S0-AKJ-179

FLOW INJECTION ANALYSIS

INSTRUMENTATION AND APPLICATIONS

FLOW INJECTION ANALYSIS

INSTRUMENTATION AND APPLICATIONS

Marek Trojanowicz

Department of Chemistry,
University of Warsaw;
Institute of Nuclear Chemistry
and Technology, Warsaw

Published by

World Scientific Publishing Co. Pte. Ltd.
P O Box 128, Farrer Road, Singapore 912805
USA office: Suite 1B, 1060 Main Street, River Edge, NJ 07661
UK office: 57 Shelton Street, Covent Garden, London WC2H 9HE

British Library Cataloguing-in-Publication Data
A catalogue record for this book is available from the British Library.

FLOW INJECTION ANALYSIS: INSTRUMENTATION AND APPLICATIONS

ISBN 981-02-2710-8

Printed in Singapore by Regal Press (S) Pte. Ltd.

Contents

Introduction

For almost half a century, flow methods of analytical measurements have had an established position in chemical analysis. Their roots can be traced back to the beginning of column-chromatographic methods and to a continuous monitoring of various physico-chemical parameters in industrial installations. The flow of the monitored medium through a suitable detector enabling the continuous recording of changes of the measured physico-chemical quantity provided the possibility of eliminating the collection of fractions in chromatographic separation or a sampling step in process monitoring, which in both cases was an evident step towards mechanisation of these operations for analytical purposes. These two streams of applications of flow measurements, the detection in column chromatography and capillary electrophoresis, and also process and environmental monitoring, evolve continuously and they are nowadays a part of analytical instrumentation of great importance.

An additional impact on the development of flow analysis in analytical laboratories was made by its new laboratory applications not connected with chromatographic detection, which appeared at the end of the fifties. Their principle was a replacement of all manipulations with a liquid sample to be analysed that was involved in the conventional manual procedure into manipulations with a segment of fluid (analysed sample) in a suitably designed flow system that ended with a flow-through detector. The main advantage of this concept is that it is the simplest way of mechanising practically all operations that need to be made with the sample in the whole analytical procedure. Then, instead of measuring the sample volume, using several pieces of glassware, transferring a sample between them (which is the main source of incidental contamination), waiting for a reaction to occur and waiting for a steady detector response, in a flow system after optimisation of the geometry of the system and a detector and also hydrodynamic conditions, the only operations to be done for each individual sample are its delivery to the flow analyser and reading or recording of transient or steady signals. So, the basic advantages of flow-analytical measurements compared to manual procedures are better reproducibility (precision) of determinations, a larger throughput and reduction of the sources of contamination. The result of determination in the flow system is not only a function of the chemistry applied, but also depends on the dynamics of the occurring process and the conditions of its observation in the time domain. These factors, together with involved on-line operations of preconcentration or separation of the analyte from the matrix, affect the selectivity of flow determinations. The flow conditions of measurement compared to static measurement may additionally enhance the analytical signal due to convection, or provide additional kinetic discrimination of interferences. Through the appropriate configuration of a flow system, a multicomponent determination can also be realised (without employing a chromatographic retention). This can also be gained sometimes by differences in the kinetics of reactions employed for a given determination.

The first successes of laboratory flow measurements have been connected with a technique of segmentation of a flowing stream with air segments, which allows one to restrict dispersion of the sample segment (reduce dilution during the flow). Such a measurement and appropriate instrumentation have been very readily accepted by overloaded clinical chemistry laboratories for routine diagnostic purposes, and later also by environmental, agricultural, and even industrial ones.

The next milestone in the evolution of flow measurements occurred in the middle of the seventies, when it was demonstrated that measurement of the flow system can be simplified and made an even more efficient instrument for chemical-analytical determinations. The main credit for this invention has to be given to Ruzicka and Hansen of the Technical University of Denmark, although at the same time, or in the same cases even earlier, in several research groups all over the world similar studies were carried out. They convincingly demonstrated the elimination of stream segmentation, and conducting measurements with waiting for steady-state analytical signal may not only worsen the flow measurement, but can simplify it and make it more efficient.

The proposed methodology and the coined name *flow injection analysis* have been very quickly accepted by the analytical community. In the first few years of its development the interest in this technique has grown exponentially, as expressed by the number of publications in scientific journals. At the beginning of the nineties its alterations named *sequential flow analysis* and *batch injections analysis* have been invented, which are discussed in a separate chapter. The number of papers published on flow injection techniques in the nineties is almost ten thousand; numerous companies provide instruments and accessories, and many developed flow injection methods appear as standard procedures in various regulations.

It would certainly be too boastful to state that flow injection methodologies have nowadays dominated chemical analysis, because chromatographic methods without doubt are most often employed in the majority of routine analytical laboratories, discrete analysers are used most frequently in clinical chemistry for diagnostic purposes, or atomic spectroscopy dominates inorganic analysis. Moreover, even in chromatographic techniques or atomic spectrometry methods the addition of flow injection sample pretreatment may significantly improve the value of numerous procedures. The biggest role of flow injection methods in contemporary analysis seems to be mechanisation of various methods with common detection methods still performed manually in numerous analytical laboratories.

Regarding the broad existing original literature this book does not pretend to be a complete review of the state-of-the-art in this field. It has been written as a result of following more or less scrupulously current publications and as a result of the author's own experiences in the design of instrumentation and development of analytical procedures almost since this methodology was invented. It has been prepared to guide through the evolution of this methodology and to illustrate its impact on chemical analysis in the twenty-five years since its invention.

This book is not only a result of my own experiences but also a result of creative contacts and collaboration with numerous partners and friends in the analytical community. My collaboration with Jarda Ruzicka and Elo H. Hansen in their laboratory at Lyngby in 1981 has significantly affected my interest in flow analysis. The long years of joint research with Mark E. Meyerhoff at the University of Michigan and Peter W. Alexander at the University of New South Wales in Sydney, and then later at the University of Tasmania, have been extremely fruitful, interesting and valuable for me. I also address my special thanks to all my students and co-workers in the Department of Chemistry, University of Warsaw, for their contributions to my knowledge and experience.

I wish also to thank Brian O'Reedy for his patient linguistic work on some of the chapters.

Marek Trojanowicz
January 1999

Chapter 1

Molecular Spectroscopy Detection

1. Visible Absorption Spectrophotometry

The absorption of visible and ultraviolet radiation by different chemical compounds results generally from excitation of bonding electrons in the absorbing molecule. In the visible region two main types of electronic transition take place, involving d and f electrons in the case of most transition-metal ions and charge-transfer electrons in the case of complex compounds, where one of the components has electron-donor and the other electron-acceptor properties. The product of radiation excitation, which corresponds to absorption of part of the radiation energy, has a very short life-time and the most common relaxation of the excited particle involves conversion of the excitation energy to heat. The amount of thermal energy produced is usually not detectable. Attenuation of a beam of radiation by an absorbing solution is expressed by transmittance T, which is defined as the ratio of the power of the beam of radiation observed after passing the absorbing solution to the initial power of the beam. The absorbance A of a solution is related to transmittance as follows:

$$A = \log T\,. \tag{1}$$

A fundamental relationship utilised for analytical purposes in molecular spectroscopy detection is the dependence of absorbance on the cell length b and concentration of an absorbing species c expressed by the Lambert–Beer law:

$$A = abc\,, \tag{2}$$

where a is a constant called the molecular absorptivity for b expressed in centimetres and c in moles per litre. The large number of known reactions producing species absorbing visible radiation, mainly with the use of organic reagents [1], is a source of very wide application of visible absorption spectroscopy in chemical analysis [2, 3]. It is commonly used in all areas of routine laboratory chemical analysis, field and clinical tests, in portable instrumentation and in process analysis.

Visible absorption spectrophotometry was already applied in pioneering works on flow injection analysis [4, 5]. Through all the twenty years of development of FIA, spectrophotometry has been and currently is the most common detection used in FIA. Spectrophotometric detectors are principal detectors of each commercial FIA instrumentation.

1.1. *Detectors*

Each detector used in FIA systems should be designed to monitor as closely as possible the events occurring in the measuring system. In photometric detectors it is facilitated by the smallest possible dead volume and the illuminated volume of the flow cell. A large volume causes poor reproducibility of height and shape of the flow injection peak, whereas a large illuminated volume results in a decrease of detection sensitivity and broadening of peaks [6]. The most often used commercial detectors are flow cuvettes with a geometry that fits conventional spectrophotometers (Fig. 1A), usually with path length 10 mm and a volume of a few to 50–60 μl. In comparative studies it was shown that unfavourable effect of an increase of cuvette volume above 25 μl is especially significant at low flow rates [7]. The optimisation of detector geometry can be made numerically by evaluation of impulse response functions for the FIA system, which show the contribution of the detector to the peak broadening [8]. The application of a capillary flow cell that utilises optical fibres to transmit light with the small illuminated volume (< 1 μl) allows one to extend the dynamic range of response and to use it in extraction systems without phase separation [6]. A design of a flow through nanocolorimeter with a cuvette working volume of 115 nl and a light path of 0.5 cm was reported [9]. FIA measurements at path length 0.1 cm were carried out with a crossed-beam thermal lens photometer, which is based on the utilisation of a single laser [10].

Flow-through photometric detectors for FIA can be made using optoelectronic components such as light-emitting diodes (LEDs) as light sources and photodiodes or phototransistors as detectors [11–29]. The commercially available LEDs cover a wide range of wavelengths, from 435 to 1300 nm [12]. The radiation from LEDs has a spectral bandwidth of about 20–70 nm, which is sufficient to substitute for the commonly-used-in-spectrophotometers combination of broadband sources and monochromators. LEDs are stable, inexpensive

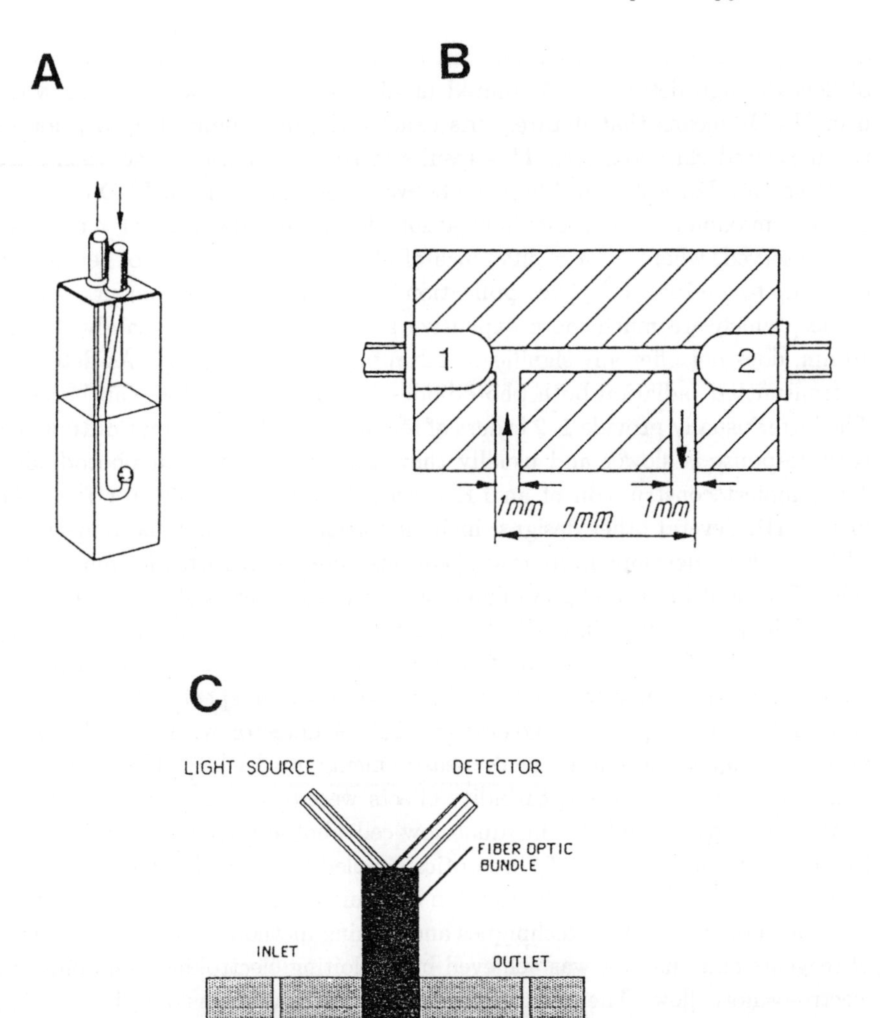

Fig. 1. Flow-through cells used in FIA with spectrophotometric detection: (A) commercial flow-through cuvette for conventional spectrophotometers; (B) flow-through cell with light-emitting diode (1) and phototransistor (2) [13]; (C) flow-through cell with bifurcate optical fibre for the light source and detector and fibrous indicator pad [42]. (*Reprinted by permission of copyright owner.*)

and easy-to-use light sources and can be easily employed in various designs of flow through detectors. A limited number of available wavelengths when using LEDs means that it rarely fits exactly the maximum of absorption of the measured chromophore. This results in some decrease of the sensitivity of detection. For a 20 nm difference between the maximum of LED emission and the maximum of chromophore absorption a 20% decrease of sensitivity was observed, whereas for a difference of about 50 nm the loss of sensitivity grows up to 40–70% [11]. The application in absorption measurements of laser diodes, which are much more expensive and exhibit the half bandwidth 1–10 nm, does not offer any significant advantage over LEDs [12]. As detectors of transmitted radiation both photodiodes and phototransistors can be used. The latter usually provide 1–2 orders of magnitude greater current output but their response is slower and usually they are more noisy than photodiodes. The simplest construction of an LED-based flow through detector is shown in Fig. 1B. Several other designs, including cells built within the body of an LED [12], were developed. Some applications were also reported for multi-LED cells. The eight LEDs and photodetectors arranged in series allow the observation of the peak formed in the FIA system and analysis of the two-component mixture with different kinetics of colour species formed [17]. An integrated multidiode light source was employed for flow injection spectrophotometry of two- and three-component mixtures [22, 24]. A detector with three LEDs in series was applied in doublet peak measurements in FIA [28]. The compensation of refractive index and turbidity effects were obtained in an LED-based dual-wavelength, double-beam, dual-flow-cell photometric detection system. Detection through the use of fibre optics coupled to the LED photodiode system was employed also in a fabrication of a micro FIA system based on glass substrates by lithographic techniques and etching methodology [29a]. Mobility of reagents and analytes was achieved by exploiting electrokinetic mobility or electro-osmotic flow. The total volume of reactants used was 0.5 μl.

Using a sufficiently strong source of radiation, the absorption measurements can also be carried out in flow cells filled with solid or gel sorbent, which integrates detection and preconcentration steps in FIA [30]. This technique, called *ion-exchange absorptiometry* [31], was successfully employed in FIA [32–37]. Although theoretical considerations indicate most favourable application for this purpose of thin-layer packed cells with a thickness of sorbent up to 0.5 mm [37], a commercially available flow cuvette such as that shown in Fig. 1A was successfully used for such measurements [32–35]. This methodology was employed for determination of various analytes with coloured reaction products retained on ion-exchangers [32–34], a hydroxypropyl derivative of a Sephadex dextran gel [35] and hydrophobic C18 sorbent [36, 37]. Such determinations were also performed by using a chromogenic ligand immobilised on a cation-exchange resin placed in the flow cell [38]. Transmittance spectrophotometry using reactions taking place at the surface of a filter paper on which a layer of dried reagent mixture had been deposited was also utilised for reactions occurring at a gas-solid interface in the FIA system for determination of bromine and chlorine in the gas phase [39].

In the case of using solid supports in the optical path, mostly for immobilisation of the chromogenic reagent, the reflectance measurement is also employed instead of absorbance measurement. The light is introduced into the flow cell through bifurcated optical fibres. This approach was applied for either the indicator dye reagents immobilised on a cross-linked styrene-divinylbenzene polymer matrix [40], or commercial indicator strips situated in the flow stream at the tip of the optical fibre (Fig. 3C) [41–44]. In reflectance measurements the reflected radiation is a much more complex function of concentration than for Lambert–Beer's law in absorptive measurements. Only for transport layers on a white opaque background may a reflection vs. concentration dependence resemble a linear relationship, but usually it is affected by radiation scattering, the nature of the reflecting medium, the geometry of illumination and the radiation collection. Such a detection with the immobilised commercial pH indicators was used for sequential determination of both acids and bases [43] and pH of rainwater [44].

Besides various ways of chemistry improvements and optimisation of hydrodynamic parameters of the FIA system, a further improvement in flow injection photometry can be achieved by the use of differential detection with two similar detectors arranged in a series, and separated with a transfer line of suitable length. It was shown that the optimum response is obtained when the dispersed sample volume is approximately equal to the volume of the transfer line between cells [45].

One of the difficulties encountered in some cases in flow injection photometry is interference due to changes of the refractive index of solutions transported through the detector, which causes deformations of the signal, the noisy response resulting in sensitivity and reproducibility deterioration. It is particularly pronounced for a large difference in concentrations between the carrier solution and the sample and in single-line FIA systems with limited dispersion. Instrumentally this effect can be eliminated by carrying measurements at two wavelengths: one at which the absorbance change is observed due to chemical reaction and another which reflects no influence of colour-forming reaction, but permits observation of the refractive index. As was mentioned above, such measurements can be made with LED detectors [27], with diode-array spectrophotometers [46, 47], or with a dedicated flow cell with optical fibre joints and different filters placed at the end of a multimode coupler [48]. The easiest way to overcome this problem is to use a large volume of the injected sample [49, 50].

1.2. *Measuring Procedures*

The most common configuration of the measuring system in flow injection photometry is the two-line manifold, where the sample is injected into the carrier stream of distilled water, buffer or a chemically inert solution with a similar matrix composition as sample, the reagents then being added by confluence [51]. For a large sample volume such a system is not interfered with by changes of the refractive index differences, but the quality of the pump, the confluence point and the method of downstream mixing are of crucial importance [49]. The insertion into both lines of the air pulse dampers and/or packed bed reactors significantly reduces the amplitude of the baseline noise. In the comparison of different manifolds for the determination of Fe(II) with 1,10-phenanthroline, the double line configuration gave the lowest detection limit [52]. Using the injection valve with the possibility of simultaneous injection of two solutions to different streams, a merging technique of FIA measurement can be employed, which is most often used to reduce the consumption of reagents [53, 54].

Reversed FIA systems with constant aspiration of the sample yield better detectability [52, 55, 56], although there are also opposite observations [57]. Such a procedure is advantageous when the volume of the injected sample is not critical, but rather the consumption of reagents [58, 59]. It can be successfully applied in multicomponent determinations, where to the same sample different reagents can be injected for the determination of different analytes (Fig. 2) [60].

The reagents needed to form a coloured product with analyte, used for sample pretreatment or elimination of interferences, are mostly used in soluble form in the continuously pumped solutions or are injected in reversed FIA systems. Several different ways have also been proposed. The determination of sulphur(IV) based on reaction with formaldehyde and pararosaniline requires one to use a three-line manifold. Utilisation of a passive cation-exchange membrane

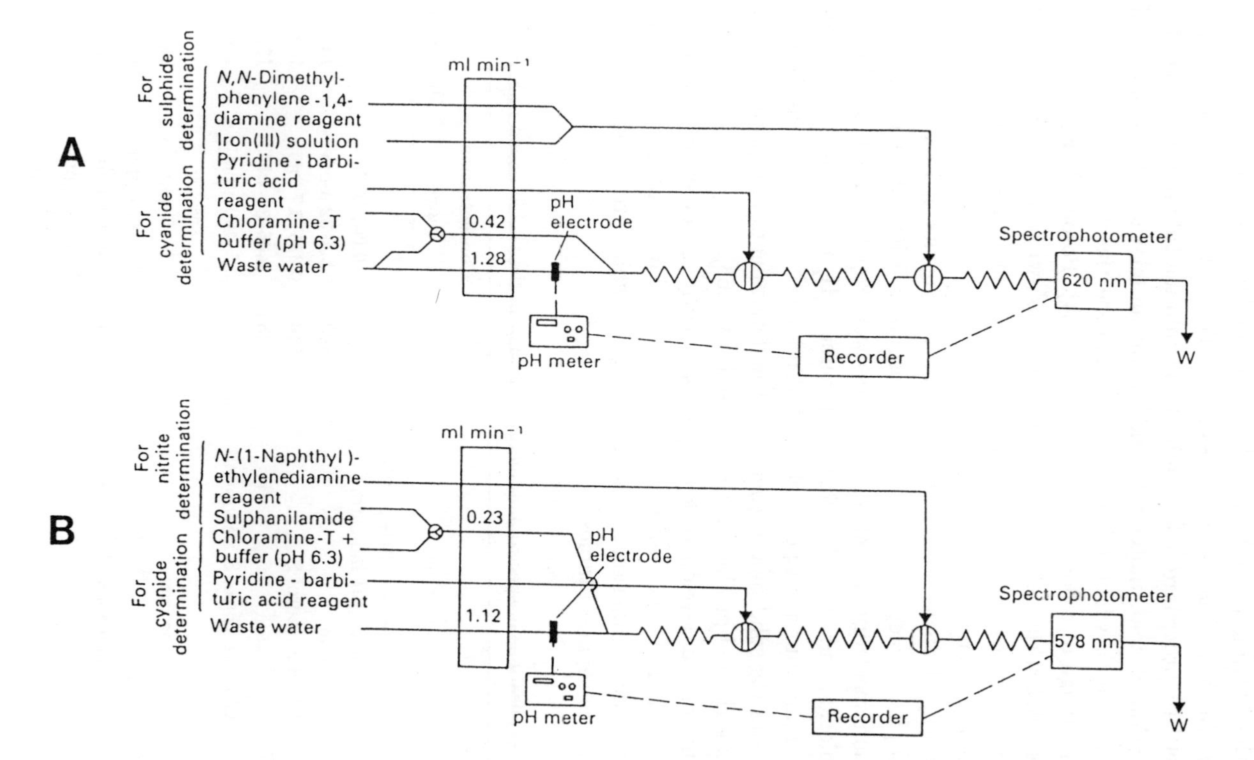

Fig. 2. Reversed flow injection manifolds used for the simultaneous determination of pollutants in waste waters [60]: (A) determination of pH, sulphide and cyanide; (B) determination of pH, nitrite and cyanide. (*Reprinted by permission of copyright owner.*)

reactor for the introduction of base and a pressurised porous membrane reactor for the introduction of acidic pararosaniline allows one to employ a single flow channel FIA procedure [63]. Similarly, membrane tubings were used for the introduction of acid, molybdate and hydrazine for the determination of phosphate in waters, which allows one to eliminate the need for confluence points in the design of a flow injection manifold [64]. Porous PTFE was used to introduce PAR in the determination of copper [65], whereas a silicon rubber membrane was employed for the introduction of bromine gas in the indirect determination of phenols [66].

Another way of dispersionless introduction of reagents in FIA is application of solid-phase reactors with the use of insoluble or immobilised reagents. Flow injection spectrophotometric procedures were reported for the determination of chloride and bromide using solid mercury(II) thiocyanate and silver thiocyanate minicolumn, respectively [67]. Cerium(IV) arsenate was used as a strongly oxidising solid-phase reactor for the determination of promethazine by monitoring the red product of the oxidised drug [68].

A long list of the developed flow injection spectrophotometric procedures is shown in Table 1. It does not contain catalytic and multicomponent procedures, and also determinations of gaseous analytes, which are discussed below and presented in separate tables.

Table 1. Applications of spectrophotometric detection in the visible region in FIA.

Analyte	Reagent	λ, nm	Concentration range	Reference
Acetoacetate	Hydroxylamine, methyl orange	520	0.1–5 mM	133
Acids, bases	Colorphast pH indicator	640	0.001–1 M	43
Ag(I)	Dithizone	500	5.4–32 ppm	134
Al(III)	Chrome azurol S	546	0.015–0.7 ppm	135
		545	0.06–30 μM	137
	Chrome azurol S, cetylpyridinium chloride	625	5–400 ppb	136
	Erichrome cyanine	535	1–15 ppm	138
			0.06–0.8 ppm	175
	Pyrocatechol violet	581	3–300 ppb	139
		583	0.045–1.0 ppm	23
	Xylenol orange	506	4.5–7.0 ppm	140

continued

Table 1 (*continued*)

Analyte	Reagent	λ, nm	Concentration range	Reference
Amines, aromatic	4-N-methylaminophenol, dichromate	530	0.05–20 ppm (as $N\text{-}NH_2$)	141
Arsenite	Ammonium molybdate, $SnCl_2$	690	5–500 ppb	142
B(III)	Azomethine-H	420	0.1–6.0 ppm	89, 143, 250
			1–200 ppb*	144
Be(II)	Xylenol orange	493	1.9–4.0 ppm	145
Bi(III)	Pyrocatechol violet	590	30–100 μM	14
	Lead tetramethyl-enedithiocarbamate	380	0.1–5.0 ppm	146
Bromide	Chloramine T, phenol red	590	0.09–100 ppm	109, 147, 148
	AgSCN, Fe(III)	465	0.01–1.2 mM	67
	$KMnO_4$, N,N-diethyl-*p*-phenylenediamine	550	0.3–20 μM	149
Bromine (g)	α-naphthoflavone, As(III)	520	0.5–100 ppm	39
Ca(II)	*o*-Cresolphthalein	580	5–144 ppm	103, 150
		570	10–250 ppm	154
	Glyoxal-*bis*(2-hydroxyanil)	520	3–500 ppm	54
	Methylthymol blue	630	40–160 ppm	151
	Zn-EGTA, PAR	505	0.8–7.2 ppm	142
	Dicyclohexano-24-crown-8, propyl orange	420	0.2–100 μM	153
Cd(II)	Dithizone	520	50–500 ppb	155
	Iodide, malachite green	690	2–200 ppb	156
	1-(2′-pyridylazo)naphthol	560	0.085–2.0 ppm	25
Chemical oxygen demand	Dichromate	445	5–2500 ppm	78, 79
			1.5–100 ppm	80
Chlorate	Iodide	370	0.083–0.83	180
Chlorate, chlorite	Iodide	360	0.1–8.3 ppm ClO_3^- 0.1–10 ppm ClO_2^-	166, 167
Chloride	$Hg(SCN)_2$, Fe(III)	480	7–500 ppm	103, 157
			0.2–15 ppm	158
			50–800 ppm	159
			0.04–1000 ppb	49
	Fe(II), Hg-tripyridyl-s-triazine	600	0.01–10 ppm	160
Chlorine	α-naphthoflavone, As(III)	520	1–100 ppm	39
	Methyl orange	510	1.4–38 ppm	161

continued

Table 1 (*continued*)

Analyte	Reagent	λ, nm	Concentration range	Reference
	o-tolidyne	438	0.08–18 ppm	161
	N,N-diethyl-*p*-phenylene diamine	515	0.3–14 ppm	162
	4,4′-tetramethyldiamin-othiobenzophenone	640	0.2–1.0 ppm	163
	4-nitrophenylhydrazine, N-(1-naphthyl)ethylene diamine	532	0.03–40 ppm	164
	3,3′-dimethylnaphtidine	535	0.03–1.0 ppm	165
	Iodide, N,N-diethyl-p-phenylene diamine	550	0.5–20 mM	149
Co(II)	4-(2-pyridylazo)resorcinol	565	0.6–500 ppb	13
	Nitroso-R-salt	516	0.5–50 ppm	168
		520	3.5–6000 ppb*	169
CO_2	Cresol red		440–1760	170
Cr(VI)	1,5-diphenylcarbazide	540	0.1–20 ppm	103, 171
			0.015–4.0 ppm	172
			0.3–2000 ppb	45
			3–700 ppb	75
			10–1000 ppb	73
Creatinine	Picrate	515	3–200 ppm	173
CS_2(g)	Diethylamine, Cu(II)	421.5	3–30 ppm	126
Cu(II)	Aquaion	805	0.8–2.4 g/l	174
		810	0.6–30 ppb*	32
	4-(2-pyridylazo)resorcinol	565	3–10 ppb	13
	Bathocuproine	485	1–3700 ppb	45
	Cuprizone	595	0.3–4.0 ppm	175
	CS_2, diethanolamine	385	0.1–20 ppm	176
	Lead diethyldithiocarbamate	436	0.04–2.0 ppm	177
	Tetramethylenedithiocarbamate	435	0.12–12 ppm	178
Cyanide	Chloramine T, pyridine,	578	0.3–5.0 ppm	60
	barbituric acid	494	0.02–4.0 ppm	108
	Isonicotinic acid, pyrazolone	548	6–1000 ppb	179
Dissolved Organic Carbon	Peroxodisulphate, phenolophthalein	552	0.1–2.0 ppm C	86
Ethylenediamine	Cu(I), pyridine-2-carbaldehyde	475	1.4–84.6 ppm	181

continued

Table 1 (*continued*)

Analyte	Reagent	λ, nm	Concentration range	Reference
Fe(II)	1,10-phenanthroline	510	0.1–30 ppm	182
			0.7–710 ppb	45
			0.05–2.0 ppm	183
			0.01–2.0 ppm	52
	Bathophenanthroline	535	0.1–3.0 ppm	175
	3-pyridyl-3′-sulphophenylmethanone 2-pyrimidylhydrazone	580	10–210 ppb	184
	2-nitroso-5(N-propyl-N-sulphopropylamino)phenol	753	1–100 ppb	185
	2-(5-nitro-2-pyridylazo)-5-(N-propyl-N-sulphopropylamino)phenol	582	1–100 ppb	50
Fe(III)	Thiocyanate	480	10–400 ppb*	34
			22–56 ppm	186
	Acetohydroxamic acid	440	0.2–10 ppm	187
Fluoride	La(III), Alizarin complexone	620	0.03–1.2 ppm	188, 189
	La(III), Alizarin complexone, sodium dodecyl sulphate	574	0.05–1.2 ppm	59
	Zr, Alizarin red S	520	0.1–10 ppm	83
Formaldehyde	5,5′-dithiobis(2-nitrobenzoic acid), sulphite	437	10–600 ppm	190
Gd(III)	1-(2-pyridylazo)-2-naphthol	560	0.9–8.8 ppm	191
Haloamines	Iodide, starch	590	1–60 ppm	192
HCl	Bromocresol green	444	0.10–0.16 M	70
HCN (g)	Chloramine T, isonicotinic acid, 3-methyl-1-phenyl-2-pyrazolin-5-one	630	0.05–65 ppmv	127
H_2O	Karl Fischer reagent	525, 546	0.01–0.2%	193
H_2SO_4	Bromophenol blue	585	20–900 g/l	106
Iodide	$KMnO_4$, N,N-diethyl-*p*-phenylenediamine	550	0.2–10 μM	149
Isoprenaline	Hexacynaoferrate(III)	585	5–50 ppm	94
Li(I)	14-crown-4-dinitrophenol	420	0.4–2.0 mM	194
Mg(II)	1-(2-Hydroxy-3-sulpho-5-chloro-1-phenylazo)-2-naphthol-3,6-disulphonic acid	510	0.2–2.4 ppm	195
	Calmagite	530	24–120 ppm	196

continued

Table 1 (*continued*)

Analyte	Reagent	λ, nm	Concentration range	Reference
Mn(II)	Formaldoxime	455	0.1–2.0 ppm	197, 198
Mo(VI)	$SnCl_2$, Fe(III), thiocyante	470	0.05–1.0 ppm	199
			100–600 ppm	88
Nd(III)	1-(2-pyridylazo)-2-naphthol	560	0.03–21 ppm	200
NH_3	Nessler reagent	660	1–200 ppb*	201
			0.2–1.5 ppm	183
	Hypochlorite, phenol, nitroprusside	620	0.05–0.5 ppm	202
		695	0.005–1.0 ppm	75
		585	0.1–1.6 ppm	183
		620	0.005–6.0 ppm	96
	Hypochlorite, salicylate, nitroprusside	620	0.03–10 ppm	203
	Phenol red	540	0.085–3.4 ppm	204
	Bromothymol blue	635	0.017–5.0 ppm	21
NH_3(g)	Bromothymol blue	520	0.04–1.0 ppm	130
N_2H_4	4-dimethylaminobenzaldehyde	460	0.02–0.3 ppm	183, 205
Ni(II)	Aquoion	410	3.6 g/l	174
	Dimethylglyoxime	445	10–70 ppm*	206
		460	2.5–30 ppm	248
Nitrate	Hydrazine, sulphanilamide, N-(1-naphthyl)ethylene diamine	520	1–10 ppm	207
	Cd, sulphanilamide, N-(1-naphthyl)-ethylene diamine	565	0.1–10 μM	16
		520	0.4–9 ppm	208
		565	0.1–50 ppm	20
		540	0.7–100 μM	209
		540, 630	2–440 μM	48
	UV reduction, sulphanilamide, N-(1-naphthyl)ethylene diamine	540	0.03–100 μM	210
	Cd, *p*-aminoacetophenone, *m*-phenylenediamine	456	0.007–2.9 ppm	211
Nitrite	Sulphanilamide, N-(1-naphthyl)ethyene diamine	565	0.1–10 μM	16
		535	0.08–0.8 ppm	53
		540	0.1–1.5 ppm	60
	Sulphanilamide, 4-N-methylaminophenol, dichromate	530	3–30 ppm	141
NO_2(g)	Sulphanilamide, 4-(1-naphthyl)ethylene diamine	540	50–250 $\mu g/m^3$	131

continued

Table 1 (*continued*)

Analyte	Reagent	λ, nm	Concentration range	Reference
Ozone	Indigo blue	600	0.03–4.0 ppm	212
	Bis(terpyridine) iron(II)	552	0.06–1.35 ppm	213
Pb(II)	Dithizone	520	0.05–8.5 ppm	214
	Dicyclohexyl-18-crown-6, dithizone 4-(2-pyridylazo) rezorcinol	512	5–200 ppb*	85
Perchlorate	Brilliant green	640	0.036–2.5 ppm	215, 216
Permanganate	Ethylenebis(triphenyl-phosphonium) bromide	545	0.58–25 ppm	215
pH	Merck ColorpHast indicators	580	2.5–7.0 pH	42, 44
	Phenol red	433, 558	8.0–8.2 pH	98
Phenol	4-aminoantipyrine, $[Fe(CN)_6]^{3-}$	510	0.1–5.0 ppm	217
	4-aminoantipyrine, peroxodisulphate	515	0.005–15 ppm*	218
	3-methyl-2-benzothiazoline hydrazone, Ce(IV)	470	0.012–15 ppm	219
	p-nitroaniline, nitrite	475	0.03–3.4 ppm	220
Phosphate	Heptamolybdate	450	30–200 ppm	70
	Heptamolybdate, ascorbic acid	660	7.5–75 ppm	4
			0.02–8.0 ppm	19, 75
	Heptamolybdate, Sb(III), ascorbic	885	0.05–4.0 μM	55
	acid	660	6–200 ppb*	36
	Heptamoybdate, $SnCl_2$	670	0.03–2.5 ppm	95
	Heptamolybdate, pyrosulphite	630	95–380 ppb	196
	Heptamolybdate, hydrazine	818	0.012–1.0 ppm	64
	Heptamolybdate, malachite green	627, 750	0.06–30 ppb*	35
Phosporus	Peroxodisulphate, heptamolybdate, malachite green	650	2–500 ppb	222
Polyphenols	Folin-Ciocalten reagent	750	100–900 ppm	93, 223
Promethazine	Ce(IV)	514	5–400 ppm	68
Propoxur	*p*-aminophenol, periodate	600	0.12–25 ppm	224
Resorcinol	*p*-aminophenol, periodate	540	0.016–8 ppm	225
Silicate	Heptamolybdate, oxalic acid, ascorbic acid	886	0.5–100 μM	226
	Heptamolybdate, oxalic acid, $SnCl_2$	695	10–100 ppb	183

continued

Table 1 (*continued*)

Analyte	Reagent	λ, nm	Concentration range	Reference
	Heptamolybdate, oxalic acid, 1-amino-2-naphthol-4-sulphonic acid	820	2–100 ppb	183
Sn(IV)	Pyrocatechol violet	576	0.3–40 ppb*	227
SO_2(g)	di-μ-hydroxo-bis[bis(1,10-phenanthroline)	510	0.5–15 ppm (v/v)	230
	Pararosaniline, formaldehyde	580	25–500 mg/m^3	128
	Fe(III); 1,10-phenanthroline	508	0.5 μg/l*	132
Sulphate	Methylthymol blue, Ba	608	0.1–6.0 ppm	81
	Dimethylsulphonazo III, Ba	662	1–30 ppm	82
			0.2–14 ppm	228
Sulphide	N,N-dimethylaniline, Fe(III)	662	1–45 ppm	229
	N,N-dimethylphenylene-1,4-diamine, Fe(III)	662	0.5–90 ppm	60
Sulphite	Pararosaniline, formaldehyde	580	2–200 ppm	232, 233
			0.00016–3 mM	63
		568	0.03–70 ppm	234
	5,5′-dithiobis(2-nitrobenzoic acid)	437	8–300 ppm	190
	Disulphide 5,5′-dithiobis(2-nitrobenzoic acid	412	0.5–20 ppm	58
	p-aminobenzene, formaldehyde	520	0.2–300 ppm	235
Surfactants, anionic	Methylene blue	660	4–360 ppm	236
			0.04–3.5 ppm	237
		652	0.08–10 ppm	238
	Ethyl violet	610	0.01–1.0 ppm	239
	Bromocresol purple	588	1–15 μM	84
	4-(4-dimethylaminophenylazo)-2-methylquinone	560	2–80 μM	240
Surfactants, cationic	Orange II	490	45–1340 ppm	241
	Bromocresol purple	588	10–50 μM	84
Surfactants, nonionic	Tetrabromophenolphthalein ethyl ester potassium salt	609	2–60 ppm	242
Terbutaline	4-aminoantipyrine, hexacyanoferrate(III)	550	12–150 ppm	243

continued

Table 1 (*continued*)

Analyte	Reagent	λ, nm	Concentration range	Reference
Ti(IV)	H_2O_2	410	9–1000 ppm	244
U(VI)	2-(5-bromo-2-pyridylazo)-5-di-ethyl-aminophenol, zephiramine	579	0.1–15 ppm	245
	2-(5-bromo-2-pyridylazo)-5-di-ethyl-aminophenol, fluoride	578	0.5–20 ppm	246
Zn(II)	Xylenol orange	568	26–64 ppm	247
	Zincon	620, 800	0.1–20 ppm	47

*with on-line preconcentration

Among numerous applications of flow injection photometry there are some very original procedures which deserve to be mentioned. Besides widely used reversed FIA, for which was used a mathematical treatment for the development of a conventional standard addition method [69], a sample-to-standard additions method was also reported [70]. The latter eliminates the need to obtain a calibration graph as is commonly practised in conventional FIA determinations. It was also shown that a pH gradient produced in the FIA system when the sample is injected of different pH than carrier solution can be exploited for multicomponent determination, if the colour-forming reagent used forms products with different analytes at different pH values [71].

In the analysis of real samples it is often necessary to adjust the analyte concentration to fall within the linear response of the detector. It can be performed by the use of an appropriate dilution system [72], or by designing a system where the sample is inserted between two reagent segments with different concentrations [73]. In such a system two simultaneous working ranges of concentration are achieved.

An especially large sampling rate can be obtained in the system, where the injected sample is placed between air segments [74]. This was applied in monosegmented continuous flow systems [75, 76]. Air bubbles were removed from the stream prior to the detector in the permeation cell with PTFE membrane [75], or resampling was employed with the removal of a fixed volume from the central part of the monosegmented sample and injection with a second injection valve into the detector line [76]. This approach was successfully used in the FIA system with solvent extraction without phase separation [25]. A satisfactory limitation of dispersion was also demonstrated in the system where the sample is injected with an air plug positioned at the tailing portion of the sample [250]. The appropriate design of the injection device allows the measurement, where the air phase can be discarded without flowing through the detector, which was described as a *relocating detectors procedure* [251]. Such a concept was illustrated in determinations of aluminium and iron.

Operating with significantly smaller sample volumes, that in conventional FIA systems (100 nl) is possible in the system with electro-osmotic flow and spectrophotometric detection [77]. The advantage of such a system is the separation of the sample matrix, typically uncharged water molecules from charged analyte ions or a charged coloured product.

Almost all applications of flow injection photometry shown in Table 1 are based on measurement of the increase in absorbance of the developed chromophore, or of the analyte itself. There are, however, also methods based on indirect measurements based on reaction of the analyte with the coloured reagent resulting in a decrease in absorbance, proportional to the analyte concentration. This procedure is employed in the determination of chemical oxygen demand [78–80]. The decoloration of barium complexes with various days is used for the determination of sulphate in waters [81,82], the decrease in absorbance of the zirconium/alizarin red S complex is used for fluoride determination [83], and the decrease in absorbance of bromocresol purple is employed for the determination of ionic surfactants [84].

As indicated in Table 1, in numerous methods various ways of on-line sample pretreatment are used in order to improve detectability or selectivity of detection. Some of them are unique and worth being mentioned. The system preconcentration on minicolumn with chelating resin was combined with subsequent selective solvent extraction with the crown ether and following by reaction with dithizone was developed for trace determination of lead [85]. The combination of the on-line UV photo-oxidation with gas diffusion was employed for the determination of dissolved organic carbon [86]. In the FIA system with an electro-osmosis-based fluid propulsion the on-line preconcentration based on the electrostacking effect as used in capillary electrophoresis was reported [87].

Several interesting solutions were developed for adaptation of FIA to the analysis of solid samples. Electrolytic dissolution of steels in a simple electrolytic chamber connected to the FIA system was used for the determination of molybdenum [88]. The determination of total boron in soils was based on the direct introduction of solid samples into the on-line cell with an ultrasonic probe [89]. The determinations of free sulphur dioxide in wine and dried apple samples can be carried out in the setup, where a gas extraction device to generate and purge gaseous SO_2 was combined with a flow reversal manifold to directly process the gaseous plug generated from the sample [90]. Operations of weighting, dissolution or extraction of the analyte from solid samples or oils were carried out also in robotised FIA systems [91–93].

The flow injection measurement based on transient analytical signal depends on numerous chemical and physical parameters, which should be carefully optimised. Most often it is carried out by single variable optimisation of all essential hydrodynamic and chemical parameters. Among the multivariable methods of optimisation most commonly employed are various versions of the simplex method. The optimisation of the flow injection spectrophotometric method of determination of isoprenaline indicated the advantage of the modified simplex over the univariate procedures in simultaneous optimisation of four variables such as flow rate, tube length, reagent concentration, and pH, or additionally sample size [94]. The univariate method of optimisation involves keeping all but one variable constant and finding the best value of this variable. In the simplex method several variables are changed simultaneously for a search of the most suitable measuring conditions. The multivariable simplex method is especially effective when interactions between the variables can occur. For the optimisation of phosphate determination two procedures of the factorial design and simplex method were compared [95]. In the optimisation of FIA procedures, of great importance is the selection of an appropriate performance criterion. The procedure can be carried out to maximise the peak height or signal-to-noise ratio or the correlation coefficient of the calibration graph. Another criterion can be the peak width. In the optimisation of phosphate determination the response used was peak width/(peak height)^2 [95]. The optimisation of ammonia determination based on the indophenol blue reaction was carried out using the modified simplex method and the multivariate Powell method with a linear combination of the peak height and residence time as the experimental target function [96]. The Powell algorithm needed fewer evaluations of the objective function than did the modified simplex method, although it does not mean generally that this method of multivariate optimisation is more efficient in absolute terms than the simplex method. A useful way of visualising experimental variables on the performance of an FIA method is to construct a three-dimensional response surface as a plot of instrumental responses vs. chemical concentrations of flow parameters. They can be obtained in an automated manner on a computer-controlled flow injection methods development system [97].

Measurements of the acidity of solutions are very common in laboratory, environmental and process analysis. With the use of colour indicators either single point pH determination or acid-base titrations can be carried out in FIA systems with spectrophotometric detection. The pH measurements in solutions of low buffering capacity and low ionic strength were realised using the reflectance spectroscopy of an immobilised pH indicator [42, 44]. A sea water pH was measured in the FIA system, where the acid-base adsorption properties of phenyl red injected into a sea water stream were measured [98]. Flow injection determinations of concentrated acids and bases were also done by the use of a buffered carrier stream containing a dissolved acid-base indicator [99, 100]. Within a limited range of linear relationship between peak height and concentration of the analyte was found, and also the simultaneous determination of acids and bases using the same carrier stream was demonstrated. A similar system was developed for a sequential determination of both dilute acids and bases using a single-line manifold with a reflectance cell with immobilised acid-base indicators.

In a single- or double-line manifold, usually with a larger dispersion than in common FIA systems with spectrophotometric detection, determinations corresponding to conventional acid-base titrations can be carried out with colour indicators. These procedures utilise the gradients of analyte concentration formed in the rising and descending parts of the signal of the injected sample and therefore they are called *gradient titrations.* In the single channel version, the titrant containing the indicator is used as the diluent in which the sample is injected. The quantitation is based on relating the time span (i.e. peak width) between the points of the same dispersion on the ascending and descending parts of the peak to the concentration of the analyte [101–103]. The relation between the time span (Δt), and the concentration of the analyte (C_A^0) and the concentration of the titrant carrier solution (C_B) is given by the equation [103]

$$\Delta t = (V_m/Q) \ln 10 \log(C_A^0/C_B) + (V_m/Q) \ln 10 \log(S/V_m n)\,, \qquad (3)$$

where S is the injected sample volume, V_m is the volume of the mixing stage (gradient tube or a mixing chamber), Q is the flow rate, and n is the stoichiometric factor between the reacting components. An example of typical signals recorded in flow injection gradient titrations is shown in Fig. 3. The set points

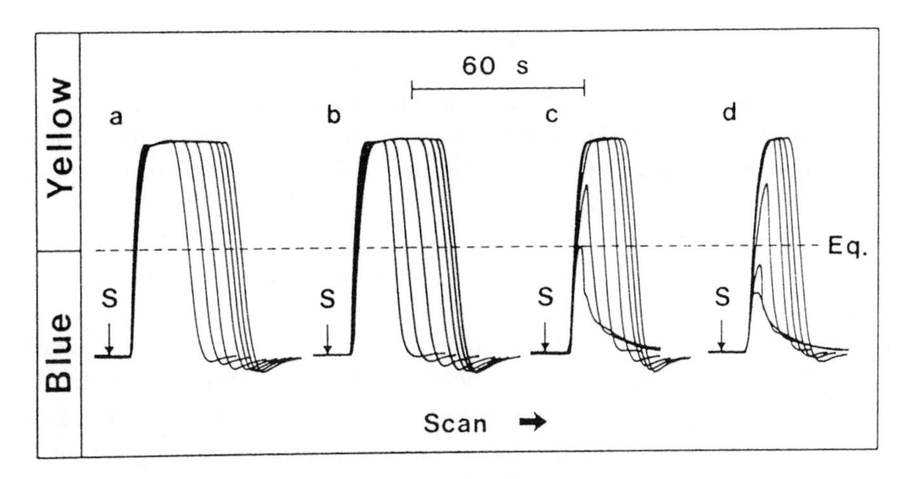

Fig. 3. Gradient titration curves obtained in the flow injection system in the titration of 5, 10, 20, 40, 60, 80 and 100 mM hydrochloric acid with (a) 0.5, (b) 1.0, (c) 5.0, and (d) 10 mM NaOH with spectrophotometric detection using a bromothymol blue indicator in the titrant solution [103]. (*Reprinted by permission of copyright owner.*)

used for peak width measurement must correspond to the equivalence point in that specific titration system. The width of signal is a logarithmic function of the analyte concentration. The effect of different parameters on the analytical signal in gradient titrations was discussed in detail for acid-base titrations [104]. It was also shown that in the single channel version it is not necessary to include a titrant in such a procedure with a large dispersion system and colour indicator [105]. A largely mechanised instrumental version of this procedure was developed for the determination of sulphuric acid in metallurgical process streams by titration with sodium acetate using a bromocresol blue indicator [106].

1.3. *Kinetic Methods of Detection*

The flow injection concept of analytical measurement based on recording of transient signal can be especially effectively applied for analytical purposes by the utilisation of kinetic properties on numerous chemical systems. Very early transient redox effect was utilised for FIA by Dutt and Mottola [107]. Into the continuously circulated reagents the analyte was injected, which in a very rapid redox reaction perturbs the concentration of the monitoring species and then it is followed by subsequent but slower reaction regenerating the monitored species. Such a procedure can be used for determination of chromium(IV) based on oxidation of ferroin with oxalic acid as activator, which can then be indirectly applied in the determination of chemical oxygen demand, and also for determination of V(V), Mn(VII) and Ce(IV) using the diphenylamine sulphonate indicator system.

The formation of the unstable intermediate product in the FIA system can also be exploited for analytical purposes, which was shown for example in the reaction of cyanide with a pyridine-barbituric acid reagent [108]. Such a system requires very detailed hydrodynamic optimisation, and in the mentioned system for cyanide determination a reversed FIA configuration was used.

Differences in the rate of reaction between various analytes and the same reagent can be utilised for designing multicomponent measuring schemes, which is discussed below. Such effects can be advantageous for the improvement of the selectivity of FIA determination in comparison to conventional steady-state measurements. The determination of bromide based on oxidation with chloramine-T to bromine, followed by reaction with phenol red, is possible under FIA conditions in the presence of a large excess of chloride, because of the much slower formation of chlorinated products [109].

The use of conventional kinetic methods of determination (initial rate, fixed and variable time) is possible in the FIA system of such a design, where the entrapment of the sample plug into a closed loop allows repetitive passage through a single detector until the sample is completely dispersed into the carrier [249]. Such a setup enables the calculating kinetic parameters and also the measurement of concentrated samples without a prior dilution.

Numerous sensitive spectrophotometric FIA methods were developed with the use of catalytic effects, mainly for the determination of trace amounts of transient metals (Table 2). The determination of Cu(II) based on catalysing the Fe(III) reaction with thiosulphate in the presence of thiocyanate can be carried out in the closed-loop system with electrochemical removal of the catalyst by reducing to Cu(0) and simultaneous regeneration of the monitored species by anodic oxidation of Fe(II) [110]. In most developed methods, however, a simple FIA measurement with sample injection is used. In the determination of selenium, which catalyses the oxidation of phenylhydrazine by potassium chlorate, the direct FIA system was compared with reversed and stopped flow configurations [111]. The best detection limit was found for the reversed system, and the widest linear range of responses for direct FIA measurement.

Table 2. Catalytic methods in FIA with spectrophotometric detection in the visible region.

Analyte	Catalysed reaction	λ, nm	Concentration range	Reference
Ag(I)+Hg(II)	$[Fe(CN)_6]^{4-}$ + α, α'-bipyridyl	536	1–64 ppb Ag 0.5–75 ppb Hg	112
Cu(II)	Fe(III)+thiosulphate with SCN-	480	5–250 ppm	110
	Fe(III)+thiosulphate	525	0.1–1.8 ppm	116
Fe(II)+Fe(III)	Leucomalachite green + $S_2O_8^{2}$- + 1,10-phenanthroline	618	0.3–15 ppm Fe(II) 0.6–15 ppm Fe(III)	114
Fe(III)	*p*-phenetidine + periodate	540	0.7–500 ppb	117
	N,N′-dimethyl-*p*-phenylenediamine + H_2O_2	554	0.08–100 ppb	117
Formaldehyde	Brilliant green + sulphite (inhibition)	615	0.02–3.0 ppm	115
Mn(II)	N,N-diethylaniline + periodate	475	0.02–1.0 ppb	118
	Malachite green + periodate	615	10–70 ppb	119
	N,N-diethylaniline + periodate	615	0.002-15 ppb*	120
	Leucomalachite green + periodate	620	0.04–12 nM*	121
Mo(VI)	Iodide + H_2O_2	350	0.7–1000 ppb	122
			1.0–40 ppb	123
Mo(VI)	Iodide + H_2O_2 with starch	580	0.04–3.2 μM Mo 0.06–3.2 μM W	113
Se(IV)	Phenylhydrazine + chlorate + chromotropic acid	360	0.15–50 ppm	111
Sulphate	$ZrOCl_2$ + methylthymol blue	586	50–500 ppm	124
V(V)	4-aminoantypyrine + N,N-dimethyl-aniline + bromate + tiron	555	0.05–2.0 ppb	125

*with on-line preconcentration

Differences in catalytic effects have found several applications in FIA systems for multicomponent determinations. Hg(II) and Ag(I) exhibit different catalytic effects on the ligand substitution reaction between hexacyanoferrate(II) and α, α'-bipyridyl with thiourea as an activator, which was utilised for simultaneous determination in the system, where flow was stopped and absorbances were determined at 100 and 200 s after sample injection [112]. Simultaneous differential rate determination of Fe(II) and Fe(III) was based on their different behaviour in the redox reaction between leucomalachite green and peroxodisulphate with and without the presence of the activator

1,10-phenanthroline [114]. For this purpose the original procedure exploiting the formation of a double peak as a result of injecting a large sample zone, sandwiched between reagent zones of appropriate composition, was developed.

As the example of application of inhibiting properties in flow injection photometric systems the determination of formaldehyde based on its inhibition of the brilliant green-sulphite reaction [115].

1.4. *Determination of Gaseous Analytes*

In this area of application of flow injection photometry, especially interesting systems from the point of view of practical applications are closed-loop FIA systems. Repetitive processing of air samples for sulphur dioxide determination was based on the reaction of SO_2 at a gas–liquid interface with dinuclear complex of iron(III) with 1,10-phenanthroline [230]. The method does not require a pretrapping of SO_2 and the main reagent was electrochemically regenerated by controlled potential electrolysis. Most often, however, determinations of gaseous components of air or process gases are carried out in the FIA systems with a cell absorbing the gaseous analyte. Such cells are usually equipped with a planar or tubular membrane of various porosity and permeability, through which the analyte diffuses into the receiver stream. The analysed gas is pumped by a fixed period of time, after which the receiver solution containing diffused gas is injected into the FIA system with an appropriate colour-forming reagent. Most favourable is to incorporate the absorbing membrane cell, called also a permeation denuder, into the sample loop of the injection valve. Two possible configurations of such a system are shown in Fig. 4. In one of them the receiver solution is also used as a carrier solution in FIA measurements, whereas in the second one the absorbing solution is injected into another carrier or reagent solution.

Hollow-membrane fibre gaseous collectors of analyte were used in determination of carbon disulphide in air [126] and trace HCN in process gas streams. CS_2 is absorbed in the ethanol stream, which then merges with reagents, forming coloured chelate of Cu(II) with dithiocarbamate. HCN is collected with a silicone rubber tube in a caustic solution, which then merges with a stream of chloramine-T and a mixture of isonicotinic acid and 3-methyl-1-phenyl-5-pyrazolin-5-one. The system with planar permeation denuder and polypropylene membrane was utilised for determination of SO_2 in air with analyte preconcentration in the recipient channel [128]. In a dual-line manifold EDTA solution was used as the absorber stream, which was merged with the

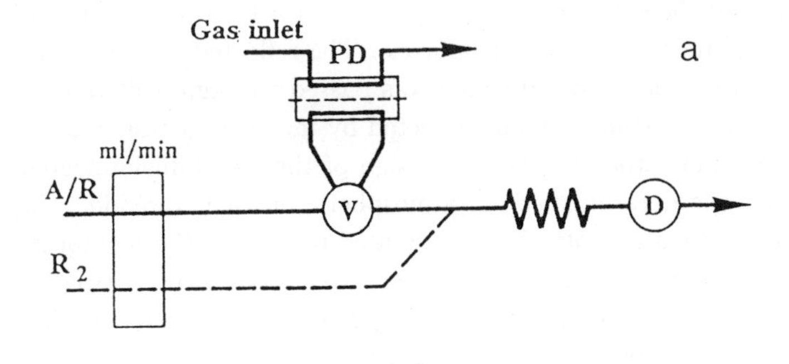

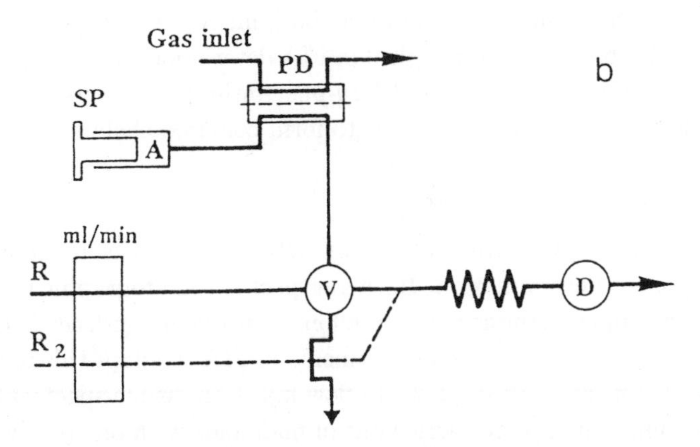

Fig. 4. Arrangements of the sampling systems for the determination of gaseous analytes in FIA with a permeation module used (a) instead of the sample loop in the injection valve, and (b) located in the loading channel of the injection valve [128]. V — valve; D — detector; A — absorber stream; R — reagent stream; SP — syringe pump; R — optional additional reagent channel. (*Reprinted by permission of copyright owner.*)

premixed pararosaniline-formaldehyde reagent stream. A planar membrane for absorbing gases was used also in flow through modules functioning simultaneously as gas absorber and spectrophotometric detector with optic fibre light transmission. The detector with reflectance measurements of the light used earlier for enzymatic determination of creatinine [129], was employed also for the determination of ammonia in air in a stopped-flow system [130]. Such a detection/reaction cell is equipped with a hydrophobic membrane, which separates acceptor and donor sides. It serves also as a reflector and light is delivered to the cell and the signal returned to the detector by a bifurcated optical fibre placed perpendicularly to the membrane. The reflected light was measured as absorbance and the collected signal was passed through a filter before reaching the detector. Ammonia was detected by monitoring the colour change of an acid-base indicator. In another design of the gas diffusion/detector module used for the determination of atmospheric nitrogen dioxide, a light beam was introduced via an optical fibre mounted parallel to the membrane surface, which allows absorptive spectrophotometry detection [131]. Nitrogen oxide was absorbed through a microporous polypropylene membrane in solution containing N-(1-naphthyl)-ethylenediamine in HCl.

The absorption of the gas analyte can be also carried out in a layer of foamed hydrophobic material containing both micropores and macropores. Biporous PTFE absorbers were applied in FIA determination of SO_2 in air [132]. An aqueous solution of Fe(III) and 1,10-phenanthroline was used as a reagent, in which Fe(III) was reduced by SO_2 to form coloured chelate.

1.5. *Multicomponent Determinations*

A very favourable feature of the analytical method is the possibility of multicomponent determination. In this respect chromatographic methods and some spectroscopic methods have a predominant role in modern chemical analysis. There are also numerous developments in FIA to adapt this methodology for multicomponent analysis [252]. In flow injection visible spectrophotometry such determinations can be carried out in manifolds with one or several detectors, or with detectors enabling simultaneous detection at several wavelengths. Basic configurations of manifolds used to achieve this are shown in Fig. 5.

In the simplest configuration the sample injected with one injection valve is split into separate branches of the manifold with different sample pretreatment and separate detectors, such as it was used for simultaneous determination of nitrate and nitrite [253]. The splitting of the sample segment can also be realised in combination with dialysis of the sample with a double-line dialyser, which was reported for simultaneous determination of chloride and calcium [254]. Multicomponent determinations can also be carried out in the system of several parallel manifolds with separate detectors and simultaneous, independent injection of the sample solution into each carrier stream by coupled injection valves [158, 255, 256]. An exceptionally complex system was developed for simultaneous determination of inorganic sulphur species in aqueous samples of importance to the petroleum industry using absorptive and

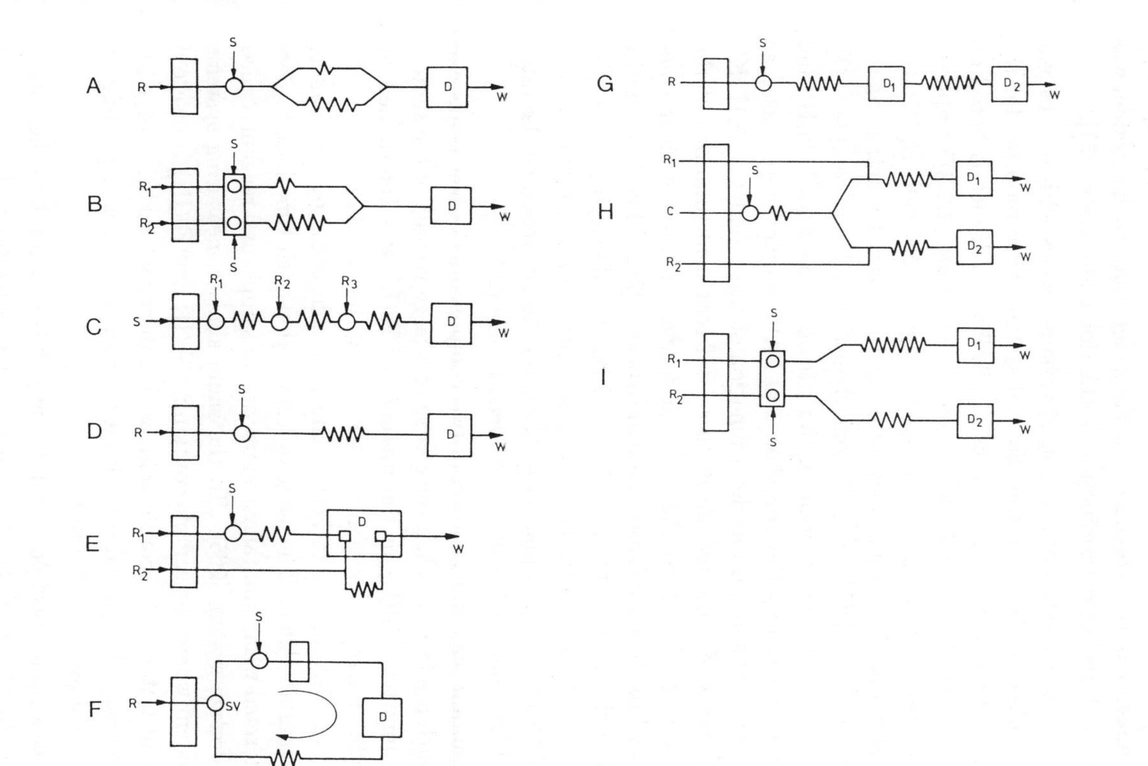

Fig. 5. Typical manifolds for multidetection flow injection systems with a single (A–F) and two (G–I) detectors [252]. (A) with splitting and confluence points; (B) with dual injection and a confluence point; (C) reversed FIA system with sequential injection of the reagents; (D) zone penetration system; (E) with two flow cells connected in series in a double-beam photometer; (F) closed-open system; (G) in-series detectors; (H,I) parallel detectors. R — reagents; S — sample injection point; D — detector; C — carrier; SV — selecting valve; W — waste. (*Reprinted by permission of copyright owner.*)

turbidimetric detectors [256]. Systems more difficult to realise due to the necessity of careful balance of back-pressure are FIA setups with sample splitting followed by confluence of the streams and a single detector. Such a system was developed for simultaneous determination of chloride and nitrate [257].

A difference in the kinetics of reaction of various species with the reagent has also been utilised for this purpose. Several systems were reported for the determination of two-component mixtures of alkaline earth metals based on the difference between the dissociation rates of the cryptand (2.2.2) complexes with phthalein complexone as the chromogenic reagent [258–260]. Hg(II) and Zn(II) were determined simultaneously utilising more rapid reaction of the Hg-Zincon complex with DCTA complexone than the Zn-incon complex [17]. The different rates of the dissociation of the citrate complexes of Co(II) and Ni(II) followed by measuring the absorbance of PAR complexes were used as a basis for the kinetic simultaneous determination of these analytes [261]. Kinetic determinations of these metals in three different configurations of the FIA system were based also on the different formation rates of their complexes with 2-hydroxybenzaldehyde thiosemicarbazide [262]. The best results were obtained in the setup with splitting up of the sample segment, then the confluence point and a single detector. A kinetic procedure for sequential determination of silicate and phosphate was based on the differences in the rates of formation of heteropolyacids and molybdenum blue [263].

For the simultaneous determination of two components doublet peaks can be utilised, which are formed when the centre of a large volume of the sample zone remains unmixed. Ni(II) in the presence of Fe(II) was determined by direct spectrophotometry of aquoion at the centre of the sample zone, whereas Fe(II) was first oxidised on-line to Fe(III) and then complexed by thiocyanate to form a red complex [264]. The same determination with time-based selectivity was performed with simultaneous injection of sample and reagent, called the *zone penetration* system [265]. To the same kind of measuring systems belongs the system where the sample containing Fe(II) and Fe(III) is inserted between zones of water and ascorbic acid, with subsequent addition of 1,10-phenanthroline [266]. The signal provides a plateau region corresponding to Fe(II) followed by a peak of total iron.

Simultaneous sample injection into different points of a non-branched manifold allows one to obtain different pretreatment of analytes in the injected sample zones. This was reported for the determination of Fe(III) and Ti(IV) with tiron [267]. Simultaneous determination of Cu(II) and Zn(II) was developed in the FIA system, where two sample plugs were injected into the same carrier [268]. One of them was for zinc determination and merged with a plug of a masking agent and another was used for the sum of copper and zinc. A similar function is obtained in the system with multicommutation, where in a single-line manifold by using several injection valves a certain sequention of sample and reagent zones is introduced. This was demonstrated for determination of iron and chromium using salicylic acid and diphenylcarbazide as chromogenic reagents [248].

There are also examples of determination of two components by successive injection of the sample solution into FIA systems of changeable configurations after each injection, or at least with a change of reagent for each injection. In such a case none of the manifolds shown in Fig. 5 was used. Cr(III) and Cr(VI) were determined with or without on-line oxidation of Cr(III) [269], Pb(II) and Bi(III) were determined with arsenazo III after pH changes of the carrier solution [270], whereas Ca and Mg were determined in the system with a buffer containing or not containing the masking agent for Ca [271].

The different stability of complexes of various metals with the same ligand at different pH can be utilised in FIA systems, where on the interface between sample zone and carrier a pH gradient is formed. This was employed for determinations of Pb(II) and V(V) with PAR [71, 272] and Cu(II) and Zn(II) with zincon [273].

Reversed FIA systems of various designs were successfully applied in simultaneous determinations of Cr(III) and Cr(VI) [61, 62, 274]. Systems with successive injection of specific reagents were developed for determinations in the same aspirated sample pH, sulphide and cyanide or pH, nitrite and cyanide (Fig. 2) [60]. Then the use of two selection valves providing easy change of buffer and chromogenic in the measuring system allows the determination of Cu(II), total iron and Al(III) in the same sample at the same wavelength using cuprizone, bathophenanthroline and eriochrome cyanine R, respectively [175]. Yet another application of the reversed FIA system was developed for determination of creatine and creatinine [278]. The sample was merged with a picrate stream. A continuous signal was proportional to the creatinine concentration, whereas peaks obtained after injection of 1-naphthol and biacetyl were proportional to the creatine concentration.

An entirely different concept of multicomponent determinations in FIA systems with spectrophotometric detection is based on simultaneous measurements at different wavelengths and appropriate processing of experimental data. It is advantageous with regard to the possible simplicity of the FIA manifold, but requires a more sophisticated detector and a larger number of calibrating solutions. The number of wavelengths used should be at least the same as the number of determined species. With the use of a flow-through detector with an integrated multidiode light source with LEDs of three wavelengths, mixtures of Al(III) and Zn(II) [22], and also Al(III), Fe(III) and Zn(II) [24] using xylenol orange were analysed. For this purpose, however, the most suitable detector is a diode array mainly used in HPLC [275]. Multiwavelength detection can be used to enhance the selectivity, as was demonstrated for the determination of the anti-cancer agent teniposide using 50 wavelengths [276]. In determinations of a single analyte the diode array detector can be employed to increase the concentration range of available responses by performing measurements at two different wavelengths corresponding to the maximum of absorption and far from it. This was applied for determination of nitrite [277]. Multicomomponent applications of diode array detectors were reported for two-component mixtures of metal ions reacting with one [277–280] or a mixture of chromogenic reagents [281–284], for three-component mixtures of cations [284], and organic compounds [285, 286]. In most cases a limited number of wavelengths were used corresponding to the number of analytes, although measurements in a very large spectrum and their interpretation were reported [279, 281]. It was also shown that by using second-derivative spectra a better resolution can be obtained than by direct absorption measurements [279]. Applications of flow injection photometry to multicomponent analysis are presented in Table 3. Some other examples are discussed in Chapter 7.

Table 3. Multicomponent determinations in FIA systems with spectrophotometric detection.

Principle of differentiation of analytes	Analytes	Manifold type*	Concentration range	Reference
Reaction kinetics	Ag(I), Hg(II)	D	1.0–64 ppb Ag 0.5–75 ppb Hg	112
	Ca, Sr	B, G	0.2–2.0 mM	259
	Ca, Mg	D	0.04–0.16 mM	260
	Co(II), Ni(II)	H	2–8 ppm	261
	Fe(II), Fe(III)	B	1–10 ppm	114
		A, G, H	0.1–1.0 mM	262
	Mg, Sr	B, G	0.2–2.0 mM	259

continued

Table 3 (*continued*)

Principle of differentiation of analytes	Analytes	Manifold type*	Concentration range	Reference
	Phosphate, silicate	C	0.25–1.5 ppm P 2.5–15 ppm Si	263
	Zn(II), Hg(II)	D	15–50 μM	17
Zone penetration	Fe(II), Ni(II)	D	2.7–5.4 mM Fe 0.17–0.24 mM Ni	264
		C	1.0–6.0 mM Fe 0.05–0.35 mM Ni	265
	Fe(II), Fe_{total}	D	0.1–12 ppm	266
pH gradient	Cu(II), Zn(II)	D	0.1–3.0 ppm Cu 0.4–12 ppm Zn	273
	Pb(II), V(V)	D	0.03–0.15 mM Pb 0.025–0.11 mM V	71
		D	6.0–10 ppm Pb 0.5–3.0 ppm V	272
Different chemistry in branched manifold	Ca, chloride	H	100–2500 ppm Ca 100–1500 ppm Cl	254
	Chloride, NH_3	I	0.2–20 ppm	158
	Chloride, nitrate	A	10–100 μM	257
	Fe(II), Fe(III)	I	0.5–120 ppm	255
	Nitrate, nitrite	H	0.1–20 mM nitrate 0.05–2 mM nitrite	253
	Sulphide, sulphite, thiosulphate	I	0.5–10 mM sulphide > 0.1 mM sulphite > 0.2 mM thiosulphate	256
Different chemistry in single-line manifold	Cr(III), Fe(III)	D	20–60 ppm Cr 25–200 ppm Fe	248
	Fe(III), Ti(IV)	C	11–105 ppm Fe 1.2–12.4 ppm Ti	267
Sample injections to altered manifolds	Bi(III), Pb(II)		1.0–11.0 ppm Bi 1.0–12.1 ppm Pb	270
	Ca, Mg		8–120 ppb Ca 1–30 ppb Mg	271
	Cr(III), Cr(VI)		0.055–4 ppb Cr(III) 0.018–2 ppb Cr(VI)	269
	Cu(II), Zn(II)		0.4–2.0 ppm	268

continued

Table 3 (*continued*)

Principle of differentiation of analytes	Analytes	Manifold type*	Concentration range	Reference
Injection of reagents	Al(III), Cu(II), Fe(III)	C	0.06–0.8 ppm Al 0.3–4.0 ppm Cu 0.1–3.0 ppm Fe	175
	Cyanide, sulphide, nitrite	C	0.3–5.0 ppm cyanide 0.5–5.0 ppm sulphide 0.1–1.5 ppm nitrite	60
	Cr(III), Cr(VI)	F	1.0–3.0 ppm Cr(III) 0.6–1.2 ppm Cr(VI)	274
		B	0.5–3.0 ppm Cr(III) 0.2–1.2 ppm Cr(VI)	62
	Creatine, creatinine	C	2–30 ppm	297
Difference in absorption maxima in the same same chemical conditions	Al(III), Zn(II)		0.2–25 ppm	22
	Al(III), Fe(III), Zn(II)		1.0–10.0 ppm	24
	Co(II), Cu(II), Ni(II)		5–50 μM	284
	o-, *m*-, *p*-cresols		0.1–100 ppm	285
	Cu(II), Fe(III)		16–25 ppm Cu 1–7 ppm Fe	277
			2–8 μM	278
			1–42μM	282
			0.4–35μM	283
	Fe(II), Fe(III)		0.2–15.5 ppm Fe(II) 3–20 ppm Fe(III)	281
	Fe(III), free acid		0.1–1.6 M Fe 0.5–6 M acid	280
	1-, 2-naphthols		1–50 ppm	285
	Ni(II), Zn(II)		1.0–7.0 ppm	279
	Nitrophenylhydrazines		0.5–30 μM	286

*according to Fig. 5

1.6. *FIA or Liquid Chromatography?*

The testing of detection conditions in FIA systems is a very often used method for the optimisation of working conditions for the chromatographic determinations [e.g. 287, 288]. The FIA system combined with a chromatographic setup can be used for effective analyte preconcentration in HPLC [289] or GC [290]. In the first of these applications transition metal ions were preconcentrated on microcolumn with Chelex-100, whereas in the second one preconcentration and ethylation of organotin species was carried out on a microcolumn with C18 sorbent.

Incorporation of a separation column in the flow system between the injection valve and the detector, and then acquisition of signals corresponding to several analytes, is a typical arrangement of the column chromatographic setup. Regardless of the size of a column, used pressure or flow rate, such a system should not be considered as a flow injection system, which sometimes can be found in the literature. Such systems are low pressure liquid chromatography systems with post-column reaction detection. They were developed, for instance, for determination of inorganic polyphosphates [291], for speciation of dissolved organic and inorganic phosphorus in environmental samples [292], for simultaneous determination of silicon and phosphorus in biological standards [293], and for determination of silicate, phosphate and arsenate [294]. Sodium and potassium were separated on a column packed with silica gel as complexes with crown ether [295]. Cobalt(II) preconcentrated on a chelating column is then separated from the excess of Mn(II) and Fe(III) on a strongly acidiccation resin, but chromatographic signals for both Co(II) and Mn(II) are recorded [296].

2. Detection in the Ultraviolet Region

Absorption in the UV region (185–400 nm) generally results from the excitation of π, α and non-bonding electrons to higher energy levels. Absorbing compounds containing these electrons mostly include organic molecules, but also some inorganic ions, e.g. nitrite, nitrate, azide or carbonate. Spectrophotometric measurements in the so-called vacuum ultraviolet ($\lambda < 185$ nm) are much more difficult, because components of the atmosphere absorb strongly in this region, and so it is not used for analytical purposes. The instrumentation for measurements in the visible and UV ranges differs in the sources of radiation used and materials for cuvettes. Deuterium or hydrogen lamps are the most common UV radiation sources, while the cuvettes used are made of quartz, fused silica or certain polymeric materials that pass UV radiation.

The similarity between the detection processes in the visible and UV regions means that their applications in FIA involve the same instrumental designs, measuring procedures and interpretation of experimental data. The majority of the applications reported involve procedures which use conventional UV/VIS spectrophotometers with flow-through cuvettes or commercial HPLC detectors. Developed flow injection procedures are based either on the direct absorption of UV radiation by the analyte or on the indirect measurement of the absorption by a product of a reaction involving the analyte. Several applications using diode array detection in the UV region for multicomponent determinations have been reported.

A direct determination of nitrate at 210–220 nm has been reported in a simple, single component system [298], as well as in a computer-controlled multichannel system, together with the spectrophotometric determination of chloride and ammonium ions [158]. In order to remove interfering organic impurities, on-line filtration is used [298]. A direct measurement of the absorbance of the chloro-complex of iron(III) at 335 nm has been employed for the determination of total iron in silicate rocks [299], and some benzodiazepines in pharmaceuticals [300]. Additional on-line solvent extraction is needed in the direct determination of bittering compounds in beer [301], while phenols and some neutral compounds can be determined using extraction with a silicon rubber membrane [302]. Matrix correction has been employed in the direct determination of ascorbic acid in soft drinks and pharmaceuticals [303]. Following a direct measurement at 245 nm, a second measurement is carried out after the decomposition of a fraction of the analyte with sodium hydroxide. The reduction in signal provides the analytical information. A similar procedure has been used in the determination of thiamine (vitamin B1) with the decomposition of the analyte carried out by on-line UV photodegradation [304].

Indirect determinations may involve the measurement of the absorbance of a product formed by a reaction of the analyte or of the decrease in the absorbance of a UV-absorbing substrate, which reacts with the analyte. Examples of the first type are the determinations of Cu(II) as a diethanolodithiocarbamate chelate [305], sulphate as $FeSO_4^+$ ion (with the effective removal of organic interferences on a charcoal column) [306], and boron, through the detection of chromotropic acid complexes [307]. The slow reaction occurring in the last example requires the use of a reactor consisting of 10 m of packed tubing connected in series with 3 m of open tubing. This allows a sample residence time of 7 min with a dispersion coefficient of about 2. Iodide can be determined by UV-FIA spectrophotometry by direct oxidation to iodine, but if iodide is first oxidised (with bromine water) to iodate, which in turn is reduced to iodine, the sensitivity of the determination can be increased six-fold [308]. In the determination of nitrite, the product of the reaction with

5,7-dihydroxy-4-imino-2-oxochroman is first extracted on-line with a mixed solvent prior to UV detection [309].

In several FIA systems with UV detection, a decrease in the absorbance of a chromophore as a result of its reaction with the analyte is measured. Sulphur dioxide can be determined by its reaction with pyridine bromide perbromide [310]. Determinations of ascorbic acid have been based on the reduction of tri-iodide ion [311] and vanadotungstophosphoric acid [312]. The reaction of water with $SnCl_4$ or $SbCl_5$ has been utilised for trace water determination in non-polar organic solvents such as benzene, 1,2-dichloroethane and n-hexane [313]. The maximum absorption of Ce(IV) is found at 320 nm. A procedure for determining chemical oxygen demand has been developed in an FIA system with a 20 m reaction tube of 0.5 mm i.d. [314]. This method gives much better detectability than FIA methods with visible spectrophotometric detection using dichromate [78] or permanganate [315].

A decrease in Ce(IV) absorbance has been utilised in the indirect catalytic determination of thiocyanate and iodide [316] using the redox reaction between cerium(IV) and arsenic(III), in a system with a double injection valve. Another catalytic FIA system with UV detection has been reported for the trace determination of Mo in manganese process solutions, based on its catalytic effect on the peroxide-iodide reaction [317]. The matrix problem in this system is minimised using a high concentration Mn(II) solution for both carrier and sample, but with a small, on-line anion exchange column to remove the molybdate in the carrier.

Diode array detectors can be used in FIA UV spectrophotometry to enhance the selectivity of a single analyte determination or for multicomponent determinations. An example of the first type of application is the previously mentioned multiwavelength detection of teniposide in blood plasma, with advanced multivariate data analysis based on partial least squares modelling [276]. Multicomponent determination has been developed for mixtures of up to four active components of pharmaceutical formulations [318], mixtures of nitrophenylhydrazines [319] and priority pollutant chlorophenols [320]. It has been shown that as the complexity of the mixture increases, third-derivative spectra generally give better results [318]. The determination of hydrazines is carried out in a 1 mm optical path length flow cell packed with C18-bonded silica, which gives a very low detection limit [319]. In a determination of chlorophenols, the analytes are first preconcentrated off-line in an XAD-4 adsorbent resin. The FIA determination is preceded by on-line extraction of ion-pairs of the chlorophenols with tetrabutylammonium ion into chloroform [320]. More details about these determinations can be found in Table 4.

Table 4. Applications of spectrophotometric UV detection in FIA systems.

Analyte	Species absorbing UV radiation	λ, nm	Concentration range	Reference
Ascorbic acid	Ascorbic acid	245	0.2–50 ppm	303
	Tri-iodide ion	350	0.1–40 ppm	311
	Vanadotungstophosphoric acid	360	1–80 ppm	312
B	Chromotropic acid complexes	361.6	0.008–6 ppm	307
Chemical oxygen demand	Ce(IV)	320	0.5–130 ppm	314
Chlorophenols	Analytes	200–430*	60–200 ppb**	320
Clotiazepam	Clotiazepam	260, 390	6–500 μM	300
Cu(II)	Diethanolodithiocarbamate	385	1–20 ppm	305
Doxylamine, etafedrine, phenylephine, theophyline	Analytes	200–300*	4–125 μM	318
Fe	Chloro-complex of Fe(III)	335	10–60 ppm	299
Iodide	Tri-iodide ion	351	0.05–15 ppm	308
Iodide, thiocyanate	Ce(IV)	254	0.1–1.0 ppm	316
Mo	Tri-iodide ion	350	7–100 ppb	317
Nitrate	Nitrate	210	0.1–40 ppm	298, 158
Nitrite	Product of reaction with oxochroman	361	5–280 ppb	309
Nitrophenylhydrazines	Analytes	300–500*	0.5–50 μM	319
Sulphate	$FeSO_4^+$	355	10–600 ppm	306
Sulphur dioxide	Pyridium bromide perbromide	300	25–140 ppm	310
Teniposide	Teniposide	200–400*	1–25 ppm	276
Thiamine	Thiamine	264	1.2–30 ppm	304
Triazolam	Triazolam	228	3–55 μM	300
Water	$SnCl_4$ or $SnCl_5$	300 or 350-420	1–100 ppm	313

*with diode array detector

**with preconcentration

3. Infrared Detection

Absorption in the infrared (IR) region can be used for the detection of molecules that exhibit small energy differences between various vibrational and rotational states. A molecule with an oscillating dipole moment as a consequence of its vibrational or rotational motion may absorb IR radiation with a wavelength ranging from 0.78 to 15 μm. When the frequency of the radiation matches a natural frequency of vibration of the molecule, a net transfer of energy takes place, resulting in a change of the amplitude of the molecular vibration, and absorption of the radiation. With few exceptions, practically all molecular species exhibit infrared absorption. IR spectroscopy is used mostly for the identification of organic compounds from absorption bands in the middle region of wavelengths from 2.5 to 25 μm. IR measurements are also finding increasing use in quantitative analysis, although they differ from UV/VIS measurements because of the greater complexity of the spectra, the narrowness of the absorption bands and the limitations of infrared instruments. The most common analytical applications are being developed for the analysis of mixtures of organic compounds of a similar structure, as well as mixtures of gases, such as in the analysis of air contaminants. When especially high resolution measurements are required, as in the determination of components in mixtures, or when fast scanning is needed in flow or chromatographic determinations, Fourier transform (FTIR) instruments are very useful.

Infrared detection is finding an increasing number of applications in FIA systems. In the simplest case, near-infrared detection is employed in the determination of reactive silicate by the formation of molybdenum blue [321]. A high output infrared LED with a maximum output at 886 nm is used as the radiation source. Dilution of the reductant solution with the sample in the developed procedure has proved to be an effective way to match the refractive indices of the carrier stream and the injected solution. Most IR applications in FIA have been developed for the middle IR spectral region, using FTIR instruments. A variable-filter IR spectrometer has been used for the determination of phenyl thiocyanate, based on absorption by the N=C=O functionality [322]. Flow cells with KBr windows have been used in FIA FTIR determinations with organic solvents [323–329], while a cell with a zinc selenide window has been used with aqueous solutions [330, 331]. In the monitoring of the decomposition of allyldiisopropylamine oxide in an FIA system, a high pressure flow cell was used when a supercritical carbon dioxide was used as the carrier [332]. Sample volumes in such determinations usually range from 100 to 300 μl.

Table 5. Applications of spectrophotometric infrared detection in FIA.

Analyte	Detection limit	Reference
Aliphatic esters	14 mM	330
Allyldiisopropylamine oxide		332
Benzene	0.02% v/v	327
tert-Butyl ether	0.035% v/v	328
Carbaryl	1.6 ppm	325
Choline compounds	0.02 pg/l	331
Ethanol	0.02% v/v	333
Ibuprafen	80 ppm	324
Phenyl isocyanate	4 ppm	322
Silicate	0.5 μM (Si)	321
Toluene	0.01% v/v	329
o-Xylene	0.02% v/v	323

Various procedures using FTIR detection have been employed in the determination of toluene in gasoline [329]. In the simplest one, the determination is based on measurements of the absorbance at 728 cm^{-1}, the position of the most intense band due to toluene, using a base-line established between 835 and 575 cm^{-1}. The determination can also be based on first-derivative spectra to avoid matrix interferences. FIA-FTIR procedures have also been developed for the determination of several other components in gasoline, such as *o*-xylene [323], benzene [327] and *tert*-butyl ether [328]. In the determination of benzene, a rapid quality control procedure has been developed, based on the on-line injection of gasoline samples into a carrier stream of a solution of benzene in hexane. Samples with a benzene content higher than the upper limit provide positive peaks, so any gasoline product with an unfavourably high content can be easily rejected.

A technique for the introduction of volatilised compounds into an FTIR gas cell has also been developed [333]. A liquid sample is injected into an electrically heated glass reactor where vapour phases are generated and then transported by carrier gas into a temperature-stabilised detector of 100 ml volume. This method has been applied in the determination of ethanol in chloroform. Examples of other applications are shown in Table 5.

4. Turbidimetric Detection

Turbidimetric and nephelometric detection methods are based on the scattering of radiation by a solution containing solid particles. The intensity of radiation observed at a particular angle in such a medium depends on the number of particles, their size and shape, and the wavelength of the radiation. Turbidimetric detection is based on the measurement of the attenuation of the power of a light beam as a result of scattering by solid particles. A relationship analogous to Lambert–Beer's law applies:

$$\log(P_0/P) = kbc\,, \tag{4}$$

where $k = 2.303\ \tau/c$, τ is the turbidity coefficient, P_0 and P are respectively the power of the light beam before and after passing through a layer of turbid solution of thickness b, and c is the concentration of solid particles. Nephelometric methods are based on the measurement of scattered radiation, usually at right angles to the incident light beam. In FIA systems, however, turbidimetric detection has only been used with simple photometers or spectrophotometers equipped with flow-through cuvettes. The value of the turbidity coefficient depends on the wavelength according to $\tau = s\lambda^{-t}$, where s is constant for a given system, and t is a constant dependent on particle size. Very frequently an ordinary white light is utilised in turbidimetric measurements.

Most of the turbidimetric FIA procedures reported in the literature have been developed for the determination of sulphate, which is difficult to determine with other detection methods. These determinations are based on the precipitation of barium sulphate [256, 334–339] or lead sulphate [340, 341]. Examples of these and several other applications are listed in Table 6.

These determinations are usually carried out in conventional spectrophotometric flow cuvettes or flow cells, with light-emitting diodes as radiation sources [335]. The sheath flow optical cell used for the turbidimetric measurement of $BaSO_4$ is shown in Fig. 6A. The light source is one arm of a bifurcated

Table 6. Applications of turbidimetric detection in FIA.

Analyte	Detected precipitate	Concentration range	Reference
Amitriptyline	Amitriptyline-bromocresol purple	30–200 ppm	344
Ammonia	$NH_{n-1}Hg_2I_n$	0.5–6.0 ppm (N)	342
Sulphate	$BaSO_4$	20–140 ppm	334
		50–200 ppm	336, 337, 339
		1–200 ppm	338
		0.25–20 mM	256
	$PbSO_4$	0.3–20 ppm	340
Total sulphur	$PbSO_4$	5–25 ppm (S)	341

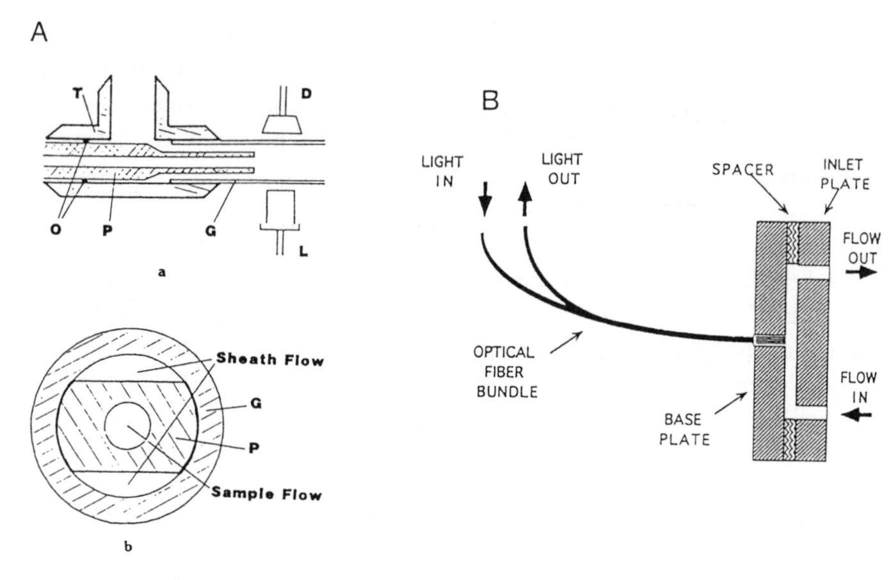

Fig. 6. Flow-through cells used in FIA systems with (A) turbidimetric detection [256] and (B) using light reflectance of the precipitate formed [343]. In A: a — longitudinal cross section; b — radial cross section; D — detector; L — light source; G — glass tube; O — o-ring; T — T-connector; P — Teflon tube. (*Reprinted by permission of copyright owner.*)

fibre optic and the source arm is connected to a miniature tungsten lamp. Such a cell can be operated over day-long periods with no indication of a rise in the baseline indicative of precipitate deposition on the optical window. In other developed procedures, polyvinyl alcohol [334, 335, 338] or gelatine [336, 337, 339] has been added to the carrier solution in order to reduce the accumulation of the precipitate on the walls of the cuvette and tubings. This can also be achieved by periodic rinsing of the FIA system with alkaline EDTA solution, or by injection of EDTA solution between sample injections [336, 337]. System performance can be improved by alternate pumping of the reagent stream and an alkaline EDTA solution at a high flow rate [338]. Another reported approach is based on the use of a reversed FIA system with continuous sample aspiration and alternating injections of barium reagent or alkaline solutions (Fig. 7A).

A filtration unit with activated carbon paper is incorporated into the conduits of the sampling line before the peristaltic pump. In another reported procedure, strongly acidic sulphate sample solutions are injected into the alkaline carrier solution containing barium and EDTA [335]. The decrease

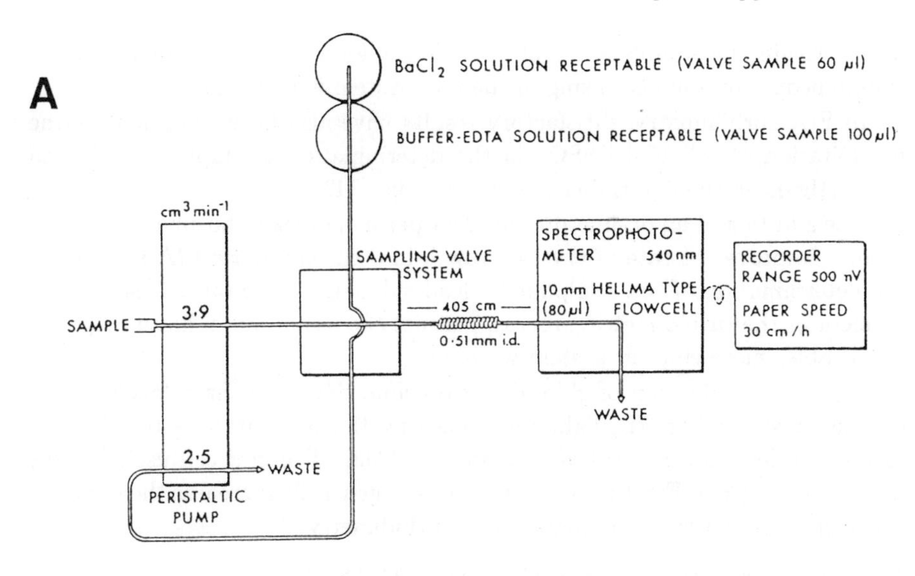

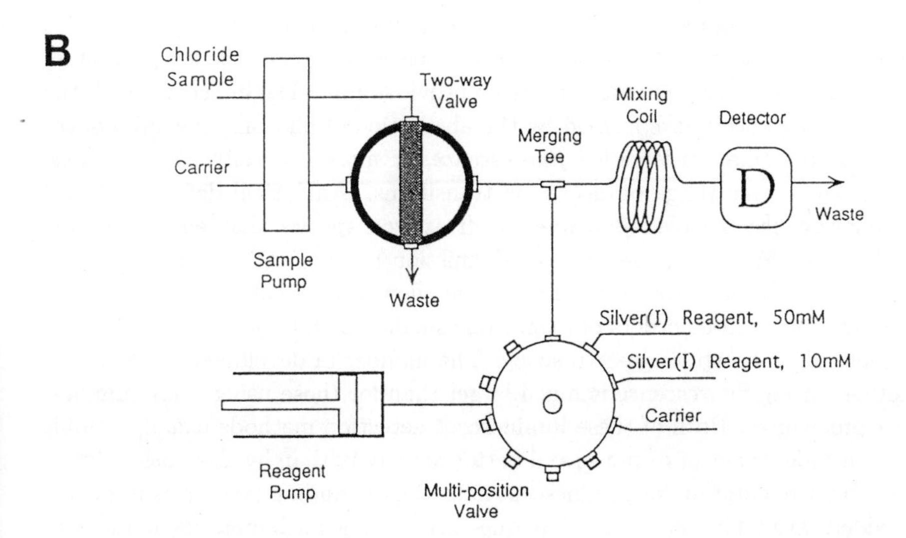

Fig. 7. Schematic diagrams of (A) a reversed FIA system with turbidimetric detection of sulphate [339] and (B) an FIA system for chloride determination using the light reflectance of the precipitate formed [343]. (*Reprinted by permission of copyright owner.*)

in pH within the sample segment is sufficient for precipitation of $BaSO_4$ at simultaneous continuous rinsing of the flow system with EDTA.

In FIA turbidimetry, satisfactory results have also been obtained for the precipitation of colloidal $PbSO_4$ in the determination of sulphate [340] and $NH_{n-1}Hg_2I_n$ in the determination of ammonia [342].

The addition of an on-line-produced suspension of lead phosphate has been used to improve the rate of crystal growth in a turbidimetric FIA system for the determination of total sulphur by lead sulphate precipitation [341]. This accelerates the turbidimetric process, permits the use of more dilute reagents, and enables more efficient system washing.

In the determination of chloride by precipitation as silver chloride in the system shown in Fig. 7B, light reflectance by the precipitate is used as the method of detection instead of turbidimetry [343]. The flow cell used for this purpose is shown in Fig. 6B. Such an approach generally increases the dynamic range of detection when compared with turbidimetry.

5. Molecular Luminescence Detection Methods

The use of luminescence detection techniques in chemical analysis is based on the formation from the analyte, or with participation of the analyte, of an excited species whose emission spectrum provides analytical information. If the excited state species are formed by the absorption of photons, the subsequent emission of radiation is called *fluorescence*, or more generally *photoluminescence*, for a wide range of electronic transitions. Analytical detection based on fluorescence is called *fluorimetry*. If excited species that emit radiation are formed chemically, the process of emission is described as *chemiluminescence*. In both cases, the emitted radiation allows analytical detection at levels that are one to three orders of magnitude smaller than those usually obtained in molecular absorption spectroscopy. The number of developed conventional methods using fluorescence is much larger than for those using chemiluminescent procedures. Both of these luminescent detection methods usually exhibit a much wider range of response and better selectivity than by absorption detection. Instrumentation for luminescent detection, including flow measurements, is widely available, so these techniques are finding an increasing number of applications in FIA [345].

5.1. *Fluorimetric Detection*

A molecule excited by the absorption of photons can return to its ground state by various deactivation steps, one of which is fluorescence, where the emission of radiation takes from 1 ns to 1 μs. Both the structure of the analyte molecule and chemical conditions affect fluorescence. It is most commonly associated with aromatic functional groups with $\pi \rightarrow \pi^*$ transitions, for example, most unsubstituted aromatic hydrocarbons exhibit fluorescence. It is favoured by increasing the rigidity of molecules; so, for example, the immobilisation of fluorescent dyes results in enhanced emission, while the fluorescent intensity of certain chelatin agents increases when they form complexes with metal ions. At sufficiently low concentration of the emitting species, the fluorescent power is a linear function of concentration. Fluorimetric detection is widely applied in the direct determination of various organic compounds. In determinations of inorganic species, the formation of a fluorescent chelate or the diminution of fluorescence resulting from the quenching action of the analyte is utilised.

In FIA fluorimetry, the direct detection of fluorescent analytes is also employed, for example in the determination of four phenothiazine derivatives, where the same excitation wavelength is used but emission is measured at wavelengths characteristic for each analyte [346]. The use of a non-selective reagent for derivatisation, such as *o*-phthaldialdehyde, gives a fluorimetric signal in an FIA system for various amines [347, 348] and amino acids [347, 349]. The use of the same reagent in reaction with sulphite allows the detection of ammonia nitrogen with considerable selectivity over amino acids [350].

In most common procedures, a selective reaction of the analyte that produces a fluorescent product is performed in a simple two- or three-line manifold. Such procedures have been developed, for instance, for the determination of formaldehyde [351], thiamine [352], boron [353], sulphate [354] and thiourea [355]. A list of the numerous other reactions employed is shown in Table 7. In certain cases, for example in the determination of sulphate based on the formation of a ternary complex between sulphate, zirconium and calcein, the method works best after separation of the analyte from its matrix [354]. In the FIA determination of gallium with lumogallion an increase in the fluorescence signal is obtained with the addition of the surfactant polyethyleneglycolmonolauryl ether [356]. Several determinations in simple systems are based on kinetic effects. Trace determination of fluoride is based on its ability to increase the rate of formation of a fluorescent Al(III)-eriochrome red B complex in the presence of hexamethylenetetramine [357]. Catalytic-fluorimetric procedures have been developed for the determination of copper(II) [358], iodide [359] and vanadium (V) [360].

Table 7. Applications of molecular fluorescence detection in FIA.

Analyte	Reagent used or catalysed reaction	λ, nm (excitation, emission)	Concentration range	Reference
Adrenaline	Iodine	350, 510	1–25 ppm	373
Al(III)	Acid alizarine garnet R	366, 460	0.013–7 ppm	362
	Lumogallion, Brij-35	484, 552	0.15–15 nM	364
	Salicylaldehyde picolinoylhydrazone	383, 468	1–90 ppb	380
Albumin	Sodium 8-aniline-1-naphthalene sulphonate	370, 470	349	
Amino acids	*o*-phthaldialdehyde	360, 550	0.01–10μM	347
Amines	*o*-phthaldialdehyde,2-mercaptoethanol	340, 432	10–50 μM	348
Ammonia	*o*-phthaldialdehyde, sulphite	360, 420	0.02–20 μM	350
B	Chromotropic acid	313, 350	0.005–100 μM	359
Be(II)	3-hydroxy-2-naphthoic acid	355, 455	0.08–0.78 ppb	363
Berberine	—	355, 517	0.004–1 μM	371
Ca, Mg	Calcein	300, 515	1–50 μM	374
Ce(III), Ce(IV)	—	260, 350	30–500 ppb	379
Coumarins	—	350, 418	1–23 μM	375
Cu(II)	2,2′-dipyridyl ketone hydrazone, O_2	350, 429	8–300 ppb	358
	2,2′-dipyridyl ketone hydrazone, H_2O_2	350, 427	0.2–300 ppb	369
C(II), Hg(II)	dipyridyl ketone phenylhydrazone, H_2O_2	350, 430	1–6 ppm	377
Cyanide	Pyridoxal	355, 435	0.1–20 ppm	368
Eu(III)	Thenoylfluoroacetone, trioctylphosphine oxide	352, 613	1.5–150 ppb	381
Fe(III), Mn(II)	Salicylaldehyde thiosemicarbazone, H_2O_2	357, 437	40–600 ppb	378
Fluoride	Zr(IV), calcein blue	323, 438	0.2–20 ppb	367
	Al(III), eriochrome red B	470, 595	0.01–3.8 ppm	357
Formaldehyde	2,4-pentanedione, ammonium acetate	410, 480	0.1–330 μM	351
Ga(III)	Lumogallion	490, 520	22–108 mM	356
			4–87 μM	370

continued

Table 7 (*continued*)

Analyte	Reagent used or catalysed reaction	λ, nm (excitation, emission)	Concentration range	Reference
Glycine	*o*-phthaldialdehyde	337, 455	0.002–8000 ppb	349
Iodide	As(III), Ce(IV)	254, 350	3–400 nM	359
Kynurenic acid	—	370, 465	0.01–6 μM	361
Metal ions	8-quinolinol-5-sulphonic acid	369, 520	0.1–10 μM	376
Nitrate	Copperized Cd, 3-amino-1,5-naphthalenedisulphonic acid	365, 470	0.01–200 μM	366
Oxalate	Zr(IV), flavonol	350, 480	10–360 μM	382
Paracetamol	$Fe(CN)_6^{3-}$, N,N'-dimethylformamide	331, 427	0.04–17.6 ppm	372
Phenothiazine derivatives	—	310–360, 382–523	0.01–4 ppm	346
Sulphate	Zr(IV), calcein	410, 505	0.8–25 ppm	354
Thiamine	Hg(II)	370, 465	0.2–7 μM	352
Thiourea	Ta(III)	227, 419	0.5–10 μM	355
V(V)	Sodium 4,8-diamino-1,5-dihydroxyanthraquinone-2,6-disulphonate	524, 582	0.2–2 ppm	360
Zn	*p*-tosyl-8-aminoquinoline	377, 495	0.01–500 nM	365

Background fluorescence in the analysis of natural samples can be eliminated by the separation of the analyte from the matrix by solid-phase extraction on an appropriate sorbent. This procedure has been employed, for example, in the determination of urinary kynurenic acid [361]. Discrimination against riboflavin in the fluorimetric kinetic determination of Al(III) has been obtained in a microcomputer-controlled FIA system, where individual peak profiles are corrected for dispersion by comparing signals in a non-reacting reference solution and a reacting sample solution [362].

On-line columns have been used in FIA fluorimetry for the preconcentration of trace amounts of Be(II) [363] and Al(III) [364], and also to separate zinc from interfering alkali and alkaline earth metal ions and to concentrate zinc from sea water prior to the reaction with the organic indicator ligand *p*-tosyl-8-

aminoquinoline [365]. A copperised cadmium column has been used to reduce nitrate on-line prior to its reaction with naphthalenedisulfonic acid to give a fluorescent product [366].

Several different manifolds, including a reversed FIA system, have been compared for the trace determination of fluoride based on the formation of a fluorescent ternary complex with Zr(IV) and calcein blue [367]. For process monitoring, a system with continuous sample aspiration was found to be most suitable. An increase in the sensitivity of FIA measurements can be obtained using a stopped-flow procedure. This has been employed in the determination of cyanide based on its reaction with pyridoxal and pyridoxal-5-phosphate [368], and in a Cu(II) determination based on its catalytic effect on the 2,2′-dipyridylketone hydrazone/hydrogen peroxide reaction [369]. This method is not free of interferences, but the use of the stopped-flow technique essentially lowers their effects. An increase in sensitivity and selectivity can also be obtained through the on-line use of solvent extraction. This has been utilised in the determination of gallium with lumogallion [370], and for determining the herbal medicine berberine, which yields a much stronger fluorescence intensity in organic solvents than in water [371].

Fluorimetric FIA determinations of pharmaceutical analytes have been satisfactorily carried out in the systems with on-line reactors packed with various reactive solid phases. Hexacyanoferrate(III) retained on an anion-exchange resin is used for the oxidation of paracetamol, whose oxidation product is reacted with N,N′-dimethylformamide in order to enhance the fluorescence [372]. In a similar manner an open-tubular PVC reactor with iodine impregnated on the inner wall is employed in the determination of adrenaline [373].

Various strategies are applied in multicomponent determinations with fluorimetric detection. The above-mentioned determinations of four phenothiazine derivatives were based on the measurement of emission at different wavelengths [346]. The determination of traces of calcium and magnesium can be based on the fluorescence of the calcein complexes after the separation of analytes from their rare earth ion matrix on an on-line ion-exchange column [374]. To discriminate between calcium and magnesium, 8-quinolinol is used as a masking agent for magnesium and EGTA for calcium. The effects of pH have been utilised in the determination of mixtures of coumarins [375] and metal ion mixtures [376]. In the former system, scopoletine and umbelliferone are determined in an FIA system at two different pH values with a double-injection valve. The analytes exhibit pH-dependent excitation spectra, so they can be determined individually at the same wavelength using simultaneous injections of analyte mixture and different buffers. Ternary and quaternary mixtures of zinc, cadmium, lead, magnesium and aluminium can be analysed through the formation of fluorescent 8-quinolinol-5-sulphonic acid complexes at different pH. A pH gradient created in an FIA system has been used for this purpose, with the fluorescence-time scans produced being processed using a partial least squares algorithm [376].

Two-component determinations of metal ion mixtures have been developed in FIA fluorimetric systems using catalytic effects. Determinations of Cu(II) and Hg(II) have been based on their catalytic effects on the oxidation of various hydrazones. The system is used in stopped-flow mode, with two sample segments being merged before the detector after undergoing different chemical reactions (Fig. 8) [377]. Sequential and differential catalytic-fluorimetric methods have also been developed for the simultaneous determination of Mn(II) and Fe(III) based on their catalytic action on the oxidation of salicylaldehyde thiosemicarbazone by hydrogen peroxide [378].

Native cerium(III) fluorescence has been exploited for the simultaneous determination of Ce(III) and Ce(IV) in an FIA system with a zinc reductor minicolumn and sample splitting between two branches of the manifold [379].

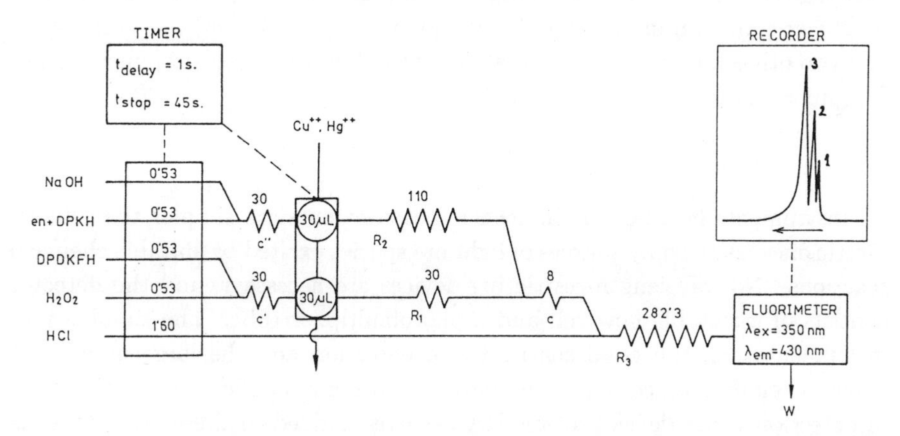

Fig. 8. Flow injection manifold used for the simultaneous catalytic-fluorimetric determination of copper and mercury [377]. The flow rates are given in ml.min^{-1}, reactor lengths in cm. DPKH — 2,2′-dipyridyl ketone hydrazone; DPDKFH — dipyridylketone phenylhydrazone; 1 — baseline; 2 — Cu(II); 3 — Hg(II). (*Reprinted by permission of copyright owner.*)

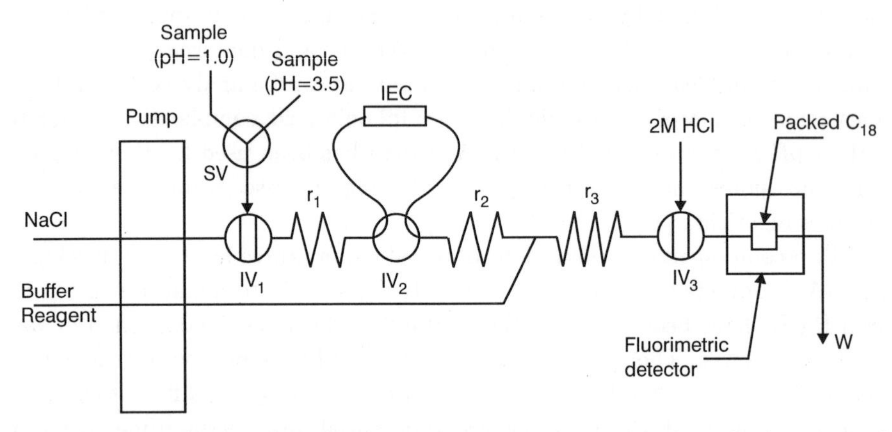

Fig. 9. Manifold for the fluorimetric speciation of aluminium based on the formation of the fluorescent Al-salicylaldehyde picolinohydrazone complex [380]. SV — switching valve; IV — injection valve; r — reactors; IEC — ion-exchange column; W — waste. (*Reprinted by permission of copyright owner.*)

A fluorimetric system with a C18 sorbent packed in a flow cell has been developed for the speciation of five different forms of aluminium [380] (Fig. 9). By adjusting the injected sample to different pH values and using an anion-exchange column, three forms of aluminium (i.e. acid reactive aluminium, total monomeric aluminium and non-labile monomeric aluminium) are determined and two other forms can be calculated. Table 7 lists developed applications of FIA fluorimetry.

5.2. *Chemiluminescence Detection*

Instrumentally, chemiluminescence measurements are simpler than fluorimetric ones, as the only sources of light are species excited by suitable chemical reactions. No wavelength-restricting devices are necessary, and the detector consists of a suitable flow cell and a photomultiplier tube. The simplicity of instrumentation, the good sensitivity of detection and the large number of known chemiluminescent systems have led to many applications of this type of detection being developed for FIA systems. Added to these advantages is the ability to design detectors and systems for solution manipulation which are specifically for chemiluminescent detection.

The most frequently used reagent for chemiluminescence detection in FIA is luminol (5-amino-2,3-dihydro-1,4-phthalazine dione), which, when oxidised

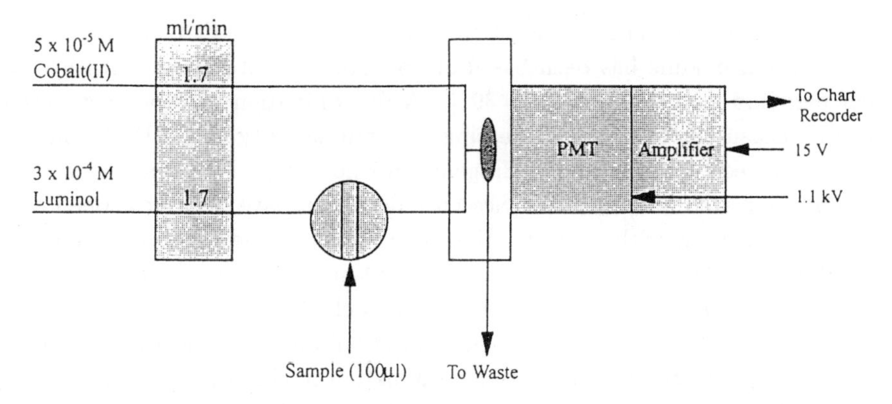

Fig. 10. Flow injection manifold for the chemiluminescence-based determination of hydrogen peroxide [383]. The distance from the injection valve to the zone merging point is 20 cm and the distance from the point of merging to the flow cell is 2.0 cm. (*Reprinted by permission of copyright owner.*)

to 3-aminophthalate, produces blue chemiluminescence. Some transition metal ions, and also a number of organic species, have a catalytic or an inhibiting effect on the luminol reaction with hydrogen peroxide or oxygen, and this is employed for analytical determination. Luminol, in the presence of Cu(II) [382] or Co(II) [383], can be used for the very sensitive detection of hydrogen peroxide. The simplest system used for the determination of H_2O_2 in sea water is shown in Fig. 10. Several procedures have been developed for the determination of transition metal ions. Determinations of Co(II) [383] and Fe(II) [384] are based on the catalytic effect of these ions on luminol oxidation by hydrogen peroxide. In the latter case, additional on-line preconcentration on 8-quinolinol column is employed. The same effect has been utilised in the determination of gold as tetrachloroaurate(III) in a reversed micellar system [385]. Determinations of Zn and Cd have been based on their inhibition of the cobalt-catalysed generation of chemiluminescence from luminol [387]. In determinations of inorganic anions, luminol has been used in procedures reported for bromide [388], nitrate [389] and nitrite [390]. The determination of bromide is based on the homogeneous catalysis of the bromine generated from the bromate-bromide-acid decomposition of hydrogen peroxide, which is necessary for luminol chemiluminescence. The nitrate determination is carried out in the system with on-line photoactivation to give peroxonitrite, which oxidises luminol. In all procedures for the determination of anions, cation-exchange columns are used for the removal of traces of transition metal ions.

The selective determination of chlorine dioxide in an FIA system with a gas-diffusion module has been based on the oxidation of luminol with hydrogen peroxide as a catalyst [212, 391]. A determination of protein is based on the measurement of the decreasing catalytic activity of Cu(II) in the reaction between luminol and hydrogen peroxide [392]. FIA systems with a gas-diffusion step and chemiluminescence detection have been developed also for determinations with reagents other than luminol. The simultaneous determinations of nitrate and nitrite are carried out in a setup where nitrogen oxide formed by the reduction of analytes produces chemiluminescence by reacting with ozone [393]. Selectivity of detection is obtained by selecting appropriate reducing agents. The selective determination of chlorine in a system with a gas-diffusion unit has also been reported, and is based on the reaction of hypochlorite ion with 2,4,5-triphenylimidazole [394].

A simpler design for chemiluminescence FIA systems can be achieved by using immobilised reagents. Two such systems have been developed for the determination of hydrogen peroxide, using the immobilised fluorophore 3-aminofluoranthene [395], or bis(2,4,6-tri-chlorophenyl)oxalate packed into a bed reactor [396]. In the first case, the detection is based on the chemiluminescent compound 1,1′-oxalyldiimidazole, and the immobilised compound serves as an active intermediate, which is excited to the first singlet state. In the second case, the immobilised peroxyoxalate is a chemiluminescent reagent, and detection is carried out in the presence of perylene as sensitiser. A lower limit of detection (6 nM) is obtained for the second system.

Apart from the above-mentioned determination of nitrate and nitrite, the speciation determination with chemiluminescence detection has also been developed for iron [397]. This detection method is based on the reaction of brilliant sulphoflavin with hydrogen peroxide and Fe(II). In the FIA system, iron is preconcentrated on a cation-exchange column, and the determination of total iron is carried out after the reduction of Fe(III) with ascorbic acid.

In addition to the use of luminol and peroxyoxalate, several applications in FIA chemiluminescence have been reported for lucigenin. Luminescence is produced by the addition of either hydrogen peroxide or an organic reducing agent to an alkaline solution of lucigenin. FIA systems have been reported for the determination of reducing sugars [398] and ascorbic acid [399]. In the determination of sugars, the chemiluminescence is significantly enhanced when the analyte is incubated with sodium periodate prior to the reaction with lucigenin. The determination of ascorbic acid is carried out in a merging zones configuration with a cation-exchange column for the removal of transition metal ions and a photoreactor. Detection is based on the chemiluminescent reaction of lucigenin with the products from the photooxidation of the analyte, sensitised by toluidine blue.

Significant changes in chemiluminescent reactions can be caused by the use of ordered surfactant assemblies such as micelles, reversed micelles or bilayer aggregates. This has been employed in the determination of total chromium using flavin mononucleotide chemiluminescence [400]. After oxidation by hydrogen peroxide, this system exhibits chemiluminescence in the presence of Cu(II) and Cr(VI). The use of cationic surfactant micelles decreases the signal for Cu(II), but not for Cr(VI), and this is exploited in the determination of chromium.

Several other examples of FIA chemiluminescence systems are included in Table 8.

Table 8. Applications of chemiluminescence detection in FIA.

Analyte	Reagents	λ, nm	Detection limit	Reference
Ascorbic acid	Lucigenin, toluidine blue	440	0.2 nM	399
Au(III)	Luminol, hexadecyltrimethylammonium chloride		0.01 ppb	386
Bromide	Luminol, bromate	417	0.0625 ppb	388
Cd, Zn	Luminol, Co(II)	425	3 (Cd); 5 (Zn) ppb	387
Chlorine	Rhodamine GG		0.1 μM	401
	2,4,5-triphenylimidazole		75 ppb	394
Chlorine dioxide	Luminol, H_2O_2		5 ppb	391
Co(II)	Gallic acid, H_2O_2	643	0.04 ppb	402
		643	8 pM	403
	Luminol, H_2O_2	425	0.01 ppb	384
Cr(III), Cr(VI)	Flavin mononucleotide, H_2O_2, tetradecyldimethylbenzylammonium chloride		50 nM	400
Cyanide	Uranine, didodecyldimethylammonium bromide		1.0 ppb	404

continued

Table 8 (*continued*)

Analyte	Reagents	λ, nm	Detection limit	Reference
Dansylalanine	Bis(2,4-dinitrophenyl)-oxalate, H_2O_2		0.5 nM	405
Fe(II)	Luminol, H_2O_2		0.1 pM	389
Fe(II), total Fe	Brillant sulphoflavin, H_2O_2		0.45 nM	397
H_2O_2	Luminol, Co(II)		10 nM	382
	Luminol, Co(II)	440	5 nM	383
	1,1′-oxalyldiimidazole, 3-aminofluoranthene		10 nM	395
	Bis(2,4,6-trichlorophenyl) oxalate, perylene		6 nM	396
Hydrazine	Hypochlorite, Ni(II)		0.5 nM	406
Morphine	Permanganate, tetraphosphate	430	0.1 nM	407
Nitrate	Luminol		70 nM	389
Nitrate, nitrite	TI(III), iodide, ozone		0.7 ppb (NO_3^-) 0.35 ppb (NO_2^-)	393
Nitrite	Luminol, H_2O_2	454	1 nM	390
Proteins	Luminol, Cu(II), H_2O_2		0.1 ppm	392
Pyrogallol	Periodate, hydroxylamine	455	0.1 μM	408
Steroids	Lucigenin			398
	Ce(IV), sulphite		13–19 ppm	409
Sugars	Lucigenin, periodate		1 μM (glucose)	398
Sulphide	Fluoresceine, hypochlorite	520	40 ppb	384
Sulphite	Permanganate, riboflavin phosphate	510	90 ppb	410
Surfactants, non-ionic	Hypochlorite, rhodamine B		3 ppm	411

6. Other Molecular Spectroscopic Detection Methods

The advantages of flow injection methodology has led to its widespread application in analytical procedures involving less common spectroscopic detection methods. These adaptations usually result in the favourable mechanisation of determinations, better precision and a higher sampling rate.

The determination of sulphur anions based on their S2 emission peaks has been achieved using a conventional atomic absorption spectrometer with flame atomisation, operated in the emission mode [412]. The spectrometer is equipped with a specially designed sample holder support device and a circular emission burner. The water-cooled steel cavity is continuously situated within the flame and the sample is introduced into it with a carrier stream of water or hydrogen peroxide. Using a 3 μl sample volume, sulphide, sulphite and sulphate can be determined with detection limits between 20 and 50 ppb of sulphur. The sequential appearance of S_2 emission peaks due to the different thermal stability of the particular sulphur anions allows the resolution of the mixture of anions within 20 s.

Refractive index measurements are very commonly used to characterise chemical species or to evaluate the composition of binary liquid or gaseous mixtures. The concentration gradient of the sample in the solvent, produced during injection, creates a refractive index gradient, which can be used for detection in FIA measurements, similar to the detection method in liquid chromatography, which is based on a refractive index gradient produced by retention processes. Fundamentally, the refractive index of a substance is determined by measuring the change in direction of collimated radiation as it passes from one solution to another. For application as a detection method in flowing streams, a conventional spectrophotometer is used and a light beam deflection (Schlieren optics) is utilised. The signal obtained is dependent on both the magnitude of the concentration change and the distance over which change occurs. When the sample zone passes through a flow cell, the light is focussed into the centre of the detection cell and then dispersed from the centre. The detector registers an absorbance decrease followed by an increase, which results in the appearance of a double peak. The height of both peaks is a function of the analyte concentration in the injected sample. Normal FIA systems with sample injection are usually employed, but it has been demonstrated that the linear working range is narrower compared with the reversed FIA method [413]. This kind of detection has been employed in the determination of glycerol/water mixtures [414], the determination of glycol in deicing/antiicing fluids [413], and the determination of sucrose [415]. In the latter case, a fibre-optics-based detector is employed with an LED as the source of radiation.

Several applications in FIA systems have been reported for optical rotation detection, commonly called polarimetry. Optical rotation measurements

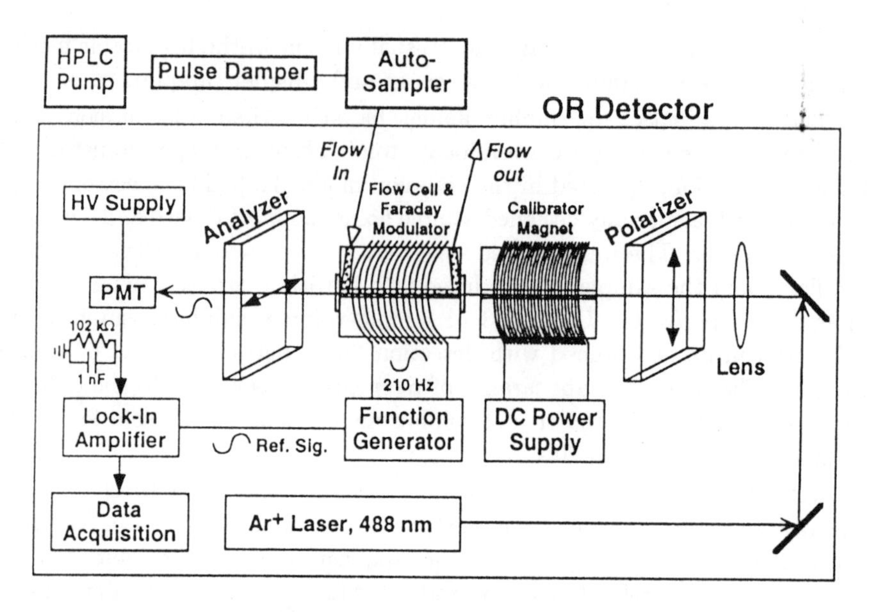

Fig. 11. Schematic diagram of a flow injection system with optical rotation detection [418]. (*Reprinted by permission of copyright owner.*)

are essential for the characterisation of chiral molecules as this is the only physical property that distinguishes the individual enantiomers. Compared with conventional polarimetry, FIA systems with optical rotation detection significantly reduce sample requirements, lower the detection limits by about two orders of magnitude, and shorten the analysis time to 1-2 min. The first report of a laser-based micropolarimeter for an FIA system describes its use in determining specific rotations from peak height data [416]. A commercially available optical rotation detector, based on a diode laser operating at 820 nm, has been used to determine the amount and enantiomeric purity of a drug in dosage form [417]. The FIA system shown schematically in Fig. 11 gives a 100-fold increase in the sensitivity of optical rotation detection compared with a high-quality conventional polarimeter [418]. It has been employed in the detection of sucrose, for measurements of the specific rotation values of various organic molecules, and for the analysis of the enantiomeric purity of $(1S, 2R)$-(+)-ephedrine.

Raman spectroscopy, based on the shifts in wavelength (from that of the incident beam) of the radiation scattered by certain molecules, is considered to be the complementary technique to infrared spectroscopy. An important advantage is that water does not cause interference, so Raman spectra can be obtained from aqueous solutions and this technique has been applied to the qualitative and quantitative analysis of inorganic, organic and biological systems. The intensity of the Raman signal is greatly enhanced at a rough surface, which is utilised in surface-enhanced Raman spectroscopy (SERS). Generally, this technique is used to study the interactions between molecules and a metal (usually silver) surface, but if the analyte can be reproducibly adsorbed and desorbed from the silver surface, SERS can be used for analytical purposes. There have already been several reports on the use of SERS as a detection method in FIA. The first applications have been developed for the detection of adsorbates such as *p*-aminobenzoic acid [419] and pararosaniline hydrochloride [420] on colloidal silver, although the adhesion of sol particles on the tubing wall is a difficult problem to overcome. When SERS is carried out at a silver electrode, an additional parameter that can be controlled is the potential applied to the electrode. This facilitates the control of adsorption and desorption of the analyte from the surface by the selection of a suitable polarisation potential. This approach has been used for the FIA detection of pyridine using a metallic silver disk electrode, which has to be mechanically polished before the beginning of a set of runs. Additionally, at the beginning of each injection, the surface is regenerated by a short oxidation-reduction cycle [421]. In order to obviate these steps, an *in situ* renewal of a silver substrate has been proposed [422]. Silver ions are introduced into the FIA system and electrodeposited on an inert glassy carbon electrode just before injection of the analyte. When the Raman spectrum of the adsorbate is recorded, the silver film is removed by anodic stripping, preparing the system for the next determination. Such a procedure has been applied in the determination of Fe(II) as its complex with 2,2′-bipyridine with a detection limit of 1 nM of Fe(II). Multicomponent determinations using this technique also seem to be possible.

7. References

1. K. Ueno, I. Imamura and K. L. Cheng, *Handbook of Organic Analytical Reagents* (CRC, Boca Raton, 1992).
2. E. B. Sandell and H. Onishi, *Colorimetric Determination of Traces of Metals* (Interscience, New York, 1978).
3. Z. Marczenko, *Separation and Spectrophotometric Determination of Elements* (Horwood, Chichester, 1985).

4. J. Ruzicka and E. H. Hansen, *Anal. Chim. Acta* **78** (1975) 145.
5. K. K. Stewart, G. R. Beecher and P. E. Hare, *Anal. Biochem.* **70** (1976) 167.
6. C. Thommen, A. Fromogeat, P. Obergfell and H. M. Widmer, *Anal. Chim. Acta* **234** (1990) 141.
7. D. S. Stone and J. F. Tyson, *Anal. Chim. Acta* 179 (1986) 427.
8. I. C. van Nugateren-Osinga, E. Hoogendam, M. Bos and E. E. van der Linden, *Anal. Chim. Acta* **239** (1990) 245.
9. J. T. Adkinson and J. C. Evans, *Anal. Chem.* **55** (1983) 2450.
10. Y. Yang and R. E. Hairrell, *Anal. Chem.* **56** (1984) 3002.
11. M. Trojanowicz, P. J. Worsfold and J. R. Clinch, *Trends Anal. Chem.* **7** (1988) 301.
12. P. K. Dasgupta, H. S. Bellamy, H. Liu, J. L. Lopez, E. L. Loree, K. Morris, P. Petersen and K. A. Mir, *Talanta* **40** (1993) 53.
13. D. Betteridge, E. L. Dagless, B. Fields and N. F. Graves, *Analyst* **103** (1978) 897.
14. S. Baban, *Anal. Proc.* (1980) 535.
15. D. Betteridge, *Fresenius Z. Anal. Chem.* **312** (1982) 441.
16. K. S. Johnson and R. L. Petty, *Limnol. Oceanogr.* **28** (1983) 1260.
17. D. J. Hooley and R. D. Dessy, *Anal. Chem.* 55 (1983) 313.
18. M. Trojanowicz, W. Augustyniak and A. Hulanicki, *Mikrochim. Acta* (1984) II, 17.
19. P. J. Worsfold, J. R. Clinch and H. Casey, *Anal. Chim. Acta* **197** (1987) 43.
20. J. R. Clinch, P. J. Worsfold and H. Casey, *Anal. Chim. Acta* **200** (1987) 523.
21. J. R. Clinch, P. J. Worsfold and F. W. Sweeting, *Anal. Chi. Acta* **214** (1988) 401.
22. M. Trojanowicz and J. Szpunar-Lobinska, *Anal. Chim. Acta* **230** (1990) 125.
23. R. L. Benson, P. J. Worsfold and F. W. Sweeting, *Anal. Chim. Acta* **238** (1990) 177.
24. M. Trojanowicz, J. Szpunar-Lobinska and Z. Michalski, *Mikrochim. Acta* (1991) I, 159.
25. I. Facchin and C. Pasquini, *Anal. Chim. Acta* **308** (1995) 231.
26. P. C. Hauser and D. W. L. Chiang, *Talanta* **8** (1993) 1193.
27. H. Liu and P. K. Dasgupta, *Anal. Chim. Acta* **289** (1994) 347.
28. M. K. Carroll, M. Conboy, A. Murfin and J. F. Tyson, *Anal. Chim. Acta* **295** (1994) 143.
29. M. Trojanowicz, E. Pobozy and J. Szpunar, *Chem. Anal. (Warsaw)* **35** (1990) 661.

29a. R. N. C. Daykin and S. J. Haswell, *Anal. Chim. Acta* **313** (1995) 155.

30. M. Valcarcel and M. D. Luque de Castro, *Analyst* **115** (1990) 699.
31. K. Yoshimura and H. Waki, *Talanta* **32** (1985) 345.
32. K. Yoshimura, *Anal. Chem.* **59** (1987) 2922.
33. K. Yoshimura, *Analyst* **113** (1988) 471.
34. F. Lazaro, M. D. Luque de Castro and M. Valcarcel, *Anal. Chim. Acta* **219** (1989) 231.

35. K. Yoshimura, S. Nawata and G. Kura, *Analyst* **115** (1990) 843.
36. D. Lacy, G. D. Christian and J. Ruzicka, *Anal. Chem.* **62** (1990) 1482.
37. J. Ruzicka and G. D. Christian, *Anal. Chim. Acta* **234** (1990) 31.
38. F. Lazaro, M. D. Luque de Castro and M. Valcarcel, *Anal. Chim. Acta* **214** (1988) 217.
39. S. M. Ramasamy, M. S. A. Jabbar and H. A. Mottola, *Anal. Chem.* **52** (1980) 2062.
40. G. K. Kirkbright, R. Narayanaswamy and N. A. Welti, *Analyst* **109** (1984) 15.
41. J. Ruzicka and E. H. Hansen, *Anal. Chim. Acta* **173** (1985) 3.
42. B. A. Woods, J. Ruzicka, G. D. Christian and R. J. Charlson, *Anal. Chem.* **58** (1986) 2496.
43. B. A. Woods, J. Ruzicka and G. D. Christian, *Anal. Chem.* **59** (1987) 2767.
44. B. A. Woods, J. Ruzicka, G. D. Christian, N. J. Rose and R. J. Charlson, *Analyst* **113** (1988) 301.
45. R. A. Leach, J. Ruzicka and J. M. Harris, *Anal. Chem.* **55** (1983) 1669.
46. E. A. G. Zagatto, M. A. Z. Arruda, A. O. Jacintho and I. L. Mattos, *Anal. Chim. Acta* **234** (1990) 153.
47. J. R. Ferreira, E. A. G. Zagatto, M. A. Z. Arruda and S. M. B. Brienza, *Analyst* **115** (1990) 779.
48. A. Daniel, D. Birot, M. Lehairte and J. Poncin, *Anal. Chim. Acta* **308** (1995) 413.
49. J. F. Tyson and A. B. Marsden, *Anal. Chim. Acta* **214** (1988) 447.
50. T. Yamane and H. Yamada, *Anal. Chim. Acta* **308** (1995) 433.
51. H. Bergamin F°., B. F. Reis and E. A. G. Zagatto, *Anal. Chim. Acta* **97** (1978) 427.
52. S. J. Chalk and J. F. Tyson, *Anal. Chem.* **66** (1994) 660.
53. E. A. G. Zagatto, A. O. Jacintho, J. Mortatti and H. Bergamin F°., *Anal. Chim. Acta* **120** (1980) 399.
54. A. O. Jacintho, E. A. G. Zagatto, B. F. Reis, L. C. R. Pessenda and F. J. Krug, *Anal. Chim. Acta* **130** (1981) 361.
55. K. S. Johnson and R. L. Petty, *Anal. Chem.* **54** (1982) 1185.
56. J. F. Tyson, *Analyst* **115** (1990) 587.
57. A. Rios, M. D. Luque de Castro and M. Valcarcel, *Talanta* **31** (1984) 673.
58. P. MacLaurin, K. S. Parker, A. Townshend, P. J. Worsfold, N. W. Barnett and M. Crane, *Anal. Chim. Acta* **238** (1990) 171.
59. M. E. Leon-Gonzalez, M. J. Santos-Delgado and L. M. Polo-Diez, *Anal. Chim. Acta* **219** (1989) 329.
60. A. Rios, M. D. Luque de Castro and M. Valcarcel, *Analyst* **109** (1984) 1487.
61. J. Ruz, A. Rios, M. D. Luque de Castro and M. Valcarcel, *Fresenius Z. Anal. Chem.* **322** (1985) 499.
62. J. Ruz, A. Torres, A. Rios, M. D. Luque de Castro and M. Valcarcel, *J. Autom. Chem.* **8** (1986) 70.
63. P. K. Dasgupta and V. K. Gupta, *Environ. Sci. Technol.* **20** (1986) 524.
64. S. J. Chalk and J. F. Tyson, *Talanta* **41** (1994) 1797.

65. P. K. Dasgupta, R. S. Vithanage and K. Petersen, *Anal. Chim. Acta* **215** (1988) 277.
66. A. Trojanek and S. Bruckenstein, *Anal. Chem.* **58** (1986) 983.
67. A. M. Almuaibed and A. Townshend, *Anal. Chim. Acta* **245** (1991) 115.
68. J. M. Calatayud and V. G. Mateo, *Anal. Chim. Acta* **264** (1992) 283.
69. Y. Israel and R. M. Barnes, *Anal. Chem.* **56** (1984) 1188.
70. Y. Israel and R. M. Barnes, *Analyst* **114** (1989) 843.
71. D. Betteridge and B. Fields, *Anal. Chem.* **50** (1978) 654.
72. D. A. Whitman and G. Christian, *Talanta* **36** (1989) 205.
73. J. Alonso-Chamarro, J. Bartroli and R. Barber, *Anal. Chim. Acta* **261** (1992) 219.
74. P. W. Alexander and A. Thalib, *Anal. Chem.* **55** (1983) 497.
75. C. Pasquini and W. A. de Oliviera, *Anal. Chem.* **57** (1985) 2575.
76. C. Pasquini, *Anal. Chem.* **58** (1986) 2346.
77. S. Liu and P. K. Dasgupta, *Anal. Chim. Acta* **283** (1993) 739.
78. T. Korenaga and H. Ikatsu, *Anal. Chim. Acta* **141** (1982) 301.
79. J. M. H. Appleton, J. F. Tyson and R. P. Mounce, *Anal. Chim. Acta* **179** (1986) 269.
80. M. L. Balconi, M. Borgarello, F. Ferraloni and F. Realini, *Anal. Chim. Acta* **261** (1992) 295.
81. B. C. Madsen and R. J. Murphy, *Anal. Chem.* **53** (1981) 1924.
82. O. Kondo, H. Miyata and K. Toei, *Anal. Chim. Acta* **134** (1982) 353.
83. T. J. Cardwell, R. W. Cattrall, M. Mitri and I. C. Hamilton, *Anal. Chim. Acta* **214** (1988) 433.
84. K. Yamamoto and S. Motomizu, *Anal. Chim. Acta* **246** (1991) 333.
85. E. A. Novikov, L. K. Shpigun and Yu. A. Zolotov, *Anal. Chim. Acta* **230** (1990) 157.
86. R. T. Edwards, I. D. McKelvie, P. C. Ferrett, B. T. Hart, J. B. Bapat and K. Kosky, *Anal. Chim. Acta* **261** (1992) 287.
87. P. K. Dasgupta and S. Liu, *Anal. Chem.* **66** (1994) 1792.
88. H. Bergamin F^{o}., F. J. Krug, B. F. Reis, J. A. Nobrega, M. Mesquita and I. G. Souza, *Anal. Chim. Acta* **214** (1988) 397.
89. D. Chen, F. Lazaro, M. D. Luque de Castro and M. Valcarcel, *Anal. Chim. Acta* **226** (1989) 221.
90. Z. Zhi, A. Rios and M. Valcarcel, *Analyst* **120** (1995) 2013.
91. J. T. Vanderslice and D. J. Higgs, *J. Micronutr. Anal.* **6** (1989) 109.
92. M. P. Granchi, J. A. Biggerstaff, L. J. Hillard and P. Gray, *Spectrochim. Acta* **B42** (1987) 169.
93. J. A. Garcia-Mesa, M. D. Luque de Castro and M. Valcarcel, *Anal. Chem.* **65** (1993) 3540.
94. D. Betteridge, T. J. Sly, A. P. Wade and J. E. W. Tillman, *Anal. Chem.* **55** (1983) 1292.
95. T. A. H. M. Janse, P. F. A. van der Wiel and G. Kateman, *Anal. Chim. Acta* **155** (1983) 89.

96. M. del Valle, M. Poch and J. Bartroli, *Anal. Chim. Acta* **241** (1990) 31.
97. A. P. Wade, P. M. Shiundu and P. D. Wentzell, *Anal. Chim. Acta* **237** (1990) 361.
98. R. G. J. Bellerby, D. R. Turner, G. E. Millward and P. J. Worsfold, *Anal. Chim. Acta* **309** (1995) 259.
99. N. Ishibashi and T. Imato, *Z. Anal. Chem.* **323** (1986) 244.
100. T. Imato and N. Ishibashi, *Anal. Sci.* **1** (1985) 481.
101. J. Ruzicka, E. H. Hansen and H. Mosbaek, *Anal. Chim. Acta* **92** (1977) 235.
102. A. Ramsing, J. Ruzicka and E. H. Hansen, *Anal. Chim. Acta* **129** (1981) 1.
103. J. Ruzicka, E. H. Hansen and A. U. Ramsing, *Anal. Chim. Acta* **134** (1982) 55.
104. J. S. Rhee and P. K. Dasgupta, *Mikrochim. Acta* (1985) III, 49.
105. H. L. Pardue and B. Fields, *Anal. Chim. Acta* **124** (1981) 39, 65.
106. T. J. Cardwell, R. W. Cattrall, G. J. Cross, G. J. Cross, G. R. O'Connel, J. D. Petty and G. R. Scollary, *Anal. Chim. Acta* **308** (1995) 197.
107. V. V. S. E. Dutt and H. A. Mottola, *Anal. Chem.* **47** (1975) 357.
108. H. Ma and J. Liu, *Anal. Chim. Acta* **261** (1992) 247.
109. W. J. M. Emaus and H. J. Henning, *Anal. Chim. Acta* **272** (1993) 245.
110. S. M. Ramasamy, A. Iob and H. A. Mottola, *Anal. Chem.* **51** (1979) 1637.
111. P. M. Siundu and A. P. Wade, *Anal. Chem.* **63** (1991) 692.
112. J. Wang and R. He, *Anal. Chim. Acta* **294** (1994) 195.
113. R. Liu, D. Liu, A. Sun and G. Liu, *Analyst* **120** (1995) 565.
114. H. Müller, V. Müller and E. H. Hansen, *Anal. Chim. Acta* **230** (1990) 113.
115. A. Safavi and A. A. Ensafi, *Anal. Chim. Acta* **252** (1991) 167.
116. S. Kawakubo, T. Katsumata, M. Iwatsuki, T. Fukasawa and T. Fukasawa, *Analyst* **113** (1988) 1827.
117. I. Ya. Kolotyrkina, L. K. Shpigun, Y. A. Zolotov and A. Malahoff, *Analyst* **120** (1995) 201.
118. Yu. A. Zolotov, L. K. Shpigun, I. Ya. Kolotyrkina, E. A. Novikov and O. V. Bazanova, *Anal. Chim. Acta* **200** (1987) 21.
119. C. Zhang, S. Kawakubo and T. Fukasawa, *Anal. Chim. Acta* **217** (1989) 23.
120. I. Ya. Kolotyrkina, L. K. Shpigun, Y. A. Zolotov and G. L. Tsysin, *Analyst* **116** (1991) 707.
121. J. A. Resing and M. J. Mottl, *Anal. Chem.* **64** (1992) 2682.
122. Z. L. Fang and S. K. Xu, *Anal. Chim. Acta* **145** (1983) 143.
123. L. C. R. Pessenda, A. O. Jacintho and E. A. G. Zagatto, *Anal. Chim. Acta* **214** (1988) 239.
124. E. A. Jones, *Anal. Chim. Acta* **156** (1984) 313.
125. S. Nakano, M. Tago and T. Kawashima, *Anal. Sci.* **5** (1989) 69.
126. I. C. van Nugteren-Osinga, M. Bos and W. E. van der Linden, *Anal. Chim. Acta* **226** (1989) 171.
127. D. C. Olsson, S. R. Bysouth, P. K. Dasgupta and V. Kuban, *Process Contr. Qual.* **5** (1994) 259.
128. W. Frenzel, *Anal. Chim. Acta* **291** (1994) 305.
129. M. T. Jeppesen and E. H. Hansen, *Anal. Chim. Acta* **214** (1988) 147.

130. P. J. Baxter, J. Ruzicka, G. D. Christian and D. C. Olsen, *Talanta* **41** (1994) 347.
131. D. Schepers, G. Schultze and W. Frenzel, *Anal. Chim. Acta* **308** (1995) 109.
132. L. N. Moskvin and J. Simon, *Talanta* **41** (1994) 1765.
133. P. Marstorp, T. Anfält and L. Anderson, *Anal. Chim. Acta* **149** (1983) 281.
134. J. Ruzicka and E. H. Hansen, *Anal. Chim. Acta* **99** (1978) 37.
135. D. Zölter and G. Schwedt, *Fresenius Z. Anal. Chem.* **317** (1984) 422.
136. B. Bouzid and A. M. G. MacDonald, *Anal. Chim. Acta* **207** (1988) 337.
137. D. J. Hawke and H. K. J. Powell, *Anal. Chim. Acta* **299** (1994) 257.
138. E. A. G. Zagatto, A. O. Jacintho, L. C. R. Pessenda, F. J. Krug, B. F. Reis and H. Bergamin, *Anal. Chim. Acta* **125** (1981) 37.
139. O. Røyset, *Anal. Chim. Acta* **185** (1986) 75.
140. T. Mochizuki, Y. Toda and R. Kuroda, *Talanta* **29** (1982) 659.
141. K. K. Verma and K. K. Stewart, *Anal. Chim. Acta* **214** (1988) 207.
142. W. Frenzel, F. Titzenhalter and S. Ebel, *Talanta* **41** (1994) 1965.
143. F. J. Krug, J. Mortati, L. C. R. Pessenda, E. A. G. Zagatto and H. Bergamin, *Anal. Chim. Acta* **125** (1981) 29.
144. I. Sekerka and J. F. Lechner, *Anal. Chim. Acta* **234** (1990) 199.
145. T. Mochizuki and R. Kuroda, *Fresenius Z. Anal. Chem.* **309** (1981) 363.
146. J. Szpunar-Lobinska, *Anal. Chim. Acta* **251** (1991) 275.
147. T. Anfält and S. Twengström, *Anal. Chim. Acta* **179** (1986) 453.
148. P. I. Anagnostopoulou and M. A. Koupparis, *Anal. Chem.* **58** (1986) 322.
149. S. Motomizu and T. Yoden, *Anal. Chim. Acta* **261** (1992) 461.
150. E. H. Hansen, J. Ruzicka and A. K. Ghose, *Anal. Chim. Acta* **100** (1978) 151.
151. J. T. Adkinson and J. C. Evans, *Anal. Chem.* **55** (1983) 2450.
152. G. Nakagawa, H. Wada and C. Wei, *Anal. Chim. Acta* **145** (1983) 135.
153. S. Motomizu, M. Oshima, N. Yoneda and T. Iwachido, *Anal. Sci.* **6** (1990) 215.
154. J. Nyman and A. Ivaska, *Talanta* **40** (1993) 95.
155. D. Klinghoffer, J. Ruzicka and E. H. Hansen, *Talanta* **27** (1980) 169.
156. J. A. G. Neto, H. Bergamin, E. A. G. Zagatto and F. J. Krug, Anal. Chim. Acta 308 (1995) 439.
157. J. Ruzicka, J. W. B. Stewart and E. A. G. Zagatto, *Anal. Chim. Acta* **81** (1976) 387.
158. J. Slanina, F. Bakker, A. Bruyn-Hes and J. J. Möls, *Anal. Chim. Acta* **113** (1980) 331.
159. J. F. van Staden, *Fresenius Z. Anal. Chem.* **322** (1985) 36.
160. B. Rössner and G. Schwedt, *Fresenius Z. Anal. Chem.* **315** (1983) 197.
161. D. J. Leggett, N. H. Chen and D. S. Mahadevappa, *Analyst* **107** (1982) 433.
162. D. J. Leggett, N. H. Chen and D. S. Mahadevappa, *Fresenius Z. Anal. Chem.* **315** (1983) 47.
163. M. Zenki, H. Komatsubara and K. Toei, *Anal. Chim. Acta* **208** (1988) 317.
164. K. K. Verma, A. Jain and A. Townshend, *Anal. Chim. Acta* **261** (1992) 233.
165. E. Pobozy, K. Pyrzynska, B. Szostek and M. Trojanowicz, *Microchem. J.* **51** (1995) 379.

166. D. G. Themelis, D. W. Wood and G. Gordon, *Anal. Chim. Acta* **225** (1989) 247.
167. A. M. Dietrich, T. D. Ledder, D. L. Gallagher, M. N. Grabeel and R. C. Hoehn, *Anal. Chem.* **64** (1992) 496.
168. M. Martinelli, H. Bergamin, M. A. Z. Arruda and E. A. G. Zagatto, *Quim. Anal.* **8** (1989) 129.
169. K. Pyrzynska, Z. Janiszewska, J. Szpunar-Lobinska and M. Trojanowicz, *Analyst* **119** (1994) 1553.
170. H. Baadenhuijsen and H. E. H. Seuren-Jacobs, *Clin. Chem.* **25** (1979) 443.
171. S. S. Jørgensen and M. A. B. Regitano, *Analyst* **105** (1980) 292.
172. J. C. de Andrade, J. C. Rocha, C. Pasquini and N. Baccan, *Analyst* **108** (1983) 621.
173. T. Sakai, H. Ohta, N. Ohno and J. Imai, *Anal. Chim. Acta* **308** (1995) 446.
174. R. Kuroda and T. Mochizuki, *Talanta* **28** (1981) 389.
175. A. Rios, M. D. Luque de Castro and M. Valcarcel, *Analyst* **110** (1985) 277.
176. R. M. Smith and T. G. Hurdley, *Anal. Chim. Acta* **166** (1984) 271.
177. J. Szpunar-Lobinska and M. Trojanowicz, *Anal. Sci.* **6** (1990) 415.
178. V. Kuban and F. Ingman, *Anal. Chim. Acta* **245** (1991) 251.
179. Z. Zhu and Z. Fang, *Anal. Chim. Acta* **198** (1987) 25.
180. K. G. Miller, G. E. Pacey and G. Gordon, *Anal. Chem.* **57** (1985) 734.
181. M. Milla, R. M. de Castro, M. Garcia-Vardas and J. A. Muñoz-Leyva, *Anal. Chim. Acta* **179** (1986) 289.
182. J. Mortatti, F. J. Krug, L. C. R. Pessenda, E. A. G. Zagatto and S. S. Jorgensen, *Analyst* **107** (1982) 659.
183. M. L. Balconi, F. Sigon, M. Borgarello, R. Ferraroli and F. Realini, *Anal. Chim. Acta* **234** (1990) 167.
184. H. Ishii, M. Aoki, T. Aita and T. Odashima, *Anal. Sci.* **2** (1986) 125.
185. N. Ohno and T. Sakai, *Analyst* **112** (1987) 1127.
186. V. V. S. E. Dult, D. Scheeler and H. A. Mottola, *Anal. Chim. Acta* **94** (1977) 289.
187. A. T. Senior and J. D. Glennon, *Anal. Chim. Acta* **196** (1987) 333.
188. H. Wada, H. Mori and G. Nakagawa, *Anal. Chim. Acta* **172** (1985) 297.
189. Z. Fang, Z. Zhu, S. Zhang, S. Xu, L. Guo and L. Sun, *Anal. Chim. Acta* **214** (1988) 41.
190. M. S. Abdel-Latif and G. G. Guilbault, *Anal. Lett.* **22** (1989) 1355.
191. J. L. P. Pavon and B. M. Cordero, *Analyst* **117** (1992) 215.
192. D. J. Legett, N. H. Chen and D. S. Mahadevappa, *Fresenius Z. Anal. Chem.* **311** (1982) 687.
193. I. Nordin-Andersson, O. Åström and A. Cedergren, *Anal. Chim. Acta* **162** (1984) 9.
194. K. Kimura, S. Iketani, H. Sakamoto and T. Shono, *Analyst* **115** (1990) 1251.
195. H. Wada, A. Yuchi and G. Nakagawa, *Anal. Chim. Acta* **149** (1983) 291.
196. J. T. Adkinson and J. C. Evans, *Anal. Chem.* **55** (1983) 2450.
197. M. F. Gine, E. A. G. Zagatto and H. Bergamin, *Analyst* **104** (1979) 371.

198. K. Oguma, K. Nishiyama and R. Kuroda, *Anal. Sci.* **3** (1987) 251.
199. H. Bergamin, J. X. Medeiros, B. F. Reis and E. A. G. Zagatto, *Anal. Chim. Acta* **101** (1078) 9.
200. J. L. Pavon, B. M. Cordero, J. H. Mendez and J. C. Miralles, *Analyst* **114** (1989) 849.
201. H. Bergamin, B. F. Reis, A. O. Jacintho and E. A. G. Zagatto, *Anal. Chim. Acta* **117** (1980) 81.
202. F. J. Krug, B. F. Reis, M. F. Gine, E. A. G. Zagatto, J. F. Ferreira and A. O. Jacintho, *Anal. Chim. Acta* **151** (1983) 39.
203. A. Cerda, M. T. Oms, R. Forteza and V. Cerda, *Anal. Chim. Acta* **311** (1995) 165.
204. G. Svensson and T. Anfält, *Clin. Chim. Acta* **119** (1982) 7.
205. W. D. Basson and J. F. van Staden, *Analyst* **103** (1978) 998.
206. R. Purohit and S. Devi, *Analyst* **120** (1995) 555.
207. B. C. Madsen, *Anal. Chim. Acta* **124** (1981) 437.
208. J. F. van Staden, A. E. Joubert and H. R. van Vliet, *Fresenius Z. Anal. Chem.* **325** (1986) 150.
209. R. Nakata, M. Terashita, A. Nitta and K. Ishikawa, *Analyst* **115** (1990) 425.
210. S. Motomizu and M. Sanada, *Anal. Chim. Acta* **308** (1995) 406.
211. S. Nakashima, M. Yagi, M. Zenki, M. Takahasi and K. Toei, *Fresenius Z. Anal. Chem.* **319** (1984) 506.
212. G. E. Pacey, D. A. Hollowell, K. G. Miller, M. R. Straka and G. Gordon, *Anal. Chim. Acta* **179** (1986) 259.
213. M. R. Straka, G. Gordon and G. E. Pacey, *Anal. Chem.* **57** (1985) 1799.
214. D. Klinghoffer, J. Ruzicka and E. H. Hansen, *Talanta* **27** (1980) 169.
215. M. Harriot and D. T. Burns, *Anal. Proc.* **26** (1989) 315.
216. D. T. Burns, N. Chimpalee and M. Harriott, *Anal. Chim. Acta* **217** (1989) 177.
217. M. Trojanowicz, E. Pobozy and J. Szpunar, *Chem. Anal. (Warsaw)* **35** (1990) 661.
218. J. Möller and M. Martin, *Fresenius Z. Anal. Chem.* **329** (1988) 728.
219. W. Frenzel, J. Oleksy-Frenzel and J. Möller, *Anal. Chim. Acta* **261** (1992) 253.
220. A. Kojlo, W. Wolyniec, H. Puzanowska-Tarasiewicz, J. Poltorak and A. Grudniewska, *Chem. Anal. (Warsaw)* **37** (1992) 253.
221. J. P. Susanto, M. Oshima, S. Motomizu, H. Mikasa and Y. Hori, *Analyst* **120** (1995) 187.
222. M. Aoyagi, Y. Yasumasa and A. Nishida, *Anal. Chim. Acta* **214** (1988) 229.
223. J. A. G. Mesa, P. Linares, M. D. Luque de Castro and M. Valcarcel, *Anal. Chim. Acta* **235** (1990) 441.
224. M. de la Guardia, K. D. Khalaf, V. Carbonell and A. Morales-Rubio, *Anal. Chim. Acta* **308** (1995) 462.
225. M. de la Guardia, K. D. Khalaf, B. A. Hasan, A. Morales-Rubio and V. Carbonell, *Analyst* **120** (1995) 231.
226. J. Thomsen, K. S. Johnson and R. L. Petty, *Anal. Chem.* **55** (1983) 2378.

227. L. F. Capitan-Vallvey, M. C. Valencia and G. Miron, *Anal. Chim. Acta* **289** (1994) 365.
228. S. Nakashima, M. Yagi, M. Zenki, M. Doi and K. Toei, *Fresenius Z. Anal. Chem.* **317** (1984) 29.
229. D. J. Leggett, N. H. Chen and D. S. Mahadevappa, *Anal. Chim. Acta* **128** (1981) 163.
230. S. M. Ramasamy and H. A. Mottola, *Anal. Chem.* **54** (1982) 283.
231. M. Yamada, T. Nakada and S. Suzuki, *Anal. Chim. Acta* **147** (1983) 401.
232. J. Ruzicka and E. H. Hansen, *Anal. Chim. Acta* **114** (1980) 19.
233. J. Möller and B. Winter, *Fresenius Z. Anal. Chem.* **320** (1985) 451.
234. Z. Zhi, A. Rios and M. Valcarcel, *Analyst* **120** (1995) 2013.
235. J. Bartroli, M. Escalda, C. J. Jorquera and J. Alonso, *Anal. Chem.* **63** (1991) 2532.
236. J. Kawase, A. Nakae and Y. Yamanaka, *Anal. Chem.* **51** (1979) 1640.
237. M. del Valle, J. Alonso, J. Bartroli and I. Marti, *Analyst* **113** (1988) 1677.
238. F. Canete, A. Rios, M. D. Luque de Castro and M. Valcarcel, *Anal. Chem.* **60** (1988) 2354.
239. Y. Hirai and K. Tomokuni, *Anal. Chim. Acta* **167** (1985) 409.
240. S. Motomizu and M. Kobayashi, *Anal. Chim. Acta* **261** (1992) 471.
241. J. Kawase, *Anal. Chem.* **52** (1980) 2124.
242. M. J. Whitaker, *Anal. Chim. Acta* **179** (1986) 459.
243. M. Strandberg and S. Thelander, *Anal. Chim. Acta* **145** (1983) 219.
244. M. Munoz, J. Alonso, J. Bartroli and M. Valiente, *Analyst* **115** (1990) 315.
245. T. P. Lynch, A. F. Taylor and J. N. Wilson, *Analyst* **108** (1983) 470.
246. E. A. Jones, *Anal. Chim. Acta* **169** (1985) 109.
247. R. Kuroda and T. Mochizuki, *Talanta* **28** (1981) 389.
248. P. B. Martelli, B. F. Reis, E. A. M. Kronka, H. Bergamin, M. Korn, E. A. G. Zagatto, J. L. F. C. Lima and A. N. Araujo, *Anal. Chim. Acta* **308** (1995) 397.
249. A. Rios, M. D. Luque de Castro and M. Valcarcel, *Anal. Chem.* **57** (1985) 1803.
250. A. R. A. Nogueira, S. M. B. Brienza, E. A. G. Zagatto, J. L. F. C. Lima and A. N. Araujo, *Anal. Chim Acta* **276** (1993) 121.
251. E. A. G. Zagatto, H. Bergamin, S. M. B. Brienza, M. A. Z. Arruda, A. R. A. Nogueira and J. L. F. C. Lima, *Anal. Chim. Acta* **261** (1992) 59.
252. M. D. Luque de Castro and M. Valcarcel, *Trends Anal. Chem.* **5** (1986) 71.
253. L. Anderson, *Anal. Chim. Acta* **110** (1979) 123.
254. J. F. van Staden, *Fresenius J. Anal. Chem.* **346** (1993) 723.
255. T. P. Lynch, N. J. Kernoghan and J. N. Wilson, *Analyst* **109** (1984) 843.
256. K. Sonne and P. K. Dasgupta, *Anal. Chem.* **63** (1991) 427.
257. A. T. Faizullah and A. Townshend, *Anal. Chim. Acta* **179** (1986) 233.
258. D. Espersen and A. Jensen, *Anal. Chim. Acta* **108** (1979) 241.
259. H. Kagenow and A. Jensen, *Anal. Chim. Acta* **114** (1980) 227.
260. H. Kagenow and A. Jensen, *Anal. Chim. Acta* **145** (1983) 125.
261. D. Betteridge and B. Fields, *Frsenius Z. Anal. Chem.* **314** (1983) 386.

262. A. Fernandez, M. D. Luque de Castro and M. Valcarcel, *Anal. Chem.* **56** (1984) 1146.
263. A. O. Jacintho, E. A. M. Kronka, E. A. G. Zagatto, M. A. Z. Arruda and J. R. Ferreira, *J. Flow Injection Anal.* **6** (1989) 19.
264. D. A. Whitman, G. D. Christian and J. Ruzicka, *Anal. Chim. Acta* **214** (1988) 197.
265. D. A. Whitman, M. B. Seasholtz, G. D. Christian, J. Ruzicka and B. R. Kowalski, *Anal. Chem.* **63** (1991) 775.
266. J. Alonso, J. Bartroli, M. del Valle and R. Barber, *Anal. Chim. Acta* **219** (1989) 345.
267. S. Kozuka, K. Saito, K. Oguma and R. Kuroda, *Analyst* **115** (1990) 431.
268. R. M. Liu, D. J. Liu and A. L. Sun, *Talanta* **40** (1993) 511.
269. J. C. de Andrade, J. C. Rocha and N. Baccan, *Analyst* **110** (1985) 197.
270. J. Martinez Calatayud, R. M. Albert and P. Camplco, *Anal. Lett.* **20** (1987) 1379.
271. H. Wada, K. Asakura, G. V. Rattaiah and G. Nakagawa, *Anal. Chim. Acta* **214** (1988) 439.
272. A. Rios, M. D. Luque de Castro and M. Valcarcel, *Anal. Chem.* **58** (1986) 663.
273. R. Liu, D. Liu, A. Sun and G. Liu, *Analyst* **120** (1995) 569.
274. J. Ruz, A. Rios, M. D. Luque de Castro and M. Valcarcel, *Talanta* **33** (1986) 199.
275. F. Lazaro, A. Rios, M. D. Luque de Castro and M. Valcarcel, *Analusis* **14** (1986) 378.
276. M. J. P. Gerritsen, G. Kateman, M. A. J. van Opstal, W. P. van Bennekom and B. G. M. Vendeginste, *Anal. Chim. Acta* **241** (1990) 23.
277. F. Lazaro, A. Rios, M. D. Luque de Castro and M. Valcarcel, *Anal. Chim. Acta* **179** (1986) 279.
278. H. Wada, T. Murakawa and G. Nakagawa, *Anal. Chim. Acta* **200** (1987) 515.
279. A. G. Melgarejo, J. M. C. Pavon and A. R. Castro, *Anal. Chim. Acta* **241** (1990) 153.
280. W. Lindberg, G. D. Clark, C. P. Hanna, D. A. Whitman, G. D. Christian and J. Ruzicka, *Anal. Chem.* **62** (1990) 849.
281. M. Blanco, J. Gene, H. Hurriaga, S. Maspoch and J. Riba, *Talanta* **34** (1987) 987.
282. V. Kuban and D. B. Gladilovich, *Coll. Czech. Chem. Commun.* **53** (1988) 1461.
283. V. Kuban, D. B. Gladilovich, L. Sommer and P. Popov, *Talanta* **36** (1989) 463.
284. P. Dolezal and V. Kuban, *Coll. Czech. Chem. Commun.* **53** (1988) 1162.
285. B. Bermudez, F. Lazaro, M. D. Luque de Castro and M. Valcarcel, *Analyst* **112** (1987) 535.
286. B. Fernandez-Band, F. Lazaro, M. D. Luque de Castro and M. Valcarcel, *Anal. Chim. Acta* **229** (1990) 177.
287. L. Dou and I. S. Krull, *J. Chromatogr.* **499** (1990) 685.
288. S. H. Han, K. S. Lee, G. S. Cha and M. Trojanowicz, *J. Chromatogr.* **648** (1993) 283.

289. P. Richter, J. M. Fernandez-Romero, M. D. Luque de Castro and M. Valcarcel, *Chromatographia* **34** (1992) 445.
290. J. Szpunar-Lobinska, M. Ceulemans, R. Lobinski and F. C. Adams, *Anal. Chim. Acta* **278** (1993) 99.
291. Y. Hirai, N. Yoza and S. Ohashi, *Anal. Chim. Acta* **115** (1980) 269.
292. I. D. McKelvie, B. T. Hart, T. J. Cardwell and R. W. Cattrall, *Talanta* **12** (1993) 1981.
293. Y. Narusawa, T. Katsura and F. Kato, *Fresenius Z. Anal. Chem.* **332** (1988) 162.
294. Y. Narusawa, *Anal. Chim. Acta* **204** (1988) 53.
295. S. Motomizu and M. Onoda, *Anal. Chim. Acta* **214** (1988) 289.
296. T. Yamane, K. Watanabe and H. A. Mottola, *Anal. Chim. Acta* **207** (1988) 331.
297. G. del Campo, A. Irastorza and J. A. Casado, *Fresenius J. Anal. Chem.* **352** (1995) 557.
298. J. Slanina, F. Bakkaer, A. G. M. Bruijn-Hes and J. J. Möls, *Fresenius Z. Anal. Chem.* **289** (1978) 38.
299. T. Mochizuki, Y. Toda and R. Kuroda, *Talanta* **29** (1982) 659.
300. R. M. Alonso, R. M. Jimenez, A. Carvajal, J. Garcia, F. Vicente and L. Hernandez, *Talanta* **36** (1989) 761.
301. Y. Sahleström, S. Twengström and B. Karlberg, *Anal. Chim. Acta* **187** (1986) 339.
302. R. G. Mechler, *Anal. Chim. Acta* **214** (1988) 299.
303. A. Jain, A. Chaurasia and K. K. Verma, *Talanta* **42** (1995) 779.
304. A. F. Danet and J. M. Calatayud, *Talanta* **41** (1994) 2147.
305. R. M. Smith and T. G. Hurdley, *Anal. Chim. Acta* **166** (1984) 271.
306. A. Kojlo, J. Michalowski and M. Trojanowicz, *Anal. Chim. Acta* **228** (1990) 287.
307. T. Lussier, R. Gilbert and J. Hubert, *Anal. Chem.* **64** (1992) 2201.
308. A. Al-Wehaid and A. Townshend, *Anal. Chim. Acta* **198** (1987) 45.
309. M. Nakamura, T. Mazuka and M. Yamashita, *Anal. Chem.* **56** (1984) 2242.
310. T. R. Williams, S. W. McElvary and E. C. Igholado, *Anal. Chim. Acta* **123** (1981) 351.
311. J. Hernandez-Mendelez, A. A. Mateos, M. J. A. Parra and C. G. de Maria, *Anal. Chim. Acta* **184** (1986) 243.
312. D. T. Burns, N. Chimpalee, D. Chimpalee and S. Rattanariderom, *Anal. Chim. Acta* **243** (1991) 187.
313. J. S. Rhee, P. K. Dasgupta and D. C. Olson, *Anal. Chim. Acta* **220** (1989) 55.
314. T. Korenaga, X. Zhou, K. Okada, T. Moriwake and S. Shinoda, *Anal. Chim. Acta* **272** (1993) 237.
315. T. Korenaga, *Bunseki Kagaku* **29** (1980) 22.
316. A. Tanaka, M. Miyazaki and T. Deguchi, *Anal. Lett.* **18** (1985) 695.
317. R. J. Hodges, *Proc. 4th Environ. Chem. Conf.*, Darwin, Australia, 1995, AO10-1.

318. M. Blanco, J. Gene, H. Iturriaga and S. Maspoch, *Analyst* **112** (1987) 619.
319. B. Fernandez-Band, F. Lazaro, M. D. Luque de Castro and M. Valcarcel, *Anal. Chim. Acta* **229** (1990) 177.
320. F. Navarro-Villoslada, L. V. Perez-Arribus, M. E. Leon-Gonzalez and L. M. Polo-Diez, *Anal. Chim. Acta* **308** (1995) 238.
321. J. Thomsen, K. S. Johnson and R. L. Petty, *Anal. Chem.* **55** (1983) 2378.
322. D. J. Curran and W. G. Collier, *Anal. Chim. Acta* **177** (1985) 259.
323. M. de la Guardia, S. Garrigues and M. Gallignani, *Anal. Chim. Acta* **261** (1992) 53.
324. S. Garrigues, M. Gallignani and M. de la Guardia, *Talanta* **40** (1993) 89.
325. M. Mallignani, S. Garrigues, A. Martinez-Vado and M. de la Guardia, *Analyst* **118** (1993) 1043.
326. S. Garrigues, M. Gallignani and M. de la Guardia, *Analyst* **117** (1992) 1849.
327. M. Gallignani, S. Garrigues and M. de la Guardia, *Anal. Chim. Acta* **274** (1993) 267.
328. M. de la Guardia, M. Gallignani and S. Garrigues, *Anal. Chim. Acta* **282** (1993) 543.
329. M. Gallignani, S. Garrigues, M. de la Guardia, J. L. Burguera and M. Burguera, *Talanta* **41** (1994) 739.
330. D. K. Morgan, N. D. Danielson and J. E. Katon, *Anal. Lett.* **18** (1985) 1979.
331. B. E. Miller, N. D. Danielson and J. E. Katon, *Appl. Spectr.* **42** (1988) 401.
332. S. V. Olesik, S. B. French and M. Novotny, *Anal. Chem.* **58** (1986) 2256.
333. E. Lopez-Anreus, S. Garrigues and M. de la Guardia, *Anal. Chim. Acta* **308** (1995) 28.
334. F. J. Krug, H. Bergamin F., E. A. G. Zagatto and S. S. Jørgensen, *Analyst* **102** (1977) 503.
335. S. Baban, D. Beetlestone, D. Betteridge and P. Sweet, *Anal. Chim. Acta* **114** (1980) 319.
336. J. F. van Staden, *Fresenius Z. Anal. Chem.* **310** (1982) 239.
337. J. F. van Staden, *Fresenius Z. Anal. Chem.* **312** (1982) 438.
338. F. J. Krug, E. A. G. Zagatto, B. F. Reis, O. Bahia, A. O. Jacintho and S. S. Jørgensen, *Anal. Chim. Acta* **145** (1983) 179.
339. J. F. van Staden, *Fresenius Z. Anal. Chem.* **326** (1987) 754.
340. R. E. Santelli, P. R. S. Lopes, R. C. L. Santelli and A. D. R. Wagner, *Anal. Chim. Acta* **300** (1995) 149.
341. S. M. B. Brienza, R. P. Sartini, J. A. Neto and E. A. G. Zagatto, *Anal. Chim. Acta* **308** (1995) 269.
342. F. J. Krug, J. Ruzicka and E. H. Hansen, *Analyst* **104** (1979) 47.
343. R. H. Taylor and J. W. Grate, *Talanta* **42** (1995) 257.
344. J. Martinez-Calatayud and C. M. Pastor, *Anal. Lett.* **44** (1990) 1371.
345. M. Valcarcel and M. D. Luque de Castro, in *Fluorescence Spectroscopy*, ed. O. S. Wolfbeis (Springer, New York, 1993).
346. B. Laassis, J. J. Aaron and M. C. Mahedro, *Talanta* **41** (1994) 1985.

347. R. L. Petty, W. C. Michel, J. P. Snow and K. S. Johnson, *Anal. Chim. Acta* **142** (1982) 299.
348. M. H. Memon and P. J. Worsfold, *Anal. Chim. Acta* **183** (1986) 179.
349. J. I. Braithwaite and J. N. Miller, *Anal. Chim. Acta* **106** (1979) 395.
350. Z. Genfa and P. K. Dasgupta, *Anal. Chem.* **61** (1989) 408.
351. S. Dong and P. K. Dasgupta, *Environ. Sci. Technol.* **21** (1987) 581.
352. C. Martinez-Lozano, T. Perez-Ruiz, V. Tomas and C. Abellan, *Analyst* **115** (1990) 217.
353. S. Motomizu, M. Oshima and Z. Jun, *Anal. Chim. Acta* **251** (1991) 269.
354. N. Chimpalee, D. Chimpalee, S. Suparuknari, B. Boonyanitchayakul and D. T. Burns, *Anal. Chim. Acta* **298** (1994) 401.
355. T. Perez-Ruiz, C. Martinez-Lozano, V. Thomas and R. Casajus, *Talanta* **42** (1995) 391.
356. N. Ishibashi, K. Kina and Y. Goto, *Anal. Chim. Acta* **114** (1980) 325.
357. V. Marco, F. Carillo, C. Perez-Cande and C. Camara, *Anal. Chim. Acta* **283** (1993) 489.
358. F. L. Boza, M. D. Luque de Castro and M. Valcarcel, *Analyst* **109** (1984) 333.
359. A. Tanaka, K. Obata and T. Deguchi, *Anal. Sci.* **2** (1986) 197.
360. R. Forteza, M. T. Oms, J. Cardenas and V. Cerda, *Analusis* **18** (1990) 491.
361. K. Mawatari, F. Iinuma and M. Watanabe, *Anal. Biochem.* **190** (1990) 88.
362. H. K. Chung and J. D. Ingle Jr., *Anal. Chem.* **62** (1990) 2547.
363. V. Kuban, J. Havel and B. Patockova, *Coll. Chech. Chem. Commun.* **54** (1989) 1777.
364. J. A. Resing and C. I. Measures, *Anal. Chem.* **66** (1994) 4105.
365. J. L. Nowicki, K. S. Johnson, K. H. Coale, V. A. Elrod and S. H. Lieberman, *Anal. Chem.* **66** (1994) 2732.
366. S. Motomizu, H. Mikasa and K. Toei, *Anal. Chim. Acta* **193** (1987) 343.
367. D. Chen, M. D. Luque de Castro and M. Valcarcel, *Anal. Chim. Acta* **230** (1990) 137.
368. P. Linares, M. D. Luque de Castro and M. Valcarcel, *Anal. Chim. Acta* **161** (1984) 257.
369. F. Lazaro, M. D. Luque de Castro and M. Valcarcel, *Anal. Chim. Acta* **165** (1984) 177.
370. T. Imasaka, T. Harada and N. Ishibashi, *Anal. Chim. Acta* **129** (1981) 195.
371. T. Sakai, N. Ohna, Y. S. Chung and H. Nishikawa, *Anal. Chim. Acta* **308** (1995) 329.
372. J. Martinez-Calatayud and C. G. Benito, *Anal. Chim. Acta* **231** (1990) 259.
373. A. Kojlo and J. Martinez-Calatayud, *Anal. Chim. Acta* **308** (1995) 334.
374. D. T. Thuy, D. Decnop-Weever, W. Th. Kok, P. Luan and T. V. Nghi, *Anal. Chim. Acta* **295** (1994) 151.
375. P. Solich, M. Polasek and R. Karlicek, *Anal. Chim. Acta* **308** (1995) 293.
376. N. Porter, B. T. Hart, R. Morrison and I. C. Hamilton, *Anal. Chim. Acta* **308** (1995) 313.

377. F. Lazaro, M. D. Luque de Castro and M. Valcarcel, *Fresenius Z. Anal. Chem.* **320** (1985) 128.
378. F. Lazaro, M. D. Luque de Castro and M. Valcarcel, *Anal. Chim. Acta* **169** (1985) 141.
379. K. H. Al-Sowdani and A. Townshend, *Anal. Chim. Acta* **179** (1986) 469.
380. P. Cañizares and M. D. Luque de Castro, *Anal. Chim. Acta* **295** (1994) 59.
381. A. Aihara, M. Arai and T. Taketatsu, *Analyst* **111** (1986) 641.
382. G. Rule and W. R. Seitz, *Clin. Chem.* **25** (1979) 1635.
383. D. Price, P. J. Worsfold and R. F. C. Mantoura, *Anal. Chim. Acta* **298** (1994) 121.
384. J. L. Burguera, A. Townshend and S. Greenfield, *Anal. Chim. Acta* **114** (1980) 209.
385. A. A. Alwarthan, K. A. J. Habib and A. Townshend, *Fresenius J. Anal. Chem.* **337** (1990) 848.
386. Imdadullah, T. Fujiwara and T. Kumamaru, *Anal. Chem.* **63** (1991) 2348.
387. J. L. Burguera, M. Burguera and A. Townshend, *Anal. Chim. Acta* **127** (1981) 199.
388. I. A. M. Shakir and A. T. Faizullah, *Analyst* **114** (1989) 951.
389. L. Renmin, L. Daojie, S. Ailing and L. Guihua, *Talanta* **42** (1995) 437.
390. P. Mikuska, Z. Vecera and Z. Zdrahal, *Anal. Chim. Acta* **316** (1995) 261.
391. D. A. Hollowell, J. R. Gord, G. Gordon and G. E. Pacey, *Anal. Chem.* **58** (1986) 1524.
392. T. Hara, M. Toriyama and K. Tsukagoshi, *Bull. Chem. Soc. Jpn.* **57** (1984) 1551.
393. T. Aoki and M. Wakabayashi, *Anal. Chim. Acta* **308** (1995) 308.
394. J. R. Gord, G. Gordon and G. E. Pacey, *Anal. Chem.* **60** (1988) 2.
395. M. Stigbrand, E. Ponten and K. Irgum, *Anal. Chem.* **66** (1994) 1766.
396. P. van Zoonen, D. A. Kamminga, C. Gooijer, N. H. Velthorst and R. W. Frei, *Anal. Chim. Acta* **167** (1985) 249.
397. V. A. Elrod, K. S. Johnson and K. H. Coale, *Anal. Chem.* **63** (1991) 893.
398. M. Maeda and A. Tsuji, *Anal. Sci.* **2** (1986) 183.
399. T. Perez-Ruiz, C. Martinez-Lozano and A. Sonz, *Anal. Chim. Acta* **308** (1995) 299.
400. H. Ohshima, M. Yamada and S. Suzuki, *Anal. Chim. Acta* **232** (1990) 385.
401. T. Nakagama, M. Yamada and T. Hobo, *Analyst* **114** (1989) 1275.
402. S. Nakahara, M. Yamada and S. Suzuki, *Anal. Chim. Acta* **141** (1982) 255.
403. C. M. Sakamoto-Arnold and K. S. Johnson, *Anal. Chem.* **59** (1987) 1789.
404. E. Yamada, C. Hamamura, K. Fukuda and M. Sato, *J. Flow Injection Anal.* **10** (1993) 48.
405. K. Honda, J. Sekino and K. Imal, *Anal. Chem.* **55** (1983) 940.
406. A. T. Faizullah and A. Townshend, *Anal. Proc.* **22** (1985) 15.
407. R. W. Abbott, A. Townshend and R. Gill, *Analyst* **111** (1986) 635.
408. N. P. Evmiridis, *Analyst* **113** (1988) 1051.
409. N. T. Defteros and A. C. Calokerinos, *Anal. Chim. Acta* **290** (1994) 190.

410. M. Yamada, T. Nakada and S. Suzuki, *Anal. Chim. Acta* **147** (1983) 401.
411. J. S. Lancaster, P. J. Worsfold and A. Lynes, *Anal. Chim. Acta* **239** (1990) 189.
412. J. L. Burguera and M. Burguera, *Anal. Chim. Acta* **157** (1984) 177.
413. A. Wijk and B. Karlberg, *Talanta* **3** (1994) 395.
414. D. Betteridge and J. Ruzicka, *Talanta* **23** (1976) 409.
415. J. Pawliszyn, *Anal. Chem.* **58** (1986) 3207.
416. P. D. Rice, Y. Y. Shao, S. R. Erskine, T. G. Teague and D. R. Bobbitt, *Talanta* **36** (1989) 473.
417. G. Liu, D. M. Goodall and J. S. Loran, *Anal. Proc.* **29** (1992) 31.
418. C. A. Goss, D. C. Wilson and W. E. Weiser, *Anal. Chem.* **66** (1994) 3093.
419. A. Berthod, J. J. Laserna and J. D. Winefordner, *Appl. Spectrosc.* **41** (1987) 1137.
420. R. D. Freeman, R. M. Hammaker, C. E. Meloan and W. G. Fateley, *Appl. Spectrosc.* **42** (1988) 456.
421. R. K. Force, *Anal. Chem.* **60** (1988) 1987.
422. V. J. P. Gouvela, I. G. Gutz and J. C. Rubim, *J. Electroanal. Chem.* **371** (1994) 37.

Chapter 2

Atomic Spectroscopy Detection Methods

Atomic spectroscopy detection methods are based on absorption, fluorescence or emission of radiation by atoms or ions. The molecular constituents of a sample are converted into atomic particles by various atomisation methods, such as flame and electrothermal techniques, inductively coupled argon plasmas, direct current plasmas, electric arcs or sparks. The spectra of gaseous atomic particles consist of narrow lines resulting from electronic transitions of the outermost electrons. The number of lines increases with the number of electrons in incomplete shells. The transition elements have an especially large number of closely spaced energy levels. Flame atomisation is used with absorption, emission and fluorescence procedures. Electrothermal atomisation is mostly used for absorption and fluorescence detection, whereas plasma sources are mainly employed for emission and fluorescence measurements. Atomic spectroscopy techniques are among the most selective and sensitive methods of detection used in chemical analysis and are widely used, particularly in the trace analysis of about 70 elements.

1. Atomic Absorption Spectroscopy

Analytical methods based on atomic absorption spectroscopy (AAS) are the most widely used of all atomic spectral methods. Their high selectivity results from unique electronic transition energies for each element. The atomised element to be determined is capable of absorbing radiation with wavelengths characteristic of electronic transitions. Atomic absorption spectra consist predominantly of resonance lines, which arise from the transition of atoms from the ground state to upper levels. The most common radiation sources for AAS methods are hollow cathode lamps, where the cathode is constructed of the element to be determined. The radiation emitted from the source is absorbed by the analyte, which has been atomised by a flame, or atomised electrothermally in a graphite furnace or a silica tube (in hydride generation techniques and cold vapour mercury measurements). The absorption signal in the detector rises to a maximum within a few seconds of ignition, then decays rapidly back to zero as the atomisation products escape into the surroundings. In flame atomisation methods, where good precision is usually attained, the most serious limitation is often insufficient detectability. With electrothermal atomisation, improved detectability is usually obtained at the expense of lower precision and a much longer time needed for one determination. In both cases, the analysis of a natural sample very frequently requires the pretreatment of the sample in order to reduce chemical interferences, to eliminate matrix effects or to preconcentrate trace amounts of the analyte.

The above-mentioned limitations can be reduced by coupling AAS measurements with flow injection sample processing. This was pioneered in 1979 by Zagatto *et al.* [1], who utilised the flow injection manifold to improve the sampling rate, and to reduce the consumption of lanthanum nitrate used as an interference suppressor. Since then, flow injection atomic absorption spectrometry (FIAAS) has been very intensively developed, as can be seen from numerous review articles in journals [2–7], and a book about the procedure [8]. Although in general FIA can be considered as a special technique for the delivery of a sample to the detector, FIAAS systems offer several advantages of a purely analytical nature. These include an improvement in nebulisation efficiency in certain circumstances and the ability to easily adjust the signal magnitude. The analytical measurements performed in FIAAS systems are less influenced by differences in the viscosity of samples or differences in matrix composition. FIAAS is also a convenient way to carry out measurements using different sample volumes and to perform on-line preconcentration operations.

1.1. *Flow Injection Systems with Flame Atomisation*

FIAAS is most frequently used in measuring systems with flame atomisation, where a solution of the sample, delivered by the flow injection manifold, is sprayed into a flame by means of a nebuliser. In the nebuliser, a liquid sample is converted into a fine spray or aerosol, which is then fed into the flame. Numerous papers on the optimisation of experimental conditions for sample injection and nebulisation in FIAAS systems have been published. Several authors indicate the advantages of using a flow rate in the flow system that is larger than the aspiration rate of the nebuliser [1, 9–11]. If the FIA flow rate is too low,

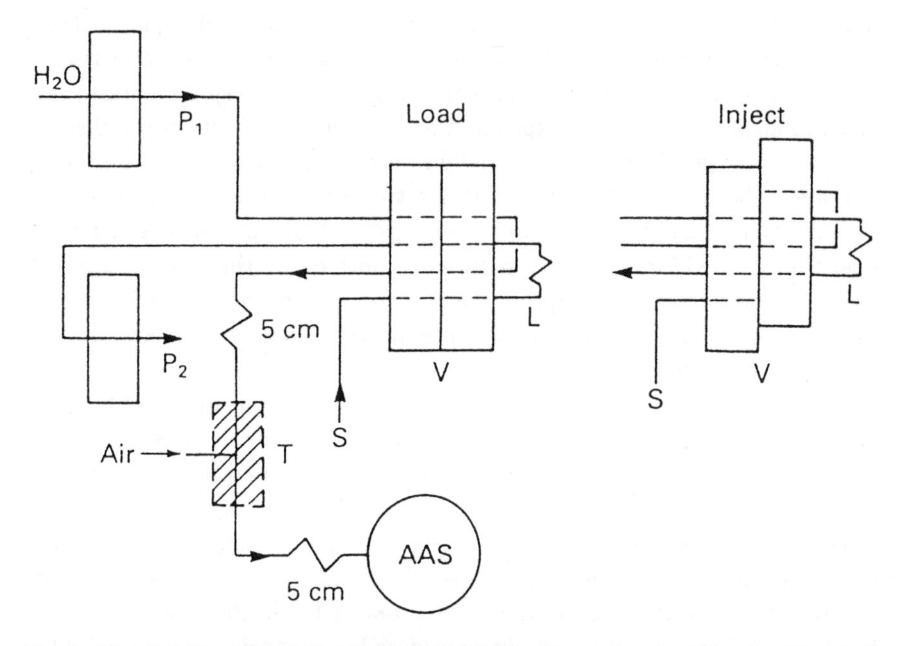

Fig. 1. Flow injection manifold with flame AAS detection and air or solvent compensation of the flow rate difference [16]. P_1 — peristaltic pump for carrier; P_2 — peristaltic pump for sample loading; L — sample loop; V — injection valve; S — sample; T — connector for air or solvent compensation; AAS — atomic absorption spectrometer. (*Reprinted by permission of copyright owner.*)

large errors in the determination may be obtained because of air suction by the nebuliser [1]. The use of a higher flow rate in the flow system enables signal enhancement in comparison with non-FIA aspiration [9]. However, one can also find many opposite opinions, arguing that the use of a flow rate smaller than the aspiration rate is advantageous [3,12–16]. In systems with a flow rate in the FIA manifold that is lower than the aspiration rate, the difference is made up by compensation aspiration of air or liquid [13, 14] (Fig. 1.). Better results obtained with air compensation were attributed to more effective liquid fragmentation [14]. An increase in the nebulisation efficiency with a lower FIA flow rate was also observed in an FIAAS system without any compensation aspiration (the flow injection valve and nebuliser being connected with a short piece of tubing) [15]. The peak signal was 12 times higher than for conventional measurements at the extremely small flow rate of 0.4 ml/min. It was, however, accompanied by a large decrease in the sampling frequency and an increase in the noise level. The results of a comparative study of FIAAS systems with FIA flow rates smaller than aspiration rates with air compensation and without compensation, showed a smaller signal in the air-compensated system than in the system without flow rate compensation [3].

The hydrodynamic conditions in the flow injection system, as well as the course of nebulisation in the FIAAS system, significantly influence the difference in analytical characteristics between FIAAS measurements and conventional flame AAS measurements. The reduction of the flow rate at a constant aspiration rate leads to a decrease in the peak height and to an increase in its area. It has been shown that thc dctcctability evaluated based on the noise level is much better for measurements of the peak area than the peak height [15]. The opinion that detectability obtained in FIAAS systems is comparable with that from conventional flame AAS measurements has been presented in several studies [10, 14, 16]. With computer data acquisition, an improvement in the detectability in FIA measurements with flame AAS detection can be obtained by appropriate digital filtering of data [17].

The use of a pump for the carrier solution in FIAAS systems is a source of additional noise and thus causes a decrease in measurement precision. Precision deteriorates with a decrease in the flow rate of the carrier solution [3, 13]. A significant improvement in precision has been observed when the volume of injected samples is increased [15]. Generally, in FIAAS systems, either with or without compensation of the flow rate difference, the precision obtained is comparable to that for conventional measurements and the RSD is close to 1% [13, 14, 16].

In the most developed FIAAS systems, the sample solution is injected using a rotary or sliding injection valve, or by continuous aspiration in time-based systems, where on-line preconcentration from a larger sample volume is carried out. Microsampling with the sequential introduction of samples and carrier through a sampling probe by a peristaltic pump, has been developed for clinical samples, with 120 μl of the sample consumed in each determination [18]. For the same purpose, air-transported sampling loading and hydrodynamic injection were applied in order to minimise dispersion and maintain the original level of flame AAS sensitivity with minimum sample consumption [19]. A variable-volume injector has been developed to avoid different dilutions of analysed samples, which differ greatly in their analyte content [20]. The injector has also been used in an FIAAS system without a peristaltic pump, with the aspiration of the nebuliser being employed to introduce the carrier stream or sample solution [21].

The simplest calibration method for FIAAS systems does not differ from that used in conventional measurements, and is performed by injecting a series of standard solutions prepared with consideration of possible matrix effect and the influence of interfering species. The use of the merging zones principle limits the consumption of the reagent used for the elimination of chemical interferences [1]. However, FIAAS systems also offer several more efficient calibration methods [22]. The analytical determinations can be based on the measurement of the peak width instead of the peak height. A large number of options are made available by the use of various configurations of branched flow injection systems and the utilisation of the concentration gradient formed as the result of the dispersion of the injected zone of the standard solution or of the sample during transport to the detector. Various procedures have been developed for standard addition methods and for sample addition to the standard solutions.

The application of multiway switches for solution streams in flow injection systems allows the transport of the same injected sample or standard solution to the detector through tubings of different length and diameter, which enables the system to have dispersion coefficients ranging from 5.9 to 38.8 [23]. Another possibility is the use of a branched flow injection system, where the injected sample or standard solution is divided into two or three lines on its way to the detector and the analytical signal obtained is in the form of two or three overlapping peaks [24, 25]. This allows the simultaneous measurement of up to five calibration plots corresponding to various concentration ranges. Several different concepts were reported for broadening the calibration range in FIAAS measurements. The peak width is a linear function of form $(C_m/C' - 1)$, where C_m is the concentration of the analyte in the injected sample and C' is the concentration corresponding to the signal magnitude at which the width of the peak is measured. Using this relationship for magnesium determination, a linear response can be obtained over three orders of magnitude [26]. Another hydrodynamic relationship used to simplify the calibration of FIAAS systems is the exponential dependence of the dispersion coefficient on the injected sample volume [27]. A variation in the dispersion coefficient can be obtained by the injection of different sample volumes, by injecting only part of the solution from the injection loop, or by controlling, with a pump, the sample volume introduced into the measuring system [18, 28].

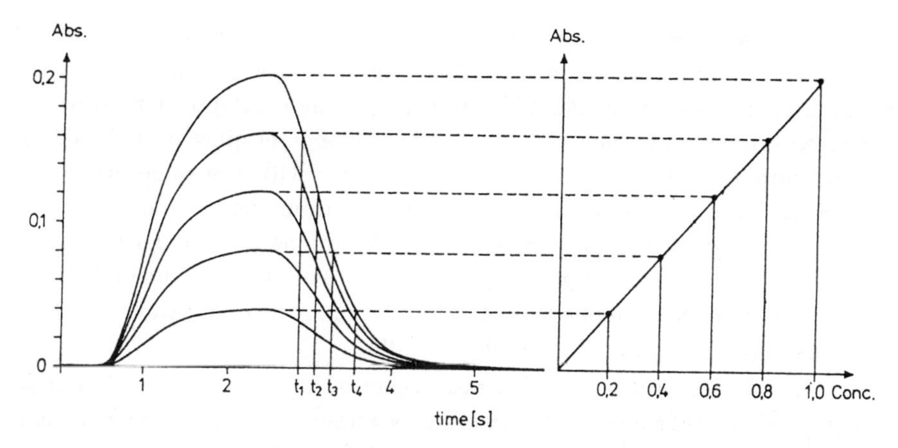

Fig. 2. Principle of the "electronic dilution" calibration method [31]. Different delays in reading of the signal correspond to different concentrations of the analyte. (*Reprinted by permission of copyright owner.*)

The calibration procedure in FIAAS systems can also be based on the reproducible formation of the concentration gradient, especially on the descending part of the flow injection peak (Fig. 2). In the so-called "electronic dilution" method, after the preliminary determination of the dispersion coefficient at different points on the flow injection peak, the calibration can be performed by the injection of a single standard solution and measurement of the signal magnitude after a given period of time [29]. The exponential increase of the signal with time can be utilised in the measuring system, where a segment of the injected standard solution flows through a mixing chamber producing a controlled increase in concentration [30]. The entire working range of the instrument can be calibrated with one reference solution by means of an algorithm based on gradient ratio evaluation and capable of utilising the additional information in the transient signal for interference correction [31]. The effectiveness of this approach has been demonstrated for the determination of calcium in the presence of phosphoric acid.

Standard addition procedures have also been utilised in FIAAS measurements, most often in reverse flow injection systems, where the sample solution is continuously pumped and standard solutions are injected into the sample stream [32]. Such a procedure has been utilised for the determination of chromium in steel [33]. The determination does not require releasing agents and uses pure chromium standard solutions. A particular version of the standard addition procedure for FIAAS measurements is the zone penetration method, where measurements are performed in the overlapping zone of the dispersed sample and standard solution [34]. The determination is based on the measurement of the absorbance at two points with the same dispersion coefficient. At the first point, only the analyte from the sample is present, while at the second point it is accompanied by the added standard amount. A zone penetration concept with injection of two sample segments of different volume has also been applied in on-line dilution, where the three different dilutions may be acquired in one injection [35].

The possibility of direct AAS measurement of liquid samples with a large content of dissolved solids in FIAAS systems without clogging of the nebuliser and accumulation of precipitates has been pointed out in many other studies [36–38]. The main advantages of FIAAS measurements are the very short time the sample spends in contact with the nebuliser and burner (typically 1-2 s for each sample), on-line sample dilution by the carrier solution and effective rinsing of the measuring system between injections of the samples. At extremely high salts content in the analysed sample, the zone sampling procedure can be used. Here, the injected sample is initially dispersed and a small part of its zone is then injected into the second part of the manifold and transported to the detector [27].

Accurate results in FIAAS systems can be obtained by flow injection nebulisation of a slurry into a flame. In the determination of heavy metals in sewage sludges, the slurry of sludge is pumped through a PTFE coil located in a microwave oven, then through a cooling bath and a gas trap. Finally, a portion of the sample is injected into a single line carrier and transported directly to the spectrometer [39]. An FIAAS system with slurry nebulisation has also been used for the rapid determination of zinc and iron in foods [40].

The use of FIAAS, instead of conventional AAS, with flame atomisation is particularly advantageous when the analysed sample requires chemical pretreatment in order to remove interferences or to preconcentrate the analyte. The elimination of the effects of phosphate and sulphate on calcium signals has been obtained by the incorporation of an anion-exchange column into the FIAAS system [41]. A buffer consisting of boric acid, triethanolamine, thiourea and acetylacetone was used for masking aluminium, iron and zinc during the preconcentration of lead on immobilised 8-quinolinol [42]. Solid-phase or solvent extraction on-line preconcentration procedures most often serve simultaneously as methods of matrix removal in FIAAS systems, but other techniques are also employed for this purpose. For the determination of certain elements in silver electrolysis solutions, a closed-loop recirculating manifold for matrix isolation has been developed, where silver is precipitated as the chloride and retained on a nylon filter [43]. The biological matrices are efficiently digested using microwave sample irradiation. This is used in conjunction with an FIAAS system for the above-mentioned analysis of sewage sludge [39], and for the determination of zinc and cadmium in biological tissues [44].

On-line precipitation in FIAAS systems can be utilised for the preconcentration of trace amounts of the analyte, for removal of interferences, and for carrying out indirect determinations. In order to remove aluminium from the solution in the determination of calcium, the latter is precipitated as the oxalate, collected on a nylon filter, a stainless-steel disc [45], or a minicolumn of glass beads [46], and then dissolved in hydrochloric acid. A continuous precipitation and filtration technique has been employed in FIAAS systems for the preconcentration of lead by precipitation with ammonia [47], and for the preconcentration of silver, calcium and iron by precipitation as the chloride, carbonate or hydroxide, respectively [48]. A solid phase with preconcentrated trace amounts can also be collected on the walls of a knotted tubular reactor, and then dissolved using an appropriate organic solvent. Such procedures have been developed for the preconcentration of lead [49] and silver [50] by coprecipitation with Fe(II)-carbamate complexes. They have also been used for the preconcentration of Cu(II)-carbamate chelate [51].

The use of organic solvents in FIAAS systems can improve the detectability of the AAS determination in two different ways. One is the more efficient combustion in the flame and easier evaporation due to lower viscosity and surface tension, giving a larger signal than in aqueous solutions [52, 53]. It has been observed that the injection of an acetone solution instead of an aqueous solution into a methyl isobutyl ketone carrier stream results in an almost nine-fold increase in sensitivity [53]. The other way is the use of appropriate solvent extraction systems, which allow not only the preconcentration of the determined species, but also their separation from matrix interferences [54–59]. Some examples of such determinations are shown in Table 1, and a typical flow injection manifold is shown in Fig. 3. In developed applications a continuously aspirated sample solution, or an aqueous carrier stream with an injected sample, is segmented with organic solution of a suitable extractant. After transport through the extraction coil the aqueous and organic phases

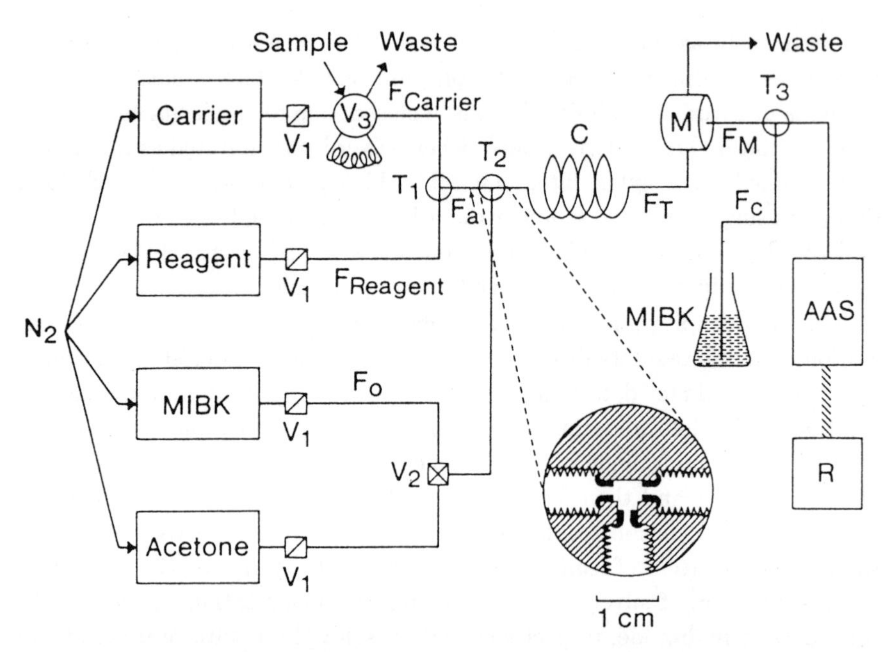

Fig. 3. Manifold of an FIAAS system with on-line solvent extraction for the determination of zinc, using extraction of zinc thiocyanate into methyl isobutyl ketone [57]. V — valves; T_1, T_3 — stream connectors; T_2 — segmentor; C — extraction coil; M — membrane phase separator. (*Reprinted by permission of copyright owner.*)

are separated and the latter is directed to the nebuliser. In the FIAAS system for the determination of zinc shown in Fig. 3, the extraction of zinc as zinc thiocyanate, after the preliminary reduction of iron to Fe(II), allowed the elimination of the iron interferences usually observed in AAS zinc determinations. The compensation for the difference between the flow rate in the flow injection system and the nebuliser aspiration rate was done with methyl isobutyl ketone, as compensation with water or air was much less effective [57].

The most common sample pretreatment method used in FIAAS systems is on-line solid-phase extraction, carried out by passing the sample solution through a microcolumn containing a suitable stationary phase [60, 61] (Table 1). The reason for the rapid development and wide application of this mode of sample pretreatment is the simplicity in its hardware and operation requirements, the availability of a wide range of sorbents, and the existence

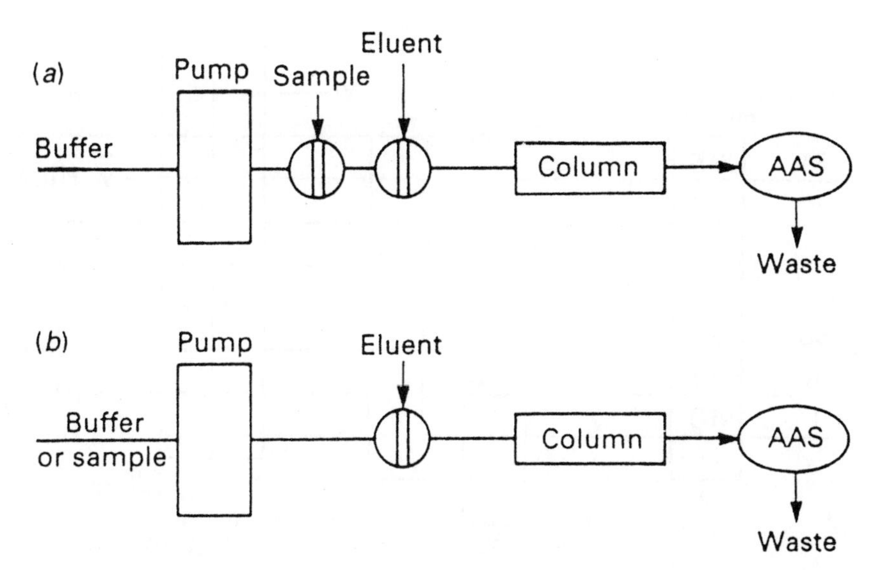

Fig. 4. Manifolds of FIAAS systems with on-line preconcentration using solid-phase extraction; a — volume-based system; b — time-based system [84]. (*Reprinted by permission of copyright owner.*)

of a wealth of knowledge on related batch procedures. FIAAS systems with preconcentration microcolumns can be clasified into two groups, based on the method of sample delivery (Fig. 4). In systems with volume-based loading, a defined volume of the sample solution is introduced using the injection valve, whereas in systems with time-based loading, the sample solution is pumped at a constant flow rate over an appropriate time interval through the preconcentration column. Both of these types of FIAAS systems (with various modifications) are widely reported in the literature (Table 1).

Two valves in system (a) allow the use of different volumes of the introduced sample and eluent, which is important for obtaining preconcentration. In the same system, under suitable chemical conditions, two columns can be incorporated in series for the preconcentration of different species. The sequential elution from each column enables the determination of various species in the same sample. A system of this type, with two different functionalised cellulose sorbents, has been developed for the simultaneous determination of Cr(III) and

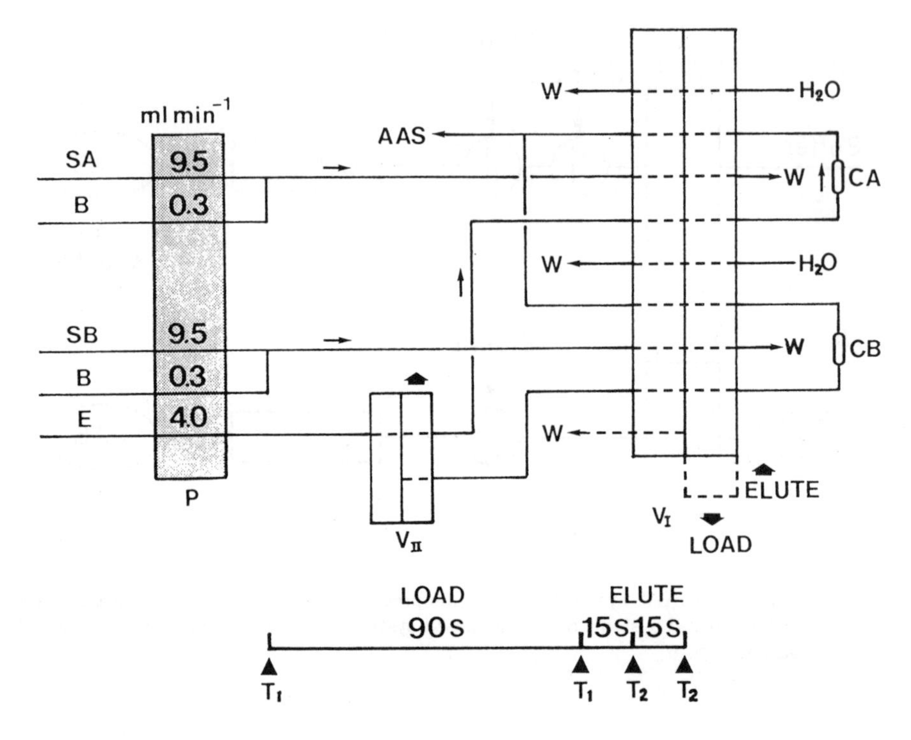

Fig. 5. Manifold of an FIAAS system with on-line preconcentration using two solid-phase extraction columns [75]. P — pump; V — valves; SA, SB — samples; CA, CB — columns; B — buffer solution; E — eluent; W — waste; T — turning points. (*Reprinted by permission of copyright owner.*)

Cr(VI) [79]. An example of an FIA system for the parallel preconcentration of two samples is shown in Fig. 5. The loading (preconcentration) and elution periods are governed by settings on the timer which controls the operation of the multifunctional valve via the pneumatic system. A simple two-way valve is included in the eluent line to switch the eluent from one column to the other to permit sequential elution of the columns during the elution stage.

As is evident from the literature data collected in Table 1, the most frequently used sorbents in preconcentration columns in FIAAS systems are commercial chelating resins such as Chelex 100 [11, 67–69, 75, 76], Muromac A-1 [70] or Spheron Oxin 1000 [83]. Many applications have been developed for 8-quinolinol immobilised on various supports [42, 56, 81, 82, 88–90] or

Table 1. Applications of flow injection AAS systems with flame atomisation in direct determinations.

Analyte	On-line processing procedure*	Enrichment factor**	Detection limit, ppb	Reference
Ag, Ca, Fe	Precipitation as chloride, carbonate and hydroxide	5–9	4 (Ag); 0.8 (Ca) 40 (Fe)	48
Ag	Coprecipitation with carbamate	26 (28)	0.5	50
Al	SPE: anionite Amberlite IRA-400	15	20	62
	SPE: cationite SCX		75	63
	SPE: Chelex 100		15	64
	SPE: Chromazurol S/anionite	30	15	65
Au	SPE: Amberlite XAD-8	(35)	2	66
Au, Cu	SE: ammonium thiocyanate in MIBK	70 (Au); 78 (Cu)	1.8 (Au); 1.0 (Cu)	59
Be	SPE: 8-quinolinol/CPG	(24)	0.1	56
Ca, K, Mg				1
Ca	Precipitation as oxalate		250	45
Ca, Fe, K, Mg, Na			70 (Ca); 50 (Fe,K); 8 (Mg); 100 (Na)	21
Cd, Pb, Zn	SPE: Chelex 100	13	10 (Pb); 1 (Cd, Zn)	11
Cd, Cu, Pb,Zn	SPE: Chelex 100	(50-100)	0.05 (Cd); 0.07 (Cu); 0.5 (Pb); 0.03 (Zn)	67
Cd, Cu, Mn	SPE: Chelex 100	60–80 (10–13)	0.09 (Cd,Cu); 0.08 (Mn)	68
Cd, Cu, Zn	SPE: Chelex 100	60–80	0.08 (Cd); 0.1 (Cu); 0.2 (Zn)	69
Cd, Cr(III), Cu, Fe(III), Mn(II) Pb, Zn	SPE: Muromac A-1	90–180 (20–30)	0.1 (Cd); 0.3 (Cr) 0.7 (Cu); 0.6 (Fe) 0.8 (Mn); 2 (Pb)	70
Cd	SPE: alumina	28	0.4	71

continued

Table 1 (*continued*)

Analyte	On-line processing procedure*	Enrichment factor**	Detection limit, ppb	Reference
Cd, Cu, Pb	SPE: C18 bonded silica	19–25	0.3 (Cd); 0.2 (Cu) 3 (Pb)	72
Cd, Co, Cu, Ni, Pb, Zn	SPE: 8-quinolinol/CPG	500 (20)		81
Cd, Cu, Fe, Pb, Zn	SPE: immobilised cyanobacteria	100-1000	0.4 (Cd); 0.3 (Cu); 0.5 (Fe); 10 (Pb); 0.2 (Zn)	73
Cd, Cu, Pb, Fe, Zn	SPE: immobilised Saccharomyces cerevisiae	125–2000	0.2 (Cd); 0.7 (Cu); 0.6 (Fe); 8 (Pb); 0.1 (Zn)	74
Cd, Zn	Microwave digestion		10 (Cd); 50 (Zn)	44
Ce, Co, Ni, V	SPE: 8-quinolinol/CPG	(24–41)	3 (Ce); 0.6 (Co); 1.3 (Ni); 0.7 (V)	56
Co	SPE: Chelex 100	48 (48)	0.2	75
	SPE: Chelex 100		20	76
	SPE: Nitroso-R-salt modified alumina	300	0.44	77
Cr				33
	SPE: poly(hydroxamic acid) resin	50	100	78
	SPE: Cellex P for Cr(III); Cellex T for Cr(VI)		0.78 (Cr^{III}); 1.4 (Cr^{VI})	79
Cu				19,43
	SE: pyrrolidine dithiocarboxylate chelate in MIBK	15–30	2	58
	Dialysis and SPE on cationite		295	80
	SPE: 8-quinolinol/ silica gel	35 (35)	1.5	82
	SPE: Spheron Oxin 1000	100	0.3	83

continued

Table 1 (*continued*)

Analyte	On-line processing procedure*	Enrichment factor**	Detection limit, ppb	Reference
	SPE: Pyrocatechol violet/XAD-2	(20)	0.06	84
	SPE: chelate sorption on PTFE	120	0.2	51
	SPE: fullerenes	185	0.3	85
Fe, Zn			600 (Fe); 300 (Zn)	40
Fe	SPE: C_{18} bonded silica		5 (Fe^{II})	86
K				27
Ni	SPE: Eriochrome blue-black R/XAD-2		0.1	87
Pb	SPE: 8-quinolinol/ C_{18} bonded silica	14	4	88
	SPE: 8-quinolinol/ silica gel	40 (13)	1.4	42, 89
	SPE: 8-quinolinol resin		10	90
	SPE: alumina	250 (50)	0.4	91
	SPE: dithizone activated carbon	(18)	2	92
	SPE: Cellex P	(10–15)	0.17	93
	SPE: fullerenes	70	5	94
	Precipitation with ammonia	700 (11)	1.0	47
	Coprecipitation with Fe(II) carbamate complex	(30)	2	49
	Donnan dialysis	100	1	98
Organolead	Oxidation with iodine		100 (Pb)	95
compounds	Oxidation with iodine		600 (tetramethyl) 800 (tetraethyl)	96
V	SPE: Anionite SAX		50	97
Zn	SE: thiocyanate complex in MIBK)	200		57

*SE — solvent extraction; SPE — solid-phase exctraction; MIBK — methyl isobutyl ketone; CPG — controlled-pore glass

**The number in parentheses is the concentration efficiency value (min^{-1}).

anion-exchange resins and non-ionic sorbents modified with chelating ligands [65, 84, 87, 92]. Several authors have also used alumina [71, 91] and other non-modified ion-exchangers [62, 63, 79, 93, 97]. Recently, microorganisms covalently immobilised on CPG [73, 74] and also fullerenes have been employed for the preconcentration of metal chelates [85, 94]. The use of C18-bonded silica [72, 86] or PTFE [51] is more common for the preconcentration of metal chelates.

The criterion most often used for evaluating a preconcentration system is the *enrichment factor.* It is usually calculated by comparing the peak height before and after preconcentration, or by comparing the slopes of the linear portion of the calibration curves before and after preconcentration. The enrichment factor values thus produced are sometimes very high (Table 1), but do not give sufficient information about the efficiency of preconcentration, as they can be achieved by long preconcentration periods of hours. The use of *concentration efficiency* has been proposed as a better criterion for the evaluation of the efficiency of preconcentration system [67]. This is defined as the product of the enrichment factor and the sampling frequency expressed as the number of samples analysed per minute. In other words, it is the number of enrichments obtained per minute in a given experimental FIA system. Some of the values reported in the surveyed literature are given in Table 1. In the most efficient systems, the concentration efficiency can reach values up to 100 [67]. Other parameters, such as the consumptive index or the retention efficiency, may also be helpful in the description and comparison of different on-line preconcentration methods in FIAAS systems [60].

Preconcentration on different sorbents in FIAAS systems has been compared in several papers. The properties of the chelating resin Chelex-100 with iminodiacetic functional groups, 8-quinolinol; immobilised on CPG, and a commercial resin with salicylate functional groups have been compared in the determination of Cd, Cu, Pb and Zn [76]. The highest concentration efficiency was obtained for immobilised 8-quinolinol; however, in terms of the recovery of retained metal ions, the most satisfactory results were obtained with Chelex-100. The common practice of eluting retained metal ions with solutions of mineral acids can be inappropriate for Chelex-100 because of resin swelling and changes in the microcolumn packing. It can be better to replace acidic eluents with solutions of suitable ligands to avoid these problems and to provide a particular elution selectivity. Quantitative elution of copper retained on Chelex-100 can be obtained with the use of an L-cysteine solution, whereas only partial elution of Cd, Pb and Zn is observed [68]. Better adsorption-desorption behaviour for Cd is observed using basic alumina than with Chelex-100 [71]. Column packing should generally exhibit a very low degree of swelling and shrinkage with any change of solvent during preconcentration or elution. It should also exhibit high mechanical resistance and have satisfactory kinetic properties. Good performance is obtained from FIAAS systems with column volumes of about 100 μl column length to diameter ratios of 10–15 [60]. The optimum particle size for obtaining good enrichment without excessive back pressure is 150–200 μm at a preconcentration flow rate of 8–9 ml min^{-1}. Figure 6 compares the signal recorded using a conventional Cu

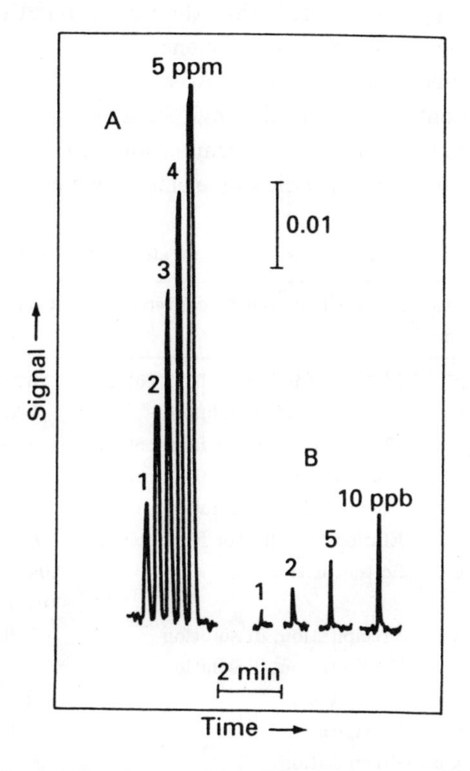

Fig. 6. Comparison of recordings from flame AAS measurements of Cu(II) made using (A) conventional aspiration and (B) in a time-based FIAAS system with 20 min preconcentration using a 70 cm microcolumn with XAD-2 loaded with pyrocatechol violet [84]. Preconcentration was carried out at 6.8 ml min^{-1}, elution with 1.0 ml nitric acid at 2.2. ml min^{-1}. Scale bar shows an absorbance of 0.01 a.u. (*Reprinted by permission of copyright owner.*)

AAS instrument with that obtained from an FIAAS system, with preconcentration using an XAD-2 column loaded with the complexing ligand pyrocatechol violet.

The preconcentration of analytes and elimination of matrix interferences in FIAAS systems can also be carried out by means of Donnan dialysis with ion-exchange membranes. In an application of this procedure for the determination of lead, a tubular cation-exchange membrane is used as the sample loop of the rotary injection valve [98]. The ion-exchange tubing is immersed in the stirred sample solution and the dialysis performed for 5 min in the optimised conditions gave a 100-fold enhancement of the signal magnitude in comparison with the conventional AAS measurement of the same sample. One disadvantage of such a procedure, however, is that the ionic strength of the sample has a significant influence on the dialysis efficiency.

On-line sample manipulations in FIAAS systems allow indirect determinations to be carried out (Table 2), although such determinations usually have poorer detection limits than direct determinations. Most are based on precipitation of the analyte with an indicator element, which is then measured by

Table 2. Applications of FIAAS systems with flame atomisation in indirect determinations.

Analyte	Indicator element	Principle of analyte conversion	Concentration range	Reference
Ammonia	Ag	Elution of indicator from column	5–100 μM	99
Chloride	Ag	Precipitation, dissolution	3–100 ppm	100
Cyanide	Cu	Elution of indicator from column	1–50 ppm	101
	Ag	Elution of indicator from column	0.5–8 μM	99
	Ag	Precipitation, dissolution	3–150 μM	102
EDTA	Cu	Elution of indicator from column	0.1–50 ppm	103
Nitrate, nitrite	Cu	Solvent extraction	0.04–2.2 (NO_3^-); 0.4–10 (NO_2^-)	104, 105
Oxalate	Ca	Precipitation, dissolution	5–90 ppm	100
Saccharin	Ag	Precipitation, dissolution	5–75 ppm	106
Sulphate	Pb	Precipitation	2–20 ppm	110
Sulphide	Cd	Precipitation	0.01–2 ppm	107
Sulphonamides	Ag, Cu	Precipitation	2.5–35 ppm	108
Tannins	Cu	Precipitation	0.7–25 ppm	109
Thiocyanate	Ag	Precipitation, dissolution	1–120 μM	102
Thiosulphate	Ag	Elution of indicator from column	0.5–9 μM	99

AAS following the dissolution of the precipitate. This method has been used for the determination of chloride and oxalate [100], cyanide and thiocyanate [102] and saccharin [106]. In methods developed for determination of cyanide [99, 101] and ammonia and thiocyanate [99], reaction with the analyte causes dissolution of a precipitate of the indicating element packed in a flow through reactor. In a determination of EDTA, the analyte elutes Cu(II) retained in a Chelex-100 column [103]. In determinations of sulphide, a colloidal precipitate of CdS is introduced into the nebuliser following the removal of excess Cd(II) on a cation-exchange column [107]. Excess Cu(II) is measured after precipitation of tannins [109] in a determination of the latter. The determination of sulphonamides has been carried out in a reversed FIAAS system with continuous sample aspiration and injection of an indicator metal ion [108]. A determination of nitrate and nitrite has been based on the formation of an ion pair between nitrate and a Cu(I) Neocuproine complex and subsequent extraction with MIBK [104,105]. The AAS measurement is performed in the organic phase. Nitrite is determined after oxidation with Ce(IV).

1.2. *Electrothermal Atomisation*

With electrothermal atomisation, the sensitivity of AAS detection is enhanced by two to four orders of magnitude due to the rapid atomisation of the entire sample. Atomisation occurs in a graphite furnace atomiser which is a cylindrical tube, open at both ends, with a hole for introduction of the sample. A small sample, usually of a few microlitres, is first evaporated at low temperature, then ashed at a higher temperature on an electrically heated surface of carbon, tantalum or some other conducting material in the shape of a tube, strip, rod or boat. After ashing, the current is increased to raise the temperature rapidly to 2000–3000°C, causing the analyte to atomise over a period of a few milliseconds to seconds. The absorbance is measured in the region above the heated conductor. AAS measurements with electrothermal atomisation are much more sensitive than those with flame atomisation, but they are generally much less precise and suffer from much more severe matrix interference problems. Several causes of this poorer selectivity have been identified including broad-band molecular absorption resulting from an incomplete breakdown of the matrix in the short atomisation time, scattering by incompletely decomposed organic particles, and the existence of stable molecular halides in atomised samples. For these reasons, several background correction techniques and procedures to minimise chemical interferences have been developed, especially for AAS measurements with electrothermal atomisation.

The graphite furnace atomiser is generally suitable for batch measurements procedures; however, the processes of preconcentration and elution carried out in FIA systems are discontinuous in character, and can be joined instrumentally with the discrete, non-flow through nature of electrothermal AAS [111]. The earliest work with electrothermal FIAAS systems has dealt with the use of solvent extraction for the on-line preconcentration of heavy metals and separation of the analyte from its matrix [112, 113]. In a system involving a two-step extraction of metal dithiocarbamates, using a set of membrane phase separators, a 50- to 100-fold signal enhancement is obtained using the injection of 13 μl of the sample extract into a graphite furnace with a low air flow stream [113]. In several of the developed systems, solid-phase extraction is performed [111, 114–116]. A typical FIAAS system with electrothermal atomisation is shown in Fig. 7. The FIA system is interfaced to the graphite furnace by connecting the transfer capillary of the FIA system to the sample introduction capillary of the autosampler arm. This particular system has been used for the preconcentration of Cr(VI) on a 15 μl C_{18}-bonded silica microcolumn using sodium diethyldithiocarbamate as the chelating agent [114]. Speciation analysis in this system requires off-line oxidation of Cr(III) to Cr(VI). A 12-fold enhancement in sensitivity compared with the direct introduction of 40 μl samples is achieved after preconcentration for 60 s.

In similar systems, diethyldithiocarbamate has been used for the preconcentration of Cd, Cu, Ni and Pb on C_{18} [111], diethyldithiophosphate has been used for the preconcentration of Cd, Cu and Pb on C_{18} [115], and an anion-exchange resin column has been used for the removal of sulphur anions in the determination of manganese [116]. A different system design has been used for the on-line preconcentration of lead on 8-quinolinol bound on methacrylate gel [117]. A sorbent column is packed in the top of the sampling arm of the autosampler and all solutions (sample, wash, eluent) are introduced by an air flow stream.

FIAAS systems with electrothermal atomisation have also been designed with microwave-assisted mineralisation. In the determination of iron and zinc in adipose tissue, a solution of mineralised samples is mixed on-line with surfactant and a Pd-Mg matrix modification, and introduced with a sampling arm assembly into the graphite furnace [118]. In determinations of cobalt in whole blood, a microwave-assisted mineralisation is carried on-line. After degassing in a gas diffusion cell and cooling in an ice chamber, 20 μl of pretreated sample is introduced into the graphite furnace [119].

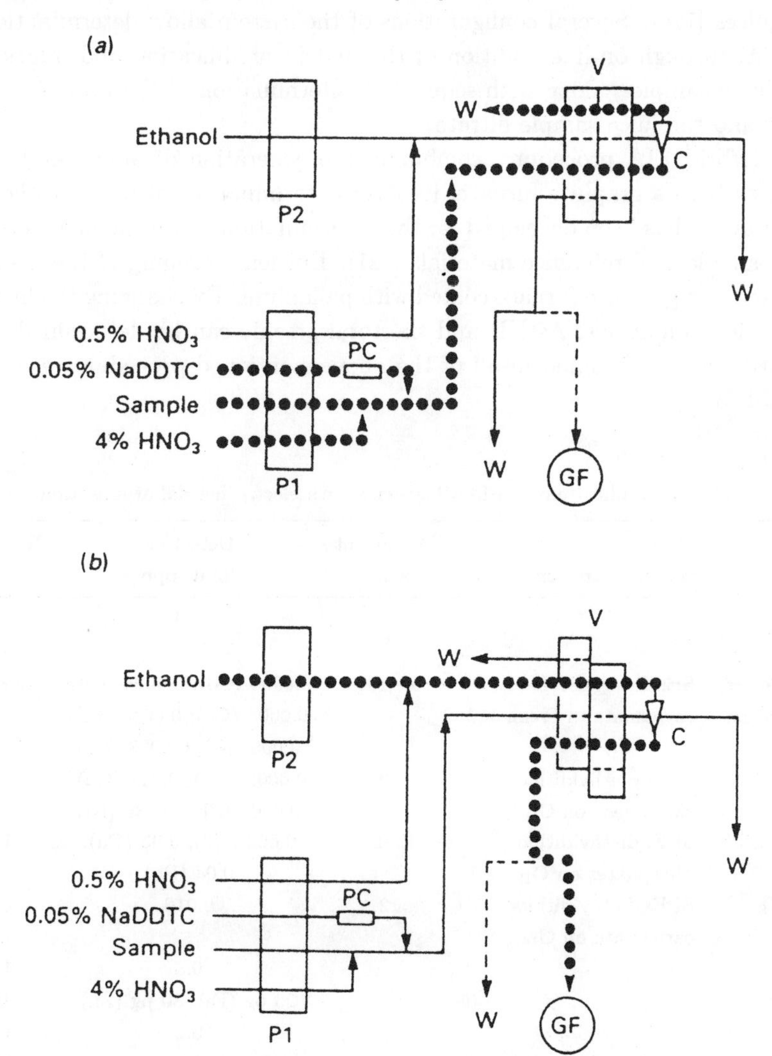

Fig. 7. Manifold of an FIAAS system with electrothermal atomisation and solid-phase extraction, and the sequence of operation for the determination of Cr(VI) [114]. P — peristaltic pumps; C — conical column packed with C18 sorbent; PC — precolumn packed with C18 sorbent; V — valve; GF — graphite furnace; W — waste; a — sample loading; b — analyte elution. (*Reprinted by permission of copyright owner.*)

An FIAAS system with electrothermal atomisation has also been developed with slurry sample introduction for the determination of aluminium in fruit and juices [120]. Several configurations of the system allow determination of total Al through on-line addition of the matrix modification and dilution of the slurry samples, along with sequential determination of the total Al in the slurry and the juice sample filtrate.

A hybrid FIA procedure, combining the generation of arsine and, after trapping it in a graphite furnace, its AAS determination with electrothermal atomisation, has been developed for the determination of arsenic in waters and various biological reference materials [121]. Efficient trapping of the arsine is achieved in a graphite furnace coated with palladium. By changing the hydride generation conditions, As(III) and the total arsenic can be determined. The detection limits obtained in all of the above-mentioned procedures are given in Table 3.

Table 3. Applications of FIAAS systems with electrothermal atomisation.

Analyte	On-line preconcentration*	Enrichment factor	Detection limit, ppb	Reference
Al			5.0	120
As			0.1	121
Cd, Co, Fe, Ni, Pb, Zn	SE: diethyldithiocarbamate in Freon 113	50–100	0.0005 (Cd); 0.0024 (Cu); 0.0008 (Co); 0.0044 (Fe); 0.005 (Ni); 0.0033 (Pb)	113
Cd, Cu, Ni, Pb	SPE: diethyldithiocarbamate on C_{18}	20	0.0008 (Cd); 0.017 (Cu); 0.0065 (Pb); 0.036 (Ni)	111
Cd, Cu, Pb	SPE: diethyldithiophosphates on C_{18}	12–13	0.003 (Cd); 0.05 (Cu); 0.04 (Pb)	115
Cr(VI)	SPE: diethyldithiocarbamate on C_{18}	12	0.016	114
Co			0.3	119
Fe, Zn			20 pg (Fe), 30 pg (Zn)	118
Mn			0.2	116
Pb	SPE: 8-quinolinol on methacrylate gel		0.1	117
	SPE: diethyldithiocarbamate on C_{18}	26	0.003	122

*SE — solvent extraction; SPE — solid-phase extraction.

1.3. *Measurements Involving Conversion of the Analyte into the Gaseous Phase*

A common method used to improve the detectability in AAS determinations of elements such as As, Bi, Ge, Se, Sn, Sb, Pb and Te is the generation of their hydrides, which are transferred into an electrically heated quartz tube furnace. This operation is usually carried out under flow conditions. The use of the flow injection technique here reduces the consumption of sample and reagent, and increases the sampling rate when compared with manual procedures or steady-state measurements with segmented flow.

A typical method for hydride generation is to mix the HCl acidified sample solution with sodium tetrahydroborate solution. In most FIAAS systems, this is carried out by the injection of 0.2–2.0 ml of the sample into the carrier HCl solution, which is then merged with a stream of tetrahydroborate solution. An example of a manifold used for the determinations of arsenic and selenium in environmental samples is shown in Fig. 8. Often another stream of reagents for a supporting reaction or for masking interferences is added. In the manifold shown in Fig. 8 1,10-phenanthroline is used as a masking agent to suppress interferences from Cu and Ni in the determination of selenium [124]. In the determination of arsenic and antimony, a stream of iodide solution is added to reduce the analyte from pentavalent to tetravalent states [125–127]. By carrying out hydride generation with different acid media it is possible to determine

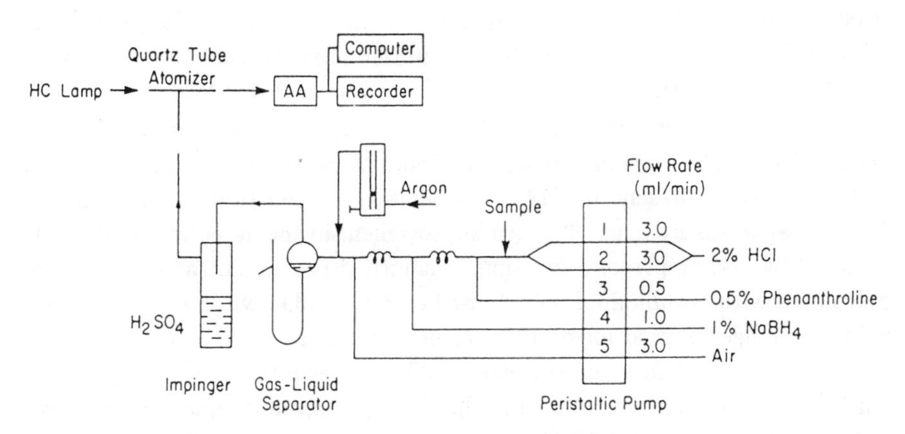

Fig. 8. Manifold of an FIAAS system with hydride generation for the determination of arsenic and selenium in environmental samples [124]. (*Reprinted by permission of copyright owner.*)

As(III) and the total arsenic in the same setup separately [128]. As(III) and total As monomethylarsonic and dimethylarsinic acids can also be determined in this way [129]. In the determination of total selenium, a preliminary reduction of Se(VI) to Se(IV) is necessary, as only the latter can be reduced to hydride. This is accomplished by off-line reduction with hydrochloric acid at an elevated temperature [124], by on-line reduction with concentrated HCl in a coil heated to 140°C [130], or by using an on-line microwave cell [131]. A chemical modifier for the suppression of interferences can also be introduced with an additional line in an FIAAS system with hydride generation. The on-line addition of tellurium(IV) efficiently suppresses interference by noble metals in the determination of As, Bi and Se [127]. Tetrahydroborate can also be added to the system using an on-line column packed with the strong anion-exchange resin Amberlyst A-26 in the tetrahydroborate form [132]. This allows a significant reduction in the dilution of the sample in the FIA system.

Various separators have been developed for the separation of gaseous and liquid phases in FIAAS systems. The most typical design is a U-shape with the excess solution being pumped away [35, 56, 107, 123, 124, 128, 133–135]. A separator with a microporous PTFE tube yields a doubling of the signal in an arsenic determination, compared with the U-tube separator [136]. A gas-liquid separator based on a cotton gauze membrane has been reported in manifolds for the generation of the hydrides of As, Bi, Sb, Se and Te [137]. A flat PTFE membrane separator has been used in a manifold for the determination of Bi [138] and also for the determination of arsenic in a system with a gas diffusion module. In the latter case, the effect of interferences was observed to be lower than in systems with a conventional gas/liquid separator [126].

The advantage of using an FIAAS system for hydride generation AAS has been confirmed in selenium determination, where the tolerated level of interfering species present in FIAAS systems is 10–100 times higher than in steady-state measurements [37]. An improvement in the detection limit can be achieved by on-line preconcentration. Bismuth is preconcentrated using a microcolumn with 8-quinolinol immobilised on CPG [133], while a strongly basic anion-exchanger is used for on-line preconcentration of Se(IV) [56, 133]. The latter analyte can also be preconcentrated by coprecipitation with a generated lanthanum hydroxide precipitate [139]. The precipitate is quantitatively collected on the inner walls of an incorporated knotted reactor, and subsequently eluted with HCl into the FIAAS system. A cation-exchange column has also been used on-line to suppress interference by copper and other metal cations in the determination of arsenic and selenium [127]. Applications of FIAAS systems with hydride generation are listed in Table 4. Many of these systems have been employed for speciation analysis (Chapter 7).

Table 4. Applications of FIAAS systems in measurements with analyte transfer into the gaseous phase.

Analyte	On-line preconcentration*	Enrichment factor	Detection limit, ppb	Reference
As			10	126
			1.0	132, 143
			0.2–0.5**	129
As, Bi, Sb, Se, Te			0.4 (As); 0.1 (Bi); 0.08 (Sb); 0.6 (Se); 0.2 (Te)	125
As, Se			0.3	124
Bi			0.08	123
	SPE: 8-quinolinol/CPG	30	0.002	133
Hg			0.06	37
			0.2	140
	SPE: 8-quinolinol/CPG	20	0.002	56
Sb			0.3	135
Sn			2.5	144
Se			0.06	37
			1.2	134
			0.7	130
	SPE: anion-exchanger	15	0.002	56,133
	Coprecipitation with $La(OH)_3$	24	0.001	139

*SPE — solid-phase extraction

**for various species

A technique widely used to improve mercury detectability in AAS determinations is the generation of a mercury cold vapour, by reducing mercury compounds to metallic mercury with Sn(II) chloride or sodium tetrahydroborate. Instrumentally, such measurements require the same FIAAS system components as those used in the hydride generation method. Various types of gravity separators for mercury vapour and solution have been developed [56, 140, 141],

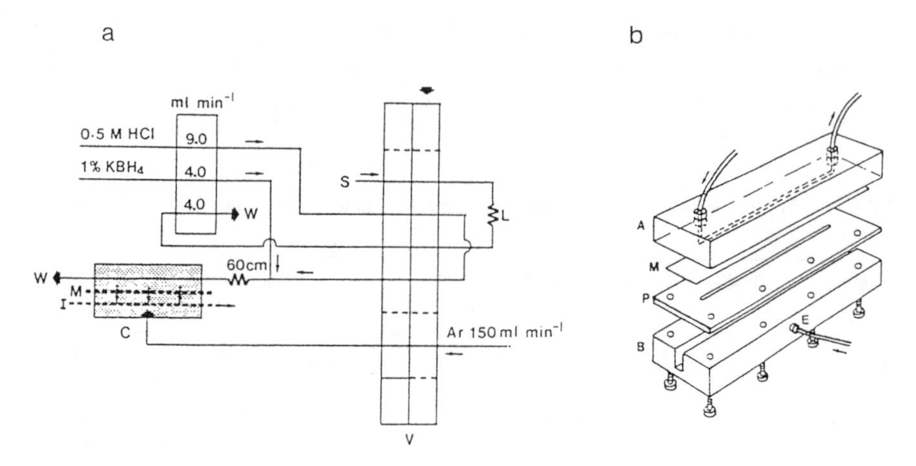

Fig. 9. Manifold of an FIAAS cold vapour system with a membrane phase separator (a) and the detailed structure of the combined phase separator/absorption flow cell (b) [37]. C — phase separator; L — 400 μl sample loop; V — 8-channel multifunctional valve; M — teflon membrane with nylon gauze backing; S — sample inlet; I — incident light from hollow-cathode lamp; W — waste; A — perspex block with 1.5×110 mm channel at the bottom; B — perspex block with a $3 \times 6 \times 170$ mm groove acting as absorption flow cell; P — 1 mm thick perspex plate with 1.5×110 mm slit in the centre; E — argon inlet. (*Reprinted by permission of copyright owner.*)

along with membrane separators that employ porous PTFE membranes that are permeable to mercury vapour [37, 142]. In the latter cases, a vapour-diffusion flow cell acts both as a membrane separator and as an absorption cell. It is shown, together with the manifold used, in Fig. 9.

In one of these systems, under optimised conditions, the detection limit was as low as 0.06 ppb at a sampling rate of 200 h^{-1} [37]. This method, however, is not free of interferences, such as sulphide, chloride, Se or noble metals [141]. The effects of some of these, and of organic species, can be eliminated by the preliminary mineralisation of samples with a mixture of sulphuric and nitric acid with potassium permanganate [140]. The lowest detection limit, 2 ppt at a sampling rate of 60 h^{-1}, has been obtained in a system where the generation of the cold vapour was preceded by mercury preconcentration on a microcolumn with immobilised 8-quinolinol [56]. Applications of FIAAS systems with hydride generation and cold vapour measurements are presented in Table 4.

2. Atomic Emission Spectroscopy

The atomisation of analyte molecules in a flame, a plasma source or in an electric arc or spark is the basis of atomic emission spectroscopic detection. The emission of radiation is caused by the excitation of outer electrons to higher orbitals by the heat of a flame or more energetic sources such as an inductively coupled plasma (ICP), a direct current plasma (DCP), a microwave-induced plasma (MIP), or an electric arc or spark. Flame emission spectroscopy, commonly called flame photometry, was the first to be applied as a detection method in FIA [145], and has found several applications, although in FIA the most commonly used techniques are those with plasma sources. Methods with excitation by an arc or spark have found widespread application in elemental analysis since the 1930's, but because of the nature of the excitation they are not used in analytical flow measurements. Since the 1970's these methods have been replaced by emission techniques using plasma excitation, which gives greater reproducibility of atomisation conditions with simpler equipment and can be used with liquid samples. These techniques can be easily adapted to FIA measurements.

2.1. *Flame Photometry*

In contrast with atomic absorption spectrometry with flame atomisation, the flame acts as the radiation source in flame photometry and therefore there is no need to use an individual lamp for each element. This technique can be used for a few tens of elements, but its most important uses have been in the determination of sodium, potassium and lithium, especially in the analysis of physiological fluids and plant materials. In analytical practice, a low-temperature flame is normally used in order to prevent the excitation of most other metals. The spectra obtained from the alkali and alkaline-earth metals are simple and the required emission line can be isolated using interference filters. In flame photometry, the emission lines are often superimposed on bands emitted by oxides or other molecular species present in the sample. A correction can be made by scanning for a few nanometers on both sides of the analyte peak, or by single measurements on both sides of the peak. The average of the two measurements is subtracted from the total peak height. Often, in order to avoid the non-controlled effects of such variables as flame temperature, fuel flow rate and background radiation, determinations are carried out with the use of lithium as an internal standard. A fixed amount of lithium is introduced into each standard and sample. In one FIA system, where flame photometry is used for the determination of sodium and potassium and AAS is used for the determination of calcium and magnesium, a carrier stream of water with the injected sample or standard solutions is merged with a lithium solution [146]. Determinations of 1–140 ppm sodium and 0.6–21 ppm potassium have been carried out using a calibration curve method. It has been shown, however, that in the FIA flame photometric determinations of sodium, potassium and calcium that a fast standard addition procedure can be more advantageous [147]. This method is based on the concentration/time profile generated by the injection of a single standard solution, which is then merged with the undispersed aspirated sample solution. Such a procedure can correct more efficiently than the calibration plot method for sudden changes in the FIA system which may occur between successive samples. This feature is especially evident for multicomponent determinations in which both matrix effects and spectral interferences may affect the result.

The simultaneous flame photometric determination of Na, K, Li and Ca has also been carried out in an FIA system with a fast scanning monochromator to scan sample emission spectra from the flame [148]. Sodium, potassium and calcium have been determined simultaneously in soil extracts and in tap water over wide concentration ranges, using lithium as an internal reference by scanning on two different levels of the dispersed sample zone for each injection. An example of the response at two different post-injection delay times is shown in Fig. 10. In the same work, a novel standardisation method based on the zone penetration technique has been proposed. The sample and standard solutions are dispersed within each other, allowing standard addition measurements at different sample/standard ratios from a single injection.

2.2. *Emission Spectroscopy Based on Plasma Sources*

The use of plasma sources in emission spectroscopy offers several advantages compared with flame or electrical excitation methods. Lower inter-element interferences result from the higher temperatures they generate. The simultaneous recording of spectra of numerous elements can be carried out because good spectra can usually be obtained for most elements under the same excitation conditions. Low concentrations of elements that form compounds highly resistant to decomposition by heat can be determined. Determinations are possible over several orders of magnitude in the concentration. In an argon plasma which is utilised for emission spectroscopy, argon ions and electrons

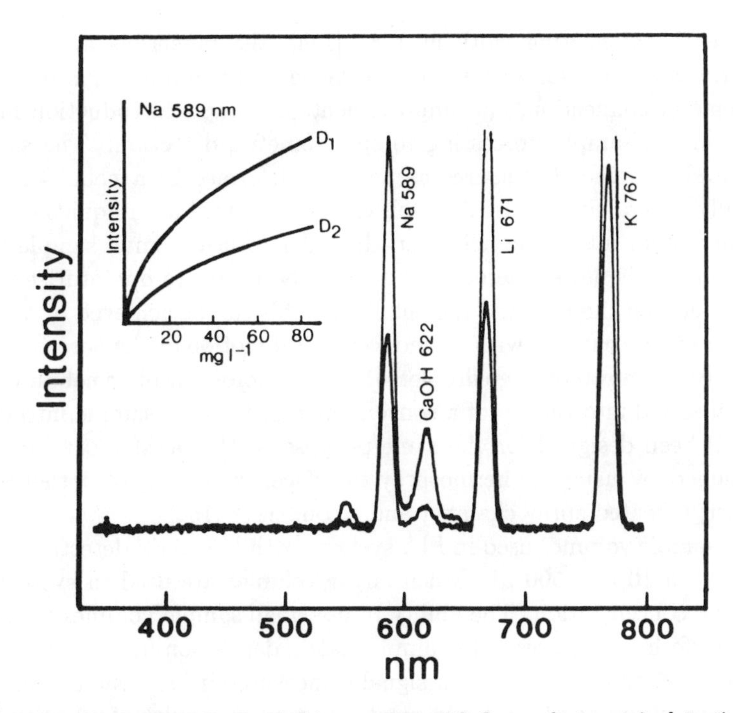

Fig. 10. Flow injection signals recorded in a system with flame photometric detection using a fast scanning monochromator at two dispersion levels scanned at 1.25 and 2.5 s after injection of 40 μl of a 20 ppm standard solution of Na, K, Ca and Li [148]. The calibration curves for Na at 589 nm at the two dispersion levels are shown in the inset. (*Reprinted by permission of copyright owner.*)

are the main conducting species. The plasma is formed by using one of a number of different sources. The most common is a high power radio-frequency generator producing an inductively coupled plasma (ICP). ICP – atomic emission spectroscopy (ICP-AES) offers better sensitivity and less interferences than other plasma-based emission techniques. Direct current plasma sources (DCP) are simple and less costly, but provide fewer emission lines than ICP and the sensitivities obtained are lower in many cases. There are two basic types of plasma emission spectrometers: sequential and multichannel. Sequential instruments are designed to measure line intensities on a one-to-one basis, whereas in multichannel instruments, an array of photomultipliers allows the simultaneous measurement of the intensity of emission lines from a large number of elements.

Since the pioneering work in 1981 [144], the numerous applications of ICP-AES developed for FIA systems have seen the optimisation of the FIA-spectrometer connection, and improvements in sample introduction methods and in on-line sample processing to spectrometric detection. The sample is introduced into the ICP source as an aerosol formed in a nebuliser or as a thermally generated vapour. Aerosols can be produced from liquids and solids by means of an ultrasonic nebuliser. In order to use a small sample volume and to obtain the lowest detection limit it is necessary to miniaturise the connection between the FIA and the nebuliser. This can be achieved by the use of a smaller cloud chamber with an induced spiral motion of the aerosol allowing the effective removal of large droplets [150]. A microconcentric nebuliser which can be inserted into the tip of a conventional, ICP torch, sample introduction tube has been designed for the same purpose [151]. Satisfactory results can be obtained by using a thermospray interface, consisting of a thermospray nebuliser, a heated spray chamber and a condenser [152].

The sample volumes used in FIA systems with ICP-AES detection are usually between 10 and 500 μl. Much larger volumes are used in systems with on-line preconcentration. The ability to use small sample volumes means that some interfering processes, for example matrix deposition in the nebuliser, can be avoided. Although a transient signal is measured it has been demonstrated that the precision of FIA measurements is about one order of magnitude better than that from conventional steady-state measurements [153]. Solid and gaseous samples can also be directly analysed in flow injection systems with this type of detection. The simultaneous determination of Cr, Ni, Mn, Si and Fe in stainless steel has been achieved in such a system after on-line dissolution in a flow though electrolytic cell. The solution from the electrolytic chamber is impelled by an air carrier stream towards a mixing-dilution chamber [154]. A similar procedure has been employed in the determination of several elements in aluminium alloys [155]. Gaseous samples containing B, C, Cl and S can be introduced directly into an ICP-AES spectrometer with Ar carrier gas [156]. In determinations of As [157–159], As, Sb and Se [160], Ge [161] and Sb [162], hydrides are formed in an FIA system and introduced into the spectrometer. A method for the determination of mercury in biological samples through the generation of a cold vapour has also been developed [163]. This procedure consists of the on-line extraction of the 1,5-bis(di-2-pyridyl)methylene thiocarbonohydrazide complex of mercury into MIBK followed by mixing of the organic phase with $SnCl_2$ in N,N-dimethylformamide to generate mercury vapour.

The large dispersion in FIA systems helps to eliminate the effects of large sample viscosity [149]. In FIA ICP-AES systems, several versions of the standard addition method have been developed [164-166]. In the last-cited work, the zone-sampling mode was used. This enables controlled dilution of the standard solution prior to its injection into the final standard carrier stream. In order to minimise the interferences due to easily ionisable elements, matrix matching of sample and standards can be employed in computerised systems [167].

Minicolumns with ion-exchangers [168,169], activated alumina [170-174] and chelating sorbents such as Chelex-100 [175-177] and iminodiacetic acid/ethylcellulose [177] have been employed for on-line preconcentration and matrix removal. The sorption of Al-8-quinolinol on a minicolumnn with Amberlite XAD-2 has been to preconcentrate aluminium [178]. A column with activated alumina has been used for the speciation of chromium [170]. Acidic alumina is used for the separation and preconcentration of Cr(VI), whereas Cr(III) is not retained. The simultaneous determination of phosphate and total phosphorus in waste-water has been carried out in an FIA system with spectrophotometric

Table 5. Detection limits (ppb) obtained in measurements with ICP-AES detection.

Analyte	Conventional determination [180]	Flow injection systems: Without preconcentration [151]	With on-line preconcentration [175]	With Donnan dialysis [185]
Al	45	–	20	–
Ba	2.3	2.5	0.04	–
Be	0.3	–	0.008	–
Cd	2.7	4.9	0.04	0.10
Co	7	27	0.1	–
Cu	5.4	–	0.2	0.04
Fe	6.2	–	0.5	–
Mn	1.4	1.6	0.04	–
Ni	15	–	0.2	0.3
Pb	42	64	2	0.8
Zn	1.8	–	0.5	0.06
Sample volume in FIA systems		0.01 ml	30 ml	350 ml

Table 6. Application of flow injection systems with ICP-AES detection.

Analyte	On-line analyte processing*	Detection limit, ppb	Reference
Al	SPE: anion-exchanger Amberlite IRA 400	3	168
	SPE: 8-quinolinol/ XAD-2	10	178
As	Hydride generation	8.2	157
		0.03	158
As, Sb, Se	Hydride generation	3.4 (As); 7.0 (Sb); 3.6 (Se)	160
Au	SPE: Amberlyst A-26	1	189
B		500	150
Be	Solvent extraction	50	183
Br, C, Cl, S	Determination of gaseous species	20–80 pg	156
Cd	Solvent extraction	0.4	182
Cd, Co, Cu, Pb	SPE: iminodiacetic acid/ethylcellulose	0.02 (Cd); 0.06 (Co); 0.09 (Cu); 0.5 (Pb)	177
Cr	SPE: alumina	0.05	171
Fluoride	Solvent extraction	30	184
Ge	Hydride generation	0.4	161
Hg	Solvent extraction and vapour generation	4	163
Iodide, iodate	SPE: anion-exchanger	0.75 (I^-); 31 (IO_3^-)	169
Mo	SPE: alumina	0.2	75
Phosphorus total		0.5	179
Sb	Hydride generation	0.19	162
Multielement		0.5–40	188
		see Table 5	151
	SPE: Chelex 100	see Table 5	175
	Donnan dialysis	0.04–1.8	185

*SPE — solid-phase extraction

(470 nm) and ICP-AES detectors in series [179]. A comparison of the detection limits obtained in conventional ICP-AES measurements with those for FIA systems, with and without solid-phase extraction on Chelex-100, is shown in Table 5. In FIA systems without preconcentration, similar or worse detection

limits were found, but preconcentration improved the detection limit by up to 75 times. The removal of sample matrix and the preconcentration of Cu and Cd has been achieved by on-line voltammetric stripping in an FIA system with ICP-AES detection [181].

On-line solvent-extraction has been employed in the direct determination of Cd [182] and Be [183], and in the indirect determination of fluoride in water [184]. In the latter case, a complex of the lanthanum/alizarin complexone/fluoride in hexanol containing N,N-diethylaniline is introduced into the plasma and the emission of lanthanum is measured.

In a system with on-line Donnan dialysis, enrichment factors of over 200 for cations with an 8 min dialysis time, and even up to 650 for silver with a 30 min dialysis time, can be obtained [185]. A list of ICP-AES applications in FIA systems is shown in Table 6.

Emission spectrometry with other plasma sources is much more rarely used in flow injection systems. Systems with MIP-AES have been developed for the determination of copper, with on-line preconcentration on C18-bonded silica [186], while chloride, bromide and iodide can be determined in a setup involving the generation of volatile halogens, which are introduced into the plasma [187].

3. Atomic Fluorescence Spectroscopy

The fluorescence of atoms can be observed when they are irradiated with an intense source containing wavelengths that can be absorbed by the element. Electrodeless discharge lamps are the most common sources for atomic fluorescence, and fluorescence spectra are most conveniently measured perpendicular to the incident light path. This method, however, has found far fewer applications than atomic absorption or atomic emission spectroscopy. It has been applied to the analysis of metals in biological substances, agricultural samples and lubricating oils. The most important application is for the determination of trace amounts of mercury in the environment. Several procedures for this type of determination have been developed for FIA systems [190–192]. Compared with atomic absorption methods, they are significantly less affected by interference from organic compounds. The equipment used is also simpler and less expensive.

In FIA systems for the cold vapour determination of mercury, an atomic fluorescence photometer consisting of an electrodeless discharge lamp and a solar-blind photomultiplier [190] or commercial mercury detectors [191, 192] have been employed. In the simplest system, mercury (II) is reduced with

Sn(II) to Hg(0) and, after gas-liquid separation, is measured fluorimetrically with a detection limit of 0.12 ppb [190]. In another system, employing a microcolumn of sulphydryl cotton, a sequential determination of unretained inorganic mercury and preconcentrated methylmercury can be carried out [191]. The limit of detection for methylmercury, based on the processing of a 0.5 ml sample, is evaluated at 0.006 ppb.

The use of on-line microwave digestion in a Teflon coil allows the determination of mercury in injected slurried samples of environmental materials [192].

4. References

1. E. A. G. Zagatto, F. J. Krug, H. F. Bergamin, S. S. Jorgensen and B. F. Reis, *Anal. Chim. Acta* **104** (1979) 279.
2. J. F. Tyson, *Analyst* **110** (1985) 419.
3. J. F. Tyson, *Anal. Chim. Acta* **214** (1988) 57.
4. J. F. Tyson, *Anal. Chim. Acta* **234** (1990) 3.
5. J. F. Tyson, *Spectrochim. Acta Rev.* **14** (1991) 169.
6. M. Trojanowicz and E. Olbrych-Sleszynska, *Chem. Anal. (Warsaw)* **37** (1992) 111.
7. B. Welz and M. Sperling, *Pure Appl. Chem.* **65** (1993) 2465.
8. J. L. Burguera (Ed.), *Flow Injection Atomic Spectrometry* (Dekker, New York, 1989).
9. M. W. Brown and J. Ruzicka, *Analyst* **109** (1984) 1091.
10. V. Carbonell, M. de la Guardia, A. Salvador, J. L. Burguera and M. Burguera, *Microchem. J.* **40** (1989) 233.
11. S. Olsen, L. C. R. Pessenda, J. Ruzicka and E. H. Hansen, *Analyst* **108** (1983) 905.
12. W. R. Wolf and K. K. Stewart, *Anal. Chem.* **51** (1979) 1201.
13. N. Yoza, Y. Aoyagi, S. Ohasi and T. Tateda, *Anal. Chim. Acta* **111** (1979) 163.
14. I. L. Garcia, M. H. Cordoba and C. Sanchez-Padreno, *Analyst* **112** (1987) 271,
15. J. M. Harnly and G. R. Beecher, *J. Anal. At. Spectrom.* **1** (1986) 75.
16. Z. Fang and B. Welz, *J. Anal. At. Spectrom.* **4** (1989) 83.
17. M. Trojanowicz and B. Szostek, *Anal. Chim. Acta* **261** (1992) 521.
18. R. A. Sherwood, B. F. Rocks and C. Riley, *Analyst* **110** (1985) 493.
19. S. Xu and Z. Fang, *Microchem. J.* **51** (1995) 360.
20. J. L. Burguera, M. Burguera, C. Rivas, M. de la Guardia and A. Salvador, *Anal. Chim. Acta* **234** (1990) 253.
21. M. de la Guardia, A. Morales-Rubio, V. Carbonell, A. Salvador, J. L. Burguera and M. Burguera, *Fresenius J. Anal. Chem.* **345** (1993) 579.
22. J. Tyson, *Fresenius Z. Anal. Chem.* **329** (1988) 663.
23. J. F. Tyson, J. R. Mariara and J. M. H. Appleton, *J. Anal. At. Spectrom.* **1** (1986) 273.
24. J. F. Tyson and S. B. Bysouth, *J. Anal. At. Spectrom.* **3** (1988) 211.

25. S. R. Bysouth and J. F. Tyson, *Anal. Proc.* **23** (1986) 412.
26. J. F. Tyson, *Analyst* **109** (1984) 319.
27. B. F. Reis, A. O. Jacintho, J. Mortatti, F. J. Krug, E. A. G. Zagatto, H. F. Bergamin and L. C. R. Pessenda, *Anal. Chim. Acta* **123** (1981) 221.
28. I. Lopez Garcia, P. Vinñs, N. Campillo and M. Hernandez Cordoba, *Anal. Chim. Acta* **308** (1995) 85.
29. S. Olsen, J. Ruzicka and E. H. Hansen, *Anal. Chim. Acta* **136** (1982) 101.
30. J. F. Tyson, J. M. H. Appleton and A. B. Idris, *Analyst* **108** (1983) 153.
31. M. Sperling, Z. Fang and B. Welz, *Anal. Chem.* **63** (1991) 151.
32. J. F. Tyson, J. M. H. Appleton and A. B. Idris, *Anal. Chim. Acta* **145** (1983) 159.
33. J. F. Tyson and A. B. Idris, *Analyst* **109** (1984) 23.
34. Z. Fang, J. M. Harris, J. Ruzicka and E. H. Hansen, *Anal. Chem.* **54** (1985) 1457.
35. Z. Fang, M. Sperling and B. Welz, *Anal. Chim. Acta* **269** (1992) 9.
36. B. D. Mindel and B. Karlberg, *Lab. Pract.* **30** (1981) 719.
37. Z. Fang, S. Xu and S. Zhang, *Anal. Chim. Acta* **179** (1986) 325.
38. Z. Fang, B. Welz and G. Schlemmer, *J. Anal. At. Spectrom.* **4** (1989) 91.
39. V. Carbonell, M. de la Guardia, A. Salvador, J. L. Burguera and M. Burguera, *Anal. Chim. Acta* **238** (1990) 417.
40. J. C. de Andrade, F. C. Strong III and N. J. Martin, *Talanta* **7** (1990) 711.
41. O. F. Kamson and A. Townshend, *Anal. Chim. Acta* **155** (1983) 253.
42. S. R. Bysouth, J. F. Tyson and P. B. Stockwell, *Analyst* **115** (1990) 571.
43. E. Debrah, J. F. Tyson and M. W. Hinds, *Talanta* **11** (1992) 1525.
44. M. Burguera, J. L. Burguera and O. M. Alarcon, *Anal. Chim. Acta* **214** (1988) 421.
45. C. E. Adeeyinwo and J. F. Tyson, *Anal. Proc.* **26** (1989) 58.
46. C. E. Adeeyinwo and J. F. Tyson, *Anal. Chim. Acta* **214** (1988) 339.
47. P. Martinez-Jimenez, M. Gallego and M. Valcarcel, *Analyst* **112** (1987) 1233.
48. F. Esmadi, M. Kharoaf and A. S. Attiyat, *Microchem. J.* **40** (1989) 277.
49. Z. Fang, M. Sperling and B. Welz, *J. Anal. At. Spectrom.* **6** (1991) 301.
50. S. Pei and Z. Fang, *Anal. Chim. Acta* **294** (1994) 185.
51. H. Chen, S. Xu and Z. Fang, *Anal. Chim. Acta* **298** (1994) 167.
52. K. Fukamachi and N. Ishibashi, *Anal. Chim. Acta* **119** (1980) 383.
53. A. S. Attiyat and G. D. Christian, *Anal. Chem.* **56** (1984) 439.
54. L. Nord and B. Karlberg, *Anal. Chim. Acta* **125** (1981) 199.
55. L. Nord and B. Karlberg, *Anal. Chim. Acta* **145** (1983) 151.
56. Z. Fang, Z. Zhu, S. Zhang, S. Xu, L. Guo and L. Sun, *Anal. Chim. Acta* **214** (1988) 41.
57. J. A. Sweileh and F. F. Cantwell, *Anal. Chem.* **57** (1985) 420.
58. V. Kuban, J. Komarek and D. Cajkova, *Coll. Czech. Chem. Commun.* **54** (1989) 2683.
59. S. Lin and H. Hwang, *Talanta* **40** (1993) 1077.
60. Z. Fang, *Spectrochimica Acta Rev.* **14** (1991) 235.

61. V. Carbonell, A. Salvador and M. de la Guardia, *Fresenius J. Anal. Chem.* **342** (1992) 529.
62. M. P. P. Garcia, M. E. D. Garcia and A. S. Sanz-Medel, *J. Anal. At. Spectrom.* **2** (1987) 699.
63. H. J. Salacinski, P. G. Riby and S. J. Haswell, *Anal. Chim. Acta* **269** (1992) 1.
64. M. P. P. Garcia, A. L. Garcia, M. E. D. Garcia and A. Sanz-Medel, *J. Anal. At. Spectrom.* **5** (1990) 15.
65. P. Hernandez, L. Hernandez and J. Losada, *Fresenius Z. Anal. Chem.* **325** (1986) 300.
66. S. Xu, L. Sun and Z. Fang, *Anal. Chim. Acta* **245** (1991) 7.
67. Z. Fang, J. Ruzicka and E. H. Hansen, *Anal. Chim. Acta* **164** (1984) 23.
68. Y. Liu and J. D. Ingle Jr. *Anal. Chem.* **61** (1989) 520.
69. Y. Liu and J. D. Ingle Jr. *Anal. Chem.* **61** (1989) 525.
70. S. Hirata, K. Honda and T. Kumamaru, *Anal. Chim. Acta* **221** (1989) 65.
71. A. Karakaya and A. Taylor, *J. Anal. At. Spectrom.* **4** (1989) 261.
72. Z. Fang, T. Guo and B. Welz, *Talanta* **38** (1991) 613.
73. A. Maquieira, H. A. M. Elmahadi and R. Puchades, *Anal. Chem.* **66** (1994) 3632.
74. A. Maquieira, H. A. M. Elmahadi and R. Puchades, *Anal. Chem.* **66** (1994) 1462.
75. Z. Fang, S. Xu and S. Zhanf, *Anal. Chim. Acta* **200** (1987) 35.
76. M. C. Temprano, J. P. Parajon, M. E. D. Garcia and A. Sanz-Medel, *Analyst* **116** (1991) 1141.
77. M. Trojanowicz and K. Pyrzynska, *Anal. Chim. Acta* **287** (1994) 247.
78. A. Shah and S. Devi, *Anal. Chim. Acta* **236** (1990) 469.
79. A. M. Naghmush, K. Pyrzynska and M. Trojanowicz, *Anal. Chim. Acta* **288** (1994) 247.
80. J. F. van Staden and C. J. Hattingh, *Anal. Chim. Acta* **308** (1995) 214.
81. F. Malams, M. Bengtsson and G. Johansson, *Anal. Chim. Acta* **160** (1984) 1.
82. M. A. Marshall and H. A. Mottola, *Anal. Chem.* **57** (1985) 729.
83. V. Kuban, J. Komarek and Z. Zdrahal, *Coll. Czech. Chem. Commun.* **54** (1989) 1785.
84. A. M. Naghmush, M. Trojanowicz and E. Olbrych-Sleszynska, *J. Anal. At. Spectrom.* **7** (1992) 323.
85. Y. P. de Peña, M. Gallego and M. Valcarcel, *Anal. Chem.* **67** (1995) 2524.
86. S. Krekler, W. Frenzel and G. Schultze, *Anal. Chim. Acta* **296** (1994) 115.
87. E. Olbrych-Sleszynska, K. Brajter, W. Matuszewski, M. Trojanowicz and W. Frenzel, *Talanta* **39** (1992) 779.
88. J. Ruzicka and A. Arndal, *Anal. Chim. Acta* **216** (1989) 243.
89. S. R. Bysouth, J. F. Tyson and P. B. Stockwell, *Anal. Chim. Acta* **214** (1988) 329.
90. R. Purohit and S. Devi, *Anal. Chim. Acta* **259** (1992) 53.
91. Y. Zhang, R. Riby, A. G. Cox, W. McLeod, A. R. Date and Y. Y. Cheung, *Analyst* **113** (1988) 125.

92. Y. P. de Peña, M. Gallego and M. Valcarcel, *Talanta* **42** (1995) 211.
93. A. M. Naghmush, K. Pyrzynska and M. Trojanowicz, *Talanta* **42** (1995) 851.
94. M. Gallego, Y. P. de Peña and M. Valcarcel, *Anal. Chem.* **66** (1994) 4074.
95. C. G. Taylor and J. M. Trevaskis, *Anal. Chim. Acta* **179** (1986) 491.
96. R. Borja, M. de la Guardia, A. Salvador, J. L. Burguera and M. Burguera, *Fresenius J. Anal. Chem.* **338** (1990) 9.
97. B. Patel, S. J. Haswell and R. Grzeskowiak, *J. Anal. At. Spectrom.* **4** (1989) 195.
98. J. A. Koropchak and L. Allen, *Anal. Chem.* **61** (1989) 1410.
99. F. T. Esmadi, M. Kharaof and A. S. Attiyat, *Anal. Lett.* **23** (1990) 1069.
100. P. Martinez-Jimenez, M. Gallego and M. Valcarcel, *Anal. Chem.* **59** (1987) 69.
101. A. T. Haj-Hussen, G. D. Christian and J. Ruzicka, *Anal. Chem.* **58** (1986) 38.
102. F. t. Esmadi, M. Kharaof and A. S. Attiyat, *J. Flow Injection Anal.* **10** (1993) 33.
103. E. B. Milosavljevic, L. Solujic, J. L. Hendrix and J. H. Nelson, *Analyst* **114** (1989) 805.
104. M. Gallego, M. Silva and M. Valcarcel, *Fresenius A. Anal. Chem.* **323** (1986) 50.
105. M. Silva, M. Gallego and M. Valcarcel, *Anal. Chim. Acta* **179** (1986) 341.
106. M. C. Yebra, M. Gallego and M. Valcarcel, *Anal. Chim. Acta* **308** (1995) 275.
107. B. A. Petersson, Z. Fang, J. Ruzicka and E. H. Hansen, *Anal. Chim. Acta* **184** (1986) 165.
108. R. Montero, M. Gallego and M. Valcarcel, *J. Anal. At. Spectrom.* **3** (1988) 725.
109. M. C. Yebra, M. Gallego and M. Valcarcel, *Anal. Chim. Acta* **308** (1995) 357.
110. J. Zorro, M. Gallego and M. Valcarcel, *Microchem. J.* **39** (1989) 1989.
111. M. Sperling, X. Yin and B. Welz, *J. Anal. At. Spectrom.* **6** (1991) 295.
112. K. Bäckström and L. G. Danielsson, *Anal. Chem.* **60** (1988) 1354.
113. K. Bäckström and L. G. Danielsson, *Anal. Chim. Acta* **232** (1990) 301.
114. M. Sperling, X. Yin and B. Welz, *Analyst* **117** (1992) 629.
115. R. Ma. W. Van Mol and F. Adams, *Anal. Chim. Acta* **293** (1994) 251.
116. J. L. Burguera, M. Burguera, C. Rivas, P. Carrero, M. Gallignani and M. R. Brunetto, *J. Anal. At. Spectrom.* **10** (1995) 479.
117. E. Beinrohr, M. Cakrt, M. Rapta and P. Tarapci, *Fresenius Z. Anal. Chem.* **335** (1989) 1005.
118. J. L. Burguera, M. Burguera, P. Carrero, C. Rivas, M. Gallignani and M. R. Brunetto, *Anal. Chim. Acta* **308** (1995) 349.
119. M. Burguera, J. L. Burguera, C. Rondon, C. Rivas, F. Carrero, M. Gallignani and M. R. Brunetto, *J. Anal. At. Spectrom.* **10** (1995) 343.
120. M. A. Z. Arruda, M. Gallego and M. Varcarcel, *Anal. Chem.* **65** (1993) 3331.
121. M. Burguera and J. L. Burguera, *J. Anal. At. Spectrom.* **8** (1993) 229.
122. Z. L. Fang, M. Sperling and B. Welz, *J. Anal. At. Spectrom.* **5** (1990) 639.
123. O. Aström, *Anal. Chem.* **54** (1982) 190.
124. C. C. Y. Chan and R. S. Sadana, *Anal. Chem.* **270** (1992) 231.
125. M. Yamamoto, M. Yasuda and Y. Yamamoto, *Anal. Chem.* **57** (1985) 1382.

126. G. E. Pacey, M. R. Straka and J. R. Gord, *Anal. Chem.* **58** (1986) 502.
127. G. D. Marshall and J. F. van Staden, *J. Anal. At. Spectrom.* **5** (1990) 681.
128. R. Torralba, M. Bonilla, A. Palacios and C. Camara, *Analysis* **22** (1994) 478.
129. T. R. Rüde and H. Puchelt, *Fresenius J. Anal. Chem.* **350** (1994) 44.
130. M. G. C. Fernandez, M. A. Palacios and C. Camara, *Anal. Chim. Acta* **283** (1993) 386.
131. L. Pitts, P. J. Worsfold and S. J. Hill, *Analyst* **119** (1994) 2785.
132. S. Tesfalidet and K. Irgum, *Anal. Chem.* **61** (1989) 2079.
133. S. Zhang, S. Xu and Z. Fang, *Quim. Anal.* **8** (1989) 191.
134. K. KcLaughlin, D. Dadgar, M. R. Smyth and D. McMaster, *Analyst* **115** (1990) 275.
135. M. B. de al Calle Guntinas, Y. Madrid and C. Camara, *Anal. Chim. Acta* **252** (1991) 161.
136. M. Yamamoto, K. Takada, T. Kumamura, M. Yasuda, S. Yokoyama and Y. Yamamoto, *Anal. Chem.* **59** (1987) 2446.
137. G. D. Marshall and J. F. van Staden, *J. Anal. At. Spectrom.* **5** (1990) 675.
138. W. F. Chan and P. K. Hou, *Analyst* **115** (1990) 567.
139. G. Tao and E. H. Hansen, *Analyst* **119** (1994) 333.
140. C. Pasquini, W. F. Jardim and L. C. de Faria, *J. Autom. Chem.* **10** (1988) 188.
141. S. E. Birnie, *J. Autom. Chem.* **10** (1988) 140.
142. J. S. De Andrade, C. Pasquini, W. Baccan and J. C. Van Loon, *Spectrochim. Acta B* **38** (1983) 1329.
143. S. Schmid and M. Bahadir, *Fresenius J. Anal. Chem.* **346** (1993) 683.
144. M. Burguera, J. L. Burguera, C. Ricas, P. Carrero, R. Brunetto and M. Gallignani, *Anal. Chim. Acta* **308** (1995) 339.
145. E. A. G. Zagatto, F. J. Krug, H. Bergamin, S. S. Jorgenson and B. F. Reis, *Anal. Chim. Acta* **104** (1979) 279.
146. W. D. Basson and J. F. van Staden, *Fresenius Z. Anal. Chem.* **302** (1980) 370.
147. M. C. U. Araujo, C. Pasquini, R. E. Bruns and E. A. G. Zagatto, *Anal. Chim. Acta* **171** (1985) 337.
148. Z. Fang, J. M. Harris, J. Ruzicka and E. H. Hansen, *Anal. Chem.* **57** (1985) 1457.
149. A. O. Jacintho, E. a. g. Zagatto, H. Bergamin, F. J. Krug, B. F. Reis, R. E. Bruns and B. R. Kowalski, *Anal. Chim. Acta* **130** (1981) 243.
150. P. L. Kempster, H. R. van Vliet and J. F. van Staden, *Anal. Chim. Acta* **218** (1989) 69.
151. K. E. Lawrence, G. W. Rice and V. A. Fassel, *Anal. Chem.* **56** (1984) 289.
152. J. A. Koropchak and D. H. Winn, *Anal. Chem.* **58** (1986) 2561.
153. M. P. Granchi, J. A. Biggerstaff, L. J. Hillard and P. Gray, *Spectrochim. Acta* **42B** (1987) 169.
154. I. G. Souza, H. Bergamin, F. J. Krug, J. A. Nobrega, P. V. Oliveira, B. F. Reis and M. F. Gine, *Anal. Chim. Acta* **245** (1991) 211.
155. A. J. Ambrose, L. Ebdon and P. Jones, *Anal. Proc.* **26** (1989) 377.

156. B. R. LaFraniere, R. S. Houk, D. R. Wiederin, D. R. Wiederin and V. A. Fassel, *Anal. Chem.* **60** (1988) 23.
157. R. R. Liversage, J. C. van Loon and J. C. Andrade, *Anal. Chim. Acta* **161** (1984) 275.
158. N. H. Tioh, Y. Isreal and R. M. Barnes, *Anal. Chim. Acta* **184** (1986) 205.
159. R. M. Barnes and X. Wang, *J. Anal. At. Spectrom.* **3** (1988) 1083.
160. G. S. Pyen and R. F. Browner, *Appl. Spectrosc.* **42** (1988) 508.
161. F. Nakata, H. Sunahara, H. Fujimoto, M. Yamamoto and T. Kumamaru, *J. Anal. At. Spectrom.* **3** (1988) 579.
162. T. Nakahara and N. Nikui, *Anal. Chim. Acta* **172** (1985) 127.
163. P. C. Rudner, J. M. C. Pavon, A. G. de Torres and F. S. Rojas, *Fresenius J. Anal. Chem.* **352** (1995) 615.
164. E. A. G. Zagatto, A. O. Jacintho, F. J. Krug, B. F. Reis, R. E. Bruns and M. C. U. Araujo, *Anal. Chim. Acta* **145** (1983) 169.
165. Y. Isreal and R. M. Barnes, *Anal. Chem.* **56** (1984) 1188.
166. B. F. Reis, M. F. Gine, F. J. Krug and H. Bergamin, *J. Anal. At. Spectrom.* **7** (1992) 865.
167. M. F. Gine, H. Bergamin, B. F. Reis and R. L. Tuon, *Anal. Chim. Acta* **234** (1990) 207.
168. M. P. Garcia, M. E. D. Garcia and A. Sanz-Medel, *J. Anal. At. Spectrom.* **2** (1987) 699.
169. J. P. Dolan, S. A. Sinex, S. G. Capar, L. Montaser and R. H. Clifford, *Anal. Chem.* **63** (1991) 2539.
170. A. G. Cox, I. G. Cook and C. W. McLeod, *Analyst* **110** (1985) 331.
171. A. G. Cox and C. W. McLeod, *Anal. Chim. Acta* **179** (1986) 487.
172. N. Furuta, K. R. Brushwyler and G. M. Hieftje, *Spectrochim. Acta* **44B** (1989) 349.
173. C. W. McLeod, I. G. Coox, P. J. Worsfold, J. E. Davies and J. Queay, *Spectrochim. Acta* **40B** (1985) 57.
174. A. G. Cox, C. W. McLeod, D. L. Miles and J. M. Cook, *J. Anal. At. Spectrom.* **2** (1987) 553.
175. S. D. Hartenstein, J. Ruzicka and G. D. Christian, *Anal. Chem.* **57** (1985) 21.
176. S. D. Hartenstein, G. D. Christian and J. Ruzicka, *Can. J. Spectrosc.* **30** (1985) 144.
177. S. Caroli, a. Alimonti, F. Petrucci and Zs. Horvath, *Anal. Chim. Acta* **248** (1991) 241.
178. B. Fairman, A. Sanz-Medel and P. Jones, *J. Anal. At. Spectrom.* **10** (1995) 281.
179. J. L. Manzoori, A. Miyazaki and H. Tao, *Analyst* **115** (1990) 1055.
180. R. K. Winge, V. J. Peterson and V. A. Fassel, *Appl. Spectrosc.* **33** (1979) 206.
181. J. R. Pretty, E. A. Blubaugh, E. H. Evans, J. A. Caruso and T. M. Davidson, *J. Anal. At. Spectrom.* **7** (1992) 1131.
182. T. Kumamaru, Y. Nitta, F. Nakata, H. Matsu and M. Ikeda, *Anal. Chim. Acta* **174** (1985) 183.

183. M. Yamamoto, Y. Obata, Y. Nitta, F. Nakata and T. Kumamaru, *J. Anal. At. Spectrom.* **3** (1988) 441.
184. J. L. Marzoori and A. Miyazaki, *Anal. Chem.* **62** (1990) 2457.
185. N. Kasthurikrishnan and J. A. Koropchak, *Anal. Chem.* **65** (1993) 857.
186. Y. Madrid, M. Wu, Q. Jin and G. M. Hieftje, *Anal. Chim. Acta* **277** (1993) 1.
187. F. Camuna, J. E. S. Uria and A. Sanz-Medel, *Spectrochim. Acta* **48B** (1993) 1115.
188. C. W. McLeod, P. J. Worsfold and A. G. Cox, *Analyst* **109** (1984) 327.
189. M. M. Gomez and C. W. McLeod, *J. Anal. At. Spectrom.* **8** (1993) 461.
190. H. Morita, T. Kimoto and S. Shimomura, *Anal. Lett.* **16** (1983) 1187.
191. W. Jian and C. W. McLeod, *Talanta* **11** (1992) 1537.
192. A. Morales-Rubio, M. L. Mena and C. W. McLeod, *Anal. Chim. Acta* **308** (1995) 364.

Chapter 3

Electrochemical Detection Methods

1. Potentiometric Detection

1.1. *Properties of Potentiometric Detectors in FIA*

Potentiometric detection is based on the measurement of the electromotive force of the measuring cell, which usually consists of a suitable indicating electrode (whose potential depends on the concentration or, more precisely, the activity of the analyte) and of a reference electrode, maintained at a constant potential during the measurement. The measurement of the potential difference between these electrodes is carried out with potentiometers of high input impedance, which prevent electrode processes that might disturb the equilibrium at the electrode-solution interface. Ion selective electrodes with various membranes are the most common indicating electrodes in modern potentiometry. The chemical composition and structure of the membrane determine the selectivity of the electrode response towards a given analyte. The potential of the membrane electrode is described by the Nikolsky–Eisenman equation:

$$E_i = E_i^0 + s \log(a_i + \Sigma K_j a_j^{z_i/z_j}), \qquad (1)$$

where E_i is the potential of the electrode measured in a solution of activity a_i of the main ion and a_j are the activities of other ions present in the solution which affect the electrode potential. z_i and z_j are electrical charges of main and other ions, respectively. E_i^0 is the standard potential of the electrode, which depends on the type of electrode and the electroactive materials used, while s is the slope of the semilogarithmic relationship between the electrode potential and the activity of the main ion. The latter depends on the ionic charge and the temperature. For a monovalent ion at room temperature, $s = 59.2$ mV per decade of activity change, while for divalent ions, $s = 29.6$ mV/decade. The value of the selectivity coefficient K_j depends on the kind of membrane used and the physico-chemical mechanisms of interaction of various ions with the membrane. From the point of view of analytical applications of membrane electrodes, the main ion is not always the one to which the electrode response is the most sensitive, and therefore the K_j values may be either greater or smaller than 1. Redox electrodes composed of inert materials such as noble metals or different forms of carbon are also employed as indicating electrodes. Their potential depends on the activity of both forms of the redox couple in solution and is given by the Nernst equation:

$$E = E^0_{\mathrm{Ox,\ Red}} + s \log(a_{\mathrm{Ox}}/a_{\mathrm{Red}})\,, \tag{2}$$

where a_{Ox} and a_{Red} are the activities of the oxidised and reduced forms of a redox couple, respectively. First kind metallic electrodes, for which the potential value is determined by the equilibrium between the electrode material phase and metal ions in the solution, as well as second kind electrodes, where the active metal surface is covered with a sparingly soluble salt of this metal, are also used as indicating electrodes. Second kind electrodes can be used for the detection of the activity of an anion forming a sparingly soluble precipitate, and they are also used as the reference electrodes. The constant potential of the reference electrode is maintained by using the electrode in a half-cell with a constant concentration of the anion, to which the electrode responds.

Potentiometric measurements are among the most common measuring methods in modern instrumental chemical analysis, mainly due to the common use of pH measurements in science and industry. The development of numerous other membrane electrodes of exceptional selectivity has greatly increased the range of analytical applications of potentiometry. Their unique feature is a semilogarithmic dependence of the potential on the activity of the detected ion (or its concentration under constant ionic strength conditions), which allows their use over a very wide range of activities (concentrations). The instrumentation necessary to carry out potentiometric measurements is usually simpler than that required for other detection methods.

These features, together with the ease of constructing electrodes and flow through potentiometric cells in most analytical laboratories, are the main reasons that potentiometric detection was employed very early [1, 2]. As evidenced by the number of published original research papers and review articles [3–6], it is one of the most frequently used FIA detection methods. Its other main attributes are the simplicity of the experimental set-up, its sensitivity and the relatively simple chemistry involved with the most frequently determined analytes. Potentiometry, like other electrochemical detection methods, is not affected by the colour or turbidity of the sample and in most cases does not need the use of expensive specific reagents. Apart from the above-mentioned semilogarithmic dependence of the potential response on the activity of the analyte, the other unique feature of potentiometric detection is that it gives information about the particular form of the element. This is essential for speciation analysis. All of these features are utilised in FIA.

Apart from the obvious mechanisation of the analytical procedure which FIA always provides, the use of this methodology creates more favourable conditions for potentiometric detection, when compared to conventional nonflow measurements [3]. In FIA measurements, it is easier to avoid an incidental contamination or that resulting from the leak of the solution from the reference electrode. A steady flow of solution past the sensing surface of the indicator electrode prevents the deactivation of the surface due to precipitation or adsorption. Because of different kinetics of the heterogeneous interaction between the main sensed ions, and interferences with the indicating electrode material, in certain cases the use of flow injection measurement with a very short time of contact between the sample solution and the electrode surface may significantly improve the selectivity of detection through the so-called mechanism of kinetic discrimination.

The concentration-time relationship at the sensing electrode surface can be described theoretically for flow injection potentiometric measurements under certain practically realisable conditions. Mass transport and kinetic models discussed in the literature [7, 8] differ with respect to the nature of the rate-determining process. This can be the transport of the primary ion to the electrode surface through a stagnant liquid layer, or a kinetic process of dehydration or/and subsequent adsorption due to the space charge region, which is formed in a precipitate exchange or a complex formation equilibrium. None of these models provide an accurate description of the potentiometric signal in FIA without taking into account the surface nonidealities. It has been shown, however, that a treatment based on the assumption that ideal mixing occurs in microcavities in the electrode surface describes fairly successfully the experimentally obtained potential-time transients. The most significant conclusion of these considerations was that the rate of response of solid-state potentiometric sensors is in principle the same as that of other electrochemical types. The distortions observed in the potential-time signal mainly result from their logarithmic character. An experimental comparison of a tubular

AgI precipitate-based electrode and a tubular conductivity detector has been made [8].

In practical analytical FIA measurements, the response time limitation of potentiometric detectors is observed mainly as a slow return of the potential value to the base-line level [9,10] or as a deformation of the peak shape [11]. The dynamic behaviour of potentiometric membrane electrodes depends on the concentration level of the main sensed ion and on the direction of the concentration change (increasing or decreasing). At higher concentrations, potential changes are faster than at lower concentrations. The efficient way to improve the dynamic behaviour of the measuring system in order to increase the sampling rate is to introduce a small concentration of the main ion into the carrier solution. Suggested initially for low concentration measurements with a copper selective electrode [12], this has became a common practice in flow injection potentiometry. This procedure provides a more stable base-line and faster return of the potential value to the base-line level, but it is accompanied by a decrease in the flow injection signal magnitude with an increase of the main ion concentration in the carrier solution [9].

In flow injection measurements with potentiometric detection, deviations of the slopes of the calibration plots from the theoretical ones predicted with Eq. (1) may occur [9, 13]. The super-Nernstian slopes can be explained by different rates of increase of the signal at low and high concentrations. The short residence time of the sample solution at the sensing surface of the electrode (a few seconds) results in the observation of different proportions of the steady state for each concentration [9]. It is, however, more difficult to explain the peak shape caused by the negative signals commonly referred to as "over-shoots", which occur mostly in measurements done in the presence of interfering ions. In measurements with iodide electrodes, their magnitude and shape significantly depend on the degree of oxidation of the electrode surface and ageing of the electrode [11]. These deformations can be explained by sudden activity changes at the electrode, resulting from the desorption of primary ions from the membrane surface that immediately follows the adsorption of excess interferent ions from solution [14].

The response time of a potentiometric detector in FIA is also affected by the ionic strength of the sample solution, which results in changes of the signal magnitude. This effect is associated with obtaining a negative prepeak, which was observed with increase of the ionic strength in determination of fluoride [15]. A decrease of the response slope in flow injection potentiometry can be expected for electrodes with insufficient response times, even when the electrode exhibits Nernstian behaviour under conventional steady state conditions. This has been found using calcium [16] and potassium [17] selective electrodes with photocured membranes containing commonly used ionophores.

Flow injection measurements with potentiometric detection are usually carried out in systems with medium dispersion (dispersion coefficient $D < 5$). In most cases, it is necessary to mix the injected sample with the carrier solution on-line in order to adjust the ionic strength and provide a constant ratio of the measured activity of the determined ion to its concentration in the injected sample. Often, it is necessary to use buffering species either in the carrier stream, or as additional reagent stream to adjust pH, along with a masking agent to eliminate interferences. In some cases, either dilution should be avoided, for example for pH determination in soil extracts [18], or very large dilution should be applied, as in the gradient titrations discussed below. When the dynamic properties of the electrode are constant in the measured concentration range, the peak height has a Nernstian relationship with the analyte concentration, but it also depends on dispersion in the flow injection system [19]. Thus,

$$H = H_0 - S \log D \,, \tag{3}$$

where H_0 is the signal magnitude for an undiluted sample of the same original concentration. The dilution of the sampling zone in FIA measurement limits the linear response range of membrane electrodes. It has been found [19] that

$$\log C_{\mathrm{FIA}} = \log C_{\mathrm{NF}} + \log D \,, \tag{4}$$

where C_{NF} is the lower limit of the linear response range of the electrode functioning in a non-flow measurement, and C_{FIA} is the lower linear detection limit in FIA. Obviously, an increase in dispersion in an FIA system leads to lower detectability. The demonstrated in certain cases improvement in detectability in FIA systems, when compared to conventional non-flow measurements [20], probably results from a more efficient elimination of various contamination effects, and better conditioning of the indicating electrode through the long-term rinsing of its surface with carrier solution, when this is used without addition of the sensed ions.

At very low concentrations, where a curvature of the semi-logarithmic relationship between peak height and concentration is observed, a linear dependence of peak height with concentration can be observed. For a second kind chloride electrode it has been found that in FIA measurements the following relationship is obeyed [21]:

$$H = k(2DK_{so}^{0.5})^{-1}C \qquad \text{for} \qquad (C/D)^2 \ll 4K_{so}\,, \tag{5}$$

where C is the concentration of the analyte in the injected sample, K_{so} is the solubility product of the membrane material and k is a constant. This relationship can be utilised for trace determinations in FIA with potentiometric detection.

Carrying out potentiometric measurements under flow injection conditions with a very short time of contact of the sample zone with the surface of the sensing electrode, may alter the observed selectivity of detection. This occurs mainly with those interfering ions which under steady-state conditions provide a potentiometric response larger than that of the analyte. The presence of such species in the solution contacting the electrode surface may cause significant changes on the sensing surface or even in the bulk of the electrode membrane. Because of the short contact time in FIA measurement, these interactions are minimised. This results in an apparent decrease in selectivity coefficient values, and is treated as kinetic discrimination of interferences. In measurements with a second kind chloride electrode at $D = 5.2$, the selectivity coefficients for bromide and iodide were only 1.9 and 1.8, respectively [21]. Similarly, in FIA measurements with a membrane chloride electrode, the value of the selectivity coefficients for bromide, iodide and thiocyanate decreases with an increase in the flow rate [22]. For a similar reason, the strong interaction of thiosulphate with an iodide electrode, which leads to membrane damage in steady-state measurements, is not harmful to the electrode in FIA measurements [11]. The same effect was observed for PVC-based nitrate [23] and 5,5-diethylbarbiturate [24] electrodes, and with cation electrodes sensitive to Ag(I) and Cu(II) [13]. Such an effect is not observed for interfering ions with a selectivity coefficient lower than 1 [25], and a deterioration in the selectivity of FIA measurements has even been observed [24].

1.2. *Construction of Detectors*

One of the important advantages of using potentiometric detection in FIA systems is the relative ease of construction of flow cells for the use of conventional electrodes or of dedicated flow through detectors. The proper design of these is of great significance to the dynamic behaviour of the measuring flow system. The flow through detector should ensure the most efficient exchange possible of the solution at the sensing surface, i.e. its dead volume should be minimised. Due to possible interference caused by the formation of the flow potential the distance between indicating and reference electrodes should be as small as possible, and in tubular cells the channel between these electrodes should be large in diameter [26]. Usually, it is advantageous to ground the solution. The reference electrode should be placed in a position that eliminates possible contamination of the sample at the sensing electrode surface. It has been shown in flow injection potentiometry that reference electrodes, with liquid junctions providing additional junction potential to the total EMF of the measuring cell, may create large transient signals with a sudden change in the electrolyte composition in flow measurements [27]. Hence the best reference electrodes are those with a constant flow of the eletrolyte.

The most common types of flow-through cells used in FIA systems with conventional ion-selective electrodes are shown in Fig. 1. In a cascade design, a layer of solution is formed at the sensing surface due to surface tension (Fig. 1A) [2, 28, 29]. More frequently used are thin-layer cells, with a strictly regulated thickness of the liquid layer at the sensing surface (Fig. 1B) [30–32]. An increasing number of applications have been found for wall-jet cells [33], which are easy to produce and handle, especially in the large volume configuration (Fig. 1C) [9, 34]. A comparison of several different flow cell configurations in flow injection potentiometry using a fluoride selective electrode has shown that the best dynamic performance and the smallest carry-over errors are obtained with a wall-jet cell [35]. In this configuration the solution spreads uniformly over the entire membrane, creating a thick diffusion layer. With the optimal flow rate and distance between the outlet nozzle and the sensing surface, the outer solution does not affect the measured signal. This design, however, cannot be used in stopped-flow measurements, and is difficult to use in multicomponent measurements, unless a miniaturised array of sensing electrodes is employed.

Among the dedicated potentiometric detectors designed for flow injection systems, the most common type is the tubular cell, which can be manufactured for solid electrodes, glass membrane electrodes, and polymer membrane electrodes. Cells with solid electroactive material can be made by drilling a channel in a pellet of sensing electrode material (Fig. 2A) [12, 36], or by deposition of an appropriate precipitate on the inside surface wall of silver tubing to form a second kind electrode [37–39]. A polymer membrane tubular electrode is made by replacing part of the wall of a small piece of PVC tubing

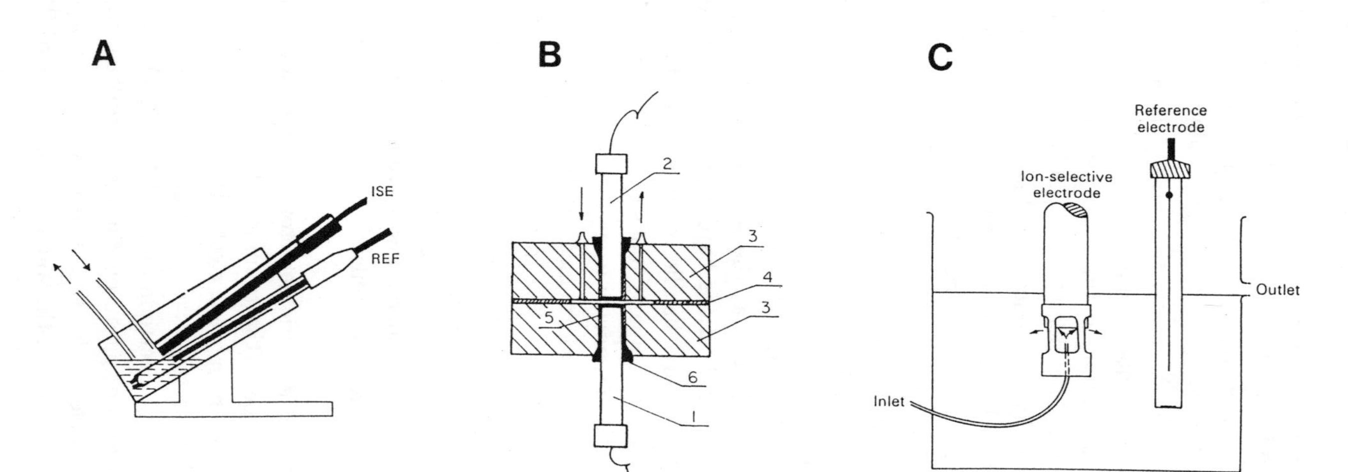

Fig. 1. Flow-through cells used with conventional ion-selective electrodes for potentiometric detection in FIA systems: (A) cascade type with ion-selective electrode (ISE) and calomel reference electrode (REF) [28]; (B) thin-layer cell: 1 — ion-selective electrode; 2 — reference electrode; 3 — cell body; 4 — spacer; 5 — rubber seal; 6 — fixing screws [31]; (C) large volume wall-jet cell [34]. (*Reprinted by permission of copyright owner.*)

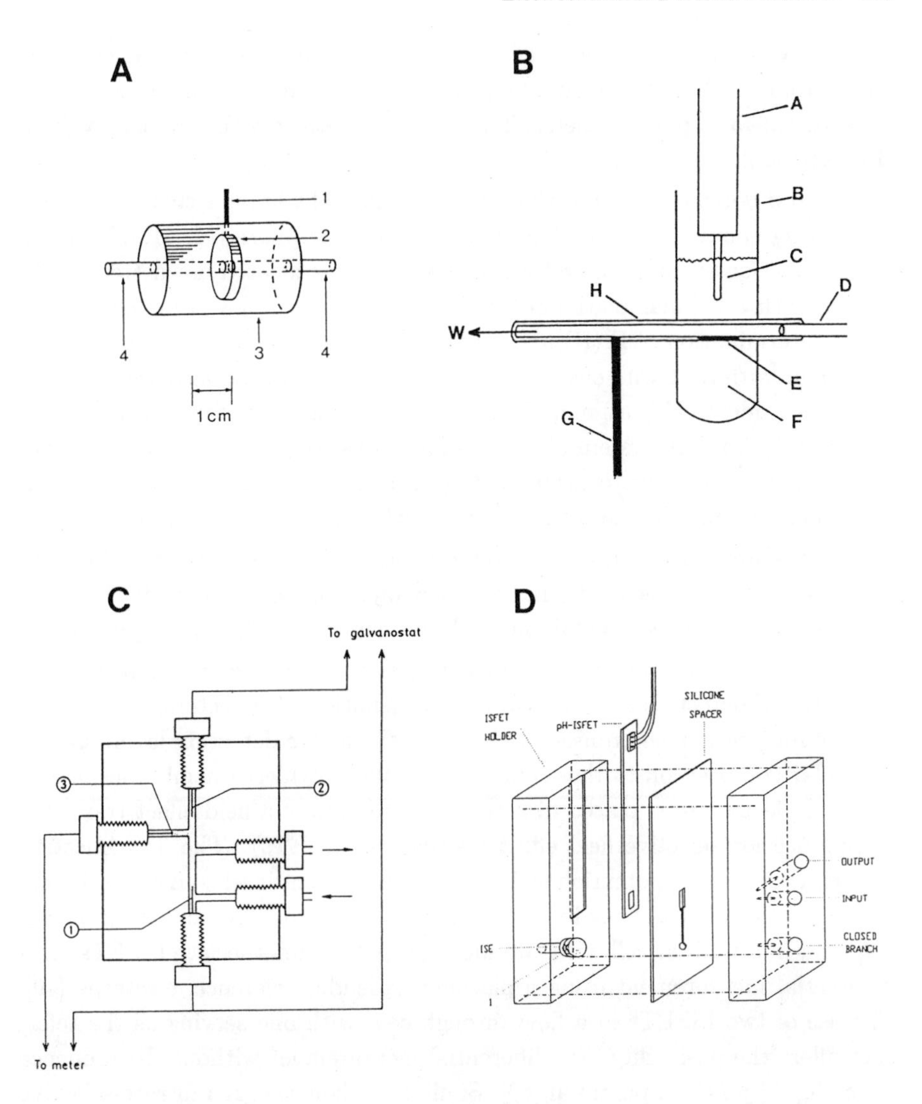

Fig. 2. Potentiometric flow-through cells designed for FIA systems: (A) disposable flow through electrode with solid-state membrane [12]: 1 — silver wire; 2 — solid-state membrane; 3 — polymer body; 4 — inlet/outlet; (B) tubular polymer detection unit [40]: A — coaxial cable; B — plastic pipette tip; C — Ag/AgCl electrode; D — stainless steel tube; E — polymer ion-selective membrane; F — internal solution; G — salt bridge; H — PVC tubing; W — waste; (C) flow cell with wire electrodes [44]: 1 — indicating wire electrode; 2 — auxiliary Pt-wire electrode; 3 — reference Ag/AgCl electrode in a compartment with agar gel containing KCl; flow-through sandwich assembly for the ISFET and the ISE [51]. (*Reprinted by permission of copyright owner.*)

with a PVC-ionophore-plasticiser ion-selective membrane. When enclosed in an appropriate housing, containing an internal solution and reference electrode, it acts as a potentiometric flow-through detector with zero dead volume (Fig. 2B) [40].

Tubular polymer electrodes have been produced by lining a channel through conducting epoxy resin with a PVC-containing ionophore [24, 41, 42]. In all these detection systems the reference electrode was located downstream. Detection with a tubular platinum electrode has been employed in the indirect redox determination of Fe(II) and ascorbic acid, providing better results than a detector with a Pt wire electrode [43]. The same work has also seen the first use of two identical platinum electrodes placed in series for potentiometric detection in an FIA system giving a signal consisting of two peaks of opposite direction [43]. This concept was later used to increase the sensitivity of potentiometric fluoride detection with ion-selective membrane electrodes [34].

A flow-through cell with a metallic copper wire electrode (Fig. 2C) has been successfully employed for indirect flow injection potentiometry of copper-complexing ligands [44], metal ions [45], inorganic ions [46] and sugars [47].

Progress in analytical potentiometry in recent years has brought a miniaturisation of detection systems via the hyphenation of detectors with a large scale of integration electronics. A chemically sensitive detection layer, usually of similar composition to those employed in the polymeric membranes of ion-selective electrodes, is placed directly onto the gate of a field effect transistor to produce ion-selective field effect transistors (ISFETs). The flow injection applications of such potentiometric sensors have been developed since the beginning of the 1980's. They have been used in a cascade flow cell configuration [48], and in a wall-jet cell with two identical potassium-selective ISFETs for a differential measurement using a platinum (pseudo) reference, electrode [49]. The use of two ISFETs in a flow through cell, with one serving as the reference offers the possibility of a differential measurement without the influence of the liquid junction potential [50]. Similarly a flow through nitrate-selective electrode with a polymeric membrane in solid contact with graphite epoxy has been used in flow injection potentiometry with pH-sensitive ISFET as the reference (Fig. 2D) [51].

The entire hydraulic part of an FIA system with potentiometric detection using a polymeric membrane electrode with a solid metallic contact can be miniaturised by designing integrated microconduits [52, 68].

1.3. *Measuring Techniques*

Flow injection potentiometry is used both for the determination of analytes, which are directly sensed by the indicating electrode used, and for the indirect determination of analytes which react with the ion that is measured directly by the electrode.

Direct FIA potentiometric measurements are most commonly carried out in a single line manifold, where the sample is injected into the carrier solution, or in two-line manifolds, where the sample is introduced into a stream of distilled water, which is then merged with the reagent solution that ensures constant ionic strength, and which very often also contains buffering components and masking agents. Analytical determination is usually based on measurements of peak height as a function of the analyte concentration in the injected sample or standard solutions. The above-mentioned addition of the main ion to the carrier solution not only stabilises the base-line and improves the dynamic properties of detection, but also allows the determination of very low concentrations of the analyte below the level of the added concentration in the carrier solution. Although in such measurements a linear response range can be found [53], it can be shown that the term $10^{H/S}$ is a linear function of analyte concentration, where H is the peak height, and S the slope of the electrode response [54]. An example of such a signal is shown in Fig. 3. In FIA measurements with a second kind Ag/AgCl electrode chloride, concentrations below the solubility of silver chloride have been measured [34].

A quantitative determination based on the measurement of a peak height for standard solutions may be erroneous, even up to 50%, in the case of heterogeneous samples such as whole blood, because of undefined dilution and dispersion [55]. It has been demonstrated in FIA determinations of potassium and pH with ISFET potentiometric detectors that such errors can be eliminated by the use of a standard addition procedure in a dedicated FIA system [56]. The most efficient procedure was the method of standard addition with multiple additions, incorporating the Gran plot technique.

The most frequent difficulties encountered in flow injection potentiometric determinations in natural matrices are caused by the limited selectivity of membrane ion-selective electrodes. These can be eliminated using numerical corrections, chemical removal of interferences by masking, or dialytic on-line removal. Numerical correction procedures have been developed for the determination of nitrate [57] and ammonia [58]. In the FIA determination of nitrate in a variety of environmental samples the nitrate content is calculated

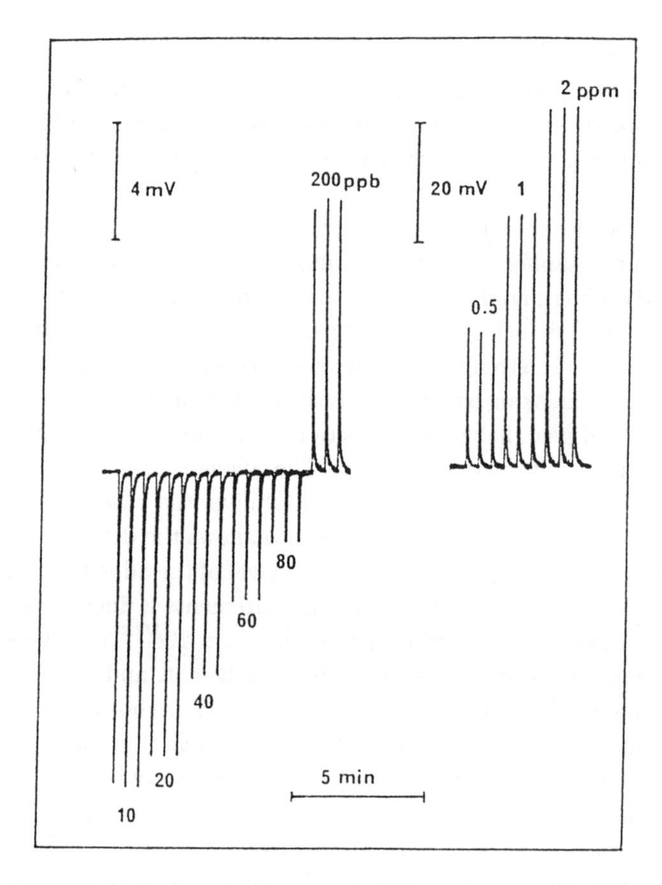

Fig. 3. Flow injection peaks recorded for potentiometric detection with a fluoride selective electrode using carrier solution of 0.2 M NaCl + 0.2 M acetic buffer pH 5.2 containing 100 μg/l fluoride [54]. Sample injection volume 100 μl. Flow rate 6.6 ml/min. (*Reprinted by permission of copyright owner.*)

by taking into account the chloride level determined by a chloride electrode, and using an experimentally determined selectivity coefficient [57]. In the determination of ammonium ion with a nonactin ammonium-selective electrode, an array of electrodes selective for ammonium, sodium, potassium and calcium ions is used, and the array response is interpreted using the Projection Pursuit regression model, which leads to improved precision and accuracy over FIA measurements with a single electrode [58].

Of the chemical methods for eliminating interferences, the simplest involves the addition of a sufficiently high level of interferent background to the carrier solution, allowing the elimination of its effect on the results for injected samples. Such a procedure has been successfully applied in the determination of iodide in iodised table salt [37], and bromide in soil extracts [38], in order to eliminate the influence of very high levels of chloride. The most common example of the masking of interferences is the use of DCTA [15, 30, 33] or EDTA [53] in the determination of fluoride. The complexones are added in order to bind alkaline earth metal, Al and Fe(III) ions, which may cause negative errors in fluoride determinations. It has been found, however, that the presence of DCTA in the carrier solution may result in a steady increase in the response time, and the addition of DCTA to the sample solution off-line, prior to injection, is recommended [59]. Another example is the elimination of sulphide interference in the FIA determination of chloride using the addition of bromate in nitric acid [60] or hydrogen peroxide in an alkaline medium to the carrier solution [61].

On-line dialysis of the analyte can be used not only for the elimination of interferences in potentiometric FIA detection, but also for an improvement in detectability. In the trace determination of copper, the on-line use of a column containing 8-quinolinol, immobilised on porous glass, enables the preconcentration of analyte and the elimination of anionic interferences [62]. The on-line preconcentration and separation of analyte from a complex matrix can be based on the volatility of the analyte. These methods have been employed in the process determination of sulphide in diisopropanolamine solutions with a gas-diffusion membrane module [63], and in the determination of ammonia with a nonactin-based ammonium electrode [64, 65]. One on-line separation of fluoride is based on its conversion to volatile trimethylfluorosilane. After a combination of continuous evaporation and gas diffusion (called pervaporation) it is absorbed in dilute NaOH solution, decomposed to free fluoride, and determined with a fluoride electrode [66]. The suppression of interferences by means of on-line dialysis has been demonstrated in the elimination of the effect of most of the inorganic and organic anions normally found in blood in the determination of salicylate in serum with a tubular polymer membrane electrode based on manganese (II) tetraphenylporphyrin [67]. In the determination of lithium in serum using coated wire electrodes, on-line dialysis increases the Li/Na selectivity by about 40% and solves the problem of protein interference [68]. In the determination of high chloride levels in electroplating baths, dialysis can be employed for effective on-line dilution of samples [69].

An improvement in the detection limit or sensitivity in flow injection potentiometry can be obtained not only by a suitable choice of the indicating electrode, optimisation of the flow-through cell and hydrodynamic conditions, or the use of on-line preconcentration; the sensitivity of detection can be significantly increased using multiple cells. This has been demonstrated for a system with air-segmented flow [70], but the same concept can be used in FIA systems. The use of real-time digital filters for signal processing gives better detection limits and an improvement in the precision of potentiometric detection [71].

FIA systems developed with direct potentiometric detection are listed in Table 1. The chemical conditions for determination depend upon the particular matrices of samples being analysed and are discussed in more detail in further chapters concerning the various practical applications of FIA.

Table 1. Applications of direct potentiometric detection in FIA.

Analyte	Type of electrode	Concentration range, pX	Reference
Ag(I)	Membrane solid-state	2.0–6.0	13
Anionic surfactants	Membrane polymeric	3.0–4.0	102
Bromide	Ag/AgBr	1–5000 ppm	10, 38
Ca	Membrane polymeric	1.0–5.0	16, 90, 91, 92, 94–97, 103-105
	ISFET	2.0–3.3	48, 101
Catechol	Pt modified	2.0–6.0	99
Cd(II)	Cd/CdS	0.25–10 g/l	106
	Membrane solid-state	2.0–5.0	13
Chloride	Membrane solid-state	1.0–4.0	69, 94, 107, 108
	Ag/AgCl	1.0–5.0	10, 21, 61, 90, 109
	Membrane polymeric	2.0–4.0	110, 111
CO_2	Membrane polymeric	2.0–4.0	112
Cu(II)	Membrane solid-state	1.0–6.0	9, 12, 13, 62
Cyanide	Membrane solid-state	3.0–6.6	31, 59
5,5-diethylbarbiturate	Membrane polymeric	1.0–4.0	24
Fluoride	Membrane solid-state	1.0–7.0	9, 15, 30, 33, 35, 53 54, 59, 66, 93, 113
H^+	Membrane glass	3.0–5.0	53
	Membrane polymeric	4.0–9.5	18
	ISFET	1.0–13.0	48, 51, 55, 56, 101

continued

Table 1 (*continued*)

Analyte	Type of electrode	Concentration range, pX	Reference
Iodide	Membrane solid-state	1.0–6.0	9, 11, 25, 30
	Ag/AgI	10–5000 ppm	10, 37
K	Membrane polymeric	1.0–5.0	2, 17, 29, 56, 90, 91 93–97, 114, 115
	ISFET	1.0–6.0	48–50, 101
Li	Membrane polymeric	2.7–3.7	68
Na	Membrane solid-state	−0.3 – 3.0	116
	Membrane polymeric	1.0–4.7	2, 93, 95–97
NH_3	Air-gap gas-sensing	2.0–3.0	1
NH_4^+	Membrane polymeric	2.0–5.0	40, 58, 64
Nitrate	Membrane solid-state	1.0–3.0	98
	Membrane polymeric	1.0–5.0	2, 28, 41, 57, 90, 92, 93, 114
Pb(II)	Membrane solid-state	1.0–5.0	13
	ISFET	3.0–5.0	50
Salicylate	Polymeric	2.0–5.7	67
SO_2	Gas-sensing	0.5–1.0 ppm	3
Sulphide	Membrane solid-state	1.0–5.0	63, 108, 117

As mentioned above, numerous FIA systems have been developed for the indirect determination of analytes which react with species that are directly measured with an indicating electrode, or components which can be chemically transformed into species sensed by the indicating electrode. The first kind of analyte can be determined using first kind metallic electrodes, membrane ion-selective electrodes, and also redox electrodes. The applications of a metallic copper electrodes have already been indicated. A metallic silver electrode has been employed for the determination of cyanide [65, 72] and sulphide [65] in systems with gas-separation units. A metallic gold electrode has been employed for the determination of cyanide [73]. Sulphite can determined using a solid-state ion-selective electrode with a mercury(II)sulphide-mercury(I) chloride membrane [74]. Sulphate orthophosphate and tripolyphosphate can be determined with a lead ion-selective electrode [75], while phosphate species in detergents are determined with a calcium ion-selective electrode [76]. In an indirect determination of sulphate in the FIA system with a lead sulphate reactor, a response function has been obtained with a linear dependence on analyte concentration in the injected samples [76a]. Applications of other membrane electrodes in indirect flow injection potentiometry are shown in Table 2.

Table 2. Applications of indirect potentiometric detection in FIA.

Analyte	Indicator electrode	Concentration range, pX*	Reference
Al(III)	Fluoride	0.015–0.04 M	3
		0.5–50 ppm	34
Ascorbic acid	Pt, Ce(III)/Ce(IV)	0.5–50 ppm	34
Borate	Tetrafluoborate	0.04–0.4 ppm B	82
Chlorine	Ag/AgCl	2.0–5.3	80
	Au, Fe(II)/Fe(III)	5.0–7.3	78
Chloroacetone	Chloride	2.0–5.0	81
Chloronitrobenzene	Chloride	2.0–3.7	81
Cyanide	Ag metallic	2.0–6.0	72
	Au metallic	0.25–70 ppm	73
Ethanol	Pt, Fe(II)/Fe(III)	3–40% (v/v)	79
Hydrazine	Iodide/Pt	50–300 ppb	84
Inorganic anions	Cu metallic	1.0–5.0	46
NO_x, nitrite	Nitrate	3.0–5.3	79
Organic ligands	Cu metallic	2.0–5.0	44
Orthophosphate	Pb membrane	2.0–5.0	75
Sugars	Picrate	25–200 g/l	77
Sulphate	Pb membrane	1–1000 ppm	75
		0.1–10 ppm	76a
Sulphite	HgS/Hg_2Cl_2	0.1–10 ppm	74
Sulphide	Ag metallic	2.0–5.0	65
Tripolyphosphate	Ca membrane	2.0–4.0	76

*If other concentration units are not shown.

A redox electrode with an Fe(III)/Fe(II) indicator couple has been used for the trace determination of chlorine in water [78], and ethanol in alcoholic beverages [79]. In the latter case, the determination is based on the oxidation by dichromate of alcohol permeating across a porous membrane, and the subsequent reduction of excess dichromate with the ferrous ion in a flow system. The concentration of the ferric ion formed is measured potentiometrically with a platinum-plated electrode.

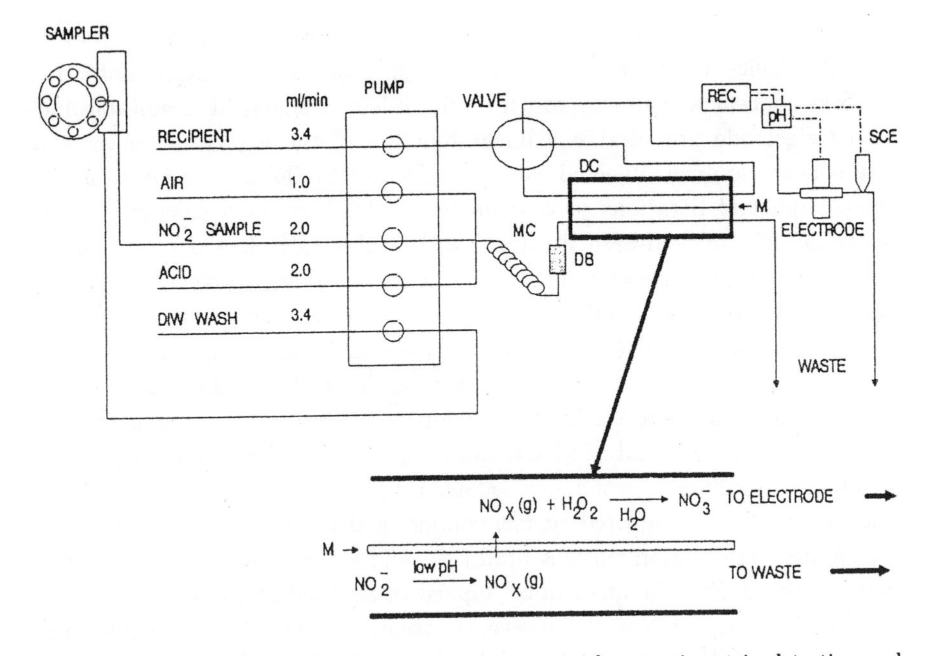

Fig. 4. Schematic diagram of a flow injection system with potentiometric detection and on-line dialysis for detection of NO_x/nitrite [79]: DC — dialysis chamber; MC — mixing coil; M — microporous membrane; pH — pH/mV meter; REC — strip chart recorder; SCE — saturated calomel electrode; DB — tubular debubbler. (*Reprinted by permission of copyright owner.*)

The chemical transformation of an analyte in an FIA system into a species detected by an indicating electrode has been adopted for the determination of NO_x [79], free chlorine [80], chloroorganic compounds [81] and boron [82]. In the system for the determination of nitrogen oxide and nitrite (Fig. 4), nitric oxide is trapped across a teflon membrane in a dialyser and converted to nitrate by a buffered recipient solution. This solution is then injected and carried to a tubular nitrate-selective electrode [79]. A double electrode system composed of an iodide-selective electrode and a platinum electrode has been employed in the indirect determination of residual chlorine [83] and hydrazine [84].

Potentiometric detection can also be used in systems developed for flow injection titrations. These can be carried out in different ways. For gradient titrations, FIA systems are designed to give a large dispersion of the injected sample, by incorporating into the system a miniature mixing chamber or a piece of tubing of larger internal diameter. Samples are passed through a gradient device and then mixed with a continuously flowing stream of titrant of fixed concentration. The width of the selected signal at a suitable indicating electrode potential is a linear function of the logarithm of the analyte concentration in injected samples. This methodology has been used in complexometric titrations of calcium with EDTA using a calcium-selective electrode [85], in acid-base titrations with a glass pH electrode [86], and in precipitation titrations of sulphate with a barium-selective electrode [87]. In so-called single-point titrations, a small volume of the sample is injected into a water stream, which then merges with an acidic or basic buffer solution, and the peak height is recorded using a glass electrode. In this titration technique, peak maxima are a linear function of the acid or base concentration [88]. In yet another kind of FIA titration procedure, a piston burette is used as a variable-volume injection device [89]. By plotting the peak height for each injection versus the injected titrant volume, a titration curve very similar to a conventional titration curve is produced, and an equivalence volume can be determined. This technique can be applied to acid-base titrations.

The sequential connection of several potentiometric detectors in flow injection systems or the construction of flow through cells with several indicating electrodes is often used for multicomponent flow injection potentiometry. This possibility was demonstrated very early with the use of a cascade detector with potassium and sodium ion-selective electrodes [2]. Such measurements are relatively simple if, with the same carrier solution, no interferences are observed in the determination of any of the analysed components of the sample. This technique has been used successfully in a flow through detector with solid-contact polymer membrane electrodes for potassium, calcium, nitrate and chloride, in the determination of these species in soil extracts. Sodium acetate with some background level of all analytes is used as the carrier solution [90]. Potassium and calcium content in wine can be determined in a similar fashion [91]. This is impossible, however, in the simultaneous determination of calcium and nitrate in waters, and it is necessary to optimise the individual carrier solution for each analyte [92]. A very interesting method for the simultaneous determination of two analytes, using a flow through cell with two indicating electrodes but no conventional reference electrode, has been proposed, and is shown in Fig. 5 [93]. In this system, a double sample loop and an asynchronous sample injection technique are used. Each electrode serves as an indicating or reference electrode, and the determination of each component

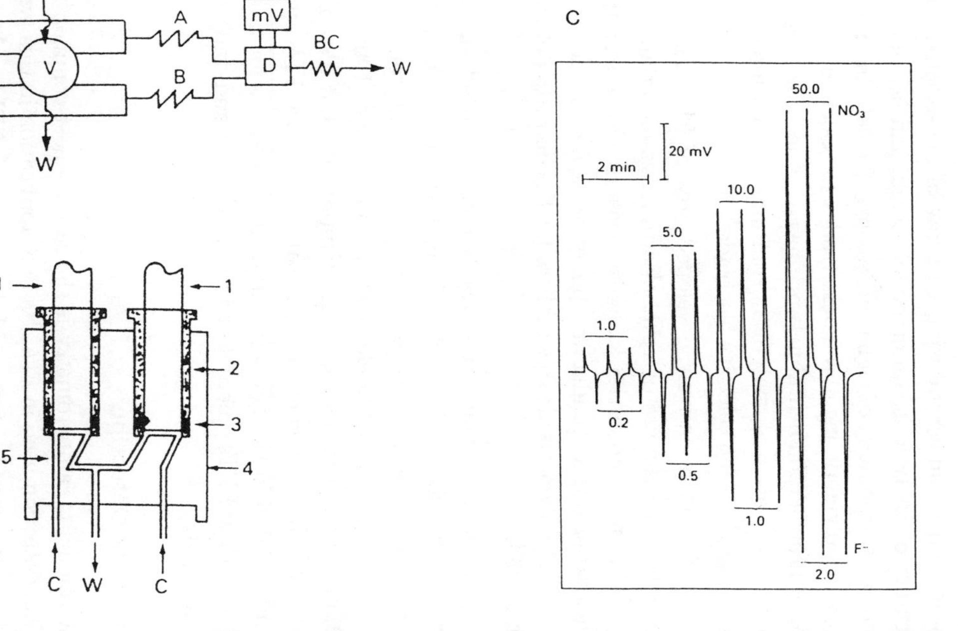

Fig. 5. Schematic diagram of a flow injection system (a), detector cell (b), and recorded signals (c) for simultaneous determination of fluoride and nitrate with potentiometric detection using ion-selective electrodes [93]. (P) peristaltic pump; (V) injection valve; (S) sample channel; (A,B) reaction coils; (D) detector cell; (BC) back-pressure coil; (1) ion-selective electrodes; (2) PTFE sleeve; (3) rubber O-ring; (4) perspex body; (C) carrier solution; (W) waste. (*Reprinted by permission of copyright owner.*)

can be carried out with the use of a different carrier solution for each given electrode. In multicomponent flow injection potentiometry of different cations in natural waters or blood serum using an array of ion-selective electrodes, a cross-interference of analytes is commonly observed, and hence it is necessary to employ advanced methods of signal processing [94–96]. In another system with an array of polymeric membrane electrodes for calcium, potassium and sodium, a back-propagation neural network has also been investigated for the detection and identification of analytes in solutions, based on response profiles from transient signals of ion-selective electrodes [97].

Potentiometry is an intensively developed area of FIA. Further progress can be envisioned with the use of novel electroactive materials in indicating electrodes, offering new properties for novel sensors. Some examples of this trend are the use of doped conducting polypyrrole for the determination of nitrate [98], and of polymerised crown ethers for the detection of catechol and catecholamines [99].

Commonly used membrane electrodes can be employed successfully for other modes of electrochemical detection in FIA. For instance, measurements of the conductance, instead of the potential of a calcium ion-selective electrode, lead to a reduction of the electrode response time [100]. Progress in miniaturisation, not only of the electronic components, but also of hydraulic devices, may result in significant miniaturisation of entire FIA setups. This has been demonstrated in the development of microsystems for flow injection potentiometry with ISFETs, piezoelectric micropumps and valveless sample injection devices [101].

2. Amperometric Detection

After potentiometry, amperometry is the second-most-common method of electrochemical detection used in FIA. Its first application in this type of measurement [118] even preceded the rapid development of FIA, which began in the middle of the 1970's. However, because of a greater difficulty in achieving selectivity of detection when compared to, for instance, potentiometry with ion-selective electrodes, amperometry in flow conditions has mainly been developed as a sensitive method of chromatographic detection of electroactive solutes or as a detection method in conjunction with a selective biocatalytic step in biosensors (Chapter 4). Amperometric detection in flow conditions has been discussed in several review articles [119–122].

2.1. *Principle of Detection*

Amperometric detection is a particular mode of electrochemical voltammetric measurement, where the current intensity in a detection cell is observed as a function of concentration of the analyte. The magnitude of the current intensity depends on the polarising potential applied to the working electrode on which electrochemical oxidation or reduction of the analyte takes place. A hydrodynamic voltammogram obtained in flow conditions allows the selection of a suitable polarising potential to be used for amperometric detection. The voltammetric measurements are usually carried out in three electrode cells, where reference and auxiliary electrodes are used along with the working electrode. The reference electrode provides a reference potential for the polarisation of the working electrode, whereas the auxiliary electrode conducts the current resulting from the electrode process of the detected species. Voltammetric detection performed using a mercury working electrode is called polarography. This name has been in traditional use since the discovery of the method and its initial development, which mainly employed a mercury dropping electrode (DME). In conventional polarographic measurements with a DME, the mass transfer of the analyte to the working electrode surface is controlled only by diffusion and the observed diffusion or limiting current, i_d, is proportional to the bulk concentration of the analyte, according to the Ilkovič equation:

$$i_d = 708nD^{1/2}m^{2/3}t^{1/6}C\,, \tag{6}$$

where n is the number of electrons involved in the electrode reaction, D is the diffusion coefficient of the electroactive species, m is the mass of mercury flowing per unit time, t is the drop time, and C is the bulk solute concentration. In voltammetry with other working electrodes, using a linear potential sweep, a peak in the current passing through the cell is observed for an electroactive analyte. Its maximum, i_p, is given by the Randles–Ševčik equation:

$$i_p = kn^{3/2}AD^{1/2}V^{1/2}C\,, \tag{7}$$

where A is the working electrode surface area, V is the potential-sweep rate and k is a constant.

Amperometric measurements are carried out at a fixed potential at or near the potential which yields the limiting current or peak current in a static system of similar chemical composition. In flow measurements, the magnitude of the observed current depends on the hydrodynamic conditions at the electrode surface, where both diffusion and convection of the analyte contribute to the current. In amperometric flow measurements with a DME polarographic detector, the limiting current, i_l is given by the equation [123]

$$i_l = knD^{1/2}FV^{1/2}(mt)^{1/2}C\,, \tag{8}$$

where F is the Faraday constant and V is the linear flow velocity. The constant k equals 0.0154 for a flow-through cell where the solution flows parallel to the axis of a vertically positioned capillary, in the opposite direction to the mercury flow. In cells with a horizontally positioned capillary and horizontal flow, k has the value 0.0178. For flow cells with solid electrodes [123]

$$i_l = knFD(\mathrm{Sc})^{1/3}\omega(\mathrm{Re}_x)^{\alpha}\,, \tag{9}$$

where Sc is the Schmidt number (equal to νD^{-1}, where ν is the kinematic viscosity), and Re_x is the modified Reynolds number (equal to $vl\nu^{-1}$, where v is the average linear velocity of the solution and l is a characteristic length of the electrode). The variable ω characterises the effective width of the working electrode. Its meaning, together with values of the constants k and α for different types of flow cells, are as follows [123]:

Cell type	k	α	α
Tubular	8.0	0.33	$l^{2/3}r^{2/3}$
Thin-layer	0.8	0.5	b
Wall-jet	0.5	0.75	a

where l is the length and r is the diameter of the tubular working electrode, b is the width of a thin-layer electrode and a is the diameter of the solution inlet in a wall-jet cell.

The relationship between the current at the top of the flow injection peak, i_p, and the magnitude of the signal observed for the flow of the sample under steady-state conditions, i_{ss}, is as follows [124]:

$$i_p = i_{ss}[3.4V_S/(\pi a^2)][D/v_f(0.5V_S + V_R)]^{1/2}\,, \tag{10}$$

where V_S is the injected sample volume, a is the radius of the tubular channel, v_f is the time-averaged flow rate and V_R is a finite retention volume.

Generally, the flow conditions in measurements with amperometric detection decrease the thickness of the diffusion boundary layer at the working electrode surface. This results in an increase in the measured current, depending on the flow rate used and the geometry of the system, especially that of the flow cell. The effects of particular parameters are described by Eqs. (8)–(10).

2.2. *Flow-Through Cells and Electrode Materials*

The satisfactory functioning of amperometric flow through detectors requires the fulfilment of a few more conditions than potentiometric detectors. Of particular importance is the electrode area, rather than its shape or the dead volume of the detector. In most amperometric flow cells, a three-electrode configuration is used, where auxiliary (counter) and reference electrodes are used along with the working electrode. The voltage applied between the working and auxiliary electrodes is always accompanied by an uncontrolled error due to the ohmic drop, iR (i =current; R =cell resistance). In order to minimise this effect, the electrodes should be placed as close together as possible, except in cases where the products of an electrode reaction at the counter electrode might interfere with detection at the working electrode. Therefore, the optimal position of the auxiliary electrode is downstream from the working electrode, but as close to it as possible.

Flow cells used in flow injection amperometry are generally the same as those discussed above for potentiometric detection. The most common are wall-jet and thin-layer arrangements. The wall-jet configuration can be used not only with solid working electrodes, but also with the DME [125] (Fig. 6A), although the advantages of the flow cell with a horizontally positioned capillary and a perpendicular solution flow have been demonstrated [126]. The effective dead volume of the wall-jet cell with DME is estimated to be less than 1 μl. A simple large-volume wall-jet detector cell has also been used with a sessile mercury-drop electrode [127]. Such cells are most frequently used with solid working electrodes, although not only as a large-volume type, but also as microflow cells [128]. In the wall-jet/thin-layer cell design the carrier stream flows parallel to the working electrode while the sample strikes its surface perpendicularly as it is injected with a microsyringe [129]. The negligible dispersion results in a very fast response and increased sensitivity in comparison to the thin-layer cell with conventional sample injection. A large-volume wall-jet detector has also been applied in a capillary FIA using a gravity flow [130]. The resulting system permits highly sensitive measurements with an extremely small sample volume (3–30 nl), when used in conjunction with ultra-low carrier flow rates. In flow injection measurements with a flow detector based on a jet of solution directed at a thin porous disk made of reticulated vitreous carbon better

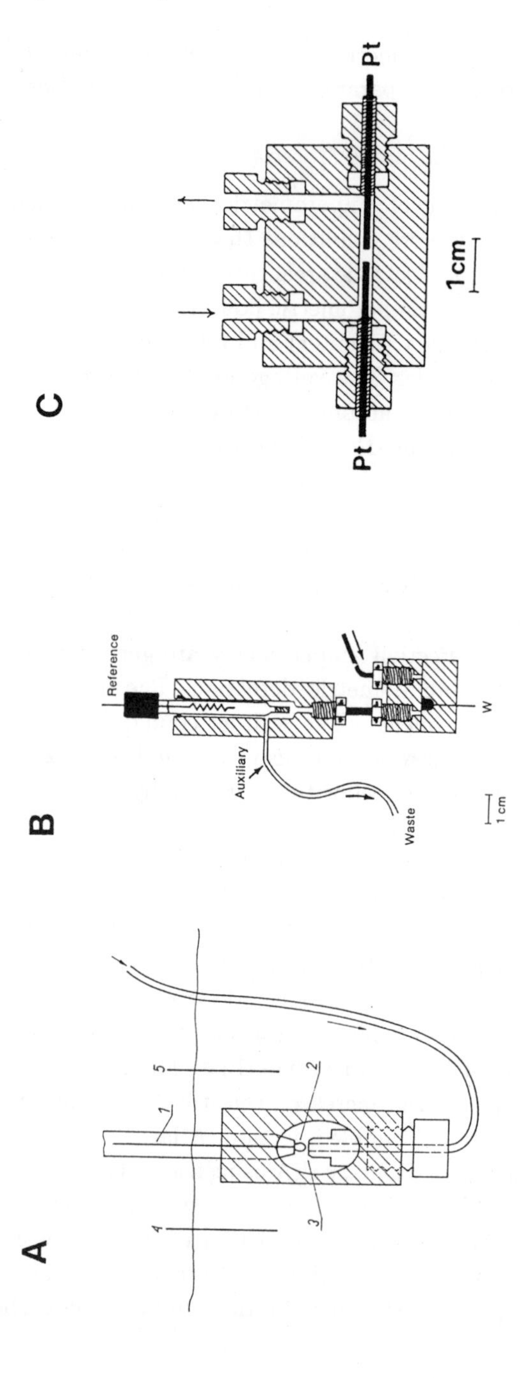

Fig. 6. Flow-through cells used in amperometric detection in FIA. (A) large-volume wall-jet detector with mercury drop electrode [125]: 1 — capillary; 2 — mercury drop; 3 — outlet; 4 — reference electrode; 5 — auxiliary electrode; (B) thin-layer cell with flat working electrode (W) [209]; (C) biamperometric detector with two platinum wire electrodes [147]. (*Reprinted by permission of copyright owner.*)

sensitivity and detectability thana for a wall-jet cell has been shown [131]. In the widely used thin-layer cell, the stream flows parallel to the working electrode, which is embedded in the channel wall (Fig. 6B) [132]. The solution layer thickness is easily controlled by the spacer. The current magnitude depends on the flow rate in various ways which are determined by the coulometric efficiency of the electrode process. If the efficiency is lower than 10%, a linear dependence on the flow rate to the power of 1/3 is obeyed [133]. In this cell type, two working electrodes [134] or an array of working electrodes can be used [135, 136]. In the dual-electrode system, the potential at the upstream electrode is scanned, while the downstream electrode can be maintained at a constant potential to monitor the reaction occurring at the up-stream electrode. The use of an array of working electrodes can give three-dimensional information (current as a function of both time and potential) by using working electrodes set at different potentials or an array of electrodes with different electrode materials. A thin-layer cell with two Pt foil electrodes has been used in flow injection biamperometry [137, 138]. Both thin-layer and wall-jet flow through cells for amperometric detection are commercially available, mainly as electrochemical detectors for HPLC.

Tubular electrodes are also used as working electrodes in flow injection amperometry [139–142]. Excellent agreement of experimental results with theory has been obtained for moderately low flow rates [141], but these cells are not often used because of difficulties in cleaning the working electrode surface and in its reproducible pretreatment for measurements. Flow cells with a wire working electrode are not often used in flow injection amperometry with a three-electrode detection system [143–145], although they have found numerous applications in biamperometric detection with two polarised electrodes [145]. A schematic diagram of a biamperometric flow through detector is shown in Fig. 6C. The wire flow-through electrode, while not adhering strictly to the published theory for annular electrodes [143], has proven to be a very reliable detector. The vibration of the working Pd wire electrode at a frequency of 400 Hz in the flow injection determination of chromium(VI) gives a 10-fold increase in the sensitivity of detection as a result of turbulent convection [144].

In flow injection amperometry, several applications using disk microelectrodes of diameter 5–50 μm in wall-jet cells have been reported [148, 149]. Their small dimensions allow an enhancement at the signal-to-noise ratio and they can also be used in highly resistive media such as dilute electrolytes or organic solvents. Along with oscillations in the flow rate, incomplete wetting of the working electrode surface and fluctuations in the polarising potential are assumed as sources of noise in flow measurements with amperometric detection [150]. In the latter two cases, a decrease in the electrode area leads to a decrease in the noise amplitude. An additional advantage is the ability to perform measurements without a potentiostat or in a two-electrode cell. In the determination of copper with a Pt disk microelectrode, it was found that the limit of detection increases with a decrease in the electrode radius below 28 μm [149]. In a thin-layer cell, a Pt disk microelectrode covered with polypyrrole has been used for the flow injection amperometry of electroinactive anions [151].

A separate and very crucial factor in the success of amperometric detection in both flow and conventional nonflow measurements and in voltammetry in general is the selection of a suitable material for the working electrode. Each material employed for this purpose has a limited range of potentials which can be applied to carry out required electrode processes. The potential limits for the most common electrode materials are as follows [152]:

	Potential limit, V vs. SCE	
Electrode material	Cathodic	Anodic
Mercury	−2.0	+0.4
Mercury film on Pt or C	−1.0	+0.4
Platinum	−0.5	+1.2
Glassy carbon	−0.8	+1.2
Graphite paste with nujol	−1.6	+1.1

The above data apply to aqueous solutions of acetate buffer at pH 4.5. All of these materials find application as working electrodes in flow injection amperometry, as well as others such as gold [128, 144, 153–155], palladium [144] and silver [156]. Each of these materials has its own advantages and drawbacks. Platinum and gold undergo oxidation to form a film of oxides, which are adsorbed and desorbed in a narrow range of potentials. The oxide layer inhibits electron transfer in electrode processes. These metals can, however, be utilised with suitable electrode polarisation in the flow injection amperometry of concentrated aqueous solutions of strong acids and bases [157]. Platinum, in spite of such behaviour, is very commonly employed for amperometric detection in FIA.

Far fewer applications have been developed using mercury electrodes in classical polarographic systems [125, 158–160], sessile mercury drop electrodes [127], or mercury film electrodes on gold [135] or copper [161, 162] supports. Mercury film electrodes generally exhibit a narrower available potential range and a poorer reproducibility than mercury drop electrodes.

The most common working electrodes used in flow injection amperometry are made of glassy carbon (GC), which is very chemically and mechanically resistant, and allows detection over a very broad range of potentials. Chemisorption and surface oxidation are much less pronounced than for platinum or gold. However, in FIA systems, the appearance of a blank response, which consists of a transient component and a steady state component, has been observed [163]. The transient response is attributed to nonfaradaic current and faradaic current from the oxidation/reduction of electroactive functional groups on the glassy carbon surface. Particularly low residual currents are observed for graphite paste electrodes, although these have found relatively few applications in FIA, with examples being the indirect determinations of sulphur dioxide [163] and nitrogen dioxide [164].

Fundamental changes in the electrochemical properties of a working electrode can be achieved by suitable conditioning or by modification with a layer of a polymer which exhibits selective permeability or electrocatalytic properties. The stability of detection of some analytes on a GC electrode can be improved by using a simple preanodization procedure to form on the surface an oxidized layer with quinoidal-like surface functionalities [166]. Such species are known for their activity as electron transfer mediators. The covering of the glassy carbon surface with a base-hydrolysed cellulosic film prevents interferences from electroinactive macromolecules adsorbed on the electrode surface [167], whereas coating with the perfluorosulphonated polymer Nafion excludes anionic and neutral interferences in the FIA detection of cationic neurotransmitters [168].

A further widening of the potential use of amperometeric detection in FIA is obtained by the modification of the working electrode surface with species of electrocatalytic behaviour. Because of a large overpotential, hydrazine compounds cannot be detected using ordinary carbon electrodes. Hydrazine can, however, be detected using a carbon-paste-modified electrode containing cobalt phthalocyanine [169], or a Prussian-blue-modified GC electrode [170]. Arsenic (III), which cannot be oxidised on carbon electrodes, has been detected on a GC electrode electrochemically modified in a mixture of $RuCl_3$ and $K_4Ru(CN)_6$

[171]. An electrode modified with poly(4-vinyl pyridine) containing $IrCl_6^{3-}$ and $IrCl_6^{4-}$ has been used for the determination of nitrite, while a polymer layer exhibiting ion-exchange properties can eliminate interferences by transition metal cations [172]. The flow injection amperometric response of a GC electrode to Fe(II), nitrite and various dithiocarbamate metal complexes can be enhanced by coating it with a ruthenium-containing polymer; however, the slow removal of the polymer from the substrate causes instability in this detection [173]. A more stable response is given by a GC electrode covered with a layer of a redox polymer containing osmium, which acts as an efficient electrocatalyst for the reduction of Fe(III) [174].

Recently, increasing attention has been concentrated on conducting polymers formed on a working electrode surface by electropolymerisation. Their repeated doping-undoping properties have been demonstrated in flow injection amperometry for the determination of electroinactive anions [145, 151, 175, 176] and cations [177, 178]. Due to the non-selectivity of the obtained amperometric signals, detection with these modified electrodes could be primarily used in ion chromatography.

2.3. *Measuring Procedures*

The majority of developed flow injection amperometric procedures for analytical determinations are carried out in a typical three-electrode system, with constant polarisation of the working electrode. A marked increase of one to two orders of magnitude in the detection limit of amperometric determination can be achieved by the use of pulse polarisation techniques due to an enhancement of the faradaic current of the electroactive analyte and a decrease in the nonfaradaic charging current. A drawback in the determination of cathodically active metal ions with polarographic detection employing a dropping mercury electrode is the necessity of oxygen removal from the analysed solution. Flow injection determinations of Pb(II) and Cd(II) have been carried out without removal of oxygen, using a constant potential pulse waveform in a solution buffered at a moderate to low pH [160]. The use of a differential pulse technique with fast pulse repetition in the flow injection amperometry of Cd(II) and Zn(II) with a copper-amalgam electrode improves the sensitivity of the determination by a factor of about ten compared with measurements at constant polarisation [162].

The use of a reverse-pulse amperometric waveform gives better sensitivity in the determination of oxidisable species with a planar GC electrode in an FIA system [179]. This involves a long application of a positive initial potential followed by a short pulse at a more negative final potential. Analytes are measured by monitoring the reduction of the oxidation product from the initial oxidation process. As with a constant-potential waveform, a reducible species can be monitored without deaeration of the solution. This method is especially favourable for analytes, for example phenol which are oxidised at potentials near that of solvent decomposition.

Numerous FIA applications have been developed using pulsed amperometric detection with a triple-pulse potential waveform [154, 157, 180–182]. A severe loss in the activity of noble metal electrodes is normally associated with the anodic detection of organic compounds by amperometry at a constant polarisation potential, because of adsorption of the products of the anodic reaction. Such a loss in response is avoided in pulsed amperometric detection by measuring the anodic signal a short time (950–250 ms) after the application of the detection potential, followed by the sequential application of large positive and negative potentials for oxidative and reductive cleaning, respectively. The frequency of the wave form can be sufficiently high to allow application in chromatographic and flow injection systems. This type of detection has been successfully employed for the determination of carbohydrates [154, 180] and numerous sulphur compounds such as thiourea, sulphide, thiocyanate, thiols and dithiocarbamates [181]. Unlike the Pt electrode, it is found that carbohydrates can be detected anodically at gold electrodes in a potential region where oxygen reduction does not occur and the formation of surface oxide is not significant. The resulting increase in the signal-to-noise ratio for gold electrodes as compared to platinum has been verified with detection limits decreased by a factor of about five for glucose, fructose, sorbitol and sucrose [154] (Fig. 7). Pulsed amperometric detection can also be utilised for the detection of electroinactive species, which when adsorbed on Pt electrodes alter the rate of surface oxide formation following a positive potential step. This has been utilised for the sensitive amperometric detection of chloride and cyanide [182]. Since the decay of the anodic current with time, corresponding to oxide formation on a Pt electrode, is a function of the hydrogen activity, this type of detection can be applied in FIA mode for the determination of highly concentrated strong acids and bases for which a glass electrode lacks sensitivity [157].

Numerous FIA applications have been developed for amperometric detection with two polarised electrodes, also called biamperometry [146]. This method, used in electroanalysis for the detection of the end-point of titrations,

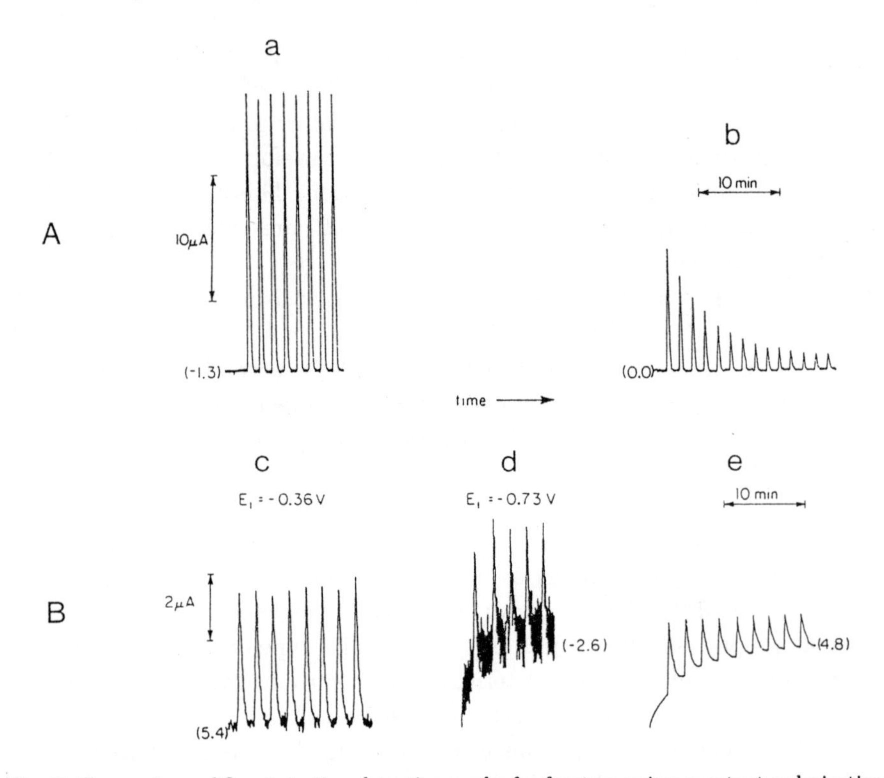

Fig. 7. Comparison of flow injection detection peaks for fructose using constant polarisation and a triple-pulse potential waveform for (A) gold and (B) platinum electrodes; injection — 50 μl of 1.0 mM fructose in 0.2 M NaOH; carrier solution: 0.2 M NaOH at 0.5 ml/min; (a) pulse mode: $E_1 = 0.15$ V (250 ms), $E_2 = 0.75$ V (100 ms), $E_3 = -1.00$ V (100 ms); (b) constant polarisation at 0.15 V; (c,d) pulse mode with E_1 shown, $E_2 = 0.48$ V (500 ms), $E_3 = -0.92$ V (250 ms); (e) constant polarisation at −0.73 V [154]. (*Reprinted by permission of copyright owner.*)

is based on the measurement of the current between two identical inert electrodes, to which a small potential difference is applied. A measurable current flows in the presence of both forms of a reversible redox couple. An increase of the applied potential difference provides a wider linear relationship between the current and the concentration of the deficit component of the redox couple [137]. This is, however, accompanied by a decrease in selectivity. Such detection can be utilised for the determination of one component of a redox couple, e.g. iron(III) [137]. Several applications of the iodine/iodide couple as the indicating system have been reported for indirect determinations.

The determination of sulphur(II) compounds including sulphide, thiosulphate, glutathione, cysteine and thiourea has been based on the induction of a reaction between sodium azide iodine, and detection of the iodide thus formed [183]. Other applications include the single-point titration of acids of various strengths [184], determination of nitrate and nitrite in waters [185], of organic solvents with the Karl Fischer reagent [138] and of residual chlorine in waters [186]. Appropriate design of the flow injection manifold (Fig. 8)

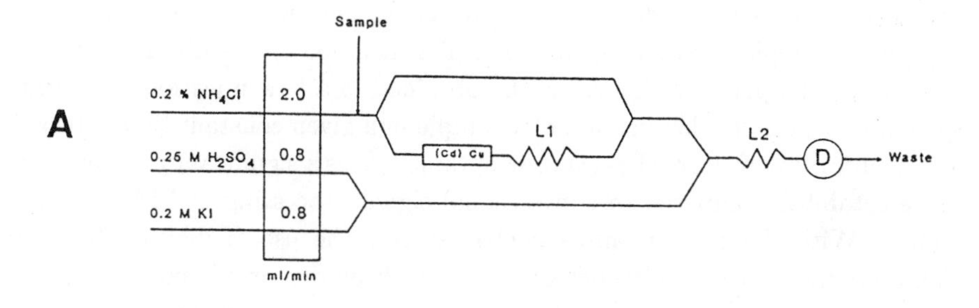

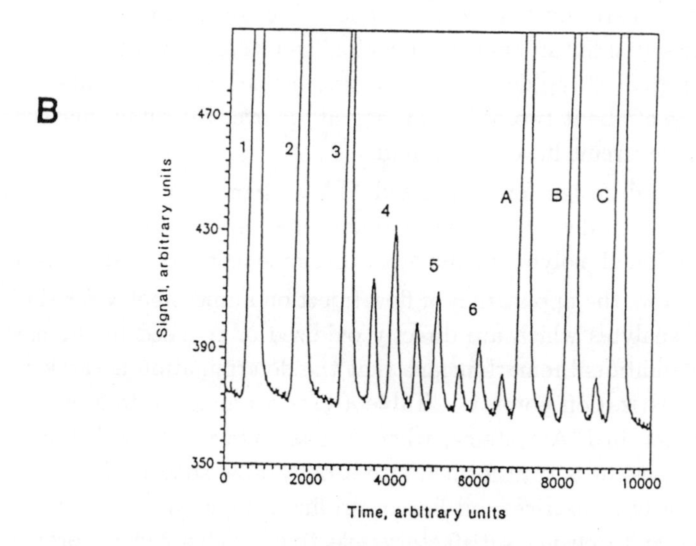

Fig. 8. Schematic diagram of a manifold (A) used for the simultaneous flow injection biamperometry of nitrite and nitrate, and display of experimental data (B) obtained for nitrate standard solutions (1–3) nitrite standard solutions (4–6), and natural water samples (A–C) [187]. (*Reprinted by permission of copyright owner.*)

allows the simultaneous determination of nitrate and nitrite in waters [197]. The Fe(III)/Fe(II) indicating couple has been employed for the indirect flow injection biamperometry of phenothiazine derivatives in pharmaceutical preparations [188], while the hexacyanoferrate(III)/hexacynaoferrate(II) couple has been used for the simultaneous determination of sucrose and reducing sugars [189]. Biamperometric FIA has also been applied in the catalytic determination of Mo(VI) in soil extracts [147] and Cu(II) in blood plasma [190]. The biamperometric detection of nitrite has also been used as an example in presenting a concept of flow injection measurements with the Fourier transform of a multiple injection signal [191]. This is based on the observation that near sinusoidal periodic signals can be obtained in a flow injection measuring system by fast multiple injection of a sample in a given constant time interval. The Fourier transform of the response obtained gives a significant improvement in detectability compared to conventional signal processing in FIA measurements. With the same biamperometric system the use of digital filters for signal processing to improve detectability has been demonstrated [71].

Cross-correlation can be employed to improve detectability in an FIA system with amperometric detection [192]. To do this, two independent flow injection systems are constructed in parallel. One flow line is used to generate the reference signal for an analog correlator circuit, and the other to generate the analyte signal. Using dopamine as a test system, improvements in signal-to-noise ratios of about two orders of magnitude over the direct measurement of the electrode circuit have been found.

A rarely used design for a reversed FIA system with continuous sample aspiration has been developed for the determination of phosphate and nitrite with the injection of molybdate and acidic bromide reagents, respectively [192].

Table 3 shows the application of flow injection amperometry for the determination of analytes which are directly oxidized or reduced in the electrode process. No additional remark means that the determination is carried out at a constant polarization potential. Table 4 lists examples of indirect amperometric detection in FIA systems, where the analyte reacts with the directly measured species, or is transformed into a detectable species. In the analysis of complex natural matrices, off-line or on-line sample pretreatment is often needed in order to ensure satisfactory selectivity. This can be provided by suitable modification of the electrode surface with permeable polymer layers [167, 168], or a mixed lipid layer [193]. The determination of L-ascorbic acid with a GC electrode in the presence of Cu(II) is an example of an unusual

Table 3. Applications of direct amperometric detection in FIA.

Analyte	Working electrode*	Concentration range	Reference
Acids, bases	Pt (t.p.)	0.25–4.0 M	157
Ammonia	Pt/polypyrrole	0.6–100 μM	249
Anions	Pt/polypyrrole	0.01–10 mM	145, 175
	Pt/polypyrrole (microelectrode, n.p.)	50 nM–10 mM	151
	Pt/polyaniline	0.1–100 ppm	176
As(III)	Pt	0.4–3.5 ppm	198
	GC modified with $Ru(CN)_6^{4-}$	5–100 μM	171
Ascorbic acid	GC	0.1–20 μM	193a
Carbohydrates	Pt (t.p.)	0.01–1.0 mM	180
	Au (t.p.)	not shown	154
Cations	Pt/polyaniline – Nafion	0.2–100 μM	17
Cd	(Cu)Hg (d.p.)	0.35–650 ppm	162
	DME	not shown	160
Chloride	Pt (t.p.)	0.1–500 μM	182
Chlorpromazine	GC (r.p.)	0.057–6.0 μM	179
Cyanide	Pt (t.p.)	0.1–1000 μM	182
	Ag	0.001–200 ppm	197
Cr(VI)	Au, Pd (vibrating)	5–100 ppb	144
o-Diphenols	Pt	0.1–1000 ppm	142
Dopamine	GC/Nafion	20–160 μM	168
	C	0.12–240 nM	192
Epinephrine	C paste	0.001–10 μM	135
Fe(II)	Pt (biamperometry)	0.1–0.2 mM	137
Fe(III)	GC modified with Os contg. polymer	0.003–1.0 mM	174
Ferrocene	Pt (microelectrode)	0.02–100 μM	148
Food colours	Hg (sessile drop)	0.1–100 ppm	127
Hexacyanoferrate(II)	C porous	0.04–15 μM	131
H_2O_2	GC	1 μM–1 M	199
Hydrazine	C paste with cobalt phthalocyanine	0.2 nM–1 mM	169
Hypochlorite	Au	0.2–290 μM	153
Iodide	Pt	10–100 μM	141, 143
Isosorbide dinitrate	DME	0.01–1.0 mM	158

continued

Table 3 (*continued*)

Analyte	Working electrode*	Concentration range	Reference
Nitrite	GC modified with Ir contg. polymer	0.01–1.0 mM	172
	GC modified with Ru contg. polymer	0.003–2 ppm	173
Penicilloic acid	DME	2–100 μM	125
Promethazine	GC/phospholipid coating	1.4–140 μM	193
SO_2	GC	0.03–6 ppm	196
Thiourea	Pt (t.p.)	0.01–1.0 mM	181
Warfarin	GC	0.005–40 ppm	200
Zn	(Cu) Hg (d.p.)	0.7–650 ppm	162

*DME — dropping mercury electrode; GC — glassy carbon; d.p. — differential pulse; n.p. — normal pulse; r.p. — reverse pulse; t.p. — triple-pulse waveform.

Table 4. Applications of indirect amperometric detection in FIA.

Analyte	Working electrode*	Detected species	Concentration range	Reference
Acids	Pt (biamperometry)	Iodine	0.1–20 mM	184
Al	Au	DASA	0.25–100 μM	128
Ammonia	GC	Bromine	0.05–0.7 mM	202
Ascorbic acid	Pt	Iodate	0.1–1.0 mM	201
Chlorine	Pt (biamperometry)	Iodine	2–1000 ppb	186
Cu(II)	Pt (microelectrode)	Cu dithiocarbamate	0.7–64 ppb	149
	Pt (biamperometry)	Fe(II)	0.2–0.8 ppm	190
Cysteine	Pt	Iodate	0.1–1.0 mM	201
Guthion	GC	Anthranilic acid	0.41–100 μM	204
H_2O	Pt (biamperometry)	Iodine	0.03–0.11% (w/w)	138
H_2O_2	GC	Iodine	0.003–6.0 mM	203
Hydrazine	GC	Bromine	0.05–0.7 mM	202
	Pt	Iodate	0.1–1.0 mM	201

continued

Table 4 (*continued*)

Analyte	Working electrode*	Detected species	Concentration range	Reference
Iodide	Au porous	Iodine	0.2–10 μM	155
Mo(VI)	Pt (biamperometry)	Iodine	1.2–1000 ppb	147
Nitrate	DME	U(VI)	4–100 μM	159
	Pt (biamperometry)	Iodine	0.07–1.5 ppm	187
Nitrite	GC	Nitrosyl bromide	0.05–0.5 mM	192
	DME	U(VI)	6–100 mM	159
	Pt (biamperometry)	Iodine	40–150 ppb	187
NO_2	Carbon paste	Ferroin	12–140 ppm	165
Mg	GC	Eriochrome black T	0.005–1.8 ppm	194
Microorganisms	Pt	Hexacyanoferrate(III)	5×10^6 – 2.4×10^9 cfu/ml	205
Phenothiazine derivatives	Pt (biamperometry)	Fe(II)	0.5–140 ppm	188
Phosphate	GC	Molybdophosphate	1–100 μM	192, 206
	Pt	Molybdophosphate	0.02–50 μM	207
Phosphorus total	GC	Molybdophosphate	0.1–30 ppm	195
Pt(VI)	Pt	Tl(I)	0.06–0.2 mM	208
SO_2	C paste	Ferroin	0.3–14 ppm	164
Sugars	Pt (biamperometry)	Hexacyanoferrate(III)	25–300 ppm	189
Thiosulphate	Pt (biamperometry)	Iodide	0.1–100 ppm	183

*cfu — colony-forming units; DASA — 1,2-dihydroxyanthraqinone-3-sulphonic acid; DME — dropping mercury electrode; GC — glassy carbon

amperometric procedure [193a]. The presence of Cu(II) allows determination at a very low polarising potential because of the formation of Cu(I) and/or the ascorbic acid radical.

Off-line sample processing can be employed in determinations carried out on physiological samples. In the determination of copper in urine, a sample cleanup and preparation was achieved using C18 cartidges and the formation of a Cu(II) dithiocarbamate complex [149], while in the determination of copper in blood plasma with biamperometric detection copper is liberated from proteins by incubation with HCl, followed by precipitation of the proteins with trichloroacetic acid [190]. On-line dialysis has been employed in the determination of magnesium in serum [194]. Gas diffusion is very often utilised in an on-line mode, as in the determination of nitrogen dioxide based on the oxidation of ferroin to its iron(III) analogue [165], in the determination of sulphur dioxide in wines [196] and of free cyanide [197]. In the determination of total phosphorus using continuous microwave oven decomposition (Fig. 9), gas diffusion is used to remove gaseous products of microwave digestion in a flow system [195]. In the so-called pneumoamperometric detection of iodide, a matallised gold membrane is used simultaneously for the conversion of iodide into iodine and as the working electrode for the detection of iodine [155].

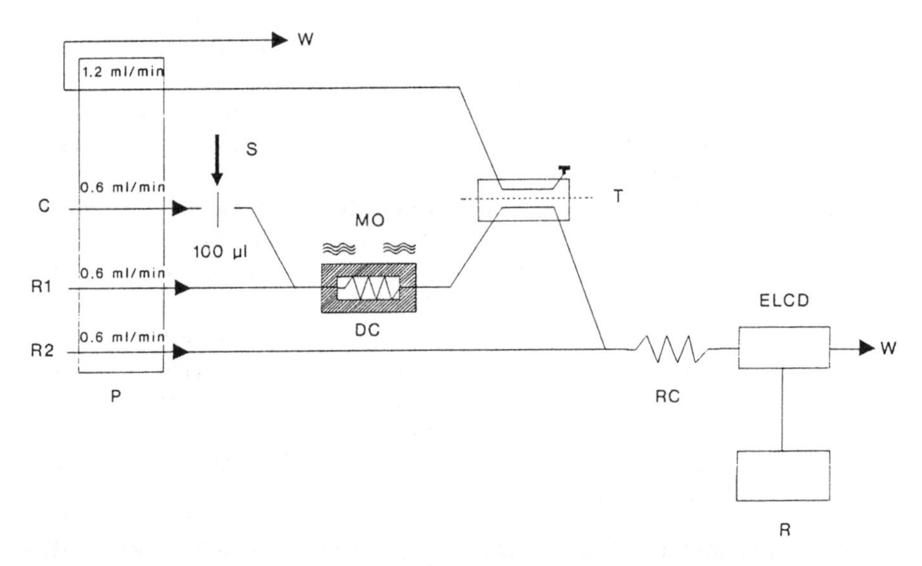

Fig. 9. Schematic diagram of a flow injection manifold for the determination of total phosphorus with amperometric detection [195]: C — carrier; R_1 — decomposition agent; R_2 — ammonium molybdate reagent; S — sample; P — peristaltic pump; MO — microwave oven; T — PTFE membrane; ELCD — electrochemical detector; R — recorder; W — waste; DC — decomposition coil (7.6 m × 0.5 mm i.d.); RC — reaction coil (2.1 m × 0.5 mm i.d.). (*Reprinted by permission of copyright owner.*)

Broader analytical information about an analysed system can be gained in a flow injection amperometric system with the use of a microelectrode array of individually controlled working electrodes [135, 136, 210]. This requires the design of a dedicated multielectrode potentiotat, but provides a three-dimensional result (potential, time, current). The working electrodes can be arranged in series in a thin-layer cell [136], or as a circular array of electrodes in a wall-jet cell [135, 210]. The working electrodes can all be made of a material such as gold [136] or glassy carbon [210] and can be used simultaneously at different polarising potentials; or, working electrodes made of different materials can be used [135]. The array can be manufactured with 8 [210], 16 [135, 136], or even 80 [136] electrodes. In a system with three working electrodes polarised at different potentials, the determination of Cd, Cu and Pb has been developed for industrial process analysis [210]. The separation of ferrocene derivatives has been demonstrated using an array of 16 microband gold electrodes [136].

3. Scanning and Stripping Methods

3.1. *Voltammetric Detection*

The advantage of voltammetric methods, in which current is measured as a function of the varying polarisation of the working electrode, is the ability to carry out multicomponent determinations. The main difficulty with the use of volammetry based on rapid potential scanning in an FIA system is the increase in the background due to the changing current and also, sometimes, distortion of the voltammetric peaks. The sweep rates used for this purpose range up to 0.8 V s^{-1} [211], which is needed for carrying out the measurement in the short period of time that the sample segment spends in the detector. In the voltammetric detection of acetaminophen and dopamine using a differential pulse mode, it has been observed that a change in the scan rate from 0.5 to 2 V min^{-1} results in some peak-height increase, while the width of peaks decreases at higher scan rates [212].

In the voltammetric flow injection systems reported in the literature the most common cell types are wall-jet flow cells used with working disk electrodes, although a thin-layer cell with a mercury micropool electrode has also been used [211]. Carbon working electrodes are used for the detection of oxidizable species [212, 213], whereas for reducible species, mostly metal cations, mercury film electrodes on a GC support [212–214] or a mercury micropool electrode [211] were employed. In order to obtain the best resolution, a fast scan rate, with differential pulse [212, 213] or rapid-scan square-wave voltammetry [214], is used. Such systems employ rather low flow rates of

0.2–2.0 ml min^{-1} and relatively large sample volumes (0.2–0.5 ml). Voltammetric detection requires a precise selection of the optimum time for initiating the potential scan.

In analysis by differential pulse voltammetry, the detection limits for oxidizable species such as acetaminophen, dopamine or chlorpromazine are similar to those for batch differential pulse voltammetry at solid electrodes [212]. These

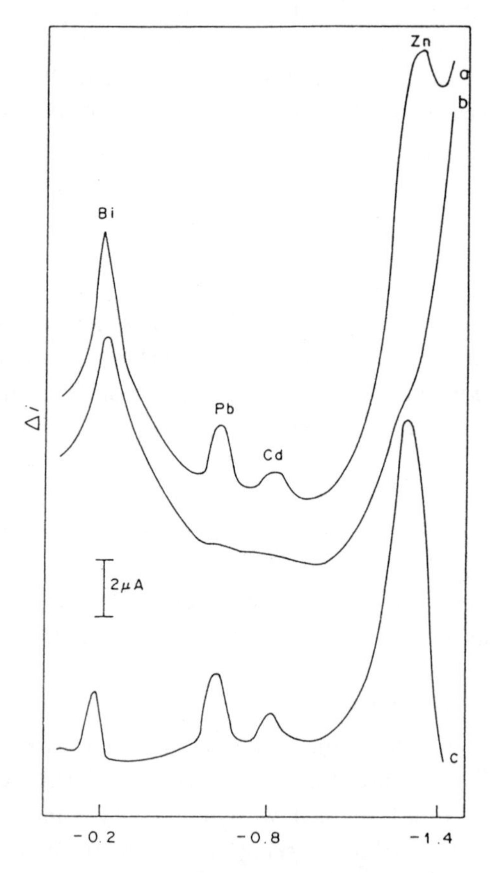

Fig. 10. Application of voltammetric detection with differential pulse mode and background-current subtraction for the injection of a sample mixture containing 20 μM Bi(III), Pb(II) and Cd(II) and 50 μM Zn(II) [213]. (a) Recorded curve without background subtraction; (b) background-current curve; (c) the subtracted response. Mercury-coated glassy carbon working electrode. Flow rate 0.3 ml/min. Scan rate 2V/min. Amplitude 50 mV; repetition time 0.5 s. (*Reprinted by permission of copyright owner.*)

measurements, especially those without deaeration, can be improved with the use of background subtraction [213]. This works well even with the differential pulse waveform which effectively corrects for nonfaradaic charging currents caused by hydrogen evolution, oxygen reduction, solvent oxidation or harmful surface processes. The success of such a procedure requires a careful matching of the electrolyte contant and pH of the samples and carrier solutions, as well as reproducible timing. Fig. 10 presents an application of the method to the analysis of a mixture of metal ions, demonstrating its effectiveness, especially for zinc and bismuth. An FIA system with deaeration of the sample has been developed for the indirect determination of magnesium, based on the adsorptive reduction current peak of the eriochrome black T-magnesium complex at a DME [214a]. The sample stream from the flow manifold is merged with the gas stream in such a way that the solution forms a thin layer on the inner wall of the tubing. This causes immediate deaeration of the solution, which then impinges on the side of the DME capillary and continues by gravity, around the mercury drop.

One disadvantage of fast scaning in flow injection voltammetry is the overlapping of voltammetric responses. The separation of overlapped signals can be achieved with satisfactory reproducibility by data processing using a Kalman filter, which fits the responses of single species to the overall response. This has been demonstrated for the separation of Pb(II) and Tl(I) signals in an FIA setup with square-wave voltammetric detection [214].

Another voltammetric technique that is employed with rapid scanning is cyclic voltammetry, in which the polarising potential is first increased linearly to a given value and then decreased to its starting point at the same rate. It is an important tool for the study of mechanisms and rates of oxidation-reduction processes but it is not used for routine quantitative analyses. Its application in FIA has been demonstrated [215, 216], but without practical analytical applications so far.

3.2. *Voltammetric stripping*

Stripping techniques are the most common type of electrochemical detection used in FIA for trace determinations [217]. In both voltammetric and, as discussed below, potentiometric stripping, low limits of detection result from the initial plating step. The analytes are first preconcentrated by electrolytic deposition on the working electrode surface, at a suitable potential. Apart from preconcentration, this stage of the procedure allows the separation of analytes from the sample matrix prior to the detection stage, which in voltammetric stripping is the anodic or cathodic reaction of the preconcentrated analyte by an appropriate potential scan. Along with the very low detection limit it provides, voltammetric stripping also enables the simultaneous determination of several analytes. In some cases, two stages of a stripping procedure can be more successfully carried out by changing the chemical conditions in the solution. The ease of changing these conditions in FIA systems is the basic advantage of perfoming stripping measurements in flow injection mode, apart from such factors as easier contamination control, which is essential in trace analysis. As the analytical stripping signal is not acquired during sample transport through the detector, the stripping current is independent of the degree of sample dispersion [218].

The most common applications of anodic stripping voltammetry are in the determination of metal cations that form amalgams with mercury. The first step is the reduction of the metal cations with preconcentration in a mercury film electrode and then their oxidation through anodic scanning with various polarisation wave forms. It has been shown that for the mercury film electrode, in linear-scan stripping mode, the stripping peak current i_p (l.s.) for laminar flow of the solution through the detector with flow rate U, is given by [218]

$$i_p(\text{l.s.}) = 1.116 \times 106 n^2 A K C_0 U^{\alpha-1} V_S v\,, \tag{11}$$

where n is the number of electrons taking part in the electrode process, A is the area of the electrode, C_0 is the initial concentration of the sample, V_S is the sample volume and v is the potential scan rate. K is a constant related to the diffusion coefficient and geometric parameters of the system. The constant α also depends on the electrode geometry. The same expression for differential pulse measurements has a slightly different form and the stripping current i_p (d.p.) also depends on the pulse width t_p [218]:

$$i_p(\text{d.p.}) = 0.138 n F A K C_0 U^{\alpha-1} V_S t_p^{-1}\,. \tag{12}$$

Examples of stripping voltammograms recorded in an FIA system are shown in Fig. 11A for several indium solutions of different concentrations. The observed peaks for Pb and Zn were due to contamination in the solutions used. In most developed procedures, voltammetric stripping measurements are carried out with differential pulse polarisation, which gives lower detection limits compared to the linear scan technique. In the determination of In, an increase in the stripping current at shorter pulse intervals in agreement with Eq. (12) has been

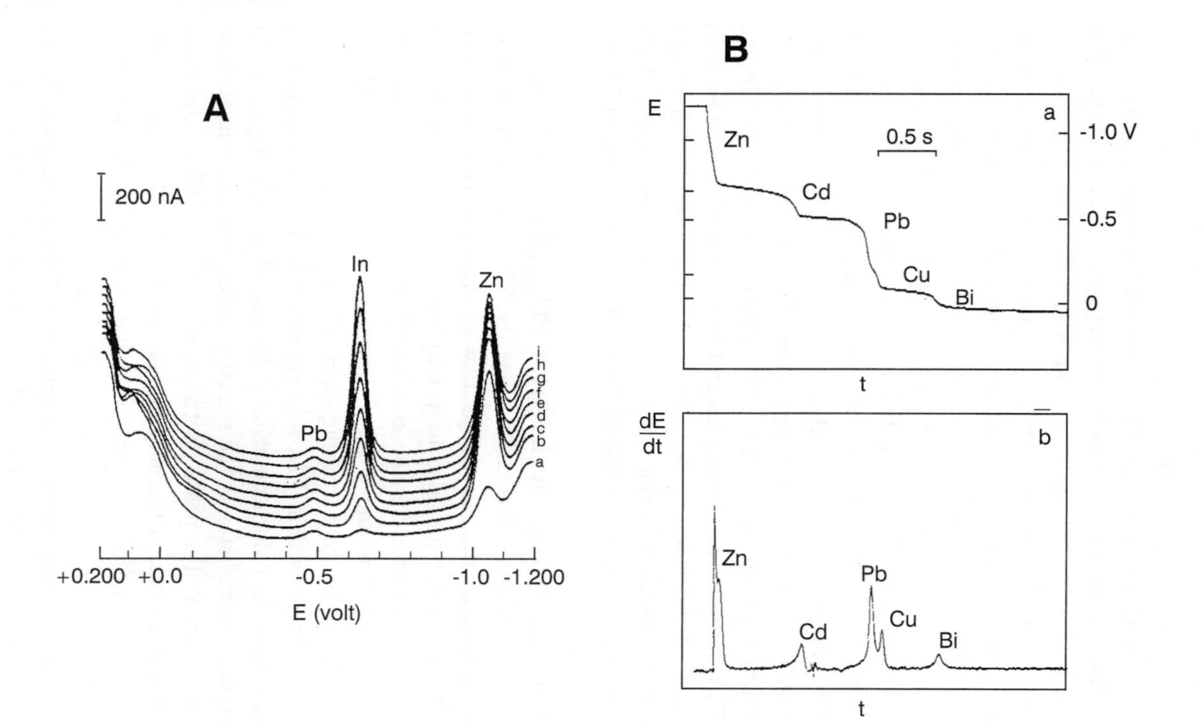

Fig. 11. Application of electrochemical stripping detections in FIA. (A) Differential-pulse anodic stripping voltammograms recorded for 1.0 ml injections of standard indium solutions: (a–i) blank, 1.0, 2.0, 3.0, 4.0, 5.0, 6.0, 7.0 and 8.0 μg/l, respectively. Lead and zinc were found as impurities in the buffer solution used to dilute the indium standard. Flow rate 0.5 ml/min, scan rate 5 mV/s, pulse amplitude 25 mV [219]. (B) Direct (a) and differential (b) recording of signal in potentiometric stripping analysis of a dissolved geological sample. Conditions: 0.1 M HCl containing as oxidant 40 mg/l Hg(II). Deposition time 150 s, deposition potential −1.15 V vs. SCE [235]. (*Reprinted by permission of copyright owner.*)

confirmed [219]. An increase in the peak height also occurs with an increase in the pulse amplitude. The conditions found for near-maximum sensitivity were a 5 mV s^{-1} scan rate, a 50 mV pulse amplitude and a 0.5 s pulse interval [219]. Under these conditions, detection limits as low as 20 pg ml^{-1} can be obtained in a sample volume of 1 ml at a sampling rate of approximately nine samples per hour. As with voltammetric detection without preconcentration, an effective correction for both faradaic and nonfaradaic background current contributions can be obtained by the subtraction of the carrier solution stripping voltammogram from the sample voltammogram, yielding a purely analytical response [220]. Generally, the removal of oxygen from the sample, prior to injection into the flow system, is not necessary because an exchange of medium takes place before the stripping step. It is important, however, to use an oxygen-free supporting electrolyte in the anodic stripping step. Trace metal contaminants should also be removed by pre-electrolysis. With subtractive anodic stripping voltammetry in FIA, non-deaerated samples and carrier solutions can be used [220]. A comparison of differential pulse stripping voltammetry with square wave voltammetry has shown that the latter provides a shorter analysis time and better sensitivity and therefore should be preferred in anodic stripping voltammetry with FIA [221].

Most applications of stripping voltammetry in FIA systems involve anodic determination of trace amounts of transition metals. These are performed on GC elecrodes with a mercury film, in wall-jet flow or thin-layer cells (Table 5). The wall-jet cell arrangement has also been employed in a setup with a Pt disk microelectrode of diameter 25 μm in a capillary flow injection system with 0.7 μl min^{-1} gravity flow and a 1.4 μl injection sample volume [221a]. In determination of Se(IV) a gold working electrode was used [222]. Carbon paste electrodes have been used in FIA adsorptive cathodic voltammetry of organic compounds such as chlorpromazine [223] and doxorubicin [224]. In these determinations, the first step is the selective adsorptive accumulation of the analyte on the surface of the paste electrode, after which differential pulse anodic stripping is carried out. A mercury film electrode on a GC support has also been used in the cathodic stripping voltammetry of chloride in an FIA system [225].

In most procedures, after appropriate preparation of the working electrode, electrolytic preconcentration is carried out when the injected sample solution reaches the working electrode, which is kept at an appropriate potential. In sample injection mode, the carrier usually serves as the washing solution, in

Table 5. Applications of electrochemical stripping detections in FIA.

Type of stripping	Analyte	Scan mode*	Concentration range	Reference
Anodic	Cd	d.p.	0.63 μM	218
	Cd, Cu, Pb, Zn	d.p.	0.1–20 μM	220
			5–30 nM	233
	Cd, Pb	d.p.	3–50 ppb	221a
	Cd, Pb, Zn	d.p.	0.01–0.1 μM	226
	Cd, Cu, Pb	d.p.	2.5–10 nM	228
	In, Pb, Zn	l.s., d.p.	0.05–300 ppb	219
	Pb	d.p.	25–120 nM	230
			0.1–0.8 μM	235
		s.w.	0.1–100 μM	221, 232
	Se(IV)	d.p.	5–100 ppb	222
	Zn	d.p.	0.05–1.0 μM	228
Adsorptive anodic	Chlorpromazine	d.p.	0.1–10 μM	223
	Doxorubicin	d.p.	0.001–1.0 μM	224
Cathodic	Chloride	l.s.	0.8–140 μM	225
Potentiometric	Bi, Cd, Cu, Pb, Sn, Tl, Zn		0.1–40 ppb	234
	Cd, Cu, Pb		0.0003–5 μM	233
	Cd, Pb, Zn		0.4–10 ppm	237
	Cd, Cu, Pb, Zn		0.2–100 ppb	238
	Cd, Pb		1–50 ppb	236
	Cu		0.1–1.0 ppm	239
	Pb		0.1–1.0 ppm	232

*d.p. — differential pulse; l.s. — linear sweep; s.w. — square wave

which stripping is carried out. The stripping step of a determination can be carried out with or without stopping the flow.

In order to avoid some interferences, an FIA stripping system can be equipped with modules for their on-line removal. Examples of such interferences include humic substances or surfactants with a strong tendency to adsorb to the electrode. They can be removed by passing the sample through a silica anion-exchanger column [226]. A chelate sorbent column containing immoblised 8-quinolinol can also be used. This retains metal ions which are then washed by the carrier solution to the detector cell. In determination of Se(IV) the use of a Chelex-100 column before loading the injection valve loop allows the removal of interfering divalent cations [222].

Porous working electrodes made of crushed, reticulated vitreous carbon and plated with mercury can be used for the determination of Cd, Cu, Pb and Zn by anodic stripping coulometry with collection in a flow system. In two-cell systems, trace metals from the injected sample are deposited in the first cell, then stripped and finally collected in the second one [227].

3.3. *Potentiometric Stripping*

Metals reduced and preconcentrated as amalgams in a mercury film can be oxidised, not only anodically by polarisation of the working electrode, but also chemically using a suitable oxidising agent in the solution. Metals will undergo oxidation sequentially with an increasing negative potential. The oxidation time is proportional to the amount of deposited metal, and it can be evalutaed from the potentiogram recorded in the coordinates potential (E) vs. time, or more easily, from the plot of the derivative dE/dt vs. time (Fig. 11B). Such a technique is called potentiometric stripping. Invented at the end of the 1970's, it soon found applications in flow measurements, including FIA [237]. The first potentiostatic preconcentration step, the working electrodes and the flow cells used are practically the same as for voltammetric stripping. The most commonly used oxidising agent, mercury(II), is dissolved, although in a system where a cathodic current is applied during stripping, dissolved oxygen can be used as an oxidising agent [232]. The use of additional current causes an increase in the signal as the stripping time is extended, due to the repetitive redeposition of the analyte into the mercury film by the cathodic current. The time taken to strip a metal from the electrode surface is given by the equation [234]

$$t_s = QH(ADC_{\text{ox}})^{-1}\,, \tag{13}$$

where Q is the amount of the analyte deposited in the mercury film, H is the diffusion layer thickness, A is the electrode surface area, D is the diffusion coefficient, and C_{ox} is the bulk oxidant concentration. In FIA measurements, Q is proportional to the sample volume and inversely proportional to the flow rate.

As with voltammetric stripping, the main advantage of performing potentiometric stripping in a flow injection system is the simplicity of matrix exchange. The sample can be transported to the detector in its own matrix. Electrolytic plating occurs in the sample matrix, but the stripping can be effected in a secondary matrix, optimized for the simultaneous determination of various metals. The air oxidation associated with manual transfer is thus eliminated. Such procedures have been successfully applied in determinations of mixtures of Cu(II) and Bi(III), Pb(II) and Sn(IV), and Pb(II) with Tl(I) [234]. The measured time periods for potentiometric stripping are in the range of a fraction of a second to a few tens of seconds, hence it is necessary to use fast strip chart recorders or computer data acquisition [234, 236].

Glassy carbon is mostly used as a support for the mercury layer, although porous carbon can also be used [237]. The film can be plated out prior to flow measurements from a volume of carrier solution to which Hg(II) and supporting electrolyte have been added, or *in situ* with Hg(II) present in the sample segment. As significant advantage of potentiometric stripping is the ability to determine metal ions in a nondeaerated solution. The calibration of the measuring system is based on the linear relationship between the stripping time and the analyte concentration in injected samples. In comparative determinations of lead, it has been observed that potentiometric stripping offers better selectivity than anodic stripping with square-wave voltammetry [232]. For determinations at the ppb level, the sample frequency is usually 10–20 samples per hour. Applications reported in the literature are listed in Table 5.

4. Conductometric Detection

The conductivity of an electrolyte solution depends only on the total number of ions in a solution and their charge, and so conductometric detection suffers from a lack of selectivity. It is very widely used in environmental and process analysis to control the salt content in waters and process streams. It is also a most common detection method employed in high-performance ion chromatography. Chromatographic detectors are also used in some flow injection systems, where the nonselectivity of detection necessitates the on-line separation of the analyte from the matrix.

Several FIA methods with conductometric detection have been developed for the determination of nitrogen compounds. These utilise gas diffusion for the separation of ammonia from the sample solution [242–244]. The change in the conductivity of the acceptor solution that results from ammonia diffusion is measured and related to the concentration of ammonium ion in the original sample. This procedure has been applied in determinations of ammonia [242, 243], in the determination of ammonia in samples obtained from Kjeldahl digestion [244] and in the determination of nitrate and nitrite [242]. In the latter case, the analytes are determined after on-line reduction to ammonia in an alkaline medium, using a column filled with metallic zinc. Speciation can be achieved by adding sulphanilic acid to remove nitrite from the sample and determining ammonia without the use of the column. The conductometric determination of ammonia in waste water samples has been compared with several spectrophotometric methods, exhibiting a comparable level of detectability [243]. The conductometric system has the simplest experimental requirements.

An FIA conductometric system has also been used to determine organic contaminants, which are photocatalytically oxidised to carbon dioxide [245]. The oxidation is carried out by near-UV illumination in a tubular Teflon reactor with an immobilised thin film of titanium oxide on its inner surface. The detection limit for methanol was found to be 1 nM with a 20 μl sample volume. This procedure has also been applied for the determination of formaldehyde, ethylene oxide and single-cell algae.

The indirect determination of chloride in waters has been accomplished in an FIA system with two ion-exchange columns in the hydrogen form and silver form [246]. The difference in the conductivity of the sample after passing through these columns was proportional to the concentration of chloride, and the limit of detection was found to be 50 ppb.

5. Flow Injection Coulometry

In conventional electroanalysis, three different techniques are based on the measurement of the charge in order to obtain quantitative information about the amount of the analyte. These are constant-potential coulometry, coulometric titration (or coulometry at constant current) and electrogravimetry. They do not need calibration and they are considered the most accurate and precise methods of chemical analysis. Their successful adaptation to the FIA system utilising full advantages of this technique is difficult and hence very few applications have been developed so far. Some applications of coulometry at constant potential and coulometric titration can be found.

In potentiostatic coulometry, the working electrode is maintained at a constant potential such that quantitative oxidation or reduction of the analyte takes place. The current decreases to zero when the analyte is removed from the solution. The charge required to oxidise or reduce the analyte is most commonly measured with an electronic integrator. As in all coulometric methods, a fundamental requirement is that the analyte be reduced in the electrode process with 100% current efficiency. Potentiostatic coulometric methods have been developed for numerous elements in inorganic compounds. The use of potentiostatic FIA coulometry has been reported for the determination of thallium(I) by means of its oxidation to thallium(III) using firmly packed platinum chips as the working electrode, with a detection limit of 0.35 μM [240]. The flow injection coulometric determination of dopamine, hydroquinone and hexacyanoferrate(II) is based on the oxidation of the analytes at a reticulated vitreous carbon working electrode with a large surface area-to-volume ratio [247]. The obtained detection limits compare quite favourably with those reported for amperometric determinations.

In constant-current coulometry, a titrant is generated electrolytically and various methods of detecting the chemical equivalence point can be used, including potentiometry, conductivity and spectrophotometry. Its principal advantages compared with volumetric processes are the elimination of standard solution preparation and the ability to perform microtitrations with small quantities of the analyte. Coulometric methods have been developed for all types of titrations, including acid-base, precipitation, complexometric and redox titrations. In one developed coulometric flow injection method, the electrolysis of water at a platinum working electrode is used to generate a reagent for an acid-base titration, while the remote spectrophotometric detection of the indicator bromothymol blue with optical fibres is used to detect the end point [241]. The flow injection arrangement is employed to deliver 19 μl samples to the integrated flow cell with generating electrode and optical fibre detection. The system has been used to titrate samples of sodium hydroxide in the range of 0.5 mM to 4 M and nitric acid ranging from 5 mM to 15 M. A commercial coulometric titrator with iodine generation has also been adopted for the flow injection Karl Fischer titration of water [248]. A manual system is automated by recirculating the anode reagent from a titrator through a mechanically controlled sampling valve. This recirculation serves as a means of mixing the sample and transporting it to the titration cell. In an automated laboratory setup, the sample inlet port of the injection valve is connected to an autosampler, whereas for on-line monitoring the sample inlet port is connected to a three-way switching valve, which is in turn connected to a circulating process sample stream. The automated method offers three advantages over manual operation: the fixed-loop operation eliminates human error and improves precision; the closed system reduces interference from atmospheric moisture; and automation increases efficiency by allowing unattended analysis.

6. References

1. J. Ruzicka and E. H. Hansen, *Anal. Chim. Acta* **78** (1974) 145.
2. J. Ruzicka, E. H. Hansen and E. A. G. Zagatto, *Anal. Chim. Acta* **88** (1977) 1.
3. K. Cammann, *Fresenius Z. Anal. Chem.* **329** (1988) 691.
4. A. Izquiredo and M. D. Luque de Castro, *Electroanalysis* **7** (1995) 505.
5. I. M. P. L. O. Ferreira and J. F. C. Lima, *J. Flow Injection Anal.* **10** (1993) 17.
6. E. Pungor, Z. Feher, G. Nagy and K. Toth, *CRC Crit. Rev. Anal. Chem.* **14** (1983) 175.
7. S. D. Kolev, K. Toth, E. Lindner and E. Pungor, *Anal. Chim. Acta* **234** (1990) 49.
8. K. Toth, E. Lindner, E. Pungor and S. D. Kolev, *Anal. Chim. Acta* **234** (1990) 57.
9. L. Ilcheva, M. Trojanowicz and T. Krawczynski vel Krawczyk, *Fresenius Z. Anal. Chem.* **328** (1987) 27.
10. J. F. van Staden, *Anal. Chim. Acta* **261** (1992) 381.
11. D. E. Davey, D. E. Mulcahy, G. R. O'Connel and R. S. C. Smart, *Electroanalysis*, **7** (1995) 461.
12. W. E. van der Linden and R. Oostervink, *Anal. Chim. Acta* **101** (1978) 419.
13. L. K. Shpigun, O. V. Basanova and Yu. A. Zolotov, *Sens. Actuators B* **10** (1992) 15.
14. M. Trojanowicz, E. Pobozy and M. E. Meyerhoff, *Anal. Chim. Acta* **222** (1989) 109.
15. W. Frenzel and P. Brätter, *Anal. Chim. Acta* **187** (1986) 1.
16. T. J. Cardwell, R. W. Cattrall, P. J. Iles and I. C. Hamilton, *Anal. Chim. Acta* **177** (1985) 239.
17. T. J. Cardwell, R. W. Cattrall, P. J. Iles and I. C. Hamilton, *Anal. Chim. Acta* **204** (1988) 329.
18. C. Hongbo, E. H. Hansen and J. Ruzicka, *Anal. Chim. Acta* **169** (1985) 209.
19. M. Trojanowicz and W. Matuszewski, *Anal. Chim. Acta* **138** (1982) 71.
20. W. Frenzel, *Fresenius Z. Anal. Chem.* **329** (1988) 698.
21. M. Trojanowicz and W. Matuszewski, *Anal. Chim. Acta* **151** (1983) 77.
22. L. Ilcheva and K. Cammann, *Fresenius Z. Anal. Chem.* **322** (1985) 322.
23. S. Alegret, J. Alonso, J. Bartroli, J. L. F. C. Lima, A. A. S. C. Machado and J. M. Paulis, *Anal. Lett.* **18** (A18) (1985) 2291.
24. J. L. F. C. Lima, M. C. B. S. M. Montenegro, J. Alonso, J. Bartroli and J. G. Raurich, *Anal. Chim. Acta* **234** (1990) 221.
25. D. E. Davey, D. E. Mulcahy and G. R. O'Connel, *Analyst* **117** (1992) 761.
26. P. Van den Winkel, J. Mertens and D. L. Massart, *Anal. Chem.* **46** (1974) 1765.
27. D. P. Brezinski, *Analyst* **108** (1983) 425.
28. E. H. Hansen, A. K. Ghose and J. Ruzicka, *Analyst* **102** (1977) 705.

29. M. Trojanowicz, Z. Augustowska, W. Matuszewski, G. Moraczewska and A. Hulanicki, *Talanta* **29** (1982) 113.
30. J. Slanina, W. A. Lingerak and F. Bakker, *Anal. Chim. Acta* **117** (1980) 91.
31. P. Petak and K. Štulik, *Anal. Chim. Acta* **185** (1986) 171.
32. M. Trojanowicz, A. Hulanicki, W. Matuszewski, A. Fuksiewicz, T. Hulanicka-Michalak, S. Raszewski, J. Szyller and W. Augustyniak, *Chem. Anal. (Warsaw)* **32** (1987) 709.
33. W. J. van Oort and E. J. J. M. van Eerd, *Anal. Chim. Acta* **155** (1983) 21.
34. W. Frenzel, *Analyst* **113** (1988) 1039.
35. W. Frenzel and P. Brätter, *Anal. Chim. Acta* **185** (1986) 127.
36. H. Müller and V. Müller, *Anal. Chim. Acta* **180** (1986) 30.
37. J. F. van Staden, *Fresenius Z. Anal. Chem.* **325** (1986) 247.
38. J. F. van Staden, *Analyst* **112** (1987) 595.
39. J. F. van Staden, *Fresenius Z. Anal. Chem.* **328** (1987) 68.
40. M. E. Meyerhoff and Y. M. Fraticelli, *Anal. Lett.* **A4** (B6) (1981) 415.
41. S. Alegret, J. Alonso, J. Bartroli, J. M. Paulis, J. L. F. C. Lima and A. A. S. C. Machado, *Anal. Chim. Acta* **164** (1984) 147.
42. J. Alonso, J. Bartroli, J. F. F. C. Lima and A. A. S. C. Machado, *Anal. Chim. Acta* **179** (1986) 503.
43. B. Karlberg and S. Thelander, *Analyst* **103** (1978) 1154.
44. P. W. Alexander, P. R. Haddad and M. Trojanowicz, *Anal. Chim. Acta* **171** (1985) 151.
45. P. W. Alexander, M. Trojanowicz and P. R. Haddad, *Anal. Lett.* **17** (A4) (1984) 309.
46. P. W. Alexander, P. R. Haddad and M. Trojanowicz, *Anal. Chem.* **56** (1984) 2417.
47. P. W. Alexander, P. R. Haddad and M. Trojanowicz, *Anal. Lett.* **18** (A16) (1985) 1953.
48. A. U. Ramsing, J. Janata, J. Ruzicka and M. Levy, *Anal. Chim. Acta* **118** (1980) 45.
49. P. D. van der Wal, E. J. R. Sudhölter and D. N. Reinhoudt, *Anal. Chim. Acta* **245** (1991) 159.
50. P. L. H. M. Cobben, R. J. M. Egberink, J. G. Bomer, R. Schouwenaar, Z. Brzozka, M. Bos, P. Bergveld and D. N. Reinhoudt, *Anal. Chim. Acta* **276** (1993) 347.
51. S. Alegret, J. Bartroli, C. Jimenez-Jorquera, M. del Valle, C. Dominguez, J. Esteve and J. Bausells, *Sens. Actuators* **B7** (1992) 555.
52. J. Ruzicka and E. H. Hansen, *Anal. Chim. Acta* **161** (1984) 1.
53. J. Fucsko, K. Toth, E. Pungor, J. Kunovits and H. Puxbaum, *Anal. Chim. Acta* **194** (1987) 163.
54. M. Trojanowicz and W. Frenzel, *Fresenius Z. Anal. Chem.* **328** (1987) 653.
55. J. J. Harrow and J. Janata, *Anal. Chim. Acta* **174** (1985) 115.
56. J. J. Harrow and J. Janata, *Anal. Chim. Acta* **174** (1985) 123.

57. E. H. Hansen, J. Ruzicka and A. K. Ghose, in *Soil Nitrogen as Fertilizer or Pollutant* (IAEA, Vienna, 1980).
58. F. J. Saez de Viteri and D. Diamond, *Electroanalysis* **6** (1994) 9.
59. O. Elsholz, W. Frenzel, C. Y. Liu and J. Möller, *Fresenius J. Anal. Chem.* **338** (1990) 159.
60. T. Krawczynski vel Krawczyk, B. Szostek and M. Trojanowicz, *Talanta* **36** (1989) 811.
61. T. Altunbulduk, H. Meier-zu Köcker and W. Frenzel, *Fresenius J. Anal. Chem.* **351** (1995) 593.
62. L. Risinger, *Anal. Chim. Acta* **179** (1986) 509.
63. W. E. van der Linden, *Anal. Chim. Acta* **179** (1986) 91.
64. G. Schultze, C. Y. Liu, M. Brodowski, O. Elscholtz, W. Frenzel and J. Möller, *Anal. Chim. Acta* **214** (1988) 121.
65. W. Frenzel, *Fresenius J. Anal. Chem.* **336** (1990) 21.
66. I. Papaefstathiou, M. T. Tena and M. D. Luque de Castro, *Anal. Chim. Acta* **308** (1995) 246.
67. Q. Chang and M. E. Meyerhoff, *Anal. Chim. Acta* **186** (1986) 81.
68. R. Y. Xie and G. D. Christian, *Anal. Chim. Acta* **58** (1986) 1806.
69. A. N. Araujo, M. B. Exebarria, J. L. F. C. Lima, M. C. B. S. M. Montenegro and R. Perez Olmos, *Fresenius J. Anal. Chem.* **351** (1995) 614.
70. D. B. Hibbert, P. W. Alexander, S. Rachmawati and S. A. Caruana, *Anal. Chem.* **62** (1990) 1015.
71. M. Trojanowicz and B. Szostek, *Anal. Chim. Acta* **261** (1992) 521.
72. W. Frenzel, C. Y. Liu and J. Oleksy-Frenzel, *Anal. Chim. Acta* **233** (1990) 77.
73. M. Sequeira, D. B. Hibbert and P. W. Alexander, *Proc. 4th Int. Environmental Chem. Conf.* (Darwin, 1995) A038-1.
74. G. B. Marshall and D. Midgley, *Analyst* **108** (1983) 701.
75. J. F. Coetzee and C. W. Gardner Jr. *Anal. Chem.* **58** (1986) 608.
76. P. W. Alexander and J. Koopetngram, *Anal. Chim. Acta* **197** (1987) 353.
76a. T. C. Tang and H. J. Huang, *Anal. Chem.* **67** (1995) 2299.
77. T. I. M. S. Lopes, A. O. S. S. Rangel, J. L. F. C. Lima and M. C. B. S. M. Montenegro, *Anal. Chim. Acta* **308** (1995) 122.
78. N. Ishibashi, T. Imato, H. Ohura and S. Yamasaki, *Anal. Chim. Acta* **214** (1988) 349.
79. G. B. Martin and M. E. Meyerhoff, *Anal. Chim. Acta* **186** (1986) 71.
80. J. F. Coetzee and C. Gunaratua, *Anal. Chem.* **58** (1986) 650.
81. L. Ilcheva and K. Cammann, *Fresenius Z. Anal. Chem.* **325** (1986) 11.
82. T. Imato, T. Yoshizuka and N. Ishibashi, *Anal. Chim. Acta* **233** (1990) 139.
83. M. Trojanowicz, W. Matuszewski and A. Hulanicki, *Anal. Chim. Acta* **136** (1982) 85.
84. M. L. Balconi, F. Sigon, R. Ferraroli and F. Realini, *Anal. Chim. Acta* **214** (1988) 367.
85. J. Ruzicka, E. H. Hansen and H. Mosbaek, *Anal. Chim. Acta* **92** (1977) 235.
86. S. F. Simpson and F. J. Holler, *Anal. Chem.* **54** (1982) 43.

87. O. Lutze, B. Ross and K. Cammann, *Fresenius J. Anal. Chem.* **350** (1994) 630.
88. O. Aström, *Anal. Chim. Acta* **105** (1979) 67.
89. J. Bartroli and L. Alerm, *Anal. Chim. Acta* **269** (1992) 29.
90. T. J. Cardwell, R. W. Catrall, P. C. Hauser and I. C. Hamilton, *Anal. Chim. Acta* **214** (1988) 359.
91. L. Ilcheva, R. Yanakiev, V. Vasileva and N. Ibekve, *Food Chem.* **38** (1990) 105.
92. J. Alonso-Chamarro, J. Bartroli, S. Jun, J. L. F. C. Lima and M. C. B. S. M. Montenegro, *Analyst* **118** (1993) 1527.
93. R. M. Liu, D. J. Liu and A. L. Sun, *Analyst* **117** (1992) 1335.
94. R. Virtanen, *Anal. Chem. Symp. Ser.* **8** (1981) 375.
95. R. J. Forster and D. Diamond, *Anal. Chem.* **64** (1992) 1721.
96. F. J. Saez de Viteri and D. Diamond, *Analyst* **119** (1994) 749.
97. M. Harnett, D. Diamond and P. G. Barker, *Analyst* **118** (1993) 347.
98. J. C. Ngila, D. B. Hibbert and P. W. Alexander, *Proc. 4th Environmental Chem. Conf.* (Darwin, 1995) AS45-1.
99. S. K. Lunsford, Y. L. Ma, A. Galal, C. Striley, H. Zimmer and H. B. Mark Jr., *Electroanalysis* **7** (1995) 420.
100. C. R. Powley, R. F. Geiger Jr. and T. A. Nieman, *Anal. Chem.* **52** (1980) 705.
101. B. H. van der Schoot, S. Jeanneret, A. van der Berg and N. F. de Rooij, *Anal. Meth. Instr.* **1** (1993) 38.
102. J. Alonso, J. Baro, J. Bartroli, J. Sanchez and M. del Valle, *Anal. Chim. Acta* **308** (1995) 115.
103. E. H. Hansen, J. Ruzicka and A. K. Ghose, *Anal. Chim. Acta* **100** (1978) 151.
104. H. Wada, T. Ozawa, G. Nakagawa, Y. Asano and S. Ito, *Anal. Chim. Acta* **211** (1988) 213.
105. N. Kolycheva and H. Müller, *Anal. Chim. Acta* **242** (1991) 65.
106. J. F. van Staden, *Fresenius Z. Anal. Chem.* **331** (1988) 594.
107. L. Ilcheva and K. Cammann, *Fresenius Z. Anal. Chem.* **322** (1985) 323.
108. J. L. F. C. Lima and L. S. M. Rocha, *Int. J. Environ. Anal. Chem.* **38** (1990) 127.
109. W. Frenzel, *Fresenius Z. Anal. Chem.* **335** (1989) 931.
110. E. Wang and S. Kamata, *Anal. Chim. Acta* **261** (1992) 399.
111. P. C. Hauser, *Anal. Chim. Acta* **278** (1993) 227.
112. J. A. Greenberg and M. E. Meyerhoff, *Anal. Chim. Acta* **141** (1982) 57.
113. W. Frenzel and P. Brätter, *Anal. Chim. Acta* **188** (1986) 151.
114. E. H. Hansen, F. J. Krug, A. K. Ghose and J. Ruzicka, *Analyst* **102** (1977) 714.
115. M. E. Meyerhoff and P. M. Kovach, *J. Chem. Educ.* **60** (1983) 767.
116. J. Bartroli, L. Alerm, F. Fabry and E. Siebert, *Anal. Chim. Acta* **308** (1995) 102.
117. M. G. Glaister, G. J. Moody and J. D. R. Thomas, *Analyst* **110** (1985) 113.
118. G. Nagy, Zs. Feher and E. Pungor, *Anal. Chim. Acta* **52** (1970) 47.
119. R. J. Rucki, *Talanta* **27** (1980) 147.

120. K. Štulik and V. Pacakova, *CRC Crit. Rev. Anal. Chem.* **14** (1984) 297.
121. M. Trojanowicz, *Chem. Anal.* (Warsaw) **30** (1985) 171.
122. D. C. Johnson, S. G. Weber, A. M. Bond, R. M. Wightman, R. E. Shoup and I. S. Krull, *Anal. Chim. Acta* **180** (1986) 187.
123. H. B. Hanekamp and H. J. Nieuwkerk, *Anal. Chim. Acta* **121** (1980) 13.
124. D. MacKoul, D. C. Johnson and K. G. Schick, *Anal. Chem.* **56** (1984) 436.
125. U. Forsman and A. Karlsson, *Anal. Chim. Acta* **139** (1982) 133.
126. H. B. Hanekamp, W. H. Voogt and P. Bos, *Anal. Chim. Acta* **118** (1980) 73.
127. A. G. Fogg and A. M. Summan, *Analyst* **109** (1984) 1029.
128. A. J. Downard, H. K. J. Powell and S. Xu, *Anal. Chim. Acta* **256** (1992) 117.
129. J. Wang and L. Chen, *Anal. Chem.* **63** (1991) 1499.
130. J. Wang and L. Chen, *Talanta* **42** (1995) 385.
131. J. Wang and H. D. Dewald, *Talanta* **29** (1982) 453.
132. P. T. Kissinger, *Anal. Chem.* **49** (1977) 447A.
133. S. G. Weber, *J. Electroanal. Chem.* **145** (1983) 1.
134. C. E. Lunte, S. W. Wong, T. H. Ridgway, W. R. Heineman and K. W. Chan, *Anal. Chim. Acta* **188** (1986) 263.
135. J. C. Hoogvliet, J. M. Reijn and W. P. van Bennekom, *Anal. Chem.* **63** (1991) 2418.
136. T. Matsue, A. Aoki, E. Ando and I. Uchida, *Anal. Chem.* **62** (1990) 407.
137. T. P. Tougas, J. M. Jannetti and W. G. Collier, *Anal. Chem.* **57** (1985) 1377.
138. C. Liang, P. Vacha and W. E. van der Linden, *Talanta* **35** (1988) 59.
139. B. G. Snider and D. C. Johnson, *Anal. Chim. Acta* **105** (1979) 25.
140. P. L. Meschi and D. C. Johnson, *Anal. Chem.* **52** (1980) 1304.
141. P. L. Meschi and D. C. Johnson, *Anal. Chim. Acta* **124** (1981) 303.
142. J. Matysik, E. Soczewinski, E. Zminkowska-Halliop and M. Przegalinski, *Chem. Anal.* (Warsaw) **26** (1981) 463.
143. J. A. Lown, R. Koile and D. C. Johnson, *Anal. Chim. Acta* **116** (1980) 33.
144. K. W. Pratt and W. F. Koch, *Anal. Chem.* **58** (1986) 124.
145. P. Ward and M. R. Smyth, *Talanta* **40** (1993) 1131.
146. M. Trojanowicz and J. Michalowski, *J. Flow Injection Anal.* **11** (1994) 34.
147. M. Trojanowicz, A. Hulanicki, W. Matuszewski, M. Palys, A. Fuksiewicz, T. Hulanicka-Michalak, S. Raszewski, J. Szyller and W. Augustyniak, *Anal. Chim. Acta* **188** (1986) 165.
148. J. W. Bixler and A. M. Bond, *Anal. Chem.* **58** (1986) 2859.
149. D. L. Luscombe and A. M. Bond, *Anal. Chem.* **62** (1990) 27.
150. S. G. Weber and W. C. Purdy, *Anal. Chim. Acta* **100** (1978) 531.
151. O. A. Sadik and G. G. Wallace, *Electroanalysis* **6** (1994) 860.
152. K. Štulik and V. Pacakova, *J. Chromatogr.* **208** (1981) 269.
153. A. N. Tsaousis and C. O. Huber, *Anal. Chim. Acta* **178** (1985) 319.
154. G. G. Neuburger and D. C. Johnson, *Anal. Chem.* **59** (1987) 203.
155. A. Trojanek and P. Papoff, *Anal. Chim. Acta* **247** (1991) 73.
156. B. Pihlar, L. Kosta and B. Hristovski, *Talanta* **26** (1979) 805.
157. J. A. Polta, I. H. Yeo and D. C. Johnson, *Anal. Chem.* **57** (1985) 563.

158. B. Persson and L. Rosen, *Anal. Chim. Acta* **123** (1981) 115.
159. R. C. Schothorst, M. van Son and G. den Boef, *Anal. Chim. Acta* **162** (1984) 1.
160. G. G. Neuberger and D. C. Johnson, *Anal. Chim. Acta* **179** (1986) 381.
161. P. W. Alexander and U. Akapongkul, *Anal. Chim. Acta* **148** (1983) 103.
162. P. W. Alexander and U. Akapongkul, *Anal. Chim. Acta* **166** (1984) 119.
163. Y. Xu, H. B. Halsall and W. R. Heineman, *Electroanalysis* **4** (1992) 33.
164. A. Rios, M. D. Luque de Castro, M. Valcarcel and H. A. Mottola, *Anal. Chem.* **59** (1987) 666.
165. F. W. Nyasulu and H. A. Mottola, *J. Autom. Chem.* **9** (1987) 46.
166. J. Wang and P. Tuzhi, *Anal. Chem.* **58** (1986) 1787.
167. J. Wang and L. D. Hutchins, *Anal. Chem.* **57** (1985) 1536.
168. J. Wang, P. Tuzhi and T. Golden, *Anal. Chim. Acta* **194** (1987) 129.
169. K. M. Korfhage, K. Ravichandran and P. R. Baldwin, *Anal. Chem.* **56** (1984) 1514.
170. W. Hou and E. Wang, *Anal. Chim. Acta* **257** (1992) 275.
171. J. A. Cox and K. R. Kulkarni, *Talanta* **33** (1986) 911.
172. J. A. Cox and K. R. Kulkarni, *Analyst* **111** (1986) 1219.
173. J. N. Barisci, G. G. Wallace, E. A. Wilke, M. Meaney, M. R. Smyth and J. G. Vos, *Electroanalysis* **1** (1989) 245.
174. A. P. Doherty, M. A. Stanley, G. Arana, C. E. Koning, R. H. G. Brinkhuis and J. G. Vos, *Electroanalysis* **7** (1995) 333.
175. Y. Ikariyama and W. R. Heineman, *Anal. Chem.* **58** (1986) 1803.
176. J. Ye and R. P. Baldwin, *Anal. Chem.* **60** (1988) 1979.
177. J. Y. Sung and H. J. Huang, *Anal. Chim. Acta* **246** (1991) 275.
178. R. C. Martinez, F. B. Dominguez, F. M. Gonzalez, J. H. Mendez and R. C. Orellana, *Anal. Chim. Acta* **279** (1993) 299.
179. J. Wang and H. D. Dewald, *Talanta* **29** (1982) 901.
180. S. Hughes and D. C. Johnson, *Anal. Chim. Acta* **132** (1981) 11.
181. T. Z. Polta and D. C. Johnson, *J. Electroanal. Chem.* **209** (1986) 159.
182. J. A. Polta and D. C. Johnson, *Anal. Chem.* **57** (1985) 1373.
183. J. Kurzawa, *Anal. Chim. Acta* **173** (1985) 343.
184. W. Matuszewski, A. Hulanicki and M. Trojanowicz, *Anal. Chim. Acta* **194** (1987) 269.
185. A. Hulanicki, W. Matuszewski and M. Trojanowicz, *Anal. Chim. Acta* **194** (1987) 119.
186. W. Matuszewski and M. Trojanowicz, *Anal. Chim. Acta* **207** (1988) 59.
187. M. Trojanowicz, W. Matuszewski, B. Szostek and J. Michalowski, *Anal. Chim. Acta* **261** (1992) 391.
188. J. Michalowski, A. Kojlo, B. Magnuszewska and M. Trojanowicz, *Anal. Chim. Acta* **289** (1994) 339.
189. J. Michalowski, A. Kojlo, M. Trojanowicz, B. Szostek and E. A. G. Zagatto, *Anal. Chim. Acta* **271** (1993) 239.
190. J. Michalowski and M. Trojanowicz, *Anal. Chim. Acta* **281** (1993) 299.

191. B. Szostek and M. Trojanowicz, *Chemometrics Intell. Lab. Systems* **22** (1994) 221.
192. A. G. Fogg and N. K. Bsebsu, *Analyst* **109** (1984) 19.
193. J. Wang and Z. Lu, *Anal. Chem.* **62** (1990) 826.
193a. A. Sano, T. Kuwayama, M. Furukawa, S. Takitani and H. Nakamura, *Anal. Sci.* **11** (1995) 405.
194. A. J. Downard, J. B. Hart, H. K. J. Powell and S. Xu, *Anal. Chim. Acta* **269** (1992) 41.
195. S. Hinkamp and G. Schwedt, *Anal. Chim. Acta* **236** (1990) 345.
196. M. Granados, S. Maspoch and M. Blanco, *Anal. Chim. Acta* **179** (1986) 445.
197. E. B. Milosavljevic, L. Solujic and J. L. Hendrix, *Environ. Sci. Technol.* **29** (1995) 426.
198. J. A. Lown and D. C. Johnson, *Anal. Chim. Acta* **116** (1980) 41.
199. H. Lundbäck, *Anal. Chim. Acta* **145** (1983) 189.
200. F. Belal and J. L. Anderson, *Mikrochim. Acta* (1985) 145.
201. S. Ikeda, H. Satake and Y. Kohori, *Chem. Lett.* (1984) 873.
202. A. G. Fogg, A. Y. Chamsi, A. A. Barros and J. O. Cabral, *Analyst* **109** (1984) 901.
203. A. Y. Chamsi and A. G. Fogg, *Analyst* **111** (1986) 879.
204. J. H. Mendez, R. C. Martinez, E. R. Gonzalo and J. P. Trancon, *Electroanalysis* **2** (1990) 487.
205. T. Ding and R. D. Schmid, *Anal. Chim. Acta* **234** (1990) 247.
206. A. G. Fogg and N. K. Bsebsu, *Analyst* **107** (1982) 566.
207. S. M. Harden and W. K. Nonidez, *Anal. Chem.* **56** (1984) 2218.
208. L. Ilcheva and A. Dakashev, *Fresenius J. Anal. Chem.* **340** (1991) 14.
209. P. T. Kissinger, *Anal. Chem.* **49** (1977) 447A.
210. P. R. Fielden and T. McCreedy, *Anal. Chim. Acta* **273** (1993) 111.
211. J. Janata and J. Ruzicka, *Anal. Chim. Acta* **139** (1982) 105.
212. J. Wang and H. D. Dewald, *Anal. Chim. Acta* **153** (1983) 325.
213. J. Wang and H. D. Dewald, *Talanta* **31** (1984) 387.
214. C. A. Scolari and S. D. Brown, *Anal. Chim. Acta* **178** (1985) 239.
214a. R. Goldnik, C. Yarnitzky and M. Ariel, *Anal. Chim. Acta* **234** (1990) 161.
215. N. Thogersen, J. Janata and J. Ruzicka, *Anal. Chem.* **55** (1983) 1988.
216. F. Canete, A. Rios, M. D. Luque de Castro and M. Valcarcel, *Anal. Chim. Acta* **211** (1988) 287.
217. M. D. Luque de Castro and A. Izquierdo, *Electroanalysis* **3** (1991) 457.
218. J. Wang and H. D. Dewald, *Anal. Chim. Acta* **162** (1984) 189.
219. J. A. Wise, W. R. Heineman and P. T. Kissinger, *Anal. Chim. Acta* **172** (1985) 1.
220. J. Wang and H. D. Dewald, *Anal. Chem.* **56** (1984) 156.
221. C. Wechter, N. Sleszynski, J. J. O'Dea and J. Osteryoung, *Anal. Chim. Acta* **175** (1985) 45.
221a. F. M. Matysik and G. Werner, *Analyst* **118** (1993) 1523.

222. D. W. Bryce, A. Izquierdo and M. D. Luque de Castro, *Anal. Chim. Acta* **308** (1995) 96.
223. J. Wang and B. A. Frelha, *Anal. Chem.* **55** (1983) 1285.
224. E. N. Chaney and R. P. Baldwing, *Anal. Chim. Acta* **176** (1985) 105.
225. M. Wasberg and A. Ivaska, *Anal. Chim. Acta* **179** (1986) 433.
226. X. Yang, L. Risinger and G. Johanson, *Anal. Chim. Acta* **192** (1987) 1.
227. E. Beinrohr, P. Tschöpel, G. Tölg and M. Nemeth, *Anal. Chim. Acta* **273** (1993) 13.
228. E. B. T. Tay, S. B. Khoo and S. G. Ang, *Analyst* **114** (1989) 1271.
229. A. Izquierdo, M. D. Luque de Castro and M. Valcarcel, *Electroanalysis* **6** (1994) 894.
230. J. Wang, H. W. Dewald and B. Greene, *Anal. Chim. Acta* **146** (1983) 45.
231. L. Almestrand, D. Jagner and L. Renman, *Anal. Chim. Acta* **193** (1987) 71.
232. J. H. Aldstadt, D. F. King and H. D. Dewald, *Analyst* **119** (1994) 1813.
233. L. Anderson, D. Jagner and M. Josefson, *Anal. Chem.* **54** (1982) 1371.
234. A. Hu, R. E. Dessy and A. Graneli, *Anal. Chem.* **55** (1983) 320.
235. M. Trojanowicz, in *Reviews in Analytical Chemistry*, ed. E. Roth (Les Editonsde Physique, Paris, 1988).
236. W. Matuszewski, M. Trojanowicz and W. Frenzel, *Fresenius J. Anal. Chem.* **332** (1988) 148.
237. G. Schultze, M. Husch and W. Frenzel, *Mikrochim. Acta* **1** (1984) 191.
238. W. Frenzel and P. Brätter, *Anal. Chim. Acta* **179** (1986) 389.
239. S. D. Kolev, C. W. K. Chow, D. E. Davey and D. E. Mulcahy, *Anal. Chim. Acta* **309** (1995) 293.
240. L. I. Ilcheva and A. D. Dakashev, *Analyst* **115** (1990) 1247.
241. R. H. Taylor, J. Ruzicka and G. D. Christian, *Talanta* **39** (1992) 285.
242. L. C. de Faria and C. Pasquini, *Anal. Chim. Acta* **245** (1991) 183.
243. A. Cerda, M. T. Oms, R. Forteza and V. Cerda, *Anal. Chim. Acta* **311** (1995) 165.
244. C. Pasquini and L. C. de Faria, *Anal. Chim. Act* **193** (1987) 19.
245. G. K. C. Low and R. W. Mattews, *Anal. Chim. Acta* **231** (1990) 13.
246. g. Lach and K. Bächmann, *Anal. Chim. Acta* **196** (1987) 163.
247. D. J. Curran and T. P. Tougas, *Anal. Chem.* **56** (1984) 672.
248. Y. Y. Liang, *Anal. Chem.* **62** (1990) 2504.
249. M. Trojanowicz, A. Lewenstam, T. Krawczyński vel Krawczyk, I. Lähdesmaki and W. Szczepek, *Electroanalysis* **8** (1996) 233.

Chapter 4

Enzymatic Methods of Detection and Immunoassays

Biochemical methods of chemical analysis have been very intensively developed in the last twenty years in numerous areas of analytical instrumentation and with applications in various fields of chemical analysis, such as clinical diagnostics, environmental, pharmaceutical, food or process analysis [1–10]. In these methods the final source of the analytical signal is one of optical, electrochemical or thermal detectors, of which use in FIA was discussed in earlier chapters. Incorporating into the analytical procedure a selective, and very often practically even specific step of chemical transformation involving biocatalysts or immunochemical interactions antigen–antibody, creates entirely new possibilities of detection with the use of mentioned detectors. A separate presentation of these detection methods is justified by the large number of developed designs and analytical applications, including in FIA, of integrated biosensors consisting of the biological component and a suitable transducer [3–5]. The biological component catalyses the chemical reaction of the analyte (enzymes, microorganisms, tissues) or specifically binds the analyte (antibodies, receptors). The transducer monitors the biochemical reaction (products or substrates). The specificity of enzymatic and immunochemical interactions results in high selectivity of biochemical detections, which are very difficult to obtain with physico-chemical methods only.

Applications of biochemical methods of detection in FIA include determination of substrates of enzymatic reactions and activity of enzymes as well as immunoassays. The simplicity of manipulation with time-dependent operations in FIA yields particularly suitable conditions for carrying out these procedures of kinetic characteristics. Either in FIA systems of commonly used dimensions, or especially in microdevices, the advantage of the utilisation of FIA in enzymatic and immunochemical determinations is the possibility of handling with very small sample volumes and limited consumption of bio-components.

1. Enzymatic Assays

All enzymes are proteins with molecular weights ranging from polypeptide chains of about 9 kDa to oligomeric complexes of up to 6.000 kDa. Some enzymes require metal cations, which are bound to specific oxygen, nitrogen or sulphur ligands of the protein. The specificity of the enzyme interaction is determined by the three-dimensional structure near the active site of the enzyme, where catalysis occurs. Enzymatic catalysis always involves complex formation between the reactant (most-often-determined substrate) and the enzyme, and it can be specific for a single substrate molecule of broad structural types. For substrate S and enzyme E,

$$E + S \Leftrightarrow E \cdot S \xrightarrow{\text{catalysis}} E \cdot P \Leftrightarrow E + P\,, \tag{1}$$

where P is the product. The observed reaction rate V depends on the substrate concentration and Michaelis–Menten constant K_M:

$$V = V_{\max}[S](K_M + [S])^{-1}\,, \tag{2}$$

where $V_{\max}$ is the maximum reaction rate, which depends on the activity of the enzyme. This equation describes the initial velocity V relative to $V_{\max}$ at a given substrate concentration and this is the basis for the analytical application of enzymes in the determination of substrates and enzyme activity.

1.1. *Systems with Soluble Enzymes*

The applications of enzymes dissolved in solution are much less frequent in FIA than those of immobilised enzymes. The use of soluble enzymes offers the possibility of design of relatively simple manifolds, especially if the product of the enzymatic reaction can be detected without the need of any additional chemical reaction. Products of enzymatic degradation of urea in the presence of soluble urease create the change of the pH of the solution, which can be measured with a potentiometric detector [11] or spectrophotometrically [12]. In a similar FIA system the bioluminescent determination of adenosine-5'-triphosphate with luciferase can be carried out [13]. An additional reaction is needed in determinations with enzymes catalysing glucose reactions with chemiluminescence detection [14], hydrogen peroxide [15] and penicillin [16] with spectrophotometric detection, or cholesterol with a potentiometric one [17]. The use of the reversal FIA system, in which the injected sample is first pumped into a detector and subsequently pumped into the same detector again by a reversed flow, provides a two-point rate assay of substrate, as was shown for glucose [18]. A multipoint determination based on monitoring of the reaction rate can be favourably carried out in the stopped-flow technique. In such a system the flow is stopped after a precisely selected time period following injection and the slope of the rate curve is calculated by measuring the absorbance increase over a fixed period of time. This technique was successfully employed for the determination of ethanol [19] and malic acid [20] with spectrophotometric detection. In the latter case the setups, both with constant aspiration of the reagent and merging zones systems with parallel injection of the sample and reagent, were used (Fig. 1). A merging zone system

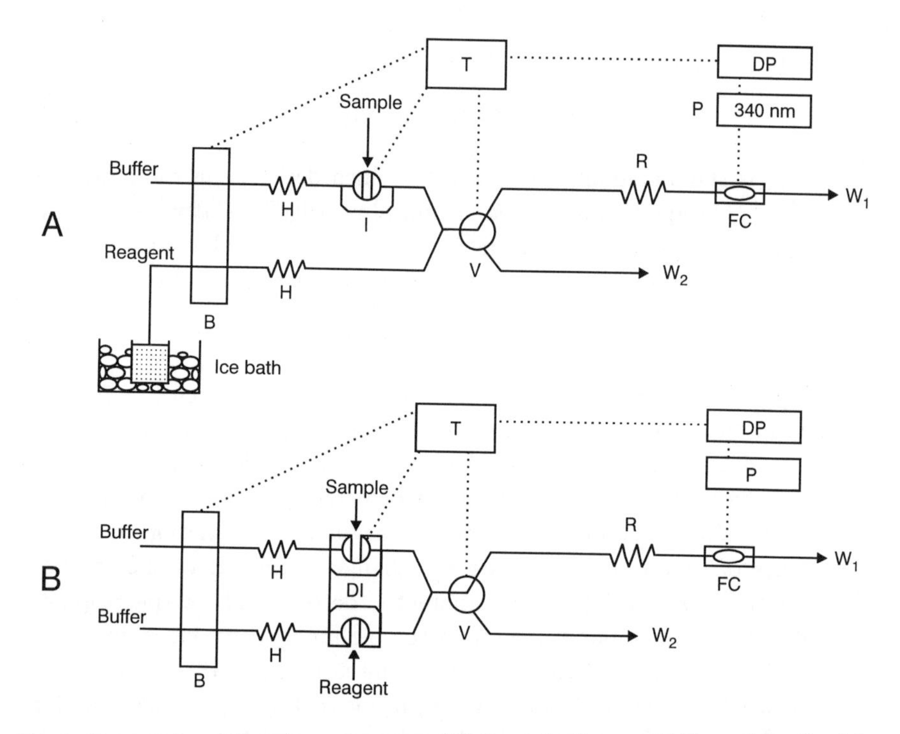

Fig. 1. Stopped-flow (A) and merging zones (B) flow injection manifolds used for the determination of free L-(-)-malic acid with soluble enzyme and UV spectrophotometric detection [20]. B — peristaltic pump; H — heating coils; I — injection valve; DI — dual injector; V — three-way valve; R — reaction coil; FC — flow cell; W — waste; P — spectrophotometer; DP — data processor; T — timer. (*Reprinted by permission of copyright owner.*)

allows the decrease of the enzyme reagent and was successfully used also for determination of triglycerides with spectrophotometric detection [21], glucose with amperometric detection [22], and ascorbic acid with spectrophotometric detection [23]. With soluble enzymes, especially advantageous are closed-loop systems, which for determination of glucose [23a] and lactose, maltose and sucrose [23b], were designed with the amperometric monitoring of the dissolved oxygen level. The consumption of oxygen in the enzymatic oxidation reaction in solution containing glucose oxidase and catalase is balanced by on-line aeration of the circulating solution. This allows analysis of more than 1000 serum samples without change of the circulating solution.

In the FIA systems where the enzymatic reaction and a reaction needed for the detection occur at different pH microporous cells can be used, where the enzyme is forced from a container through a microporous membrane into the analyte stream and where a pH gradient is formed. Then close to the membrane there is appropriate pH for enzymatic detection, whereas away from the membrane suitable pH for detection is created. This approach was demonstrated for determination of glucose [24] and sucrose [25] with chemiluminescence detection. In determination of Hg(II) based on inhibition of soluble urease a separation unit with a PTFE gas-permeable membrane was used to separate the ammonia produced from other sample components, which might react with the *o*-phthalate reagent used for fluorescence detection [26].

1.2. *Systems with Immobilised Enzyme Reactors*

As was mentioned in FIA systems, much more often enzymes are used in immobilised form, using physical or chemical binding to appropriate support, within the electroactive material of the working electrode used for detection, or in the layer through which the substrate and/or product of the enzymatic reaction can diffuse [27–29]. The immobilised enzyme can be an integral component of the biosensor employed in FIA, or, most commonly, it can be used in a separate flow-through reactor of various designs. In comparative studies of various systems for chemiluminescence determination of sucrose employing invertase, mutarotase and glucose oxidase, it was found that the use of immobilised invertase and mutarotase in controlled-pore glass (CPG) reactors is superior to the use of non-immobilised enzymes in working range, detection limit, sensitivity and time analysis [25]. Similarly, in comparison of spectrophotometric determinations with soluble or immobilised aldehyde dehydrogenase on Sepharose, it was found that the immobilised enzyme method has a lower blank value, a longer linear range and is more economic and simpler than with the soluble enzyme [30].

Immobilisation of the enzymes generally improves the stability of enzyme activity in time. An increased rigidity of the structure of the immobilised enzyme results in preserving the native configuration of the protein and making unfolding less probable. In the use of the enzyme reactors most often is used the covalent binding through functional groups in the enzyme, which are not involved in the catalytic activity such as amino, hydroxyl, and thiol groups. In certain cases the binding of these groups can be responsible for a loss of enzyme activity as a result of immobilisation. The advantage of immobilisation of the enzyme in the reactor in comparison to an integrated biosensor is the possibility of using a much larger amount of the enzyme. This allows a more efficient conversion of the substrate into a detectable product. The general expression, which enables the estimation of packed-bed enzymatic reactor $V_x(\mu\mathrm{l})$ in order to obtain 99.9% of the substrate conversion, is as follows [31]:

$$E_0 V_x V_0^{-1} \geq Q(C_x + 6.9K_\mathrm{M} + 5.9K_\mathrm{M} C_x K_\mathrm{EP}^{-1})\,. \qquad (3)$$

where E_0 is the enzyme activity (M min^{-1}) for a reactor of volume V_0 (μl), Q is the flow rate (l min^{-1}), C_x is the maximum concentration of the analyte (M), K_M is the Michaelis–Menten constant for the immobilised system, and K_EP is the dissociation constant of the enzyme-product complex. The values of K_M and K_EP can be evaluated from kinetic plots [31]. Most often, the enzyme binding is achieved by a nucleophilic group of the enzyme reacting with the activated functional group of the support.

Flow-through enzyme reactors used in FIA have various geometry and are produced with different supporting materials. Most common are packed-bed reactors, usually of 1–3 mm internal diameter and length from 10 to 50 mm. The most often employed support is alkylaminated CPG, especially aminopropylated CPG, and the procedure of enzyme immobilisation with glutaraldehyde is based on formation of the Schiff base. Although the chemistry of this process is very complex [32], the obtained reactors are very frequently used in FIA (Tables 1–3). For the immobilisation of penicillinase, functionalisation of glass beads was performed with different silanes and the best results were reported for (*p*-aminophenyl) trimethoxysilane [33]. For immobilisation of amyloglucosidase, twelve different inorganic supports were compared, including the variety of CPGs and alumina using functionalisation and glutaraldehyde immobilisation procedures [34]. The highest efficiency was obtained on CPG

Table 1. Enzymatic flow injection determination of substrates with electrochemical detections.

Substrate	Enzyme	Mode of enzyme use	Detection*	Concentration range	Reference
Ascorbic acid	Ascorbate oxidase	Reactor	Amp. Amp.	4–400 ppb	40
Cephalosporin	Cephalosporinase	Biosensor	Pot.	0.1–15 g/l	104
Cholesterol	Cholesterol oxidase and esterase	Soluble	Pot.	0.03–3.0 mM	17
		Reactor	Amp.	0.0026–31 mM	107
Creatinine	Creatinine deiminase, Leucine dehydrogenase, *L*-amino acid oxidase	Reactor	Amp.	0.2–5.0 mM	65
	Creatinine iminohydrolase	Reactor	Pot.	0.2–30 μM	63
Ethanol	Alcohol oxidase	Reactor	Amp.	0.0006–60% (v/v)	108
Fructose	Fructose 5-dehydrogenase	Reactor	Amp.	0.02–2.0 mM	67
Glucose	Glucose dehydrogenase	Biosensor	Amp.	0.5–40 mM	88
	Glucose oxidase	Biosensor	Amp.	0.01–100 mM	77
				0.01–3.0 mM	78
				0.01–70 mM	80
				0.2–25 mM	82
				1.0–40 mM	83
				3–100 mM	87
				0.1–1.0 mM	90
				0.001–8 mM	91
				0.5–20 mM	92,94
				0.02–30 mM	97
			Pot.	0.05–0.4 g/l	104
		Reactor	Amp.	1–25 mM	35
				3–30 mM	38

continued

Table 1 (*continued*)

Substrate	Enzyme	Mode of enzyme use	Detection*	Concentration range	Reference
				0.01–0.8 mM	50
				0.0023–1.0 mM	53
				0.01–20 mM	109
			Pot.	0.1–10 mM	111
		Soluble	Amp.	0.01–14 mM	22
				0.1–5.0 g/l	23a
L-glutamate	Glutamate oxidase, glutamic-pyruvic transaminase	Reactor	Amp.	0.002–2.0 mM	37
	Glutamate oxidase	Reactor	Amp.	0.01–0.3 mM	112
Glutamine	Glutaminase	Biosensor	Pot.	0.2–10 mM	100
		Reactor	Pot.	0.5–4.5 mM	71
				0.01–4.0 mM	113
	Glutaminase, glutamate oxidase	Reactor	Amp.	10–200 ppm	114
H_2O_2	Peroxidase	Biosensor	Amp.	0.01–1.0 μM	84
3-hydroxy-butyrate	3-hydroxybutyrate	Reactor	Amp.	1–50 μM	115
D-lactate	D-lactate dehydrogenase	Reactor	Amp.	0.1–500 μM	115
L-lactate	L-lactate oxidase	Biosensor	Amp.	1–10 mM	83
				2–200 μM	95
				0.2–5.0 μM	96
		Reactor	Amp.	8–64 μM	50
				0.1–500 μM	115
				0.4 μM–40 mM	117
	Lactate oxidase, lactate dehydrogenase	Reactor	Amp.	0.002–10 μM	116
Lactose	Lactase, glucose oxidase	Soluble	Amp.	0.25–2.5 g/l	23b
	β-galactosidase, glucose oxidase	Reactor	Amp.	0.01–2.0 mM	36
Maltose	Maltase, glucose oxidase	Soluble	Amp.	0.5–5.0 g/l	23b
Oxalate	Oxalate oxidase	Reactor	Amp.	5.7–900 μM	118

continued

Table 1 (*continued*)

Substrate	Enzyme	Mode of enzyme use	Detection*	Concentration range	Reference
Penicillin G	Penicillinase	Biosensor	Cond.	0.05–8 mM	102
			Pot.	0.1–20 g/l	104
		Reactor	Pot.	0.25–1.0 mM	52
				0.05–5.0 mM	33
				0.5–100 mM	48
Penicillin V	Penicillinase	Biosensor	Pot.	0.5–30 g/l	101
Phenol	Tyrosinase	Biosensor	Amp.	0.003–5 mM	89
Phenols	Laccase, tyrosinase	Biosensor	Amp.	0.002–1.0 mM	85
	Peroxidase	Biosensor	Amp.	0.5–20 μM	86
	Tyrosinase	Reactor	Amp.	10^{-9}–10^{-3} M	70
Phosphate	Nucleoside phosphorylase, xanthine oxidase	Biosensor	Amp.	1.25–100 μM	119
		Reactor	Amp.	0.02–50μM	69
Pyruvate	Lactate oxidase, lactate dehydrogenase	Reactor	Amp.	0.01–100 μM	116
Sulphite	Sulphite oxidase	Reactor	Amp.	1–10 ppm	39
				0.1–0.8 mM	54
				0.01–0.8 mM	150
Starch	Amyloglucosidase,	Biosensor	Amp.	10^{-4}–10^{-1}% (m/v)	79
Sucrose	Invertase, glucose oxidase	Soluble	Amp.	0.1–2.5 g/l	23b
	Invertase, mutarotase, glucose oxidase	Reactor + biosensor	Amp.	1–40 mM	93
Urea	Urease	Biosensor	Pot.	0.5–10 mM	100
				1–40 mM	103
		Reactor	Cond.	0.01–10 mM	120
			Pot.	0.1–100 mM	110
				0.02–2.0 mM	122
			Amp.	0.5–10 mM	121
		Soluble	Pot.	4–30 mM	11

*Amp. — amperometric; Pot. — potentiometric; Cond. — conductivity.

Table 2. Enzymatic flow injection determination of substrates with spectrophotometric detection.

Substrate	Enzyme	Detected species (λ, nm)	Concentration range	Reference
Acetaldehyde	Aldehyde dehydrogenase	NADH (340)	0.018–0.77 mM	30
L-arginine	L-arginase, urease	Ammonia as indophenol blue (629)	0.038–7.5 mM	41
L-asparagine	L-asparaginase	Ammonia as indophenol blue (629)	0.060–15 mM	42
Cholesterol	Cholesterol oxidase, peroxidase	H_2O_2 with *p*-anisidine (458)	1–180 μM	43
	Cholesterol esterase, cholesterol oxidase	H_2O_2 with ABTS (423)	0.11–8.6 mM	123
Creatinine	Creatinine iminohydrolase	Ammonia by pH indicator	25–200 μM	64
Ethanol	Alcohol dehydrogenase	NADH (340)	6.4–32 ppm, 0.05–0.4% (v/v)	19
			2–500 ppm	57
			0.73–40 ppm	124
Glucose	Glucose oxidase	H_2O_2 with Bindscheller's green (725)	0.02–2.5 ppm	125
	Glucose oxidase, peroxidase	H_2O_2 with 4-aminophenazone and HBS (505)	5–25 mM	49
		H_2O_2 with MBTH and 3-dimethylamino benzoic acid (590)	0.2–8.0 g/l	47
		H_2O_2 with 4-aminophenazone and MEHA (565)	10–200 ppb	58
		H_2O_2 with 4-aminophenazone and N,N-dimethylaniline (550)	0.16–16 mM	110
	Glucose oxidase, mutarotase, peroxidase	H_2O_2 with 4-aminoantipyrene and phenol (254)	0.5–5.0 g/l	18
H_2O_2	Peroxidase	Condensation product of 4-aminoantipyrene	0.15–40 μM	15

continued

Table 2 (*continued*)

Substrate	Enzyme	Detected species (λ, nm)	Concentration range	Reference
		with ALPS (560)		
		Product of reaction with ABTS (414)	0.04–200 μM	126
L-malic acid	Malate dehydrogenase	NADH (340)	0.1–41 ppm	20
p-nitrophenyl phosphate	Alkaline phosphatase	*p*-nitrophenol (410)	0.16–160 μM	31
Oxalate	Oxalate decarboxylase, formate dehydrogenase	NADH (340)	0.1–3.0 mM	127
Penicillin G	Penicillinase	pH changes with Merck indicator 9582 (605)	0.3–50 mM	56
Penicillin V	Penicillinase	Molybdenum blue (670)	68–680 ppb	16
Sulphite	Sulphite oxidase, peroxidase	H_2O_2 with 4-aminophenazone and MEHA (565)	3–1000 ppb	59
Sucrose	Invertase, mutarotase, glucose oxidase	H_2O_2 with 4-aminoantipyrene and DCPS (514)	0.1–500 μM	128
Triglycerides	Lipase, glycerol kinase, glycerol-1-phosphate dehydrogenase, diaphorase	Reduced form of *p*-iodonitrotetrazolium violet (503)	1.8 mM	21
Urea	Urease	Ammonia by pH indicator (610)	1–10 mM	12, 55
				61
		(590)	0.1–500 mM	
		Ammonia as indophenol blue (700)	25–500 μM	44

Abbreviations used:
ABTS — 2, 2′-azinobis (3-ethylbenzthiazoline-6-sulphonate)
ALPS — N-ethyl-N-(sulphopropyl)aniline
DCPS — 2,4-dichlorophenol-6-sulphonate
HBS — disodium 3,5-dichloro-2-hydroxybenzene sulphonate
MBTH — 3-methyl-2-benzothiazolinone hydrazone
MEHA — 3-methyl-N-ethyl-N'-(β-hydroxyethyl)aniline
NADH — reduced nicotinamide adenine dinucleotide

Table 3. Enzymatic flow injection determination of substrates with luminescence detections.

Substrate	Enzyme	Detection*	Reagent	Concentration range	Reference
Ascorbic acid	Laccase	Flu.	*o*-phenylenediamine	0.025–1.0 ppm	23
Adenosine-5′-triphosphate	Luciferase	Bio.	Luciferin	0.001–100 μM	13
Ethanol	Alcohol dehydrogenase	Flu.	NAD^+	0.1–30 ppm	129
Fructose	Mannitol dehydrogenase	Flu.	NADH	6–600 μM	46
Glucose	Glucose oxidase	Chem.	Luminol, hexacyano-ferrate(III)	50–500 ppm	14
				5–25 mM	49
				0.1–40 mM	51
			Luminol, Cu(1,10-phenanthroline)$_3$	0.01–20 μM	24
H_2O_2	Peroxidase	Chem.	Luminol	0.01–1000 μM	62
L-lactate	*L*-lactate oxidase	Flu.	Decacyclene	0.02–60 mM	105
Lactose	β-galactosidase, mutarotase, glucose oxidase	Flu.	Luminol, hexacyano-ferrate (III)	20–200 mM	130
Methanol	Alcohol oxidase, formaldehyde dehydrogenase, catalase	Flu.	NAD^+	1.2–80 μM	131
L-malate	Malate dehydrogenase, aspartate aminotransferase	Flu.	NAD^+	0.005–50 mM	132
Penicillin V	Penicillin-G-amidase	Flu.	Aminofluorescein	0.1–12 g/l	106
Sucrose	Invertase, mutarotase, glucose oxidase, peroxidase	Chem.	Luminol, hemin	0.005–1 mM	25
L-tyrosine	Tyrosinase	Flu.		0.02–1000 μM	133
Urea	Urease	Flu.	Aminofluorescein	3–10 mM	106

*Bio. — bioluminescence; Chem. — chemiluminescence; Flu. — fluorescence.

of a pore size 170 Å and a surface area of 150 m^2 g^{-1}. The aminopropylation together with binding using glutaraldehyde was applied in the immobilisation of glucose oxidase on the wall of the glass capillary in a micro-FIA system. In enzymatic determination of cholesterol in non-aqueous solutions with spectrophotometric detection, cholesterol oxidase and peroxidase were immobilised by adsorption on CPG from aqueous solution [43]. As other inorganic supports for enzyme immobilisation, there were used alkylamino-bonded silica [36, 37] and carbon fibre after anodic oxidation of the surface resulting in formation of COOH groups and activation with carbodiimide [38].

Among organic supports used in packed-bed enzyme reactors, the most common is cyanogen bromide activated Sepharose [31, 31, 39, 40]. Others include epoxy resins [41, 42], poly (glycidyl methacrylate)-coated porous glass [44], agarose [45], poly (vinyl alcohol) [46] or activated cellulose gels [47]. In determination of penicillins with potentiometric detection, the reactors containing preparations obtained by physical entrapment of enzymes in porous beds of poly-hydroxyethyl methacrylate matrix formed by γ-irradiation were used [48].

Another type of reactors employed in FIA are open tubular reactors with enzymes bound covalently to the internal wall of the reactor. Spectrophotometric and chemiluminescence glucose determinations were carried out with glucose oxidase covalently bound to the inner surface of a nylon tube [49]. Particularly good results are reported when the tubular reactor is filled with inert beads of a diameter equal to about 90% of the tube diameter due to minimised contributions to dispersion [50]. In comparison of a 50 cm nylon reactor with a CPG packed bed in determinations of glucose, better stability was found for the open tubular reactor, but an increase of dispersion results in a decrease of the sampling frequency [51]. Satisfactory functioning in potentiometric determination of penicillin G was reported for the system with an open tubular reactor with CPG particles embedded in the inner walls of plastic tubing [52].

In a rotating bioreactor used in FIA systems with amperometric detection, the enzyme was immobilised on CPG which was deposited on a sticking tape affixed to a disk that could be rotated [53, 54]. Several systems were developed, where enzymatic reaction was integrated with spectrophotometric detection. In urea determination urease was cross-linked with albumin into a cellulose pad, with an acid-base indicator dye covalently bound to the surface of the cellulose [55]. A similar system was described for determination of penicillin [56]. Photometric detection was also carried out with the cell, where to the inner walls the enzyme was attached through CPG as a support [57]. This concept was further developed in the design of the detector, where the packing material of the flow cell consists of two physically distinct layers: an upper layer of CPG with immobilised enzyme, and a lower layer of a suitable exchanger where the product of enzymatic reaction can be retained and preconcentrated. Such systems were developed for spectrophotometric determination of hydrogen peroxide and glucose [58] as well as for sulphite in environmental samples [59].

An alternative to the most common packed-bed reactors are membrane reactors [60–62]. Owing to a decreased pressure build-up, membrane reactors proved to be of some superiority to packed beds of CPG [60]. In the determination of urea with spectrophotometric detection, urease was immobilised by addition of perfluoroalkyl chains to the free amine groups of the enzyme and then adsorption of this modified enzyme on a PTFE gas-diffusion membrane [61]. Different peroxidases were covalently immobilised on affinity membranes and used in a flow cell with optical fibre transmission of light in chemiluminescence detection of hydrogen peroxide [62].

A separate role that can be played by the enzyme reactor in the FIA system is the on-line removal of interfering constituents of the sample. The CPG reactors with glutamate dehydrogenase can be utilised for removal of endogenous ammonia in potentiometric [63], spectrophotometric [64] and amperometric [65] determinations of creatinine in physiological fluids, and also in fluorimetric determination of potassium based on enzyme activation [66]. A reactor with ascorbate oxidase was used for removal of ascorbic acid in amperometric determination of fructose [67].

The applications of FIA systems with various detections and enzyme reactors are listed in Tables 1–3. Several unique designs are worth mentioning. The concept of amplification of the response of the enzymatic detection system by substrate recycling [68] was employed to enhance the sensitivity of FIA enzymatic determination of L-glutamate [37], phosphate [69] and polyphenols [70] in the set-ups with enzymatic reactors and amperometric detection. The principle of this approach is enzymatic or chemical reproduction of substrate in the measuring system in the course of detection, which results in an increase of the measured signal. In determination of phosphate it was based on the use of co-immobilised purine nucleoside phosphorylase (PNP) and xanthine oxidase (XO) and as a result of the reactions

$$\text{Inosine} + \text{phosphate} \xrightarrow{\text{PNP}} \text{hypoxanthine} + \text{ribose-1-phosphate}\,,$$

$$\text{Hypoxanthine} + 2O_2 \xrightarrow{\text{XO}} \text{uric acid} + 2H_2O_2\,,$$

uric acid is produced, which is detected amperometrically [69]. Additional co-immobilisation of alkaline phosphatase (AP) in the enzyme reactor yields a 12-fold enhancement of the amperometric signal for phosphate as result of the reaction

$$\text{Ribose-1-phosphate} \xrightarrow{\text{AP}} \text{phosphate} + \text{ribose}\,.$$

This is associated with change of the selectivity of the system. When two enzymes PNP and XO are used the system is selective for orthophosphate, whereas addition of AP makes it also sensitive to pyrophosphate and nucleotides.

In potentiometric detection of substrates, which are enzymatically degraded to ammonia, the presence of endogenous ammonia or alkali metal ions often unfavourably affects detection. The effect of these interferences can be eliminated in a branched FIA system with enzyme reactor and potentiometric detection in the sub-Nernstian range of linear response of the potentiometric detector. Such systems were developed for the determination of L-glutamine in bioreactor media [71] and creatinine in urine [72] (Fig. 2). The first peak of the obtained FIA signal corresponds to the endogenous sample ammonia, and the second one to the total ammonia present after passing through the enzyme reactor. Two linear equations can be solved to obtain the concentration of analyte.

The application of three CPG enzyme reactors with immobilised alcohol dehydrogenase (ADH), aldehyde dehydrogenase (AlDH) and L-glutamate dehydrogenase (GlDH) enables the construction of a closed-loop FIA system for the determination of ethanol with UV spectrophotometric detection (Fig. 3) [73]. In the presence of coenzyme nicotinamide adenine dinucleotide (NAD^+) determination is based on reactions

$$\text{Ethanol} + NAD^+ \xrightarrow{\text{ADH}} \text{acetaldehyde} + \text{NADH}\,,$$

$$\text{Acetaldehyde} + NAD^+ \xrightarrow{\text{AlDH}} \text{acetate} + \text{NADH}\,,$$

whereas GlDH causes NADH conversion back to NAD^+ according to the scheme

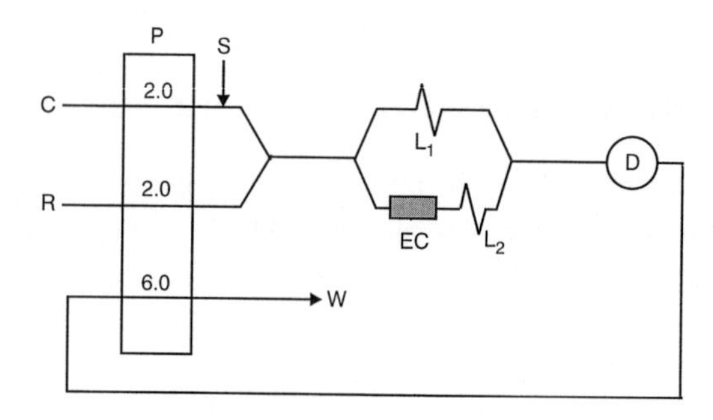

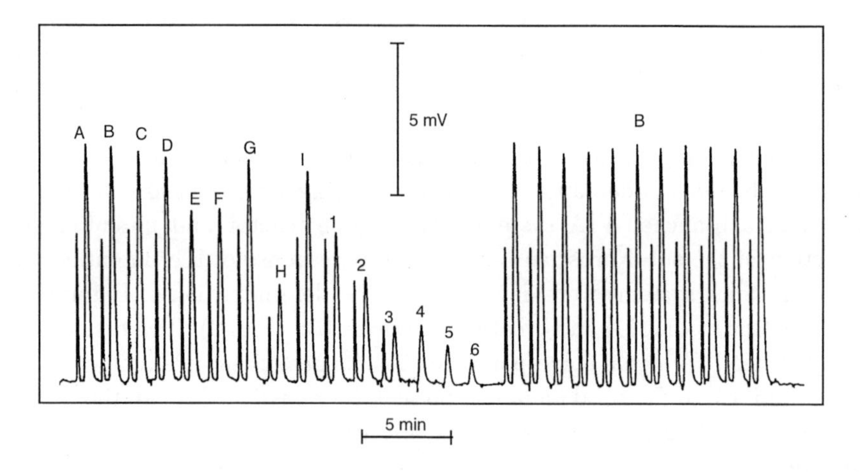

Fig. 2. Flow injection manifold used for the determination of creatinine in urine with potentiometric detection using the sub-Nernstian linear response range and flow injection peaks recorded in an optimised flow system [72]. C — carrier stream of distilled water; R — reagent stream of phosphate buffer; S — sample injection point; ER — enzyme reactor; L — teflon coils; P — peristaltic pump; D — wall-jet detector; W — waste; A–I — natural urine samples; 1–6 — standard solutions of (1) 30, (2) 20 and (3) 10 mM ammonium chloride and (4) 9, (5) 6 and (6) 3 mM creatinine. Injection volume 20 μl. (*Reprinted by permission of copyright owner.*)

$$NH_4^+ + NADH + \alpha - \text{ketoglutarate} \xrightarrow{\text{GIDH}} \text{glutamate} + NAD^+ .$$

Such a system maintains for three days the sensitivity to ethanol detection in the range of 2.4–340 mM without significant changes.

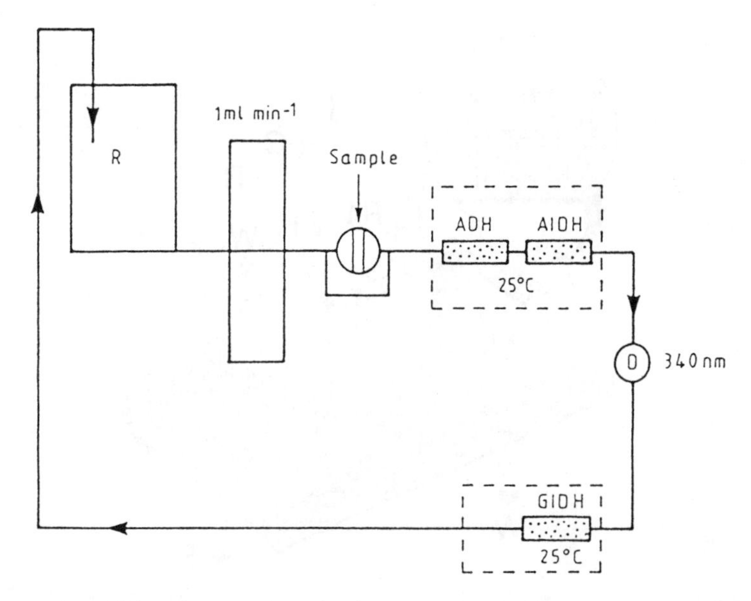

Fig. 3. Schematic diagram of the cyclic flow injection system for the determination of ethanol with immobilised alcohol dehydrogenase (ADH), aldehyde dehydrogenase (AlDH) and glutamate dehydrogenase (GlDH) reactors [73]. R — carrier; D — detector. (*Reprinted by permission of copyright owner.*)

A miniaturised FIA system with enzyme CPG reactor and chemiluminescence detector employing two photodiodes was developed for determination of glucose, creatinine and lactic acid [74] (Fig. 4). For each analyte the final product of a single or a few enzymatic conversions was hydrogen peroxide, which was determined by chemiluminescence with an alkaline reagent containing luminol and hexacyanoferrate (III). By an exchange of the enzyme reactors the system can be adjusted for the determination of four different substrates.

Systems with enzyme reactors find broad application in multicomponent determinations, which are discussed in a further part of this chapter.

1.3. *Biosensors as Detectors in FIA*

Immobilisation of enzyme or a component of the immunochemical system at the surface of the electrochemical or piezoelectric detector, and also within the electrode material, allows one to design new detection devices for analytical purposes — biosensors. Similarly combining the fibre optics with an element

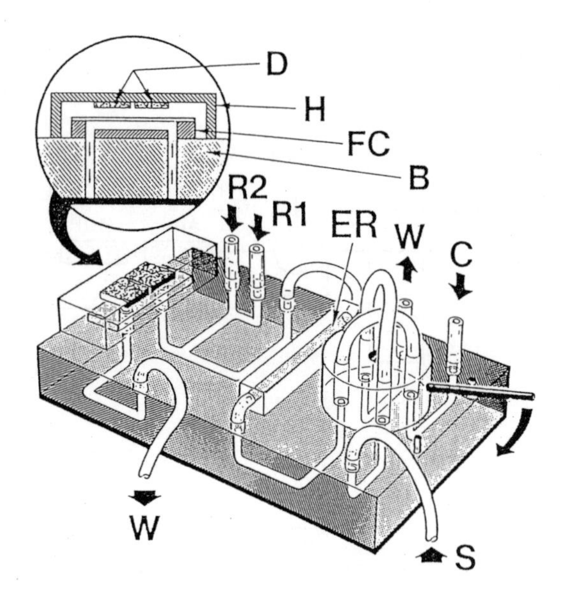

Fig. 4. Integrated flow injection microconduit for enzymatic determination with chemiluminescence detection by means of a light detector consisting of two photo discs and comprising injection valve and enzyme reactor (ER) [74]. C — carrier stream; R1 — reagent 1 (luminol); R2 — reagent 2 (hexacyanoferrate (III)); S — sample; W — waste. (*Reprinted by permission of copyright owner.*)

of molecular recognition leads to optodes, or with biocomponents to bioptodes, i.e. biosensors based on optical detections. Biosensors have been developed for many years, both as probes for non-flow measurements [3–6] and as detectors in flow measurements. Some of them are commercially available [75]. The main aim of a well-designed biosensor is selective and sensitive determination of the target analyte within complicated matrices without sample preparation [76]. Its application in the FIA system should require a relatively simple set-up of the sample delivery in comparison to similar systems with soluble enzyme or flow-through reactors.

The largest number of applications in the FIA system were developed for integrated amperometric biosensors with a wide variety of different possibilities of enzyme immobilisation. These immobilisation methods obey entrapment within the electrode material, physical attachment or chemical binding to the electrode surface or immobilisation on the membrane placed at the electrode surface. In membrane biosensors the enzyme was immobilised in a polyuretane layer sandwiched between two cellulose dialysis membranes [77], on a nylon mesh [78–81], or on a polyester membrane [82]. Also, commercially available membranes loaded with enzymes were used in glucose and lactate amperometric biosensors for FIA measurements [83]. In model theoretical considerations on the signal magnitude of the amperometric biosensor it was found that the FIA response depends both on the concentration time profile in the flow cell compartment and on the diffusion time of the membrane. It was concluded that it should be always advantageous to use a membrane with a short diffusion time provided the enzyme activity is sufficient [77]. In biosensors for the determination of glucose and lactate in blood serum it is necessary to use additional outer membranes reducing the effect of electroactive interferences [82,83]. A similar result can be obtained by lowering the detection potential. With the use of glucose oxidase this can be achieved by the use of an internal graphite paste electrode containing ferrocene [80]. Satisfactory results of FIA starch determination were reported for an amperometric membrane biosensor with three enzymes immobilised on methylated nylon mesh amyloglucosidase, mutarotase and glucose oxidase [79].

Biosensors for flow injection amperometry can also be produced by the immobilisation of enzymes on the surface of carbon electrodes. In the simplest procedure the enzyme was immobilised by adsorption from the solution on the glassy carbon surface [84] or graphite [85, 86] which was used for peroxidase [84, 86] and a mixture of laccase and tyrosinase [85]. Glucose oxidase was immobilised on the surface of the paste electrode containing mediator tetrathiafulvalene by evaporation of the enzyme solution at 4°C, then deposition of albumine cross-linked with glutaraldehyde and covering these two layers with a polycarbonate membrane [87]. Enzymes were also bound covalently to the electrode surfaces modified with phenoxazinum ion [88] or activated with a carbodiimide [89]. This procedure was employed for glucose dehydrogenase [88] and tyrosinase [89]. Yet another immobilisation method is electrochemical codeposition in the presence of platinum [90] or palladium [91] complexes. This was used for glucose oxidase immobilised on a carbon fibre ultra-microelectrode [90] or glassy carbon electrode [91]. Especially for biosensors with a glassy carbon electrode based on codeposition of palladium and glucose oxidase, a satisfactory sensitivity as well as long-term stability was found. The outer Nafion layer was efficient in elimination of interferences from ascorbate and urate.

The amperometric biosensors can also be fabricated by the electrodeposition of organic conducting polymers on the surface of a working electrode. The immobilisation of the enzyme can be obtained by its entrapment in the polymer layer resulting from the electrodeposition from the monomer solution containing the soluble enzyme [92, 94–96], or by covalent binding to a functionalized polymer layer on the electrode surface [93]. In FIA systems biosensors with polypyrrole [92, 93] and poly(*o*-phenylenediamine) [94–96] were used. The advantage of the use of electrodeposited polymer layers is relatively effective elimination of the effect of several interferences of electroactive components of natural matrices [94, 95], although their complete removal, for instance in physiological fluids, requires the use of a multilayer of electrodeposited polymers [96]. A unfavourable feature of such biosensors, observed for glucose and lactate devices, is rather quick inactivation, whereas advantageous is the possibility of very simple formation of a polymer biosensing layer even *in situ* in the flow injection system [95].

Much better stability in time and good dynamic properties in FIA measurements characterise biosensors with enzyme immobilised in the graphite paste, used as electrode material of the working electrode. This method was used for immobilisation of glucose oxidase [97], peroxidase [29, 86] and cytochrome oxidase [98]. The biosensors with immobilised peroxidase were employed for the determination of phenols, but for such a purpose better results were reported for the electrode with enzyme adsorbed on the surface of a graphite electrode [86]. The cytochrome oxidase biosensor was used for indirect determination of cyanide [98].

The immobilisation of the enzyme on the surface of ion-selective membranes is also used to obtain potentiometric biosensors, which can be used in FIA. The enzymes urease and glutaminase were immobilised on the surface of hydroxylated asymmetric ammonium ion-selective cellulose triacetate membranes, providing biosensors for urea and L-glutamine [99, 100]. Their application in FIA systems for the analysis of blood serum and in hybridoma media required, however, the use of additional ionomer membranes to enhance the selectivity. In the preferred configuration, a thin hydrophilic anion-exchange membrane was incorporated within a flow-through dialysis unit upstream from the enzyme-electrode detector. The enzyme electrode for on-line monitoring of penicillin V during fermentation was obtained by immobilisation of β-lactamase in a very thin film on a conventional combined pH glass electrode [101]. The conductivity changes with changes of pH exhibit a polypyrrole layer. The conductometric biosensor for flow injection determination of penicillin was prepared by coating the polypyrrole-deposited array device with a cross-linked penicillinase membrane [102]. Another approach used in potentiometric biosensors is to cover the ion-selective membrane with a membrane with covalently immobilised enzyme, which was used in a urea biosensor for FIA determination of urea in undiluted whole blood [103]. Complete elimination of potassium interference was achieved by measuring the actual potassium concentration in the sample separately and then a mathematical correction. Enzymes can also be immobilised by covering the pH-sensitive gate areas of ion-selective filed-effect transistors with enzyme membranes. This was utilised for obtaining enzyme-modified bio-field-effect transistors for the flow injection determination of glucose, urea, penicillin and cephalosporin C [104].

The fibre-optic biosensors can also be used for detection in FIA systems. The enzymes and the appropriate optical sensing system are immobilised on the tip of the optical fibre. The fibre-optic lactate biosensor used in fermentation process analysis is based on an oxygen optode with immobilised lactate oxidase [105]. The consumption of oxygen was determined via quenching of the fluorescence of decacyclene in the film of silicone rubber. The bioptodes for urea and penicillin were based on the pH-sensitive fluorescence dyes co-immobilised with enzymes by cross-linking with glutaraldehyde on the tip of the fibre [106]. Applications of various biosensors in FIA systems are also presented in Tables 1–3.

1.4. *Multicomponent Enzymatic Assays*

The selectivity of the biocatalytic activity of enzymes and the possibility of their immobilisation in various points of the FIA system is widely exploited for construction of the measuring systems for simultaneous multicomponent determinations. Three main types of such systems are used. The multicomponent manifold is formed by a set of parallel, independent systems with separate detectors, or in the system the injected sample is split between different lines to perform different sample processing, and then detection occurs in the same detector after the confluencing point, and as the third option, an array of different selective biosensors can be used.

The FIA multicomponent systems reported so far in the literature were designed for simultaneous determination of 2–4 components. A system with sample splitting, amperometric determination of glucose and potentiometric measurement of L-glutamine is shown in Fig. 5A [134]. A very similar system was developed for simultaneous determination of ammonia nitrogen and L-glutamine [135]. In the system for determination of glucose, urate and cholesterol in blood serum with potentiometric detectors, the sample was

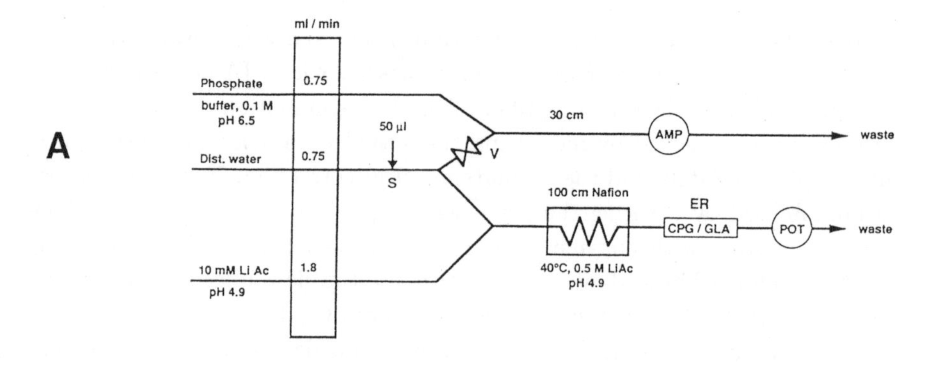

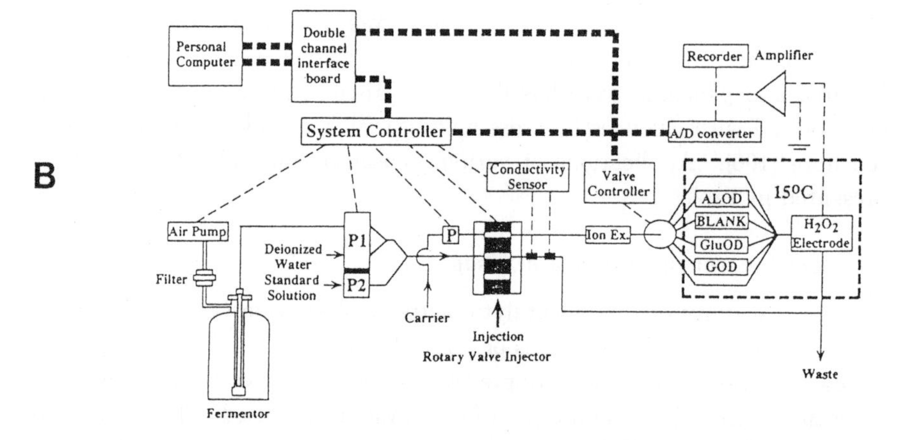

Fig. 5. Flow injection manifolds used for simultaneous determination of glucose and glutamine (A) [134] and for simultaneous determination of ethanol, glucose and glutamate (B) [144]. In (A): AMP — amperometric glucose biosensor; POT — potentiometric ammonium ISE detector; S — sample injection point; ER — enzyme reactor with immobilised glutaminase; V — flow restricting valve. In (B): P — pumps; Ion Ex. — ion-exchange column; ALOD - alcohol oxidase reactor; GluOD — glutamate oxidase reactor; GOD — glucose oxidase reactor. (*Reprinted by permission of copyright owner.*)

introduced by using a 16-way switching valve [136]. For the use in fermentation monitoring, FIA systems for determination of four components were designed, where besides ammonia nitrogen also glucose, maltose and L-amino acids or glucose, lactate and lactose can be simultaneously determined [137, 138]. In all enzymatic determinations the amperometric detection with oxygen electrode was employed and enzymes were immobilised in flow-through reactors.

FIA systems based on the splitting of the flow after sample injection and subsequent confluence before reaching the detector find main applications in two component determinations. Glucose and sucrose were determined with amperometric detection and reactors with invertase, mutarotase and glucose oxidase [119]. The determinations of ethanol and acetaldehyde with reactors containing alcohol dehydrogenase and aldehyde dehydrogenase were carried out photometrically at 340 nm [140]. The amperometric detection was also employed for simultaneous determination of polyamines and hypoxanthine with putrescine oxidase and xanthine oxidase reactors [141], and also for the determination of sulphite and phosphate with a sulphite oxidase reactor and with a coimmobilised purine nucleoside phosphorylase-xanthine oxidase reactor [142]. This type of FIA system can also be utilised for simultaneous determination of analytes, including those which are not substrates of enzymatic reaction, e.g. glucose and ascorbic acid [143].

The example of the FIA system for sequential monitoring of three analytes with a single detector and several enzyme reactors is shown in Fig. 5B [144]. The column with ion-exchange resin was used for baseline stabilisation, and the system was employed for the amperometric determination of glucose, ethanol and glutamate for on-line control of the fermentation process. Glucose and fructose were photometrically determined in the FIA system with automatic implementation of the standard addition method [145]. Two component determinations of ethanol and methanol with one enzyme reactor with immobilised alcohol oxidase can be carried out utilising different kinetics of biocatalytic reaction of these two substrates [146].

The simplest systems can be used with detection employing several selective biosensors, but then a much more difficult task is to design the detector with the layers of immobilised enzymes. A dual biosensor amperometric detector with paste electrodes and layers of immobilised enzymes was used for the determination of glucose and sucrose [147]. Between the biosensors in the thin channel a catalase membrane was formed, where hydrogen peroxide was decomposed while diffusing between electrodes. In another FIA system a biosensor array was employed where enzymes were immobilised on the pH-sensitive gates of an eight-channel field-effect transistor [148]. This was employed for the determination of glucose, maltose, sucrose, lactose and ethanol.

1.5. *Determination of Enzyme Activity*

The changes of substrate or product concentrations for enzyme-catalysed reaction can be also employed for the determination of the enzyme activity, which is a very important factor in clinical diagnostics and biotechnology. Determination of enzyme activity is widely utilised in fundamental studies of mechanisms of enzyme functioning, their activation and inhibition, and in design and fabrication of biosensors, enzyme tests, and preparations used in various clinical diagnostic instrumentation.

Such measurements can be carried out in the FIA systems as discussed above by reversing the roles of substrate and enzyme. The enzyme, which is now the analyte, is introduced into the solution of fixed concentration of substrate and monitored changes of product concentration depend on the activity of the enzyme used. By a simple measurement of the peak height the activity of trypsin with spectrophotometric detection [149], and the activity of amylase [23b] and sulphite oxidase [150] with amperometric detection were determined. More sophisticated systems were also used in order to limit the reagent consumption or to improve detectability. The merging-zones manifold was used for the determination of alcohol dehydrogenase in a bioluminescent assay with an additional enzyme reactor enabling the monitoring of concentration of the reaction product [151]. A favourable way to improve the detection limit in the FIA system is the use of the stopped-flow technique, where the sample injection is followed by a delay time for the transport of a selected section of the dispersed sample zone to the detector. Monitoring takes place during a subsequent stopped-flow period, after which the pump is reactivated. During the stopped-flow period the slope of the response curve is evaluated as the analytical signal. In the determination of the activity of lactate dehydrogenase with UV spectrophotometric detection, pyruvate and NADH were used in the reagent stream [152]. In the determination of trace amounts of a proteolytic enzyme endoproteinase, a stopped-flow procedure in the reversed FIA system was employed with additional on-line preconcentration of the reaction product *p*-nitroaniline on a C18 column [153]. In a system with spectrophotometric detection and a 30 min stopped-flow reaction time the detection limit 0.1 μU ml^{-1} protease has been achieved.

Flow injection UV spectrophotometry was applied for the assay of the activity of chymotrypsin immobilised to the surface of an IrO_2-coated titanium electrode, which is a potentiometric biosensor to certain ester substrates. The assay was based on measurements of the change in absorbance of substrate

N-benzoyl-L-tyrosine ethyl ester on hydrolysis by chymotrypsin [154]. The efficient flow-rate gradient technique was also developed for the testing of modifiers of enzyme activity and demonstrated for the acetylcholine/acetylcholinesterase system [155].

1.6. *Determinations Based on Inhibition of Enzyme Activity*

The inhibition of the activity of enzymes has been extensively investigated for biochemical processes [156, 157]. It may restrict the usefulness of the biosensors and other biochemical methods of analysis but it can also be employed for the indirect determination of the inhibiting species. Numerous such methods have been developed for FIA methodologies using various detection methods, especially for the determination of reversible inhibitors.

Metal ions inactivate urease by reaction with a sulphydryl group and this inhibition has been shown to be relatively selective for mercury (II) and Ag (I) [158]. In the FIA system with urease immobilised on a cellulose pad with an acid-base indicator and spectrophotometric detection, a reversible sensing of mercury was observed [55]. At low levels of Hg (II) in the carrier, injections of urea showed an exponential decrease in the peak height with an increase in the time of exposure, which can be used to determine the concentration of Hg (II). In an FIA system with soluble urease, inhibition of enzyme activity by trace mercury and silver was monitored by fluorimetric detection of ammonia with *o*-phthalaldehyde [26]. Detection limits are 0.1 ppb for Ag (I) and 2 ppb for Hg (II).

A number of metal ions, but also metal-chelating agents, orthophosphate and other multicharged ions inhibit the activity of alkaline phosphatase. In an FIA system with this enzyme immobilised in a CPG reactor and spectrophotometric detection of the enzymatic reaction product *p*-nitrophenol, a non-selective determination of several metals and EDTA is possible [159]. With the same enzyme a hydrolysis of 4-methylumbelliferone was investigated with fluorimetric detection. The inhibition of enzyme activity was utilised in this case for the determination of theophylline in serum in the concentration range of 0.1–200 μM without interferences from the components of natural serum samples [160].

Fluoride inhibits the activity of urease [161], acid phosphatase [162], cholinesterase [163] and liver esterase [164]. Flow injection fluoride determinations with the latter enzyme were carried out with ethyl butyrate as the substrate. Ethanol being produced as a hydrolysis product was detected spectrophotometrically in the presence of coenzyme NAD^+ with alcohol dehydrogenase immobilised on CPG [164]. Fluoride was selectively detected over the range of 15–150 ppb. Very sensitive inhibitive detection was also reported for cyanide based on its effect on cytochrome oxidase immobilised in the matrix of a lipid-cytochrome *c* modified carbon paste electrode [98]. The reversible inhibition allows the reproducible flow injection amperometric determination of cyanide at concentrations as low as 0.5 μM.

For the detection of organophosphorus pesticides and carbamates the inhibition of activity of cholinesterases is commonly exploited. Organophosphorus compounds inactivate the enzyme by phosphorylation whereas carbamates block the active centre of the enzyme by competition with the substrate. Flow injection determinations can be carried out both with spectrophotometric and with electrochemical detections. In flow injection photometric systems inhibition of the activity of acetylcholinesterase by paraoxon was monitored in the catalysed hydrolysis reactions of α-naphthyl acetate [165, 167] and acetylcholine iodide [166]. In the systems with enzyme immobilised on CPG the effective reactivating of enzyme activity by pyridine-2-aldoxime [166] and 1,1′-trimethylene-bis(4-formylpyridinium bromide) dioxime [167]. In a flow injection stopped-flow system with spectrophotometric detection of 1-naphthol based on reaction with *p*-nitrobenzenediazonium terafluoroborate (Fast Red GG Salt) (Fig. 6), a detection limit of 4 nM was obtained for paraoxon [167] and a sample throughput frequency of 10 h^{-1}.

More common, however, are electrochemical detections in the determination of pesticides based on inhibition of cholinesterase activity [168]. Potentiometric detection is based on measurement of the pH change which is produced by the formation of an organic acid in the enzymatic hydrolysis of the ester. The flow injection determinations were carried out either with CPG acetylcholinesterase immobilised in a single bead string reactor and pH detection with a glass electrode [169], or with a biosensor with enzyme immobilised on the pH glass electrode surface in the layer of an acrylamide-methacrylamide copolymer [170]. The detection limits of the insecticides ranged from 0.5 to 275 ppb [169]. In non-flow measurements with amperometric detection most often an additional enzyme choline oxidase is employed, which catalyses the oxidation of choline. The amount of formed hydrogen peroxide or consumed oxygen the activity of inhibited cholinesterase can be determined [168]. In the developed FIA methods as substrate for the determination of the activity of inhibited acetylcholinesterase acetylthiocholine [171] or 4-aminophenyl acetate

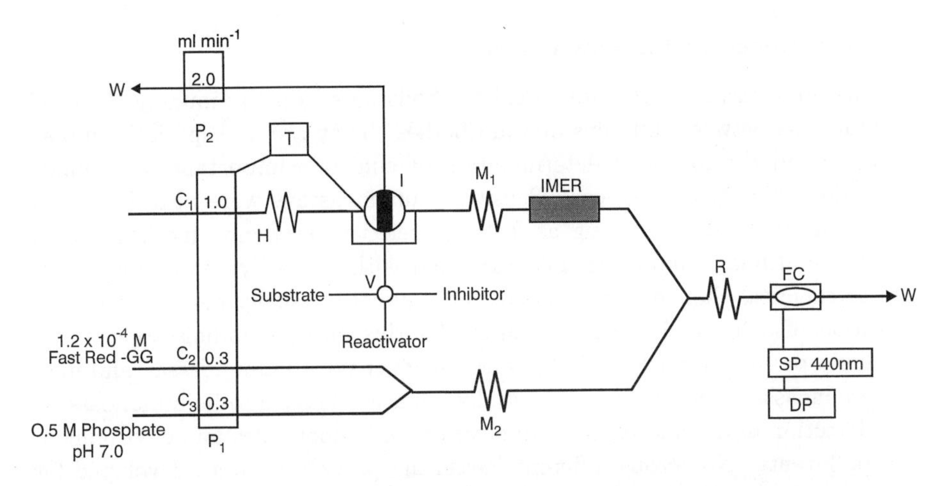

Fig. 6. Flow injection manifold used for the spectrophotometric determination of paraoxon based on the inhibition of immobilised acetylcholinesterase [167]. P — peristaltic pump; I — rotary injection valve; V — selector valve; M — mixing coils; R — indicator reaction coil; H — thermostatic coil; IMER — immobilised enzyme reactor; FC — flow cell; SP — spectrophotometer; DP — data processor; T — timer; W — waste. (*Reprinted by permission of copyright owner.*)

[81] was used and detection was based on direct oxidation of the reaction products thiocholine and 4-aminophenol, respectively. An amperometric biosensor for FIA detection of pesticides was validated with various HPLC techniques in natural waters [171]. It was found that the total amount of pesticide determined with the biosensor can be used as an indicator of the presence of the target pesticides at a level of 3.9–10.3 ppb.

In the procedures based on the inhibition of the enzyme activity, very often a difficult step is a regeneration of the enzyme. A very advantageous solution of this difficulty proposed in FIA determinations of pesticides both with spectrophotometric [172] and with amperometric [173] detections is the immobilisation of acetylcholinesterase on magnetic particles. Using a magnetic reactor with the electromagnet these particles with enzyme can be held in the reactor and then released after assay and loaded with a new portion. The result of determination of pesticides was validated with various HPLC techniques in natural waters [171]. It was found that the total amount of pesticide determined with the biosensor can be used as an indicator of the presence of the target pesticides at a level of 3.9–10.3 ppb.

2. Flow Injection Immunoassays

Immunoassays are the analytical methods based on the immunochemical interactions between antigens and antibodies. They are an especially important tool in the low-level determination of clinically important compounds in various biological matrices. Sensitive immunoassays were also developed for environmental monitoring as fast and selective screening methods [174]. Antigen-antibody binding is not associated with an easily detectable physical or chemical change, hence a detectable signal is usually generated by the addition of a labelled species. Usually for this purpose radioisotopes, fluorophores or enzymes are used. In recent years enzyme immunoassay and fluoroimmunoassay methods have essentially replaced classical radioimmunoassays for detection of trace levels of a wide variety of biomolecules and environmental pollutants. Numerous different measuring procedures were developed for homogeneous and heterogeneous assays, including both competitive and noncompetitive formats [7–10]. Many of them were successfully employed in FIA, providing a valuable alternative to reducing the use of expensive and sophisticated instrumentation [175].

2.1. *Immunoprecipitation*

Interaction between a protein and a specific antibody leads to the formation of large insoluble supramolecular structures, which significantly disperse the light. This behaviour was utilised in the serum immunoglobulin G (IgG) determination with goat anti-human IgG antibody in the merging-zones procedure based on rate turbidimetry [176]. The same principle was involved in flow injection determination of specific proteins in plasma, such as urine albumin, plasma transferrin and plasma haptoglobin [177]. A merging-zone FIA manifold with a stopped-flow procedure was used for the determination of antithrombin III [178] and monoclonal antibodies produced in fermentation of mouse–mouse hybridoma cells [179]. The analyte concentration range measured in reported FIA assays is given in Table 4.

2.2. *Fluoroimmunoassays*

In the fluorescent immunoassay techniques the labelling is a fluorescent marker. In numerous cases this technique provides enhanced sensitivity and shorter reaction times compared to the most commonly used enzyme immunoassay, therefore, it spreads rapidly in clinical analysis for the measurement of hormones, drugs and steroids. Both homogeneous and heterogeneous procedures are used in FIA. In a competitive format the analyte exists in various concentrations in the sample in contact with the specific antibody and competes with an amount of the analyte previously bound to a biological binder (antibody, binding protein, receptor or lectin). In a noncompetitive format, the analyte in the sample reacts proportionally with a biological binder. The FIA fluoroimmunoassay of serum albumin was based on the transfer of excitation energy from the labelled antigen to the labelled antibody and Fluorescein and Rodamine are used as the donor and acceptor labels [180]. The assay results in quenching of the fluorescence of the label bound to the antigen and enhancement of the fluorescence of the label bound to the antibody. A convincing illustration of the advantageous application of FIA conditions for this purpose was given in determination of the drug propanolol, where a static fluorescence analysis required several hours, whereas in a gradient FIA system it took about 5 min [181]. A fluorescent analogue of the analyte acted as a fluorescent probe and its fluorescence was enhanced on binding to orosomucoid; however, this effect is reversed by the presence of the analyte that displaces the fluorescent probe from the protein binding sites. A merging-zones flow injection system was used for a homogeneous fluorophore-linked binding assay of biotin (vitamin H) and biotin-containing compounds widely used in biotechnology and biomedical research [182].

The flow injection fluoroimmunoassay can also be employed in heterogeneous procedures using liposome-based techniques. Liposomes are spherical membrane structures that consist of a phospholipid bilayer, which can entrap water-soluble fluorescent molecules. It was demonstrated that liposomes may provide two orders of magnitude more sensitivity than enzymes in a solid-phase immunoassay [183]. Phospholipid molecules can be derivatised with the antigen and inserted into the membrane of the liposome. The antigen bound to the liposome may compete with the free analyte antigen in the sample for binding sites on the antibodies covalently bound to a solid support. Liposomes which were not bound to antibody sites in the reactor are lysed by surfactant and the released dye is measured in the detector. Liposomes containing carboxyfluorescein were used for the flow injection immunoassay of a clinical analyte theophylline and for the quantitation of anti-theophylline [184].

Non-enzymatic assays can also be performed in FIA systems with chemiluminescence detection. Using acridinium ester-labelled antibodies a flow injection immunoassay for mouse IgG was developed [185]. The light emission was collected directly from the transparent immunoreactor with immobilised anti-mouse IgG. Further details of these assays are given in Table 4.

Table 4. Flow injection immunoassays.

Analyte	Type of assay*	Detection	Concentration range	Reference
α_1-acid glycoprotein	Enz.	Potentiometry	3–15 g/l	198
α-(Difluoromethyl)ornithine	Enz.	Fluorimetry	0.02–2.5 nM	189
α-fetoprotein	Enz.	Amperometry	0.316–100 ppb	201
		Chemiluminescence	2.5–50 ppb	197
Albumin, serum	Flu.	Fluorimetry	10^{-7}–10^{-4} M	180
Anti-theophyline	Flu.(L)	Fluorimetry	0.006–0.4 mM	184
Atrazine	Enz.	Fluorimetry	0.03–1 ppb	194, 195
Biotin	Flu.	Fluorimetry	0.002–4.1 μM	182
Digoxin	Enz.	Amperometry	0.5–5.0 ppb	190
Immunoglobulin G,	Prec.	Turbidimetry	8.9–17.8 g/l	176
human	Enz.	Potentiometry	5–400 ppb	198
mouse	Flu.	Fluorimetry	0.01–2.0 μM	185
	Enz.	Fluorimetry	1.4–25 g/l	186
			1–100 ppm	193
	Enz.	Amperometry	0.81–6 ppt	200
rabbit	Enz.	Spectrophotometry	50–400 ppm	192
Insulin	Enz.	Spectrophotometry	1–250 ppm	199
Insulin, 17-α-hydroxyprogesterone	Enz.	Chemiluminescence	1.5–30 pM	197
Mouse antibovine IgG	Enz.	Chemiluminescence	0.04–0.3 μM	202
Phenytoin	Enz.	Amperometry	2.5–30 ppm	188
Theophyline	Enz.	Amperometry	0.08–20 ppm	191
		Potentiometry	0.025–0.2 μM	199
		Spectrophotometry	2.5–40 ppm	21
	Enz.(L)	Potentiometry	0.2–4000 ppb	203
	Flu.	Fluorimetry	0.5–30 ppm	187
	Flu.(L)	Fluorimetry	0.03–30 μM	184
Thyroxine	Enz.	Chemiluminescence	10^{-11}–6×10^{-11}	196
Valproic acid	Flu.	Fluorimetry	5–140 ppm	187

*Enz. — enzyme immunoassay; Enz.(L) — liposome-based immunoassay; Flu. — fluoroimmunoassay; Flu.(L) — liposome-based fluoroimmunoassay; Prec. — immunoprecipitation

2.3. *Enzyme Immunoassays*

The enzymatic procedures have been the most often used and rapidly growing immunoassays in recent years. They use enzyme labels, which are easy to handle, inexpensive and most often sufficiently stable markers to visualise the antigen–antibody binding reaction. Similarly to fluoroimmunoassays, enzyme immunoassays can also be used in homogeneous and heterogeneous systems, of which the latter one is more widely employed.

Homogeneous enzyme-multiplied immunoassay techniques are commonly used in clinical analysis for the determination of low molecular weight compounds (drugs, hormones). The assay is based on the use of a probe molecule, which show changes of some properties with a progress of antigen–antibody binding. The procedure does not involve any separation steps and can be carried out in FIA systems with various detection methods (Table 4). A homogeneous FIA immunoassay developed for serum IgG was based on the use of peroxidase as the enzyme label and fluorescence detection with H_2O_2 peroxidase catalysed oxidation of diacetyldichlorofluorescein [186]. The binding of the labelled antibody to the antigen inhibits enzyme activity, which results in a decrease in the fluorescence intensity. Fluorimetric detection was also utilised in homogeneous enzyme immunoassays of theophylline and valproic acid [187]. For the determination of theophylline, spectrophotometric detection at 340 nm was also used, where for FIA assay a commercial Syva EMIT kit in a two-reagent merging-zone system was applied [21]. The FIA enzyme immunoassay with amperometric detection was developed for the determination of anticonvulescant drug phenytoin also employing a commercially available EMIT kit [188]. The homogeneous assay was based on the competition of the analyte with phenytoin labelled with glucose-6-phosphate dehydrogenase for a limited amount of the specific antibody. The unbound enzyme-labelled phenytoin converts NAD^+ to NADH, which is detected by amperometry after reaction with 2,6-dichloroindophenol.

The heterogeneous procedures are much more often used in FIA enzyme immunoassays than homogeneous ones. They are based on the competition between the analyte and the enzyme-labelled analyte for a limited number of primary antibody binding sites immobilised on solid particles. A noncompetitive format is much less often used than the competitive one. Such a method was used to assay small haptens, for example an anticancer drug α-(difluoro-methyl) ornitine with fluorimetric detection [189]. In this assay the sample containing the hapten is incubated with an excess peroxidase-labelled antibody and the excess antibody is then separated from the bound antibody by eluting through an antigen-immobilised immunoaffinity column.

In competitive format, the analyte in the sample competes with a constant amount of the analyte immobilised onto the solid phase. In the FIA system with amperometric detection hapten, a steroidal cardiac glycoside digoxin in the sample and digoxin labelled with alkaline phosphatase competed for the solid-phase antibody. After unbound digoxin and labelled digoxin were rinsed from the reactor, the bound labelled digoxin was determined by incubation with the enzyme substrate phenyl phosphate and the produced phenol was detected amperometrically [190]. The same enzyme label and amperometric detection was employed for the immunoassay of theophylline with *p*-amino phenyl phosphate as the enzyme substrate [191].

Peroxidase labelling of the antigen was used in the FIA immunoassay of IgG with UV spectrophotometric detection with phenol, 4-aminoantipyrine and H_2O_2 as substrate mixture [192]. In this case protein A-IgG interactions were used instead of real immunochemical interaction. Several procedures were developed also for enzyme immunoassay with fluorimetric detection. IgG in the concentration range of interest in hybridoma cell fermentation was assayed using an antibody-peroxidase conjugate with membrane or magnetic particle immunoreactors [193]. Immunochemical determinations of trace levels of pesticide are possible in environmental samples using an appropriate hapten conjugate labelled with peroxidase. The pesticide atrazine derivative was used for hapten-albumin production necessary for the immunisation rabbits to obtain antiserum [194, 195]. The polyclonal antiserum used shows significant reactivities with atrazine, propazine and simazine; therefore, determination of the species equivalent of atrazine is possible. A sensitive chemiluminescent immunoassay involving peroxidase-labelled antibodies was developed for thyroxine [196]. A more complex competitive FIA assay with two antibodies and chemiluminescence detection was reported for insulin and 17-α-hydroxyprogesterone [197]. Free and bound fractions present after immune reaction were separated by an immobilised second antibody.

For the enzyme immunoassay of bivalent or polyvalent antigens, highly precise and sensitive sandwich-type procedures can be used. As illustrated in Fig. 7, in this assay the unlabelled antibody is immobilised on a solid support and is exposed to the analyte (valve 2). Then, after injection of an enzyme-antibody conjugate also via valve 2 unbound species are washed away from the reactor by the flow injection carrier buffer and via valve 1 a continuously

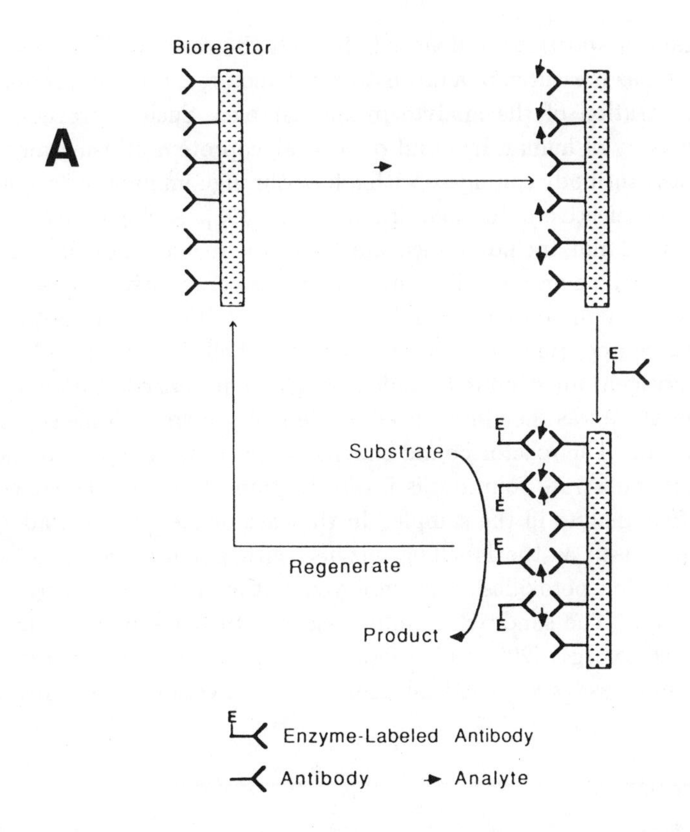

B

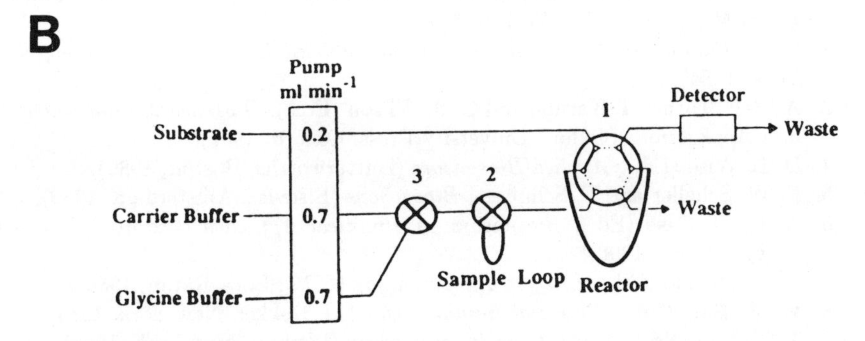

Fig. 7. Schematic diagrams illustrating the principles of the enzyme-linked flow injection sandwich immunoassay system (A) and the flows and arrangement used for the measurements (B) [198]. (*Reprinted by permission of copyright owner.*)

flowing stream of substrate is diverted through the reactor. The amount of bound activity, as measured in a downstream detector, is directly proportional to the concentration of the analyte in the sample. Such a procedure was used for the assay of human IgG and α_1-acid glycoprotein utilising an adenosine deaminase-antibody conjugate with a flow-through immunoreactor and an ammonium ion-selective potentiometric detector [198]. A single protein assay takes less than 12 min, including regeneration of the reactor. For the sandwich assays of theophylline and insulin, a more complex competitive procedure was developed, using immobilised secondary antibodies [199]. The method was based on the competition between the enzyme-labelled antigen and analyte (unlabelled antigen) for a limited number of soluble primary-antibody binding sites. This mixture was then introduced in the FIA system into the secondary-antibody reactor. The reactor bound enzyme activity, as measured by flowing an appropriate substrate solution, is inversely proportional to the concentration of the free analyte in the sample. In the case of theophylline, adenosine deaminase was used as the labelling enzyme with potentiometric detection, whereas for insulin, peroxidase was employed as the label with spectrophotometric detection. The sandwich immunoassays with amperometric detectors were developed for IgG [200] and α-fetoprotein [201]. Other applications of sandwich immunoasays with chemiluminescence detections are indicated in Table 4.

3. References

1. P. W. Carr and L. D. Bowers, *Immobilized Enzymes in Analytical and Clinical Chemistry* (Wiley, New York, 1980).
2. G. G. Guilbault, *Analytical Uses of Immobilized Enzymes* (M. Dekker, New York, 1984).
3. A. P. F. Turner, I. Karube and G. S. Wilson (Eds.), *Biosensors, Fundamentals and Applications* (Oxford University Press, Oxford, 1987).
4. D. L. Wise (Ed.), *Applied Biosensors* (Butterworths, Boston, 1989).
5. F. W. Scheller and F. Schubert, *Biosensors* (Elsevier, Amsterdam, 1989).
6. A. E. G. Case (Ed.), *Biosensors: A Practical Approach* (Oxford University Press, Oxford, 1990).
7. E. T. Maggio (Ed.), *Enzyme Immunoassay* (CRC, Boca Raton, 1980).
8. W. R. Butt (Ed.), *Practical Immunoassay* (M. Dekker, New York, 1984).
9. T. T. Ngo (Ed.), *Nonisotopis Immunoassay* (Plenum, New York, 1988).
10. C. P. Price and D. J. Newman (Eds.), *Principles and Practices of Immunoassay* (M. Stockton, Basingstoke, 1991).
11. J. Ruzicka, H. H. Hansen, A. K. Ghose and H. A. Mottola, *Anal. Chem.* **51** (1979) 199.

12. J. Ruzicka and E. H. Hansen, *Anal. Chim. Acta* **173** (1985) 3.
13. P. J. Worsfold and A. Nabi, *Anal. Chim. Acta* **171** (1985) 333.
14. C. Ridder, E. H. Hansen and J. Ruzicka, *Anal. Lett.* **15** (1982) 1751.
15. B. C. Madsen and M. S. Kromis, *Anal. Chem.* **56** (1984) 2849.
16. I. Schneider, *Anal. Chim. Acta* **166** (1984) 293.
17. S. Thahir, M. Situmorang and P. W. Alexander, *Proc. 4th Environmental Chemistry Conference*, Darwin, Australia, 1995, A012-1.
18. J. Toei, *Analyst* **113** (1988) 475.
19. P. J. Worsfold, J. Ruzicka and E. H. Hansen, *Analyst* **106** (1981) 1309.
20. C. De Maria, T. M. Moñoz, A. A. Mateos and L. G. De Maria, *Anal. Chim. Acta* **247** (1991) 61.
21. B. F. Rocks, R. A. Sherwood and C. Riley, *Analyst* **109** (1984) 847.
22. X. Wen, H. K. J. Powell, G. D. Christian and J. Ruzicka, *Anal. Chim. Acta* **249** (1991) 451.
23. H. Huang, R. Cai, Y. Du and Y. Zeng, *Anal. Chim. Acta* **309** (1995) 271.

23a. Ch. M. Wolff and H. A. Mottola, *Anal. Chem.* **50** (1978) 94.

23b. D. P. Nikolelis and H. A. Mottola, *Anal. Chem.* **50** (1978) 1665.

24. D. Pilosof and T. A. Nieman, *Anal. Chem.* **54** (1982) 1698.
25. C. A. Koerner and T. A. Nieman, *Anal. Chem.* **58** (1986) 116.
26. D. Narinesingh, R. Mungal and T. T. Ngo, *Anal. Chim. Acta* **292** (1994) 185.
27. J. Ruz, F. Lazaro and M. D. Luque de Castro, *J. Autom. Chem.* **10** (1988) 15.
28. E. H. Hansen, *Anal. Chim. Acta* **216** (1989) 257.
29. L. Gorton, E. Csöregi, J. Emneus, G. Jönsson-Pettersson, G. Marko-Varga and B. Persson, *Anal. Chim. Acta* **250** (1991) 203.
30. A. M. Almuibed and A. Townshend, *Anal. Chim. Acta* **198** (1987) 37.
31. Y. Shan, I. D. McKelvie and B. T. Hart, *Anal. Chem.* **65** (1993) 3053.
32. D. R. Walt and V. I. Agayn, *Trends in Anal. Chem.* **13** (1994) 425.
33. R. Guanasekaran and H. A. Mottola, *Anal. Proc.* **57** (1985) 1005.
34. J. Emneus and L. Gorton, *Anal. Chim. Acta* **276** (1993) 303.
35. Y. Murakami, T. Takeuchi, K. Yokoyama, E. Tamiya, I. Karube and M. Suda, *Anal. Chem.* **65** (1993) 2731.
36. T. Yao, R. Akasaka and T. Wasa, *Electroanalysis* **1** (1989) 413.
37. T. Yao, N. Kobayashi and T. Wasa, *Electroanalysis* **2** (1990) 563.
38. K. P. Ang, H. Gunasingham, B. T. Tay, V. S. Herath, P. Y. T. Teo, P. X. Thiak, B. Kuah and K. L. Tan, *Analyst* **112** (1987) 1433.
39. K. Matsumoto, H. Matsubara, H. Ukeda and Y. Osajima, *Agric. Biol. Chem.* **53** (1989) 2347.
40. G. M. Greenway and P. Ongoma, *Analyst* **115** (1990) 1297.
41. A. Alonso, M. J. Almendral, M. D. Baez, M. J. Porras and C. Alonso, *Anal. Chim. Acta* **308** (1995) 164.
42. M. J. Almendral, M. J. Porras, A. Alonso, M. D. Baez and C. Alonso, *Anal. Chim. Acta* **308** (1995) 170.
43. L. Braco, J. A. Daros and M. de la Guardia, *Anal. Chem.* **64** (1992) 129.

44. P. Solich, M. Polasek, R. Karlicek, O. Valentova and M. Marek, *Anal. Chim. Acta* **218** (1989) 151.
45. K. Kashiwabara, T. Hobo, E. Kobayashi and S. Suzuki, *Anal. Chim. Acta* **178** (1985) 209.
46. N. Kiba, Y. Inoue and M. Furusawa, *Anal. Chim. Acta* **243** (1991) 183.
47. D. Narinesingh, V. A. Stoute, G. Davis, F. Shaama and T. T. Ngo, *Anal. Lett.* **24** (1991) 727.
48. C. Macca, L. Solda and G. Palma, *Anal. Lett.* **28** (1995) 1735.
49. P. J. Worsfold, J. Farrelly and M. S. Matharu, *Anal. Chim. Acta* **164** (1984) 103.
50. R. Q. Thompson, H. Kim and C. E. Miller, *Anal. Chim. Acta* **198** (1987) 165.
51. B. A. Petersson, *Anal. Lett.* **22** (1989) 83.
52. M. C. Gosnell, R. E. Snelling and H. A. Mottola, *Anal. Chem.* **58** (1986) 1585.
53. K. Matsumoto, J. J. B. Baeza and H. A. Mottola, *Anal. Chem.* **65** (1993) 636.
54. M. O. Rezende and H. A. Mottola, *Analyst.* **119** (1994) 2093.
55. T. D. Yeridan, G. D. Christian and J. Ruzicka, *Anal. Chim. Acta* **204** (1988) 7.
56. T. D. Yeridan, G. D. Christian and J. Ruzicka, *Anal. Chem.* **60** (1988) 1250.
57. P. Linares, M. D. Luque de Castro and M. Valcarcel, *Anal. Chim. Acta* **230** (1990) 199.
58. J. M. Fernandez-Romero and M. D. Luque de Castro, *Anal. Chem.* **65** (1993) 3048.
59. M. D. Luque de Castro and J. M. Fernandez-Romero, *Anal. Chim. Acta* **311** (1995) 281.
60. S. Cliffe, C. Filippini, M. Schneider and M. Fawer, *Anal. Chim. Acta* **256** (1992) 53.
61. T. L. Spinks and G. E. Pacey, *Anal. Chim. Acta* **237** (1990) 503.
62. F. Preuschoff, U. Spohn, G. Blankenstein, K. H. Mohr and M. R. Kula, *Fresenius J. Anal. Chem.* **346** (1993) 924.
63. F. Winquist, I. Lundström and B. Danielsson, *Anal. Chem.* **58** (1986) 145.
64. M. T. Jeppesen and E. H. Hansen, *Anal. Chim. Acta* **214** (1988) 147.
65. C. S. Rui and Y. Kato, *Biosens. Bioelectron.* **9** (1994) 429.
66. J. M. Fernandez-Romero, M. D. Luque de Castro and R. Quiles-Zafra, *Anal. Chim. Acta* **308** (1995) 178.
67. K. Matsumoto, O. Hamada, H. Ukeda and Y. Osajima, *Anal. Chem.* **58** (1986) 2732.
68. A. A. Malinauskas and J. J. Kulys, *Biotechnol. Bioeng.* **21** (1979) 513.
69. T. Yao, N. Kobayashi and T. Wasa, *Anal. Chim. Acta* **238** (1990) 339.
70. Y. Hasebe, K. Takamori and S. Uchiyama, *Anal. Chim. Acta* **282** (1993) 363.
71. W. Matuszewski, S. A. Rosario and M. E. Meyerhoff, *Anal. Chem.* **63** (1991) 1906.
72. W. Matuszewski, M. Trojanowicz, M. E. Meyerhoff, A. Moszczynska and E. Lange-Moroz, *Electroanalysis* **5** (1993) 113.
73. A. M. Almuaibed and A. Townshend, *Anal. Chim. Acta* **214** (1988) 161.

74. B. A. Petersson, E. H. Hansen and J. Ruzicka, *Anal. Lett.* **19** (1986) 649.
74a. I. Satoh and T. Ishii, *Anal. Chim. Acta* **214** (1988) 409.
75. M. Alvarez-Icaza and U. Bilitewski, *Anal. Chem.* **65** (1993) 525A.
76. M. Thompson and U. J. Krull, *Anal. Chem.* **63** (1991) 393A.
77. B. Olsson, H. Lundbäck, G. Johansson, F. Scheller and J. Neutwig, *Anal. Chem.* **58** (1986) 1046.
78. G. J. Moody, G. S. Sanghera and J. D. R. Thomas, *Analyst* **111** (1986) 605.
79. J. A. Hamid, G. J. Moody and J. D. R. Thomas, *Analyst* **115** (1990) 1289.
80. S. K. Beh, G. J. Moody and J. D. R. Thomas, *Analyst* **116** (1991) 459.
81. C. La Rosa, F. Pariente, L. Hernandez and E. Lorenzo, *Anal. Chim. Acta* **308** (1995) 129.
82. W. Matuszewski, M. Trojanowicz and A. Lewenstam, *Electroanalysis* **2** (1990) 607.
83. B. A. Petersson, *Anal. Chim. Acta* **209** (1988) 231.
84. P. D. Sanchez, P. T. Blanco, J. M. F. Alvarez, M. R. Smyth and R. O'Kennedy, *Electroanalysis* **2** (1990) 303.
85. A. I. Yaropolov, A. N. Kharybin, J. Emneus, G. Marko-Varga and L. Gorton, *Anal. Chim. Acta* **308** (1995) 137.
86. T. Ruzgas, J. Emneus, L. Gorton and G. Marko-Varga, *Anal. Chim. Acta* **311** (1995) 245.
87. H. Gunasingham and C. H. Tan, *Analyst* **115** (1990) 35.
88. T. Buch-Rasmussen, *Anal. Chim. Acta* **237** (1990) 405.
89. F. Ortego, E. Dominguez, G. Jönsson-Pettersson and L. Gorton, *J. Biotechnol.* **31** (1993) 289.
90. J. Wang, R. Li and M. S. Lin, *Electroanalysis* **1** (1989) 151.
91. Q. Chi and S. Dong, *Anal. Chim. Acta* **278** (1993) 17.
92. M. Trojanowicz, W. Matuszewski and M. Podsiadla, *Biosens. Bioelectron.* **5** (1990) 149.
93. W. Schuhmann and R. Kittsteiner-Eberle, *Biosens. Bioelectron.* **6** (1991) 263.
94. D. Centonze, A. Guerrieri, C. Malitesta, F. Palmisano and P. G. Zambonin, *Ann. Chim. (Rome)* **83** (1992) 219.
95. F. Palmisano, D. Centonze and P. G. Zambonin, *Biosens. Bioelectron.* **9** (1994) 471.
96. T. Krawczynski vel Krawczyk, M. Trojanowicz, A. Lewenstam and A. Moszczynska, *Biosens. Bioelectron.* **11** (1996) 1155.
97. W. Matuszewski and M. Trojanowicz, *Analyst* **113** (1988) 735.
98. A. Amine, M. Alafaudy, J. M. Kauffmann and M. Novak Pekli, *Anal. Chem.* **67** (1995) 2822.
99. G. S. Cha and M. E. Meyerhoff, *Talanta* **36** (1989) 271.
100. S. A. Rosario, G. S. Cha, M. E. Meyerhoff and M. Trojanowicz, *Anal. Chem.* **62** (1990) 2418.
101. M. Carslen, C. Johansen, R. W. Min, J. Nielsen, H. Meier and F. Lantreibecq, *Anal. Chim. Acta* **279** (1993) 51.
102. M. Nishizawa, T. Matsue and I. Uchida, *Anal. Chem.* **64** (1992) 2642.

103. B. A. Petersson, *Anal. Chim. Acta* **209** (1988) 239.
104. U. Brand, B. Reinhard, F. Rüther, T. Scheper and K. Schügerl, *Anal. Chim. Acta* **238** (1990) 201.
105. B. A. A. Dremel, W. Yang and R. D. Schmid, *Anal. Chim. Acta* **234** (1990) 107.
106. Th. Scheper, W. Brandes, H. Maschke, F. Plötz and C. Müller, *J. Biotechnol.* **31** (1993) 345.
107. A. Carpenter and W. C. Purdy, *Anal. Lett.* **23** (1990) 425.
108. W. Künnecke and R. D. Schmid, *Anal. Chim. Acta* **234** (1990) 213.
109. M. Masoon and A. Townshend, *Anal. Chim. Acta* **166** (1984) 111.
110. L. Gorton and L. Ögren, *Anal. Chim. Acta* **130** (1981) 45.
111. W. Matuszewski, M. Trojanowicz and M. E. Meyerhoff, *Electroanalysis* **2** (1990) 525.
112. K. Matsumoto, K. Sakoda and Y. Osajima, *Anal. Chim. Acta* **261** (1992) 155.
113. S. A. Rosario, M. E. Meyerhoff and M. Trojanowicz, *Anal. Chim. Acta* **258** (1992) 281.
114. Y. L. Huang, S. B. Khoo and M. G. S. Yap, *Anal. Lett.* **28** (1995) 593.
115. G. Marazza, A. Cagnini and M. Mascini, *Electroanalysis* **6** (1994) 221.
116. T. Yao, M. Satomura and T. Nakahara, *Electroanalysis* **7** (1995) 395.
117. M. Trojanowicz, W. Matuszewski and T. Krawczynski vel Krawczyk, *J. Flow Injection Anal.* **10** (1993) 207.
118. A. M. Almuaibed and A. Townshend, *Anal. Chim. Acta* **218** (1989) 1.
119. K. B. Male and J. H. T. Luong, *Biosens. Bioelectron.* **6** (1991) 581.
120. D. Taylor and T. A. Nieman, *Anal. Chim. Acta* **186** (1986) 91.
121. M. Trojanowicz, W. Matuszewski, B. Szczepanczyk and A. Lewenstam, in *Uses of Immobilized Biological Compounds*, eds. G. G. Guilbault and M. Mascini (Kluwer, Dordrecht, 1993).
122. T. Krawczynski vel Krawczyk, M. Trojanowicz and A. Lewenstam, *Talanta* **41** (1994) 1229.
123. A. Krug, R. Göbel and R. Kellner, *Anal. Chim. Acta* **287** (1994) 59.
124. G. Maeder, J. L. Veuthey, M. Pelletier and W. Haerdi, *Anal. Chim. Acta* **231** (1990) 115.
125. M. Akiba and S. Motomizu, *Anal. Chim. Acta* **214** (1988) 455.
126. B. Olsson, *Mikrochim. Acta* (1985) II 211.
127. J. A. Infantes, M. D. Luque de Castro and M. Valcarcel, *Anal. Chim. Acta* **242** (1991) 179.
128. B. Olsson, B. Ståalbom and G. Johansson, *Anal. Chim. Acta* **179** (1986) 203.
129. J. Ruz, M. D. Luque de Castro and M. Valcarcel, *Microchem. J.* **36** (1987) 316.
130. R. Puchades, A. Maquieira and L. Torro, *Analyst* **118** (1993) 855.
131. C. G. de Maria, T. Manzano, R. Duarte and A. Alonso, *Anal. Chim. Acta* **309** (1995) 241.
132. G. C. Chemnitius and R. D. Schmid, *Anal. Lett.* **22** (1989) 2897.
133. N. Kiba, M. Ogi and M. Furusawa, *Anal. Chim. Acta* **224** (1989) 133.

134. M. E. Meyerhoff, M. Trojanowicz and B. O. Palsson, *Biotech. Bioeng.* **41** (1993) 964.
135. B. O. Palsson, B. Q. Shen, M. E. Meyerhoff and M. Trojanowicz, *Analyst* **118** (1993) 1361.
136. T. Yao, M. Satomura and T. Nakahara, *Electroanalysis* **7** (1995) 143.
137. K. Schügerl, L. Brandes, T. Dullau, K. Holzhauer-Rieger, S. Hotop, U. Hübner, X. Wu and W. Zhou, *Anal. Chim. Acta* **249** (1991) 87.
138. K. Schügerl, *J. Biotechnol.* **31** (1993) 241.
139. M. Masoom and A. Townshend, *Anal. Proc.* **22** (1985) 6.
140. F. Lazaro, M. D. Luque de Castro and M. Valcarcel, *Anal. Chem.* **59** (1987) 1859.
141. T. Yao, M. Satomura and T. Wasa, *Anal. Chim. Acta* **261** (1992) 161.
142. T. Yao, M. Satomura and T. Nakahara, *Talanta* **41** (1994) 2113.
143. W. Matuszewski, M. Trojanowicz and L. Ilcheva, *Electroanalysis* **2** (1990) 147.
144. L. C. Chen and K. Matsumoto, *Anal. Chim. Acta* **308** (1995) 145.
145. M. Agudo, A. Rios and M. Valcarcel, *Anal. Chim. Acta* **308** (1995) 77.
146. A. Macquieira, M. D. Luque de Castro and M. Valcarcel, *Microchem. J.* **36** (1987) 309.
147. X. Zhang and G. A. Rechnitz, *Electroanalysis* **6** (1994) 361.
148. T. Kullick, M. Beyer, J. Henning, T. Lerch, R. Quack, A. Zetz, B. Hitzman, T. Scheper and K. Schügerl, *Anal. Chim. Acta* **296** (1994) 263.
149. K. K. Stewart, G. R. Beecher and P. E. Hare, *Anal. Biochem.* **70** (1976) 167.
150. M. Masoom and A. Townshend, *Anal. Chim. Acta* **179** (1986) 399.
151. A. Nabi and P. J. Worsfold, *Analyst* **111** (1986) 531.
152. S. Olsen, J. Ruzicka and E. H. Hansen, *Anal. Chim. Acta* **136** (1982) 101.
153. T. Gübeli, J. Ruzicka and G. D. Christian, *Talanta* **38** (1991) 851.
154. J. A. Osborn, A. M. Yacynych and D. C. Roberts, *Anal. Chim. Acta* **183** (1986) 287.
155. J. Marcos, A. Rios and M. Valcarcel, *Anal. Chim. Acta* **308** (1995) 152.
156. M. K. Jain, *Handbook of Enzyme Inhibitors* (Wiley, New York, 1982).
157. H. Zollner, *Handbook of Enzyme Inhibitors* (VCH, Weinheim, 1989).
158. L. Ogren and G. Johansson, *Anal. Chim. Acta* **96** (1976) 1.
159. J. Marcos and A. Townshend, *Anal. Chim. Acta* **299** (1994) 129.
160. M. Sanchez-Cabezudo, J. M. Fernanez-Romero and M. D. Luque de Castro, *Anal. Chim. Acta* **308** (1995) 159.
161. C. Tran-Minh and J. Beaux, *Anal. Chem.* **51** (1979) 91.
162. F. Schubert, R. Renneberg, F. W. Scheller and L. Kirstein, *Anal. Chem.* **56** (1984) 1677.
163. C. Tran-Minh, P. C. Pandey and S. Kumuran, *Biosens. Bioelectron.* **5** (1990) 461.
164. J. Marcos and A. Townshend, *Anal. Chim. Acta* **310** (1995) 173.
165. M. E. Leon-Gonzalez and A. Townshed, *Anal. Chim. Acta* **236** (1990) 267.
166. I. A. Takruni, A. M. Almuibed and A. Townshend, *Anal. Chim. Acta* **282** (1993) 307.

167. C. G. de Maria, T. M. Muñoz and A. Townshend, *Anal. Chim. Acta* **295** (1994) 287.
168. M. Trojanowicz and M. L. Hitchaman, *Trends Anal. Chem.* **15** (1996) 38.
169. S. Kumaran and C. Tran-Minh, *Anal. Biochem.* **200** (1992) 187.
170. C. Tran-Minh, *Anal. Proc.* **30** (1993) 73.
171. J. L. Marty, N. Mionetto, S. Lacorte and D. Barcelo, *Anal. Chim. Acta* **311** (1995) 265.
172. R. Kindervater, W. Künnecke and R. D. Schmid, *Anal. Chim. Acta* **234** (1990) 113.
173. A. Günther and U. Bilitewski, *Anal. Chim. Acta* **300** (1995) 117.
174. M. Vanderlaan, B. E. Watkins and L. Stanker, *Environ. Sci. Technol.* **26** (1988) 247.
175. R. Puchades, A. Maquieira, J. Atienza and A. Montoya, *Crit. Rev. Anal. Chem.* **23** (1992) 301.
176. P. J. Worsfold, A. Hughes and D. J. Mowthorpe, *Analyst* **110** (1985) 1303.
177. J. Andersen, *J. Autom. Chem.* **12** (1990) 53.
178. R. Freitag, T. Scheper and K. Schügerl, *Enzyme Microb. Technol.* **13** (1991) 969.
179. R. Freitag, C. Fenge, T. Scheper, K. Schügerl, K. Spreinat and C. Antranikian, *Anal. Chim. Acta* **249** (1991) 113.
180. C. S. Lim, J. N. Miller and J. W. Bridges, *Anal. Chim. Acta* **114** (1980) 183.
181. J. R. Miller, *Anal. Proc.* **26** (1989) 317.
182. T. Smith-Palmer, M. S. Barbarakis, T. Cynkowski and L. G. Bachas, *Anal. Chim. Acta* **279** (1993) 287.
183. A. L. Plant, M. V. Brizgys, L. Locascio-Brown and R. A. Durst, *Anal. Biochem.* **176** (1989) 420.
184. L. Locascio-Brown, A. L. Plant, V. Horvath and R. A. Durst, *Anal. Chem.* **62** (1990) 2587.
185. C. Shellum and G. Gübitz, *Anal. Chim. Acta* **227** (1989) 97.
186. T. A. Kelly and G. D. Christian, *Talanta* **29** (1982) 1109.
187. P. Allain, A. Turcant and A. Premel-Cabic, *Clin. Chem.* **35** (1989) 469.
188. H. T. Tang, H. B. Halsall and W. R. Heineman, *Clin. Chem.* **37** (1991) 245.
189. P. C. Gunaratna and G. S. Wilson, *Anal. Chem.* **65** (1993) 1152.
190. K. P. Wehmeyer, H. B. Halsall, W. R. Heineman, C. P. Volle and I. W. Chen, *Anal. Chem.* **58** (1986) 135.
191. E. P. Gil, H. T. Tang, H. B. Halsall, W. R. Heineman and A. S. Misiego, *Clin. Chem.* **36** (1990) 662.
192. M. Nilsson, H. Håakanson and B. Mattiasson, *Anal. Chim. Acta* **249** (1991) 163.
193. W. Stöcklein and R. D. Schmid, *Anal. Chim. Acta* **234** (1990) 83.
194. P. Krämer and R. D. Schmid, *Biosens. Bioelectron.* **6** (1991) 239.
195. C. Wittmann and R. D. Schmid, *Sens. Actuators B* **15–16** (1993) 119.
196. A. A. Arefyev, S. B. Vlasenko, S. A. Eremin, A. P. Osipov and A. M. Egorov, *Anal. Chim. Acta* **237** (1990) 285.

197. M. Maeda and A. Tsuji, *Anal. Chim. Acta* **167** (1985) 241.
198. I. H. Lee and M. E. Meyerhoff, *Anal. Chim. Acta* **229** (1990) 47.
199. I. H. Lee and M. E. Meyerhoff, *Mikrochim. Acta* (1988) III, 207.
200. Y. Xu, H. B. Halsall and W. R. Heineman, *J. Pharm. Biomed. Anal.* **7** (1989) 1301.
201. Y. Xu, H. B. Halsall and W. R. Heineman, *Clin. Chem.* **36** (1990) 1941.
202. H. Liu, J. C. Yu, D. S. Bindra, R. C. Gives and G. S. Wilson, *Anal. Chem.* **63** (1991) 666.
203. T. G. Wu and R. A. Durst, *Mikrochim. Acta* (1990) 187.
204. W. Brandes, H. E. Maschke and T. Scheper, *Anal. Chem.* **65** (1993) 3368.

Chapter 5

Other Detection Methods Used in FIA

Apart from the numerous applications with common optical and electrochemical detection methods, which were discussed in previous chapters, the attractive methodology of FIA can be adapted well to other unique instrumental setups. As emphasised earlier, the main advantages of these adaptations are the ability to eliminate troublesome manual procedures, for sample transport and processing, and the ability to operate with small sample volumes. Reported developments have often involved rather preliminary approaches without strict analytical application, as in the case of the optimisation of radiometric detection for FIA [1]. However, they do add to the variety of mechanised instrumentation necessary for chemical analysis. They create a starting point for the design of new analytical instruments and the development of new, competitive analytical procedures.

1. Photoacoustic Spectroscopic Detection

When a gaseous or liquid medium absorbs radiation with a wavelength corresponding to an absorption maximum of the analyte molecule, it causes periodic heating of the medium, which results in a regular increase in temperature and the generation of an acoustic wave. A pulsed, tunable laser with a suitable wavelength and repetition rate is considered the best source of irradiation in order to observe the photoacoustic effect. The wave-front of the acoustic wave generated is cylindrically coaxial to the laser beam. An acoustic transducer (a sensitive microphone) is set perpendicular to the laser beam direction. The amplitude of the compression and the rarefaction pulse of the acoustic wave are proportional to the extent of absorption and to the energy of the laser beam. Radiation reflected or scattered by the sample has no effect on the acoustic signal.

The use of this type of detection in FIA systems has been developed for the determination of several inorganic species, such as nitrate, nitrite, iron, ammonia and phosphate in natural waters, under chemical conditions like those used for standard absorptive spectrophotometric detection [2]. These determinations are performed in a conventional Helma flow cell with a laser source selected according to the spectral characteristics of the particular species. A piezoelectric transducer is coupled to the bottom of the cell in order to detect a fraction of the acoustic wave generated inside the cuvette. The performance of the system depends on numerous factors, including the repetition rate of the laser source, the laser excitation wavelength and the geometry of the cuvette-transducer system. In the developed system, improved detection limits are obtained for iron and ammonia when compared with absorptive FIA spectrophotometry. An advantage of photoacoustic spectroscopic detection is its insensitivity to the matching of the refractive indices of the carrier and the standards.

2. Flow Injection Inductively Coupled Plasma Mass Spectrometry

Mass spectrometry is based on the conversion of the analyte into gaseous ions and resolving them on the basis of their mass-to-charge ratios. The sample molecules of the sample are ionised and fragmented by collision with streams of electrons, ions, fast atoms, or photons. Ionisation can also be achieved thermally, by a high electrical potential or in an inductively coupled plasma. The latter systems (ICP-MS) have developed rapidly since the middle of the 1980's [3] and they have already found numerous applications in combination with flow injection sample processing systems [4]. Although ICP-MS is considered a very powerful technique for trace multielement and isotopic analysis, it still has some limitations, such as a low tolerance to dissolved solids and the spectral overlapping of some polyatomic ions with certain analyte ions. Another limitation arises when concomitant elements cause nonspectroscopic interferences that result in a suppression of analyte signals.

Examples of developed FIA systems with ICP-MS are shown in Table 1. The use of on-line preconcentration on solid sorbents provides detection limits at the sub-ppb level [6, 15, 18]. Especial low detectability ($\approx 10^{-15}$ M) has been reported for determinations of gold with off-line preconcentration from 4 l samples of the cyanide complex on an anion-exchanger [5]. A decrease in the detection limit and a reduction of spectral interferences can be achieved by the generation of hydrides and their introduction into the nebuliser [8, 9, 13].

Table 1. Applications of flow injection inductively coupled plasma mass spectrometry.

Analyte	On-line analyte processing	Detection limit, ppb	Reference
Au	Preconcentration as $Au(CN)_2^-$ (off-line)	10 fM	5
	Preconcentration on sulphydryl cotton fibre	0.00019	6
Au, Cu, Zn		0.2	7
As	Hydride generation	0.003	8
Bi	Hydride generation	1.8	9
Bi, Pb, Tl		1.0	10
Cd, Cu		0.54 (Cd); 0.027 (Cu)	11
Ir, Pb, Re	Preconcentration as anions	5 pg (Re); 6 pg (Ir); 14 pg (Pt)	12
Organomercury		2.7	16
Pb	Hydride generation	40	13
Pt		0.1	14
Re	Preconcentration as ReO_4^-	0.00027	15
Th, U		0.2	17
Multielement	Preconcentration on immobilised 8-quinolinol	0.006 (Cd);0.05 (Co); 0.2 (Cu); 0.01 (Mn); 0.03 (Pb)	18
		1–2	19

In the determination of arsenic with hydride generation the replacement of hydrochloric acid with L-cysteine for the reduction of arsenic to As(III) reduced $ArCl^+$ molecular interference, significantly [8]. In the determination of 51 elements in highly concentrated solutions of phosphoric acid, sodium phosphate and sodium nitrate, the influence of matrix-induced polyatomic species has been investigated by calculating all possible polyatomic species present in the argon plasma and by determining their interference equivalent concentrations using a synthetic sample [19]. The detection limits in the actual matrix were in the order of 1–2 ppb; determinations were performed using a standard addition method. Many procedures listed in Table 1 have been used for water analysis [5, 6, 8, 15, 18], but FIA systems combined with ICP-MS have also been applied in the trace analysis of serum [7], urine [11], biological [16] and geological [9, 13] materials, and in metals and alloys [10, 17].

3. FIA in the Gaseous Phase

FIA is commonly performed in the measuring system with a liquid carrier which transports the liquid sample through the manifold to the flow-through detector. The sample can undergo various on-line operations, either homogeneous or heterogeneous. The appropriate design of an FIA system with reproducible dispersion conditions and reproducible sample processing operations enables precise and accurate analysis with small sample volumes. A similar system can be designed using a gaseous carrier for the injection of gaseous samples, and by employing detection devices that operate in the gas phase. Such systems have been developed with piezoelectric detectors, semiconductor gas sensors and an electron-capture detector.

Piezoelectric detection of gaseous species is based on measurement of a change in the oscillation frequency of a quartz crystal covered with an appropriate coating material. The frequency changes as a result of the absorption of the analyte by the coating material, causing an increase in the mass of the coated crystal. The most important step in development of a piezoelectric sensor for a given species is to find a suitable nonvolatile coating material which ensures the reversible and selective sorption of the molecules to be detected. Piezoelectric determinations of gaseous species are usually carried out by continuous aspiration of the gaseous sample and measurement of the new frequency of oscillation of the quartz crystal. Determinations of ammonia and sulphur dioxide have also been reported using an FIA system in which 10 ml of the gaseous sample is injected by septum valve into a stream of purified air or nitrogen with a flow rate of 50 ml min^{-1} [20, 21]. In the flow cell used the coating material is deposited on both electrodes of the quartz crystal and then the two surfaces of the crystal are simultaneously rinsed by the carrier gas segment of the injected sample. It has been shown that the concentration range in an FIA system is much broader than in a system with constant aspiration of the sample and then desorption of the analyte from the coating material layer [21]. In a determination of ammonia, with pyridoxine hydrochloride used as a coating material, a linear detector response was found over the range of 1 ppb to 103 ppm, with a sensor lifetime of up to four months. For sulphur dioxide detection, the best results are obtained for N,N,N′,N′-tetrakis(2-hydroxypropyl)ethylenediamine (EDTP) as coating material, with a linear response from 0.1 to 103 ppm. A stable response was observed for at least two months. Fig. 1 shows examples of FIA signals recorded in these measurements.

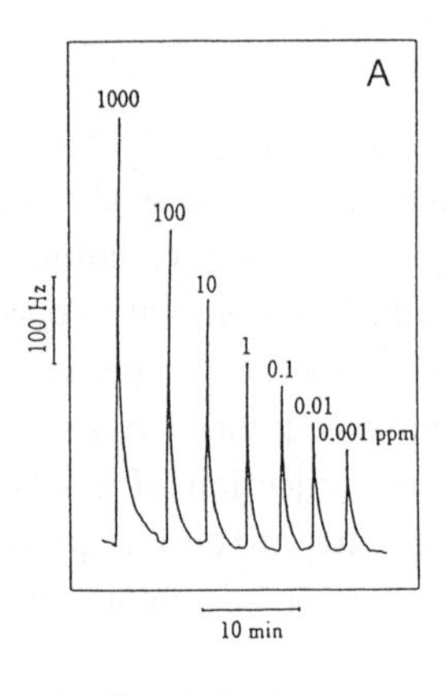

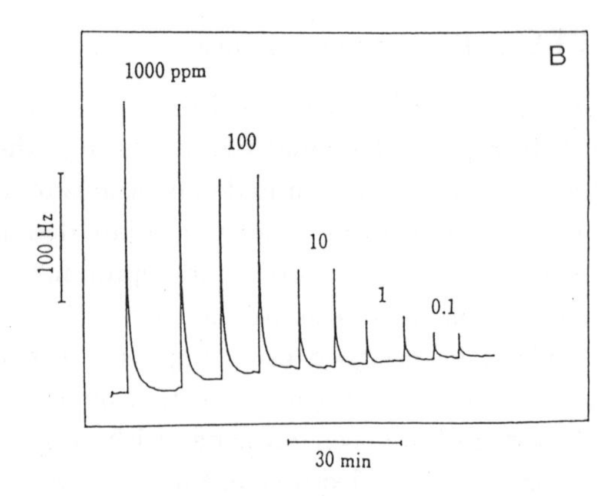

Fig. 1. Flow injection response obtained in a gaseous phase system with piezoelectric detection for (A) ammonia with pyridoxine hydrochloride coating and (B) for sulphur dioxide, with EDTP coating [21]. Air was used as the carrier gas with a flow rate of 50 ml min^{-1}. Injection sample volume was 10 ml. (*Reprinted by permission of copyright owner.*)

Another type of detector for gaseous analytes which can be employed in flow injection measurements of the gas phase are semiconductor tin oxide gas sensors. They are known to respond to various gases and vapours, including hydrogen, carbon monoxide, methane, alcohols and hydrocarbons. The response of the tin oxide semiconductor is dependent on chemisorbed oxygen on the surface reacting with reducing gases or vapours on the oxide layer of the semiconductor. The electrical resistance due to the adsorption of the analyte on the surface of the sensor is inversely proportional to the concentration of the analyte [22]. Two different commercial tin oxide sensors have been used in the design of an FIA system for monitoring alcohol concentrations in liquid samples [23]. This detection method is based on the head-space analysis of vapour above aqueous samples containing ethanol. The system operates with an air carrier stream pumped through a twin sensor cell. The sample inlet tube is introduced into the head-space of the sample container, and vapour samples are pumped into the flow-through detector for a fixed period of time, usually 5 s. Both sensors show rapid response with peak widths of approximately 30 s at a 1 l min^{-1} flow rate. Such detection can be used for an ethanol solution concentration ranging from 0.1 to 20%. The results for ethanol determination in commercial liquors show good correlation with gas chromatography analysis.

The electron-capture detector is one of the most widely used detectors in gas chromatography. Its operation is based on the ionisation of the carrier gas and the measurement of ionisation current, which decreases in the presence of molecules that tend to capture electrons. Its response is very sensitive for molecules containing electronegative functional groups such as halogens, peroxides, quinones and nitro groups. This kind of detector has been applied in the determination of volatile analytes in an FIA system with a two-phase separation using PTFE membranes [24]. In the separator, the gas diffusion of the volatile analyte from the sample to the gaseous nitrogen carrier takes place and thus the sample is transported to the detector. It has been shown that by the use of various reagents to release different volatile products, chlorine, sulphur dioxide and hydrogen cyanide can be determined. In the determination of chlorine with a sample volume of 100 μl, a detection limit of 50 ppb can be obtained.

4. References

1. K. Grudpan, D. Nacapricha and Y. Wattanakanjana, *Anal. Chim. Acta* **246** (1991) 325.
2. I. Carrer, P. Cusmai, E. Zanzottera, W. Martinotti and F. Realini, *Anal. Chim. Acta* **308** (1995) 20.
3. R. S. Houk, *Anal. Chem.* **58** (1986) 97A.
4. M. D. Luque de Castro and M. T. Tena, *Talanta* **42** (1995) 151.
5. K. K. Falkner and J. M. Edmond, *Anal. Chem.* **62** (1990) 1477.
6. M. M. G. Gomez and C. W. McLeod, *J. Anal. At. Spectrom.* **10** (1995) 89.
7. S. G. Matz, R. C. Elder and K. Tepperman, *J. Anal. At. Spectrom.* **4** (1989) 767.
8. M. F. Huang, S. J. Jiang and C. J. Hwang, *J. Anal. At. Spectrom.* **10** (1995) 31.
9. T. Akagi, T. Hirata and A. Masuda, *Anal. Sci.* **6** (1990) 397.
10. T. Mochizuki, A. Sakashita, H. Iwata, Y. Ishibashi and N. Gunji, *Anal. Sci.* **6** (1990) 191.
11. J. R. Pretty, E. A. Blubaugh, E. H. Evans, J. A. Caruso and T. M. Davidson, *J. Anal. At. Spectrom.* **7** (1992) 1131.
12. D. C. Colodner, E. A. Boyle and J. M. Edmond, *Anal. Chem.* **65** (1993) 1419.
13. X. Wang, M. Viczian, A. Lisztity and R. M. Barnes, *J. Anal. At. Spectrom.* **3** (1988) 821.
14. H. Mukai, Y. Ambe and M. Morita, *J. Anal. At. Spectrom.* **5** (1990) 75.
15. M. B. Shabani and A. Masuda, *Anal. Chim. Acta* **261** (1992) 315.
16. D. Beauchemin, K. W. Siu and S. S. Berman, *Anal. Chem.* **60** (1988) 2587.
17. P. Van de Weijer, P. J. M. G. Vullings, W. L. H. Baeten and W. J. M. De Laat, *J. Anal. At. Spectrom.* **6** (1991) 609.
18. D. Beauchemin and S. S. Berman, *Anal. Chem.* **61** (1989) 1857.
19. H. Kinkenberg, T. Beeren and W. van Borm, *Spectrochim. Acta* **49B** (1994) 171.

20. M. Trojanowicz and T. Krawczynski vel Krawczyk, *Sens. Actuators B* **9** (1992) 33.
21. T. Krawczynski vel Krawczyk and M. Trojanowicz, *Anal. Sci.* **8** (1992) 329.
22. H. V. Shurmer, J. W. Gardenr and H. T. Chan, *Sens. Actuators* **18** (1989) 361.
23. L. T. Di Benedetto, P. W. Alexander and D. B. Hibbert, *Anal. Chim. Acta* **321** (1996) 61.
24. M. Novic, L. Zupancic-Kralj and B. Pihlar, *Anal. Chim. Acta* **243** (1991) 131.

Chapter 6

On-Line Sample Processing in FIA Systems

1. Irradiation Procedures

The on-line photochemical sample pretreatment is carried most often in order to transform the analyte into the component which is directly detected by the detector used. In most cases it is a convenient step which does not require designing dedicated complex modules. Irradiation is usually performed in the tubing coiled around the lamp which is a source of radiation. Most frequently for this purpose UV radiation is employed. UV-decomposed chloro-organic compounds have been determined with potentiometric detection of liberated chloride ions with ion-selective electrode [1]. Organoarsenic compounds were photo-oxidated and generated arsenate was subsequently reduced to arsine and continuously detected by atomic absorption spectrometry [2]. A sensitive flow injection technique for the determination of dissolved organic carbon in natural and waste waters was developed based on in-line UV photo-oxidation to carbon dioxide with spectrophotometric detection at 552 nm (Fig. 1) [3]. The obtained detection limit was about 0.1 mg $C.l^{-1}$ at sampling rate 45 $samples.h^{-1}$. The results obtained were in good agreement with those obtained by a high-temperature combustion commercial carbon analyser.

The use of a photochemical reaction in FIA, with unstable compounds such as phenothiazines under UV radiation [4], and the simultaneous determination of chlorpromazine and promethazine using different configurations for implementation of the photochemical reactions [5] were reported. The determination of thiamine hydrochloride (vitamin B_1) was carried out by UV photodegradation [6]. The analytical signal was the difference between the two peaks at 264 nm, corresponding to the absorbance of the irradiated and the nonirradiated sample.

The determination of nitrate in natural waters was reported in two different FIA systems with on-line irradiation. Photoreduction of nitrate to nitrite in

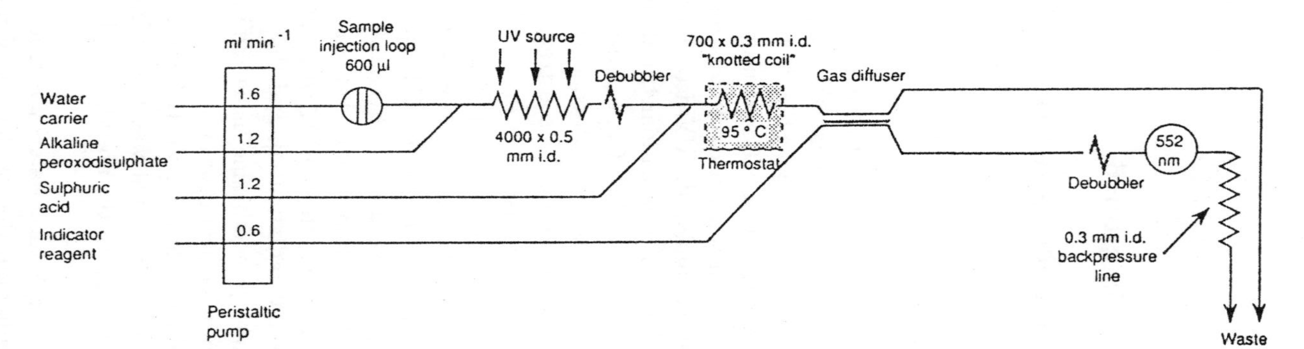

Fig. 1. Schematic diagram of the FIA manifold for the determination of dissolved organic carbon in natural and waste waters based on in-line UV photo-oxidation to carbon dioxide with spectrophotometric detection [3]. (*Reprinted by permission of copyright owner.*)

the PTFE tubing in a buffered carrier containing DTPA complexone as activator was employed in the FIA system where the nitrite formed was determined spectrophotometrically on the basis of a diazotization-coupling reaction [7]. The detection limits for nitrate and nitrite were about 30 nM and the bifurcation of the carrier stream after a sample injection valve permits simultaneous determination of these two analytes. In an acidified nitrate solution irradiated with UV light, species like peroxonitrite are formed, which are oxidants of luminol. This was the base for the development of the FIA system for the determination of nitrate with chemiluminescence detection with a linear response range of 70 nM to 0.1 mM [8].

The FIA system with a spectrophotometric system based on the photoreduction of the iron(III)-phenanthroline complex was developed for the simultaneous determination of iron(II) and iron(III) [9]. A single line system was equipped with the dual six-way injection valve. One injected sample passes the photochemical reactor, whereas the other does not. The carrier solution contained 1,10-phenanthroline and tartrate in acetate buffer. The first peak corresponded to iron(II), the second one to total iron.

The ultraviolet mineralisation in the presence of TiO_2 slurry as a catalyst was also employed as an additional detoxification step in the spectroscopic FIA method for the determination of resorcinol providing a non-polluting method of analysis [10].

Digestions in FIA systems are most effectively performed using microwave ovens. The application of microwave energy allows the reduction of the time of mineralisation of biological matrices from several hours to several minutes. This was employed in numerous non-flow analytical procedures in various biological and environmental materials, and it is finding an increasing number of applications in FIA. The procedure of determination with off-line microwave digestion was developed for determination of Zn and Cd with flame AAS detection in human kidney and liver tissue [11], and for Fe and Zn determination in adipose tissue with graphite furnace AAS [12].

More effective from the point of view of total automation of the analytical FIA procedure is a system with on-line microwave digestion. Such systems can be designed for liquid or slurry samples. Determinations of Cu, Fe and Zn in blood were carried out in the FIA setup, where the sample was mineralised on passage through the microwave reactor [13]. On-line microwave digestion with flow injection-cold vapour-AAS has also been applied for determinations of mercury in water and urine samples [14]. Rapid and efficient extraction of mercury was obtained for a range of environmental materials, in slurried form, using focused microwave heating in a 4 m PTFE digestion coil at relatively low power [15]. In combination with atomic fluorescence detection the developed procedure permits the determination of mercury with a throughput of 15 samples per hour in a range of certified environmental reference materials.

For the determination of total phosphorus in waters by FIA, an on-line microwave sample decomposition was employed with subsequent amperometric detection of orthophosphate based on reduction of the molybdophosphate formed to molybdenum blue [16]. With potassium peroxodisulphate decomposition the recoveries of phosphorus varied from 91 to 100% for organic phosphorus, and with perchloric acid decomposition the recoveries varied from 60 to 70% for inorganic polyphosphates. The microwave oven was also used to maximise the rate of the oxidation step in the determination of chemical oxygen demand in natural waters and waste waters [17]. In the FIA system with spectrophotometric detection the oxidation by a potassium dichromate-sulphuric acid mixture was carried out in a 10 m PTFE coil placed in a microwave oven.

2. Dialysis

In separation techniques employed on-line in FIA systems the membrane separation processes play an important role. Among various membrane processes such as filtration, reverse osmosis, dialysis and gas-diffusion separation, the last two have found the largest number of applications in FIA [18].

Dialysis is the process of separation of chemical species between two solutions through a semipermeable membrane. Although it is usually a slow process, a wide analytical application has been found in flow analysis with segmented streams in clinical determinations [19]. In fast flow injection systems the first application was reported for determination of inorganic species in blood serum [20] and since then numerous other applications have been developed [21]. The main areas of its applications include preconcentration of trace analytes and separation of the analyte from interfering matrices.

The nature of the used membrane determines the type of transport through the membrane. In passive processes a neutral membrane is a highly efficient molecular filter allowing species within a given range of molecular mass to diffuse across the membrane. The dialytic process is usually not very efficient but the diffusion rate can be increased by increasing the ratio of membrane surface to sample volume by appropriate design of the flow-through dialyser. With such membranes the analyte is usually transferred to the recipient stream, whereas interfering species and impermeable constituents of the sample are left in the sample stream.

In active Donnan dialysis, ions are transported across an ion-exchange membrane and their transport occurs under the influence of an ionic strength gradient. If the volume of the receiver solution of high ionic strength is significantly smaller than that of the sample solution, enrichment of the sample ions in the receiver solution results. The type of membrane chosen, anion or cation exchange, determines which ions are preconcentrated. Ion-exchange membranes used for this purpose may be flat as used in conventional flow-through dialysers (Fig. 2), or tubular.

The theoretical models for mass transfer across membranes in on-line dialysers in FIA systems were developed in several papers [22–24]. It was found, for plate dialyses with co-flow between sample and receiver stream, that in dimensions typical for analytical dialyzers the differences between flow and plug flow models were negligible [22]. The permeation of the analyte was found to be proportional to the concentration differences between the two sides of the membrane. Theoretical results were compared with experiments carried out using plate dialysers with cellulose acetate dialysis membranes for the dialysis

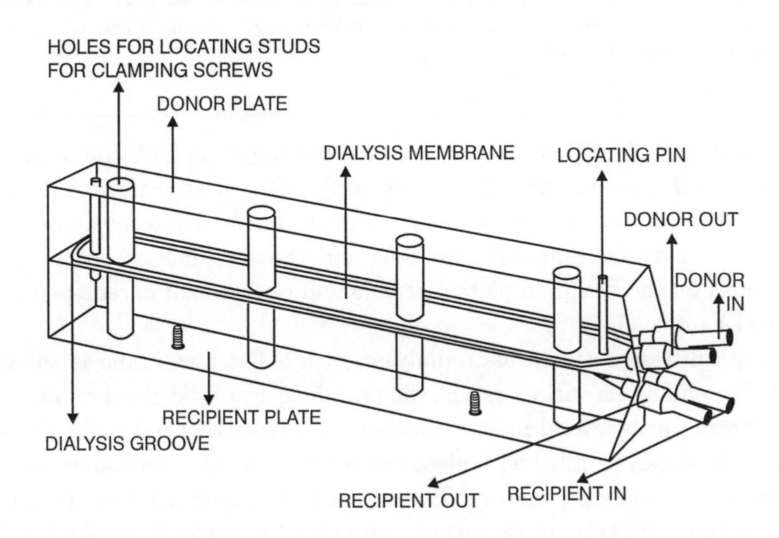

Fig. 2. Typical design of the flow-through dialyser, where a flat dialysis membrane is placed between the two plates, which are clamped by screws with a semi-tubular, triangular or square groove [21]. (*Reprinted by permission of copyright owner.*)

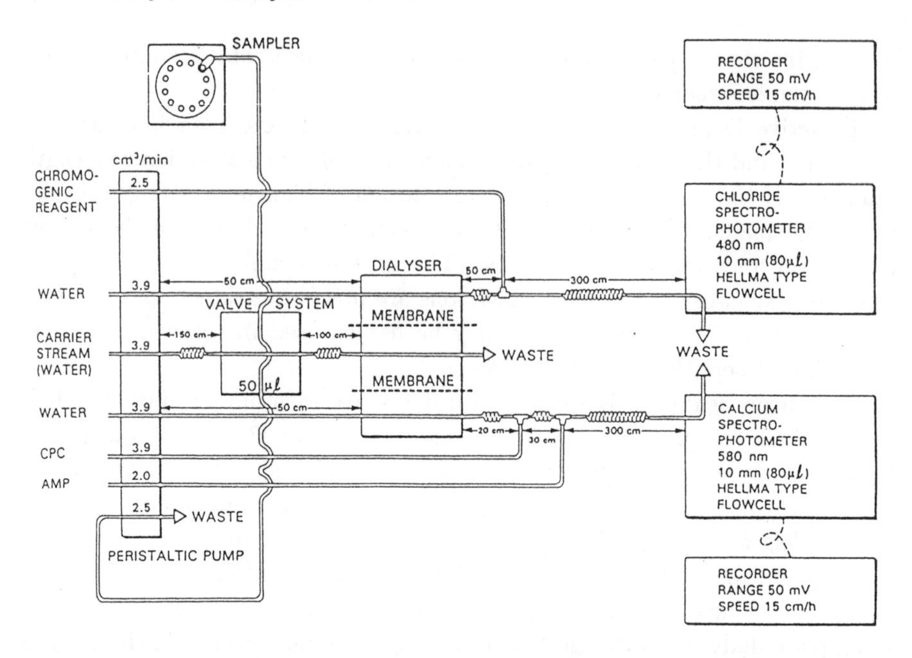

Fig. 3. Manifold for the simultaneous determination of chloride and calcium in industrial waste water with a double on-line dialyser [43]. Tubing length shown in cm. (*Reprinted by permission of copyright owner.*)

of zinc ions [22]. Detailed model treatment of dialysis in FIA conditions was discussed by Kolev and van der Linden [24]. The mass transfer in parallel plate dialysers with concurrent laminar flow in both channels was described using two convective diffusion equations and the second Fick law providing conclusions on the design of plate dialysers and operational parameters for its use in FIA mode.

A typical design of the plate dialyser with a flat membrane is shown in Fig. 2. Depending on various needs, the path lengths of the channels may vary from 10 mm up to several metres. In order to reduce dispersion the channels of the dialyser can be filled with glass beads [25]. The dialysis membrane used in flow-through dialysers are usually made of hydrophilic polymers (e.g. cellophane, cellulose acetate) of various porosity (1–10 nm) which should not carry fixed charges, and thickness from 0.015 to 0.03 mm. The molecular weight cut-off is usually in the range of 3.500–15.000 daltons. Ion-exchange membranes used for Donnan dialysis have a submicroporous structure with charged centres fixed to the pore walls. The mass transfer occurs by a combined mechanism of ion-exchange and diffusion, and membrane does not allow the penetration of ions with the opposite charge. For various purposes in FIA systems both flat anion-exchange membranes [26] and cation-exchange tubings [27–30] were used.

The dialyser unit can be placed in different positions of the FIA manifold. It can be located in the prevalve position before or after the pump. It can be placed within the sampling loop of the injection valve or, most often, as a separate module between the injection loop and the detector. In some cases the dialysis membrane is an integral part of the detector.

Dialysers with neutral semi-permeable membranes are most often employed for the separation of the analyte from interferences present in real samples. In the determination of glucose and urea with enzyme reactors employing spectrophotometric and potentiometric detections, respectively, proteins and other interfering species were removed in an on-line dialyser [31]. The same step was used in the FIA system for enzymatic amperometric determination of glucose in blood [32], determination of phosphate and chloride in blood serum with spectrophotometric detections [20] and determination of glucose in blood serum with glucose dehydrogenase and UV detection [33]. A dialyser unit with semi-permeable membranes was used to remove coloured matrices interfering in spectrophotometric determination of dialysable calcium in milk [34] and blood [35], and in fluorimetric detection of calcium with Calcein [36]. Interferences from colour of the real samples were also effectively removed by on-line dialysis in the turbidimetric determination of sulphate in urine [37] and in water [38]. Flow injection stripping analysis using a cellulose triacetate membrane modified mercury film electrode was successfully applied for the determination of lead in urine and whole blood matrices without interferences from proteins and peptides [39].

Another application developed for on-dialysis in FIA systems is enhancing the selectivity of the potentiometric response of membrane electrodes [41, 41]. The concept of the use of a dialyser unit as the injection loop was applied in determination of NO_x or nitrite [40] and salicylate in serum [41]. In both systems the analyte was dialysed into the recipient channel of the flow-through unit as the injection loop in the FIA system.

Dialysis through a semi-permeable membrane can be used to promote dilution of the concentrated sample which was demonstrated for the determination of chloride in electroplating baths with potentiometric detection [42]. A double membrane dialyser was designed for simultaneous dialysis of analytes in two acceptor streams where they were processed and channelled to various detectors for the determination of calcium and chloride in industrial effluent water [43].

An increasing number of applications in FIA systems have been observed also for Donnan dialysis through the ion-exchange membranes [26–30]. Preconcentration of cations using Nafion 811 cation-exchange tubing was employed in FIA systems with flame AAS [27] and ICP AES [29] detections. In the first system the coiled, tubular membrane was used as a direct replacement for the injection loop in a simple injection valve. A 5 min dialysis provided 100-fold enrichment in determination of lead in drinking water [27]. In the system with ICP AES detection, a signal enhancement factor of 650 for silver(I) cation was demonstrated using a 30 min dialysis [29].

Ionomer membranes were also successfully employed to enhance the selectivity of electrode-based biosensors in FIA procedures. In the systems for enzymatic determination of urea and glutamine, a nonactin-based ammonium-sensitive polymeric membrane was employed as the detector, while a thin hydrophilic anion-exchange membrane was incorporated into a flow-through dialysis unit upstream from the enzyme-electrode detector [26]. As the injected sample passes through the dialysis unit, neutral or anionic analyte molecules move through the membrane while permeation of endogeneous ammonium ions and other cations in the sample is retarded. Cation-exchange units made of Nafion tubings can also be used to reduce positive errors by endogeneous cationic interferences in flow injection enzyme-based potentiometry when using an ammonim ion-selective electrode as detector for determination of L-glutamine in bioreactor media [26,30]. Interferent cation species within the sample plug are exchanged for other cations, for example Li^+ contained within a reservoir solution surrounding the ion-exchanging tubing.

3. Gas Diffusion

Another membrane separation process which besides dialysis finds numerous on-line applications in FIA systems is diffusion of volatile species through permeable membranes. Such processes are utilised for determination of analytes in the gas phase and in solutions. In the first case the analyte is absorbed in a suitable acceptor solution from the a gas phase through the membrane in gas diffusion cell. In the second case the dissolved volatile analyte (or produced in the donor solution by an appropriate reaction) diffuses selectively through the membrane to the acceptor solution. The theoretical treatment of the membrane gas diffusion process was developed using a tank-in-series approach for the system with two flowing streams which were separated by a hydrophobic membrane permeable for gas only [44, 45]. Stopping for a certain time the acceptor flow while the donor continues to flow allows the preconcentration of the analyte in absorbing solution. Similarly to dialysis, both flat and tubular membranes are employed for gas diffusion processes.

Several FIA systems with the gas diffusion step have been developed for determination in the gas phase. In the system for determination of nitrogen dioxide in air samples the gaseous sample was introduced into a liquid carrier [46]. The detection was based on the NO_2 oxidation of tris(1,10-phenanthroline)-iron(II) to its iron(III) analogue at a gas–liquid interface in a planar gas diffusion cell. The extent of oxidation was followed amperometrically. A planar gas diffusion cell was also used for determination of ambient levels of sulphur dioxide in the FIA system with spectrophotometric detection in EDTA as the absorber stream which merged the premixed pararosaniline-formaldehyde reagent [47]. The best collection efficiency was obtained for PTFE and polypropylene membranes. In the same work determination of atmospheric ammonia was developed using a gas diffusion cell with a tubular membrane employing both spectrophotometric detection with bromocresol purple as the indicator and potentiometric detection with an ammonium selective electrode. A gas permeation unit with silicone tubing, installed as the sample loop of the injection valve, was employed in a flow injection analyser for monitoring trace HCN in process gas streams [48]. Spectrophotometric detection was based on reaction of HCN absorbed in a caustic stream with chloramine-T and a mixture of isonicotinic acid and 3-methyl-1-phenyl-2-pyrazolin-5-one. The enrichment capability of the gas diffusion unit gives the capability of measuring very low levels of HCN.

A gas diffusion unit equipped with a hydrophobic Celgard 2500 membrane was used for flow injection determination of ammonia both in the gas and the liquid phase employing an integrated fibre-optic detector [49]. The membrane in the diffusion unit serves also as a reflector for the light, where the signal is returned to the detector by a bifurcated optical fibre (Fig. 4). In the acceptor solution a bromothymol blue indicator was used. The limits of detection were 40 ppb for gas phase analysis and 1.0 ppm for aqueous phase analysis.

A different design of the gas diffusion detection device for spectrophotometric detection employing a microporous polypropylene membrane was

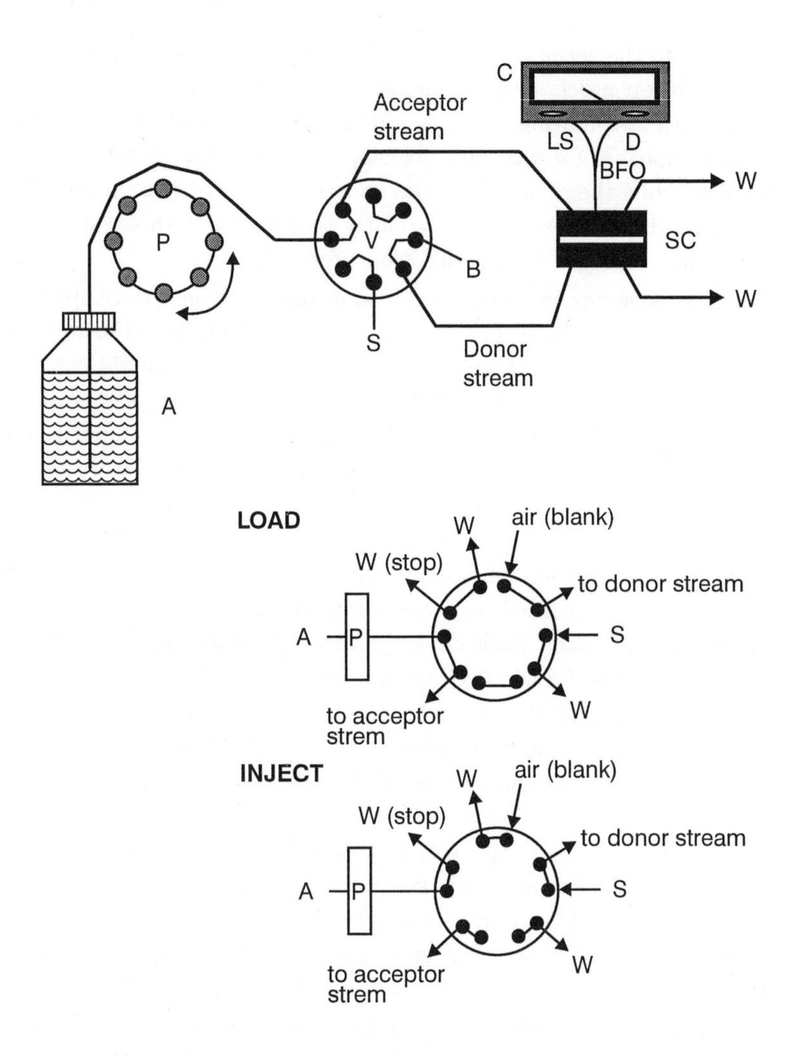

Fig. 4. *Top*: Schematic diagram of the manifold for ammonia determination in a gas stream using a gas diffusion sandwich cell (SC). P — peristaltic pump; V — 14-port two-position injection valve; LS — light source; A — acceptor; W — waste; S — sample; B — blank; C — colorimeter; BFO — bifurcated fibre-optic bundles. *Bottom*: Detail of the valve configuration, showing the flow in the load and inject positions [49]. (*Reprinted by permission of copyright owner.*)

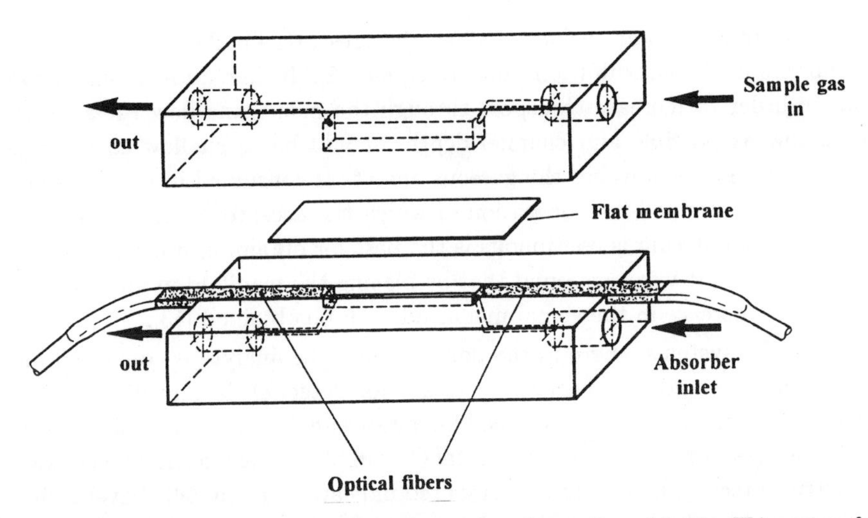

Fig. 5. Schematic diagram of the gas diffusion detection cell used in the FIA system for spectrophotometric determination of atmospheric nitrogen dioxide [50]. (*Reprinted by permission of copyright owner.*)

developed for the determination of atmospheric nitrogen oxide (Fig. 5) [50]. As absorber the solution of sulphanil-amide and *N*-(1-naphthyl)-ethylenediamine hydrochloride in hydrochloric acid was used. The working range covered in the developed system was 50-250 $\mu g/m^3$ NO_2 with a time resolution of a few seconds.

A different approach was employed for the determination of SO_2 in the gas phase by the design of a chromatomemebrane cell for use in the FIA system [51]. A chromatomembrane generated from porous hydrophobic PTFE with macro- and micropores was applied in the system with spectrophotometric detection based on formation of an Fe(II)-1,10-phenanthroline complex as a result of reduction of Fe(III) with SO_2 in the presence of 1,10-phenanthroline.

Many more applications of gas diffusion systems have been developed for determination of analytes dissolved in solutions. In most typical applications the analyte dissolved in donor solution is converted into a volatile compound which diffuses through a gas-permeable membrane into the acceptor stream, where a new reaction may take place and the resulting product is detected. With respect to the peak height response in FIA systems, the most important variables with gas diffusion steps are injection volume, flow rate, flow ratio, channel depth and shape and membrane type and area [52, 53]. The best transport properties are achieved when the pressures on the two sides of the membrane in the gas diffusion unit are equal [52]. In the study it was shown that in order to increase transport through the membrane flow rates should be as low as possible and channel depths should be as shallow as possible. For reactive systems where the gaseous analyte is consumed in the absorbing solution, the stopped-flow operation in which the acceptor stream is stopped but the donor stream is continuous is the best for obtaining preconcentration of the analyte. It was also found that by placing the manifold in an ultrasonic bath some increase in the signal magnitude can be obtained [53].

The FIA systems where in the donor stream the analyte is converted into volatile species and the reaction necessary for detection is carried out in the acceptor stream have been developed for various analytes. First, such a system was developed for determination of total CO_2 in blood plasma with spectrophotometric detection in a buffered Cresol red indicator stream [54]. Several similar systems have been reported for determination of ammonia with spectrophotometric [53, 55–58], potentiometric [56, 59] and conductimeric [58] detections. In comparison with several spectrophotometric detections the conductimetric method has advantages of the simplest experimental requirements and uses a single reagent (NaOH) which is stable and cheap. In ammonia determination in whole blood and plasma it was found that the developed FIA technique is suitable for use in clinical laboratories as "stand-by" instrumentation, and when used it is possible to analyse ammonia continuously at a sampling rate of about 60 samples per hour [55]. In comparison with several spectrophotometric detection methods, potentiometric detection with an ammonium-selective electrode showed improved base-line stability and better detectability [56].

In SO_2 determinations in the donor stream in acidic medium sulphur (IV) is converted to SO_2 and in the acceptor stream spectrophotometric detection with pararosaniline [59] or *p*-aminobenzene [60], and amperometric one using a glassy carbon electrode [61] has been used. In wines this procedure has been employed for both total and free sulphur dioxide using additionally for total SO_2 a prehydrolysis unit consisting of a well-stirred chamber where the sample was continuously mixed with a stream of NaOH [61].

In determination of free cyanide the analyte is also converted in the donor stream into volatile HCN. In the developed systems a potentiometric detection with a metallic silver electrode [63], or a pulsed amperometric detection with a silver working electrode [64], has been employed. The apparent selectivity coefficients in the system with potentiometric detection and gas diffusion step were significantly better than those reported for the common cyanide-selective electrode. In the system with amperometric detection, in order to release cyanide from complexes with transition metal ions the sample was pretreated with a mixture of ligand exchange reagents. The only potential interference of the sulphide ion was precipitated by Pb(II) using $PbCO_3$ and filtration prior to the sample injection. A method for total cyanide determination in waste waters has been developed with gas diffusion separation and spectrophotometric detection with isonicotinic acid and 3-methyl-1-phenyl-2-pyrolin-5-one, where 1,10-phenanthroline was used to eliminate interferences [65].

FIA systems with gas diffusion steps have also been developed for determination of inorganic forms of nitrogen. In the system with conductivity detection ammonia was determined as mentioned above by merging the injected sample with NaOH-EDTA solution, whereas nitrate and nitrite were determined after reduction to ammonia using a column filled with metallic zinc [66]. Simultaneous determination of nitrate and nitrite can be carried out using gas-phase chemiluminescence detection [67]. A microporous PTFE membrane was used to transfer NO to the gas phase after reducing nitrate and nitrite with Ti(III) or nitrite alone by iodide in acidic media. Chemiluminescence signals have been produced by the reaction of NO with ozone supplied from a generator. Ammonia, amines and amino acids present in natural waters did not interfere.

A gas diffusion unit with a tubular PTFE membrane was used in the FIA system for spectrophotometric determination of iodide and bromide coupled with oxidation to free halogens by permanganate in the donor stream [68]. The permeated halogens reacted in the acceptor stream with N,N-diethyl-p-phenylenediamine. The same unit can also be used for determination of residual chlorine with spectrophotometric detection or potentiometric employing a coated-wire ion-selective electrode based on quaternary ammonium salt. A tubular gas diffusion unit and potentiometric detection with a silver wire electrode were used also for the determination of sulphide after conversion in the donor stream to hydrogen sulphide [59].

In numerous developed FIA systems gas diffusion is used to separate the volatile analyte from the complex matrix of the natural sample. In determination of acetone in milk, samples are injected without pretreatment, while for determination of total oxidized ketone the acetoacetate is first converted to acetone by perchloric acid at 100°C and then spectrophotometric detection in the acceptor stream can be carried out [69]. In determination of ethanol in alcoholic beverages the alcohol permeates across a porous membrane. Determination is based on oxidation with dichromate and the reduction of excess dichromate with ferrous ion is monitored potentiometrically with a redox potentiometric electrode [70]. In the system with a gas diffusion cell fitted with a microporous Teflon membrane, residual aqueous ozone was determined using the redox reagents potassium indigo trisulphonate and bis(terpyridine)iron(II) [71]. The membrane in the gas diffusion unit significantly reduces potential interferences such as chlorine or oxidized forms of manganese.

Two different FIA systems employing gas diffusion steps have been developed for determination of free chlorine. In the system with potentiometric detection chlorine after diffusion through the gas-permeable hydrophobic membrane reacted with the acceptor stream, forming chloride which was detected using a Ag/AgCl electrode [72]. In the second system chemiluminescent reaction of hypochlorite with lophine(2,4,5-triphenylimidazole) was used for detection [73]. The determination of aqueous chlorine demonstrated excellent selectivity towards transition metals, ionic and oxy chlorinated species, and also monochloramine and dichloramine.

Chemiluminescence was also employed in a gas diffusion flow injection system of chlorine dioxide with luminol where membrane separation was used to remove ionic interferences of iron and manganese compounds [74].

A gas dialysis unit was employed in determination of ammonia nitrogen in blood not only for potentiometric detection of permeated ammonia but also to remove elevated ammonia background in commercial control sera [75]. In the latter case the gas diffusion unit had a 10-times-larger acceptor chamber than the donor compartment.

The use of the gas diffusion unit is also advantageous in FIA systems with atomic absorption spectrometric detection. The use of a dual phase gas diffusion cell with a Teflon membrane in a hydride generation AAS system for determination of arsenic(III) significantly decreased transition metal interferences [76]. It was also shown that the developed system exhibits improved selectivity towards the other hydride forming elements except for selenium (IV). The flow injection system with a vapour diffusion system using a membrane separator proved also to be convenient for a cold vapour atomic absorption determination of mercury providing a detection limit of 0.04 ppb [77].

Hydride generation can also be employed for determination of lead in the FIA system with a conductometric and gas diffusion unit [78]. Plumbane produced in the donor stream by reaction with borohydride is passed through the PTFE membrane and then reacts in the acceptor stream with bromine. The resulting ionisation is detected by conductivity measurement.

A different approach in the use of the gas diffusion step compared to separate gas diffusion units was employed in pneumoamperometric determination of iodide in the FIA system [79]. The principle of this detection is the oxidative conversion of the analyte into the electroactive volatile product iodine, which is then detected on a gas-porous gold metallised membrane.

4. Solvent Extraction

The solvent extraction based on the law of distribution of substance between two immiscible liquid phases plays a significant role among separation methods of modern chemical analysis. The analytical literature provides a large number of examples of the use of chelate complexes, ion-association compounds with high-molecular amines, acidocomplexes and salts by coordination of phospho-organic compounds in liquid–liquid extraction systems. The solvent extraction separation steps are included in many standard analytical procedures. First applications of solvent extraction in FIA have been developed by Karlberg and Thelander [80] and Bergamin *et al.* [81], and since then numerous methods have been reported and reviewed [82–84]. Appropriate design of extraction modules with thorough chemical and hydrodynamic optimisation of the flow system performance, enables one to preserve in FIA systems with on-line solvent extraction all the basic advantages of FIA. They include a large sample throughput and small sample consumption, whereas due to on-line solvent extraction the improvement in detectability and selectivity can usually be obtained.

In most common FIA systems with on-line solvent extraction, shown schematically in Fig. 6, the sample to be analysed is injected into the stream of aqueous carrier solution and then one of the following processes can occur. The chemical reaction takes place in the aqueous phase and its product after segementation of the sample zone with extracting solvent is extracted to the organic phase or the aqueous solution serves as a carrier stream only, and after segmentation of the sample zone the analyte is extracted by the appropriate reagent supplied in the organic phase.

The extraction process takes place in the extraction coil. From the extraction coil the segmented aqueous-organic stream is usually directed to the phase separator, although systems have been developed already where

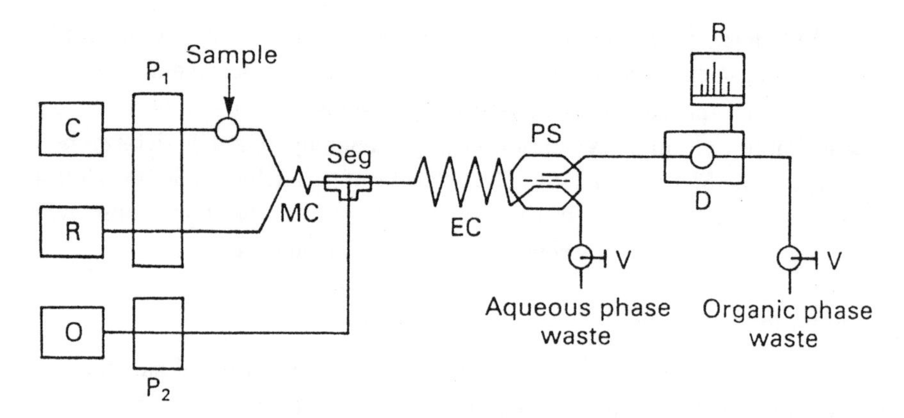

Fig. 6. Typical manifold for the FIA system with on-line solvent extraction [85]: P1, P2 — peristaltic pumps; C — carrier (aqueous phase); R — reagent stream (aqueous phase); O — organic phase; S — injection valve; MC — mixing coil; Seg — segmentor; EC — extraction coil; PS — phase separator; D — detector; R — recorder; V — waste. (*Reprinted by permission of copyright owner.*)

separation of organic and aqueous phases was not necessary, and they will be discussed below. Most frequently the separated pure organic phase is directed to the detector, whereas the aqueous phase, usually with some excess of the organic phase, is disposed of.

Segmentation and separation of the two immiscible phases are of significant importance for the effectiveness of the solvent extraction step in the FIA system. The segmentation process concerns the formation of a mixed liquid stream composed of regularly sized segments of the aqueous phase divided by segments of the organic phase. The purpose of liquid segmentation is to ensure optimum contact between two phases, and numerous practical and theoretical contributions on this subject have been published and reviewed [86]. Segementation is carried out in segmentors which should provide conditions for merging streams of aqueous and organic solutions to form regular segments of a given size. The reproducibility of segmentation depends mostly on the quality of machining and smoothness of the inner surfaces which are in contact with both solutions. The size of segments is governed mainly by the geometry of segmentor and flow rates. Most frequently segmentation is carried out in T-piece connectors made of glass, plastic or stainless steel, where aqueous and organic solvents are merged at a 90° or 180° angle. Sometimes they are additionally equipped with special inserts [87, 88]. For segmentors without Teflon inserts irregular segmentation was observed, due to penetration of the organic phase to the arm which was used as the aqueous phase inlet [87]. Less often used are segmentors where the two streams are merged at 45° [85], 60° [85, 89] or 120° [98] (Fig. 7A). As segmentor a modified Technicon A-8 connector was also used where a glass capillary was used as an aqueous phase inlet and a perpendicularly mounted platinum capillary as an organic phase inlet [90]. A detailed study of the mechanism of the segmentation process has been reported for a reversed T-connector, where the organic phase was introduced by a vertical capillary, whereas the aqueous solution, in which organic segments are formed, is transported by a horizontal capillary [87]. A very regular and reproducible segmentation has been also obtained when as segmentor a small volume mixing chamber equipped with a magnetic stirrer was applied [91].

For high flow ratios and high total flow rates, especially suitable are dropping segmentors. In a falling drop segmentor a glass capillary is mounted in Perspex housing [92] (Fig. 7C). The organic flow is connected to the capillary

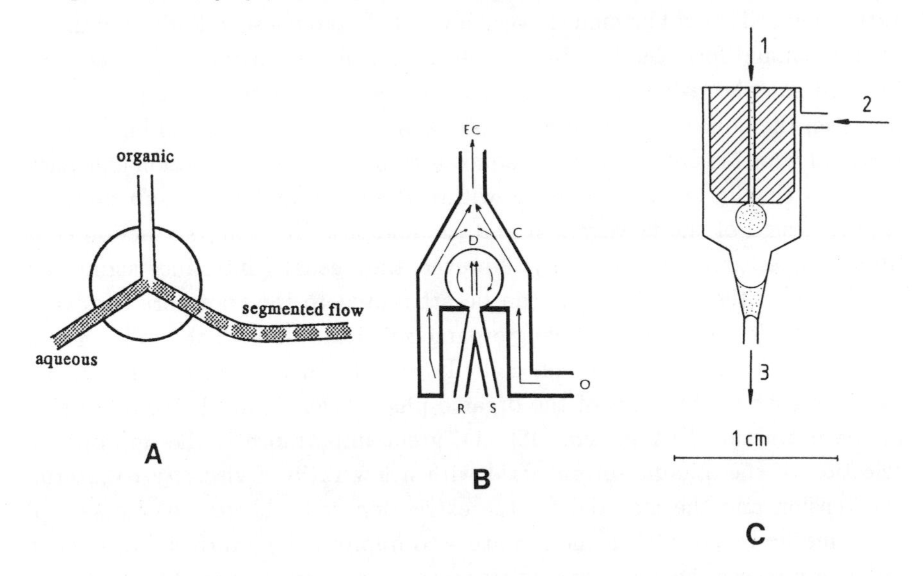

Fig. 7. Examples of construction of segmentors used in solvent extraction FIA systems: A — Teflon trisection confluence connector [98]; B — multichannel dropping segmentor with a dual-channel inlet capillary system for the organic phase being heavier than the aqueous phase [93]; R — reagent solution; S — sample solution; O — organic phase; C — conical compartment; D — reaction mixture droplet; EC — extraction/equilibrating coil; C — falling drop segmentor [92]: 1 — organic phase; 2 — aqueous phase; 3 — segmented outflow. (*Reprinted by permission of copyright owner.*)

and the segmentor works on the principle of the formation of organic droplets at the end of capillary. The droplets fall through a small chamber (130 μl) filled with the flowing aqueous phase, which conveys them through the outflow funnel, where the segments develop. The size of the organic segments depends on the inner diameter of the glass capillary. A dual-channel dropping segmentor is based on the principle of the coaxial segmentor [93] (Fig. 7B). The aqueous solution of a sample and an organic analytical reagent are simultaneously introduced into a continuous flow of an immiscible organic solvent. Droplets of precisely defined volume of the homogeneous reaction mixture are formed at the junction of the inlet capillary system.

A main process of interfacial transport of the analyte from the aqueous to the organic phase takes place in the extraction coil. In glass coils, their inner wall is wetted by water and the organic phase forms bubbles. In extraction coils made of organic polymer (most often PTFE) the organic phase wets the inner wall, whereas the aqueous phase forms bubbles. The choice of material for the extraction coil should be made in such a way that the phase initially containing analyte should form bubbles in order to diminish the carry-over between the injected samples [94]. So, for most common systems with extraction carried out from the aqueous to the organic phase, extraction coils should be made of material which is not wettable by aqueous solutions. The extraction efficiency observed usually in FIA systems is within 70–90% and depends substantially on the design of the flow system and its modules. Too long extraction coils improve the extraction efficiency; however, they usually introduce significant dispersion of the sample. This can be attributed to the transport backward of the organic phase into following organic segments through the wetting film along the wall of the extraction coil. It was shown that all factors leading to a decrease in thickness of the organic phase layer diminish the dispersion in the extraction FIA system [95]. Of great importance is the appropriate selection of the organic solvent used with a low ratio of viscosity to interfacial tension and the material for the extraction coil. There are also several other methods reported in the literature to improve the interfacial contact in order to increase the extraction efficiency, such as using a special insert to the extraction coils, elevating temperature, ultrasonication or vibrations. Precautions should be made to avoid the formation of emulsions which makes difficult phase separation.

In solvent extraction FIA systems an aqueous solution is commonly delivered using a peristaltic pump with ordinary PVC tubing. Pumping of organic solvents requires the use of pumping tubings inert to solvents made of silicone rubber, flexible polyuretane or fluoroplastics. It can also be carried out using PVC tubings by applying the displacement bottle technique. An aqueous stream is pumped into a closed bottle completely filled with an organic solvent, which is replaced at a given flow rate of the aqueous stream and delivered into the FIA system [96]. For the delivery of the organic solvent HPLC pumps can also be used, or a constant inert gas overpressure, delivering the organic solvent from a closed container.

The separation of phases is carried out in order to partition the segment after the extraction in the extraction coil in such way that the unwanted phase is directed to waste while the other phase is pumped to the detector. The phase separator is usually the most complex and most important module of the solvent extraction FIA manifold. The separation of the aqueous and the organic phase in reported devices varies from 80 to 95%. Different types of phase separators described in the literature are based on differences in density and different wetting of various materials by solutions to be separated, on selectivity of permeation through different membranes or on selective sorption of one phase by an appropriate sorbent.

The simplest device based on density differences has been made from a commercial Technicon A4 connector equipped with Teflon inserts facilitating the phase separation due to difference in wetting [90] (Fig. 8A). An all-glass gravitational phase separator has been designed, consisting of two glass bulbs connected with a narrow neck, upper of 1 ml capacity for the organic phase and lower of 10 cm capacity for the aqueous solution [97]. Another design is shown in Fig. 8B.

Phase separation in membrane separators is mostly based on the permeability of the organic phase through a Teflon membrane of 0.7–0.9 μm pore size. Because of some overpressure formed in the separator and the flexibility of the membrane it is advantageous to use some inert mechanical support of the membrane [98]. Also in this case the appropriate Teflon insert can be helpful for the separation due to differences in wetting. The membrane separators have different shapes of the separation compartment divided by the hydrophobic membrane. Most often this compartment is manufactured in the shape of a shallow straight [99–103] or coiled [104, 105] groove symmetrically on both sides of the membrane. The example of design of the membrane separator with a coiled unsymmetric groove on both sides of the membrane is shown in Fig. 8C. In order to increase the contact area between the solutions and the

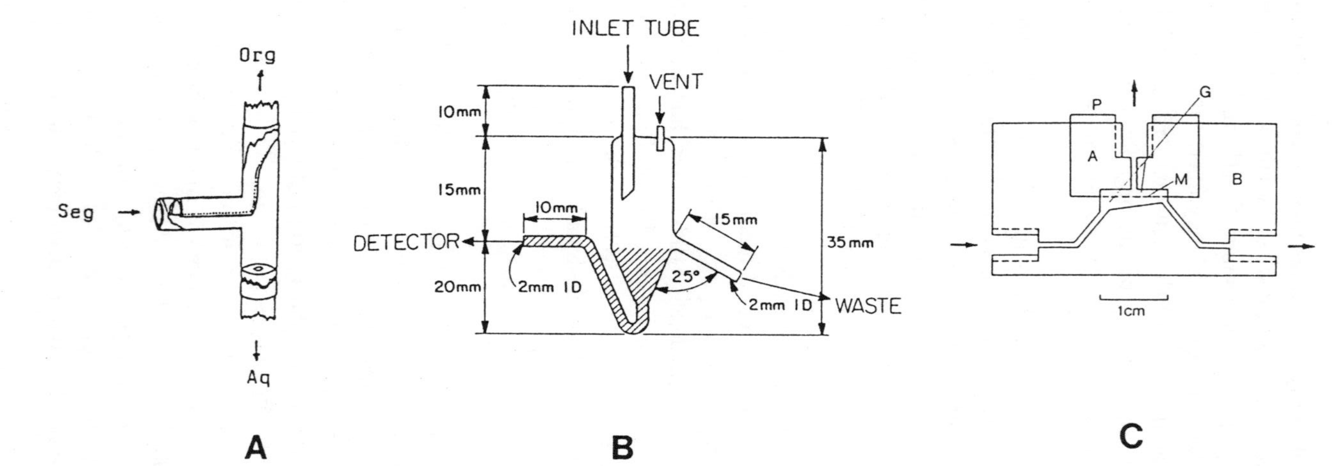

Fig. 8. Examples of constructions of phase separators used in solvent extraction FIA systems: A — separator made of T-connector with Teflon insert [90]; B — gravitational glass-blown phase separator with shaded area of total volume 250–350 μl [134]; C — membrane separator [77]: A,B — Teflon blocks; M — microporous Teflon membrane; G — grooves; P — metal plate. (*Reprinted by permission of copyright owner.*)

membrane it is advantageous to decrease the thickness of the solution layer and increase the width of the groove providing a larger contact between the membrane and flowing solution [99]. Cylindrical separators are designed with different geometry of the solution inlet and outlet, and with different volume and shape of the separation chamber. The comparison of their functioning to groove separators [100] and the comparison of the efficiency of cylindrical separators of various designs were reported by several authors [97, 100, 106]. In order to remove the remaining traces of the aqueous phase, two membrane separators or a double-membrane phase separator can be employed [107].

A completely different concept of phase separation is based on absorption of the aqueous phase in a flow-through column filled with appropriate hydrophilic loading. Such a separation device can be employed in the FIA systems where a relatively small volume of the aqueous phase is used. The choice of absorbents is very critical with regard to its physical stability. In the extraction of caffeine, when the gel-type absorbent was used, the column pressure gradually increased with the number of injections; however, with cotton-type absorbent, pretreated with a methanol-chloroform mixture, a satisfactory functioning of the system was observed [108].

Membrane modules have also been developed to perform on-line solvent extraction in FIA systems without the need for stream segmentation and phase separation. Instead of partition between segments of two immiscible liquids the extraction module contains a PTFE porous membrane, and each phase is fed only one side of the membrane [99, 104, 109]. The extraction cell has a similar construction to a dialysis cell, where the contact area between the aqueous and the organic phase is reduced to a stagnant zone in the cell. The organic solvent groove is filled with a porous support of polyethylene. Such a system is suitable for coarse liquid-liquid extraction of concentrated samples.

The phase separation in FIA extraction systems can also be avoided by the introduction of a large amount of surfactant. This allows the formation of stable microemulsion in which detection can be carried out [110, 111]. FIA systems enabling measurements without phase separation have been developed by appropriate design of spectrophotometric detection cells. The small illuminated volume of the detector ($< 1\ \mu l$) with fibre-optic coupling to the photometer made it possible to measure directly the segmented stream [112]. The same has also been achieved using a glass capillary flow cell without optic fibre visible absorption and fluorescence detection [113].

Systems without the phase separation have also been developed with iterative change of the flow direction [114, 115]. They consist of a single plug of the organic phase inserted into the carrier stream of the aqueous solution which contains the analyte. The flow is subjected to an iterative reversal, and the gradual enrichment of the organic phase with the solute is monitored photometrically. The same approach has also been reported for reversed solvent extraction into the aqueous phase [116]. In another concept of the FIA extraction system without the phase separation a segment of the sample is introduced into a glass tube between two air bubbles [117]. The segment of the organic phase containing an extracting reagent is introduced after the second air bubble and the transference of the analyte from the aqueous to the organic phase is achieved by adsorption on the surface of the tube followed by desorption into the organic phase.

Differential flow velocities in two-phase segmented flow developed between the aqueous and organic phases have been utilised for flow injection preconcentration [118]. The pseudostationary phase on the tubing wall results from the wetting by the organic solvent. When a solute extracts into the organic phase, the solute band migrates through the tubing more slowly than the aqueous segments. Introducing a segment of aqueous eluting reagent allows the rapid back-extraction of the analyte with preconcentration factors in excess of 50-fold.

FIA measurements with off-line extraction and organic carrier solution reported for the determination of anionic surfactants do not require segmentation or phase separation [119].

During the most common type of solvent extraction process in FIA systems, in the extraction tubing or coil, a solute extracts from an aqueous segment through the ends of the segment into the adjacent organic segments and also radially through the sides of the aqueous segment into the wetting film of the organic phase along the walls of the tube. If no chemical reaction is involved, the extraction process depends on the mass transfer to and from the aqueous-organic interface and follows an exponential behaviour which can be treated similarly to a first-order chemical reaction. The extraction rate in straight tubes is affected by two major factors, namely the ratio of interfacial area to volume and mass transfer to and from interfaces [120]. The extraction rate is increased by increasing the interfacial area to volume ratio, which can be most effectively achieved by decreasing the tubing diameter. The extraction rate increases rapidly with decreasing segment length for short segments and is almost constant for longer segments. Increasing the linear flow velocity increases the extraction rate with respect to time, but not in a linear proportion. Also, a significant contribution of secondary flow to convection in coiled extraction tubes has been demonstrated [121]. In the coiled tube the extraction rates are much higher than in a comparable straight tube. The tightening of coiling of the tubing increases the extraction rate with respect to time to a greater extent for long than for short segments. The signal magnitude is affected also by the length of the extraction coil and injected sample volume [122]. Both flow injection peak area and peak width at half of its height increase with an increase in the length of the extraction coil until the equilibrium values are obtained at a certain length of the extraction coil. In the effect of the sample volume two different regions have been observed. For smaller sample volumes, with a Gaussian shape of signal, a constant peak width was observed, whereas its surface depends on the sample volume. For large sample volumes the plateau of the signal value in the centre of the signal is observed. In this region the peak width becomes influenced by the sample volume.

The use of recirculation of solution in a closed loop with multiple measurement of analytical signal can be employed in the FIA system to improve the detectability [123, 124]. With the organic phase circulating in the loop it can be contacted with a fresh aqueous stream, which continuously enters the extraction coil and leaves to waste. It has been found in the system shown in Fig. 9 for extraction of copper with chloroform solution of zinc diethyldithiocarbamate that high preconcentration factors can be reached depending on flow rates, loading time, extraction efficiency and organic phase loss through the phase separator. A unique feature of the system is the possibility of performing several washing steps on the organic phase trapped in the loop to remove interferences or to strip the analyte. This has been demonstrated for the determination of the trace level of uranium in nuclear waste with spectrophotometric detection [123].

A certain improvement of detectability and sampling frequency in the FIA system with AAS detection has been observed when the sample was injected into the segmented organic-aqueous stream instead of the aqueous stream [125]. A multiple repetition of the solvent extraction has been developed in a three-stage extraction procedure where three single-step extractions are linked together by multichannel pumping and resampling [90, 94].

Spectrophotometric absorptive detection is most often employed in FIA systems, including those with on-line solvent extraction (Table 1). Several methods have been developed for fast and sensitve determination of anionic surfactants in natural waters and wastes [88, 102, 113, 114, 119, 126, 128] which

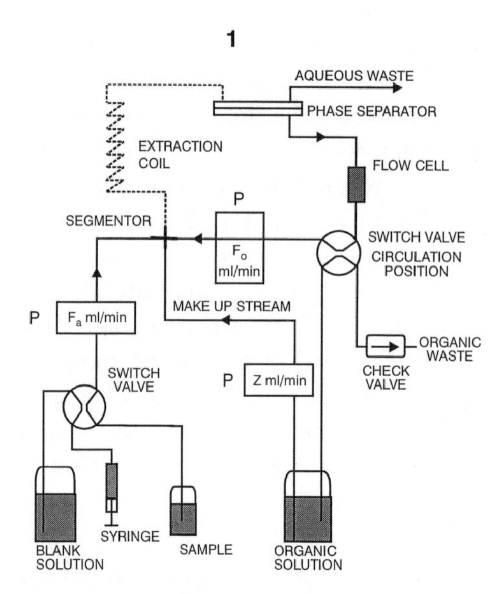

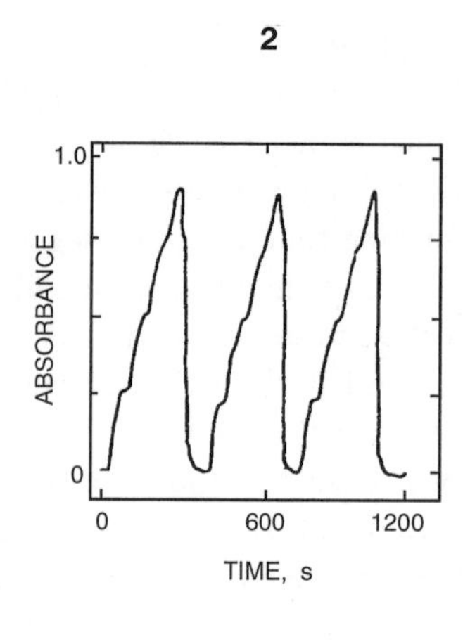

Fig. 9. Flow injection solvent extraction closed-loop system for determination of copper with zinc diethyldithiocarbamate [124]. 1 — general arrangement of the system (P — peristaltic pump); 2 — signal recorded for triplicate measurement of a 1.0 ppm Cu(II) solution. (*Reprinted by permission of copyright owner.*)

are based on extraction of appropriate ion pairs. Among other organic species determined were phenol [135], organophosphorus pesticides for further HPLC analysis [89], codeine [133] and caffeine [80]. Among inorganic species trace determination of heavy metals predominates; however, several anions have also been determined. Extraction of crown ether complexes has been applied in determination of lithium in blood serum [149,150], potassium in surface waters [145, 146] and fruits and beverages [147] and simultaneous determination of sodium and potassium in waters employing on-line separation on a silica gel microcolumn [162].

In FIA systems with solvent extraction and AAS detection, organic solvents not only allow preconcentration and separation of the determined species but, due to more efficient combustion in the flame and easier evaporation, additionally improve the detectability of the detection with flame atomisation. For continuous sample aspiration from 15 to 20 [142] an up to 60-fold [156] improvement of detectability has been observed. The extraction of zinc thiocyanate after preliminary reduction of iron to Fe(II) allowed the elimination of the iron

Table 1. Applications of on-line solvent extraction in FIA systems.

Analyte	Reagent	Solvent	Detection	Detection limit, ppb	Reference
Anionic surfactants	Methylene blue	Chloroform	VIS	4	126
				100	114
				0.015 mM	
		1,2-dichloro-benzene	VIS	3	88
		1,2-dichloro-benzene+ benzene	VIS	5	102
	1-methyl-4-(4-diethylaminophenylazo)-pyridinium ion	Chloroform	VIS	0.01 mM	128
	Ethyl violet	Toluene	VIS	10	119
	4-(4-dimethylaminoph-enylazo)-2-methylq-uinoline	Chloroform	VIS	0.002	113
Berberine	Perchlorate	1,2-dichloro-ethane	FLU	—	107
				0.8 nM	129
Bitterness		Isooctane	UV	—	131
Caffeine		Chloroform	VIS	—	80
Cationic surfactants	Orange II reagent	Chloroform	VIS	0.1 mM	132
Codeine	Picrate	Chloroform	VIS	20000	133
Non-ionic surfactants	Tetrabromophenolphthalein ethyl ester K salt	1,2-dichloroethane	VIS	2000	134
Phenol	4-aminoantipyrine	Chloroform	VIS	50	135
Pesticides		*n*-heptane	VIS	40–90	89
Polyphenols	Folin-Ciocalteu reagent	Water	VIS	—	116
Vitamin B_1		Chloroform	FLU	—	58
Ag	Dithizone	Carbon-tetrachloride	VIS	0.05 mM	136
Au	Chloride	MIBK	AAS	1.8	98

continued

Table 1 (*continued*)

Analyte	Reagent	Solvent	Detection	Detection limit, ppb	Reference
Bi	Lead tetramethylene-dithiocarbamate	Chloroform	VIS	100	152
	Iodide, tetrametylenebis (triphenylphosphonium) bromide	1,2-dichlromethane	VIS	240	157
Ca	Propyl orange	Chlorobenzene + benzene	VIS	0.2 μM	137
Cd	Dithizone	Chloroform	VIS	0.2	138
	PAN	Chloroform	VIS	85	117
Cd, Co, Cu, Ni, Pb	—	Freon 113	AAS	20	139
Chlorate	Cu(I)-6-methylpicoline-aldehyde azine	MIBK	AAS	70	140
Co	Ethylene-bis-(triphenylphosphonium) bromide	Chloroform	VIS	230	91,157
Cu	Ammonium pyrolidine dithiocarboxylate	4-methyl-2-pentanone	AAS	2	103
	Lead diethyl-dithiocarbamate	Chloroform	VIS	40	141
Cu, Ni, Pb, Zn	Ammonium pyrolidine-dithiocarbamate	MIBK	AAS	0.8–9.0	142
Dichromate	Tetramethylene bis-(triphenylphosphonium) ion	Chloroform	VIS	440	143
Fluoride	Alizarin complexone + lanthanum chloride	Hexanol +*N*, *N*-diethylaniline	ICP-AES	30	105
Ga	Lumegallion	Isoamyl alcohol	FLU	70	101
Hg	1,5-bis (di-2-pyridyl) methylene thiocarbon-ohydrazide	MIBK	HG-ICP-AES	4	144
K	Dibenzo-18-crown-6	Chlorobenzene + benzene	VIS	40	145

continued

Table 1 (*continued*)

Analyte	Reagent	Solvent	Detection	Detection limit, ppb	Reference
		Benzene	VIS	—	146
		Chloroform	VIS	—	147
		1,2-dichloro-ethane	FLU	—	148
Li	Dodecyl-14-crown-4	Chloroform	VIS	16000	149
	Aminobenzo-14-crown-4	1,2-di-chloroethane	VIS	14000	150
	14-crown-4-dinitrophenol	Chloroform	VIS	0.1 μM	155
Mn	Ethylene bis-(triphenylphosphonium) bromide	Chloroform	VIS	580	151,157
Mo	Thiocyanate	Isoamyl alcohol	VIS	50	81
Nitrate	Cu(I) Neocuproine	MIBK	AAS	400	153,154
Nitrite	Cu(I) Neocuproine	MIBK	AAS	40	153,154
Pb	Dicyclohexyl-18-crown-6, dithizone	Chloroform	VIS	50	130
	Iodide	MIBK	AAS	20	156
Perchlorate	Brillant green	Benzene	VIS	36	91,157
Phosphate	Malachite green, molybdate	Benzene + 4-methyl-pentan-2-one	VIS	0.1	85
	Molybdate	Isobutyl acetate	VIS	10	158
	–	Dioctyltin dinitrate	VIS	0.3 as P	159
U	Aluminium nitrate	MIBK	VIS	50	123
		Tributyl phosphate in heptane	VIS	100	160
Zn	Thiocyanate	MIBK + acetone	AAS	200	161

*MIBK — methyl isobutyl ketone

**Detections: AAS — atomic absorption spectrometry; FLU — fluorescence; ICP-AES — inductively coupled plasma atomic emission spectrometry; HG-ICP-AES — hydride generation inductively coupled plasma atomic emission spectrometry; UV — UV spectrophotometry; VIS — spectrophotometry in visible region.

interferences observed usually for the AAS zinc determination [161]. In the determination of lead in the FIA system with on-line solvent extraction, the improvement of the detection limit has been obtained by the use of additional solid-phase extraction with a microcolumn containing a Chelex-100 chelating agent [130]. The extraction FIA systems with AAS detection can also be utilised for indirect determination, where the analytical signal of a metallic element is proportional to the concentration of anions coextracted into the organic phase [140, 153, 154]. In the determination of ortho- and pyrophosphates, anions have been extracted to an organotin extractant immobilised on an inert support and then eluted for further processing in the FIA system [159].

5. Solid-Phase Extraction

In spite of the above-shown numerous applications of solvent extraction in FIA systems, the predominating method that has been rapidly growing in recent years is preconcentration on solid sorbents, commonly named in the literature as *solid-phase extraction* (SPE). Most often it is employed by the use of microcolumns packed with an appropriate sorbent, but also with the use of membrane devices, or by utilising the sorption on the inner walls of open tubings. The most essential advantage of this method compared to solvent extraction is the elimination of the use of toxic organic solvents, although in certain SPE procedures small amounts of organic solvents are used for elution of the retained analyte, mainly when it is proconcentrated as chelate. Another significant advantage is simplification of the FIA system, which does need a segmentor, separator and extraction coil. Additionally, in SPE FIA systems the sample can not only be injected with a limited volume but it can also be continuously aspirated by an unlimited period of time that enables one to achieve a much larger preconcentration than in solvent extraction systems.

The applications of this method pioneered by Olsen *et al.* [162] are especially frequent in FIA systems with atomic spectrometry detection methods reviewed by several authors [163–165], although, as already indicated in earlier chapters, it is also widely used in FIA systems with other detection methods. Most often this step of sample pretreatment is employed for selective preconcentration of the analyte or a group of species if the used sorbent is not sufficiently selective. Several other developed applications are shown below, where SPE is used for removal of interfering components of matrices.

The sorption process is carried out by pumping the sample solution in a suitable medium through the microcolumn with an appropriate sorbent. The time-based sampling with continuous aspiration of the sample allows higher loading efficiencies and a simpler manifold. A disadvantage of this mode in the case of some sorbents may be fluctuations of the flow rate due to a change of volume of the sorbent bed, which cause errors of determination. The flow of the sample solution through a long tubing instead of injection from the loop of the injection valve may also be a source of carry-over errors; therefore, the length and diameter of tubing used for continuous aspiration of the sample solution should be minimised. For various reasons it is also advantageous to design manifolds where the solution flowing through the column during the preconcentration phase does not enter the detector but is directed towards waste.

Of great importance for effectiveness of preconcentration is appropriate design of the microcolumn, which should depend on the breakthrough capacity of the analyte, the particle size of the packing material and the sample loading rate and volume. The higher enrichment factors are usually obtained for a larger ratio of column length to column diameter [166]. This is limited, however, by back pressure generated at high sample loading rates. The inner diameters of microcolumns used in FIA systems are usually from 1.5 to 7.5 mm, the length from 2 to 10 cm, and the bed volume from 50 to 500 μl. In order to decrease the dispersion of the eluted analyte, conical microcolumns are also used [167–169]. The decrease of particle size of the column packing results in an increase of breakthrough capacity and sharper elution peaks; however, it also causes an increase of the back pressure which limits the flow rate used in sample loading. A reasonable range of a particle size is 150–200 μm at a loading rate of 8–9 ml.min^{-1}. [164]. Examples of construction of microcolumns are shown in Fig. 10. A design with especially good tolerance of the high sampling rate is the one with threaded fittings.

Equally important as the preconcentration phase is the elution step of the preconcentrated analyte. The used eluent should ensure a fast elution of the analyte with a possibly small volume of the solution and should not attack the packing material. Most often, as eluents are used, concentrated solutions of acids and bases or solutions of complexing agents, which for example with chelating sorbents prevent swelling and shrinking of the resin when acids are used. Organic solvents are used for elution of heavy metal complexes preconcentrated on nonpolar sorbents. Reversal of the flow direction between loading and elution steps allows one to avoid the column becoming more and more tightly packed, which results in restriction of the flow. It is often advantageous in decreasing the dispersion of the analyte during elution.

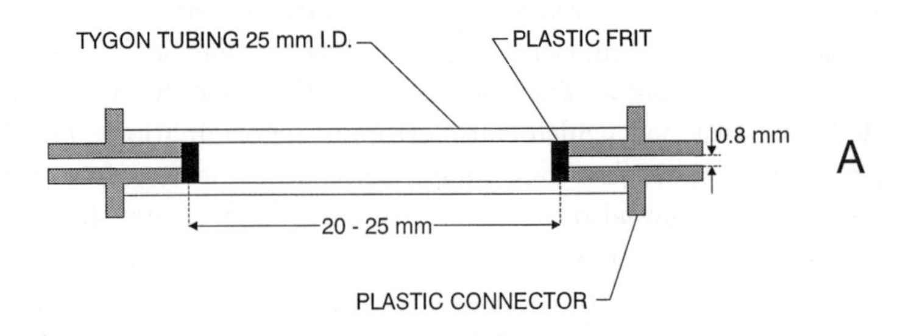

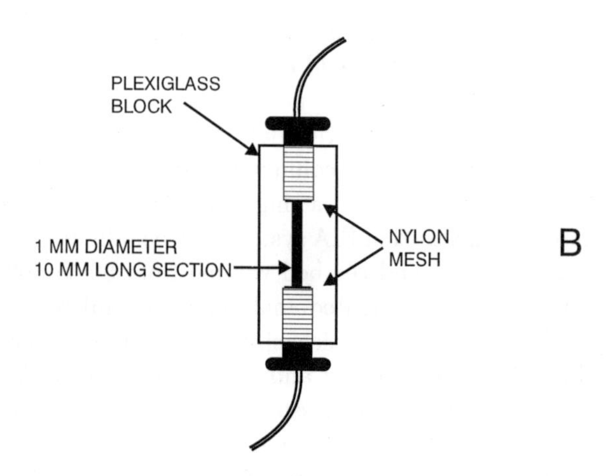

Fig. 10. Examples of construction of microcolumns for solid-phase extraction used in FIA systems: A — column with push-in connectors [171]; B — column with threaded connectors [244]. (*Reprinted by permission of copyright owner.*)

In order to evaluate the efficiency of on-line preconcentration in the SPE process, the values of the enrichment factor (EF) or concentration efficiency (CE) are used. EF is a measure of the signal amplification as a result of preconcentration without taking into consideration the time period of the preconcentration or sample volume used. Hence EF values are sometimes very high; however, they do not provide adequate information about the efficiency of the procedure. The CE value can be determined by comparison of the peak height before and after the preconcentration [167], or by comparing the slopes of the linear part of the calibration curves before and after the preconcentration [172]. As a more adequate criterion for the evaluation of the efficiency of preconcentration in FIA systems, concentration efficiency can be used, which was defined as the product of EF and the sampling frequency in the number of samples analysed per minute [173]. CE values as high as 100 EF.min^{-1} have been reported [173].

As packing materials for on-line SPE in FIA several different groups of sorbents are used, including chelating resins, sorbents with coupled complexing ligands, ion-exchange resins, non-polar sorbents employed for preconcentration of metal chelates and sorbents loaded with complexing ligands (Table 2).

Table 2. Packing materials used in on-line solid-phase extraction for preconcentration of ionic species in FIA systems with microcolumns.

Type of sorbent	Sorbent	Analyte	Detection	Reference
Chelating resins	Chelex-100	Cd, Pb, Zn	AAS	162
		Cd	AAS	174
		Cd, Cu, Mn	AAS	183
		Pb	VIS	175
		Cu	POT	175
		Heavy metals	ICP-AES	184
	Muromac A-1	Cd, Cr(III), Cu, Fe(III), Mn, Pb, Zn	AAS	176
		Cd	ICP-AES	185
		Al, Cr, Fe, Ti, V	ICP-AES	186
	Spheron Oxin 1000	Cu	AAS	177
	Resin with 8-HQ	Pb	AAS	178
		Ni	VIS	179
	Poly(hydroxamic acid) resin	Cr(III)	AAS	180
	N-methylglucamine resin	B	VIS	171
Sorbents with coupled ligands	8-HQ on silica	Cu	AAS	166
		Co	VIS	181
		Pb	AAS	170
		Heavy metals	GF-AAS	198
	8-HQ on controlled-pore glass	Cu	POT	192
		Cd, Cu	AAS	163
		Cd, Co, Cu, Ni Pb, Zn	AAS	187
		Co	AAS	77

continued

Table 2 (*continued*)

Type of sorbent	Sorbent	Analyte	Detection	Reference
		Be, Ce, Co, Ni, V	ICP-AES	77
		Fe(II)	CHLU	188
		Bi	HG-AAS	199
	8-HQ on vinyl polymer gel	Mn	VIS	189
		Al	FLU	190
		Zn	FLU	191
	8-HQ on methacrylate gel	Pb	GF-AAS	195
	Sulphydryl cotton	Hg, MetHg	CV-AFS	192
		Au	ICP-MS	193
	Fibrous cellulose with diethylenetriamine groups	Fe	VIS	194
	Fibrous cellulose with diethylenetriamine tetracetic groups	Mn	VIS	200
Ion-exchange sorbents	Cation-exchanger	NH_4^+	VIS	201
	Amberlite IR-120	Se(IV)	HG-AAS	77, 199
	Anion-exchanger D-201	V(V)	AAS	202
	Anion-exchanger SAX	Zn	VIS	203
	Anion-exchanger	Au	ICP-MS	204
	BioRad AG 1-x2	Au	AAS	168
	Amberlite XAD-8	Al	ICP-AES	205
	Amberlite IRA 400	Mo, V, W, Zn	ICP-AES	206
	Amberlite IRA 93	Ag	AAS	207
	Alumina	As, B, Cr, Mo, P, Se, V	ICP-AES	208
		Cr(III)	ICP-AES	209
		Cr(III), Cr(VI)	ICP-AES	210
		Co	VIS	211
		Mo	ICP-AES	212
		P	ICP-AES	213
		Sulphate	ICP-AES	214
	Cellulose with triethylamino groups	Cr(VI)	AAS	215

continued

Table 2 (*continued*)

Type of sorbent	Sorbent	Analyte	Detection	Reference
	Cellulose with	Cr(III)	AAS	215
	phosphonic acid groups	Pb	AAS	216
Sorbents	Anion-exchanger with	Al	AAS	217
loaded	chromazurol			
with	XAD-2 with	Ni	AAS	218
ligands	eriochrome blue-black R			
	XAD-2 with	Cu	AAS	219
	pyrocatechol violet			
	Alumina with nitroso-R-salt	Co	AAS	220
	C_{18} with ferrozine	Fe(II)	VIS	221

*8-HQ — hydroxyquinoline.

Among chelating resins most applications have been reported for resins with iminodiacetic acid functional groups (Chelex-100, Muromac A-1) and 8-hydroxyquinoline (Spheron Oxin 1000). Chelex-100 is widely used in non-flow preconcentrations and also in the earliest FIA procedures; however, it is not advantageous due to swelling properties, which are not shown by resin Muromac A-1 with the same functional groups [176, 185, 186]. An especially large number of applications in FIA systems with various detections have been reported for different sorbents with 8-hydroxyquinoline (8-HQ) immobilised on various supports, such as silica gels, controlled-pore glass (CPG) and polymer gels. To silica gel 8-HQ can be attached by the diazo coupling [196], similarly to many other ligands, such as 4-(2-pyridylazo)resorcinol or pyrocatechol violet [170]. Azoimmobilisation procedures are also used for ligand coupling to CPG and to vinyl polymers [197]. Using the latter sorbent the spectrophotometric method for manganese determination in sea water has been developed, which was shown to be less expensive than methods requiring the use of a GFAAS, ICP or ICP-MS [189], and also a method for the shipboard determination of Al in sea water with fluorimetric detection [190]. The use of a microcolumn of sulphydryl cotton allows the rapid sequential determination of inorganic mercury and methylmercury at the ng.l^{-1} level with cold vapour-atomic fluorescence spectrometric detection [192] and also the determination of gold in waters using for detection inductively coupled plasma mass spectrometry [193]. 8-HQ bonded on methacrylate gel and packed in the tip of the sampling arm of the sampler was used for preconcentration of lead in the FIA system with graphite furnace AAS detection [195].

In spite of the worse selectivity than numerous sorbents with complexing functional groups, wide application in FIA systems is reported for ion-exchange resins. Among them, most frequently are used anion-exchange resins for preconcentration of trace elements that occur as anions or trace cations that can form negatively charged complexes, which were employed for preconcentration of aluminium [205], gold (III) [168, 204] and zinc [203, 206]. For this purpose, also widely are used ion-exchange properties of activated alumina, which, depending on activation, acts as an ion-exchanger for anion or cations. When an acid carrier stream was used Cr(III) did not deposit and Cr(VI) deposition was quantitative. This allowed simultaneous determination of the two oxidation states of chromium with ICP-AES detection using ammonia solution for stripping Cr(VI) [210]. Another FIA procedure for determination of Cr(III) and Cr(VI) with AAS detection has been developed using two functionalized cellulose sorbents with cation- and anion-exchange properties [215].

Chelating agents can also be sufficiently stable when sorbed on various supports as a result of hydrophobic or electrostatic interactions. Such prepared sorbents have found several applications for on-line SPE in FIA systems. As supports for such loaded chelating sorbents are used both non-polar sorbents such as Amberlite XAD-2 [218, 219] and C_{18} [221] and ion-exchange resins [217, 220]. A suitable choice of support and ligand allows one to obtain sorbents that can be used reversibly numerous times. This method enables one to prepare specific sorbents without the necessity of synthesis of appropriate resins.

Another possibility of the effective on-line SPE in FIA systems in determination of trace elements is sorption of earlier-produced chelates on non-polar sorbents and their elution after preconcentration with organic solvents (Table 3). The advantage of this method is elimination of the need to use the selective stationary phase. Most often for this purpose is used the most common HPLC stationary phase octadecyl bonded silica, and as retained chelates carbamates are used. An example FIA system used for this purpose is shown in Fig. 11. In this case in the FIA system with AAS detection lead was preconcentrated as pyrrolidinedithiocarbamate on an activated activated carbon [228]. The chelate was eluted with methyl isobutyl ketone, which is water-immiscible and therefore the analyte is not dispersed on transfer to the

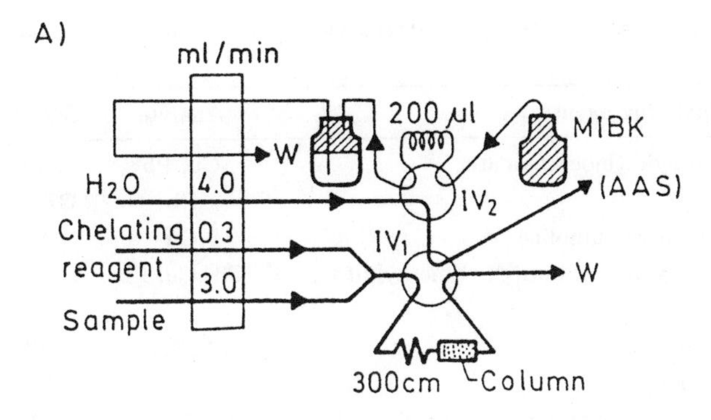

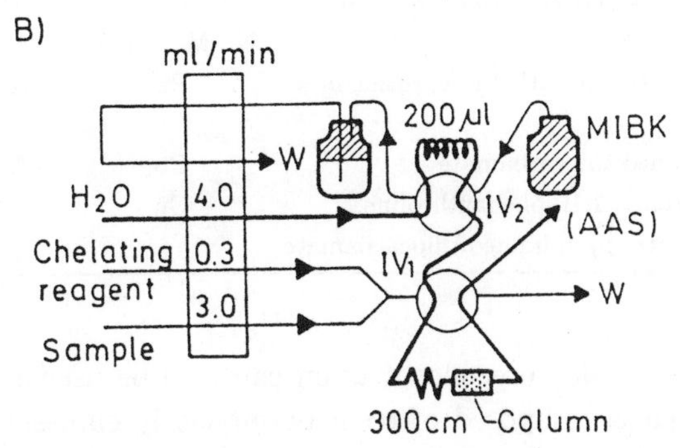

Fig. 11. Manifold of the FIA system with flame AAS detection used for the determination of lead with preconcentration by solid-phase extraction of pyrrolidinedithiocarbamate on an activated carbon [228]: A — preconcentration step; B — elution step; IV — injection valve; W — waste; MIBK — methyl isobutyl ketone. (*Reprinted by permission of copyright owner.*)

instrument nebuliser. Very promising results were reported in such systems, when instead of C_{18} or activated carbon C_{60} fullerenes were used, which provided a better preconcentration factor and the highest selectivity [229].

The C_{18} solid phase can also be used in the FIA system for preconcentration of typical organic analytes, which has been demonstrated in the simultaneous determination of carbaryl and 1-naphthol with Fourier transform infrared spectrometric detection [231].

Solid-phase extraction of metal chelates has also been demonstrated without the use of a preconcentration column. The copper diethyldithiocarbamate was adsorbed on the walls of a PTFE knotted reactor [232] and a sorbed species was eluted with methyl isobutyl ketone.

Table 3. Preconcentration of heavy metals as chelates in FIA systems with on-line solid-phase extraction.

Sorbent	Chelating agent	Analyte	Detection	Reference
C_{18}	Diethyldithiocarbamate	Cu, Pb	AAS	222
bonded		Pb	GF-AAS	223,226
silica	8-hydroxyquinoline	Cu	AAS	222
	Diethylammonium -N,N-diethyldithio-carbamate	Cd, Cu, Pb	AAS	169
	Ferrozine	Fe(II)	AAS	224
Silica gel	EDTA, tartrate	Be	FLU	225
XAD-4	Bis(carboxymethyl)-dithiocarbamate	Co, Cu, Cr, Ni, Mo, Pt, V	ICP-MS	227
Activated carbon	Pyrrolidinedithiocarbamate, dithizon	Pb	AAS	228
Fullerenes	Pyrrolidinedithiocarbamate	Pb	AAS	229
	Neocuproine, 1,10-phenanthroline, thiocyanate, pyrrolidinedithiocarbamate	Cu	AAS	230

Ion-exchange resin for preconcentration can also be used in the shape of a membrane which is composed of resin permanently enmeshed in a PTFE membrane. This was employed for preconcentration of iodine as iodide and iodate in the FIA system with ICP-AES detection [233].

As efficient sorbent for preconcentration of trace metals, certain microorganisms covalently immobilised on CPG can also be used. It was demonstrated in FIA systems with AAS detection for preconcentration of Cd and Cu on the yeast *Saccharomyces cerevisiae* [234], and Cd, Cu, Fe(III), Pb and Zn on cyanobacteria *Spiruline platensis* [235].

In FIA systems with spectrophotometric and fluorimetric detections, optical measurement is possible directly on the sorbent bed. Described as an ion-exchanger absorptiometric detection [236], it has found numerous applications [237, 238]. In FIA systems this concept leads to integration of the retention stage with the detection. For this purpose, on a sorbent located in the flow cell the analyte can be immobilised in a transient manner. Copper can be determined by the measurement of absorbance of copper ions retained on a cation-exchange resin [236]. The analysis of a mixture of amines was carried out using a multichannel UV-VIS diode array detector with analytes retained on C_{18} silica beads [239]. Most often a sorbent located in the flow cell is used to preconcentrate the product of reaction of the analyte with a suitable color-forming reagent in a coil located before the flow cell. Trace amounts of bismuth, which form iodo-complexes, were retained on anion-exchanger Sephadex A-25 [240]. Determination of Cr(VI) was based on the reaction with 1,5-diphenylcarbazide and the retention of the reaction product in a cation-exchange resin [241]. Traces of iron(III) in water and wine can be determined by measuring the absorbance of the anion-exchange Dowex 1-X2-200 which retains the previously formed iron-thiocyanate complex [242]. Two different methods were developed for trace determination of phosphate. The ion associate formed by the reaction of molybdophosphate with malachite green was adsorbed on a hydroxypropyl derivative of a Sephadex dextran gel and direct absorptiometric measurements were made in the gel phase [243]. In the second method the heteropoly complex formed with heptamolybdate was retained on C_{18} sorbent [244]. In the determination of tin the complex formed between Sn(IV) and pyrocatechol violet was concentrated on a Sephadex QAE A-25 gel packed in a flow cell [245]. Monitoring of the fluorescent complex retained on the C_{18} material packed in the flow cell was reported in flow injection speciation of aluminium [246]. The immobilisation of the biocatalyst and a reaction product in two layers in a flow cell was proposed for determination of hydrogen peroxide and glucose in the FIA system with spectrophotometric detection [247]. The enzyme was immobilised permanently in a layer of CPG, while the product of the reaction between H_2O_2, 4-aminophenazone and aniline derivative was temporarily retained in a Sephadex resin layer.

The main purpose of the use of SPE in both mentioned applications was preconcentration of the analyte or the product of its reaction in order to amplify the analytical signal and improve the detection limit of the determination. There were also numerous FIA procedures where SPE was successfully applied for removal of matrix components interfering in detection. For instance, a cation-exchange resin column was used to remove on-line interfering ions in catalytic determination of molybdenum with spectrophotometric detection [248]. Polyvalent cations were removed on a cellulose phosphate column in spectrophotometric determination of manganese [249], while a Chelex-100 column was used to remove cationic interferences in determination of cadmium with malachite green based on formation of stable cadmiumchloro-complexes which are not sorbed [250]. In voltammetric determination of trace metals a silica anion-exchange column was employed to remove the compound with a strong tendency to adsorb to the electrode [251], whereas a cation-exchange

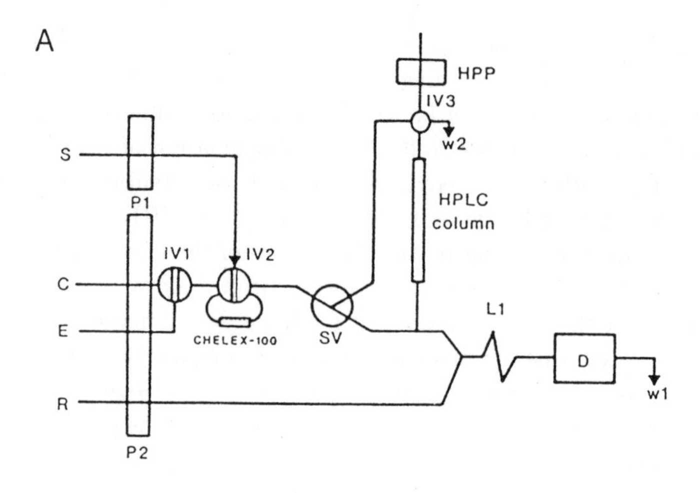

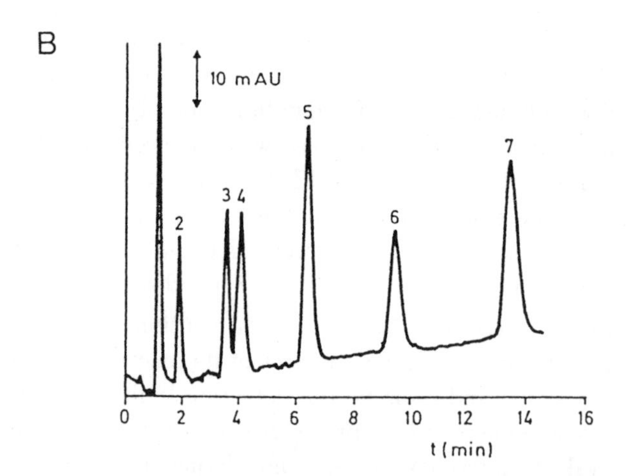

Fig. 12. FIA manifold (A) and recorded chromatogram (B) in an FIA/HPLC integrated system for preconcentration on Chelex-100 and determination of transition metal ions [256]. S — sample; C — carrier stream of distilled water; E — eluent 0.1 M nitric acid; R — derivatizing solution 0.1 mM PAR in ammonium acetate buffer; P — low pressure pump; HPP — high pressure pump; IV — injection valves; W — waste; D — diode array spectrophotometer. Chromatographic conditions for B: C_{18} Ultrabase column (25 × 4.6 mm, 5.0 μm), mobile phase 50 mM tartaric acid with 2 mM sodium octanesulphonate pH 3.5, flow rate 1.3 ml.min^{-1}. Signal corresponds to 5 min preconcentration at 3.2 ml.min^{-1} of solution containing (1) 25 ppb Cu; (2) 70 ppb Pb; (3) 10 ppb Zn; (4) 15 ppb Ni; (5) 25 ppb Co; (6) 25 ppb Cd and (7) 15 ppb Mn. (*Reprinted by permission of copyright owner.*)

column was used to remove interfering divalent cations in stripping voltammetric determination of Se(IV) [252]. A strong anion-exchanger microcolumn located in the aspiration line of the injection valve was employed for separation of anionic interferences in spectrophotometric determination of arsenite [253]. In numerous FIA procedures SPE is also used off-line to remove interferences from the sample solution prior to the injection. This was used, for instance, with a cation exchanger for removal of cations in spectrophotometric determination of sulphate [254]. A Chelex-100 column was used to collect cationic interferences in determination of As and Se by hydride generation AAS [255].

Combining in one measuring setup the FIA manifold with preconcentration on a Chelex-100 column together with an ion-pair chromatographic system with a C_{18} column was reported for the determination of transition metal ions in waters (Fig. 12) [256]. The FIA system allows the preconcentration of analytes prior to the chromatography but can also be used as a screening system for the determination of the total concentration of heavy metals.

6. Flow-Through Reactors

The use of microcolumns packed with sorbent is the most frequently used method of on-line preconcentration and removal of interferences in FIA systems; however numerous applications find microcolumns packed with materials which react with the analyte, providing appropriate chemical transformation of the analyte convenient for its further detection [257, 258]. Such flow-through reactors can be manufactured directly from sparingly soluble salts, metals, minerals or ion-exchange resins with immobilised ion which can react with the analyte. The physical entrapment of the reagent in various matrices or immobilisation on the surface of various sorbents is also used for this purpose. The basic task of the use of packed-bed reactors is minimisation of dispersion and dilution of the sample while transforming into another detectable species or integration of this transformation with the detection [238]. The reaction taking place in the reactor can also be employed to develop indirect determinations.

The simplest example of the application of flow-through reactors is the oxidation or reduction of the analyte. Copper-coated metallic cadmium packed in flow-through reactors is commonly used in determination of nitrate for its reduction to nitrite. This was employed in FIA systems with spectrophotometric [259–264], fluorimetric [265] and biamperometric [266] detections. For the same purpose copperised cadmium tubes were used [267]. A microcolumn packed with a Jones reductor (amalgamated zinc shot) was employed in

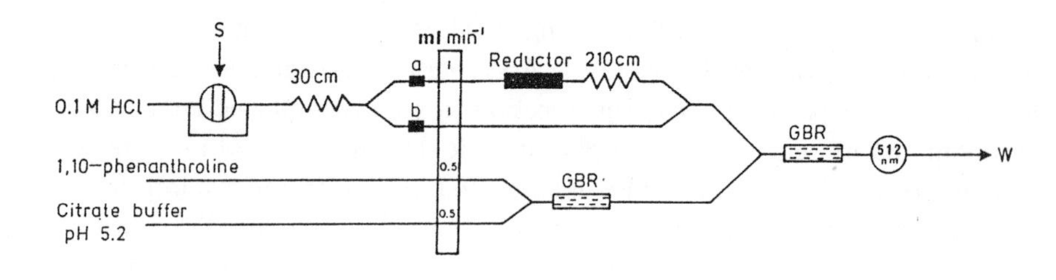

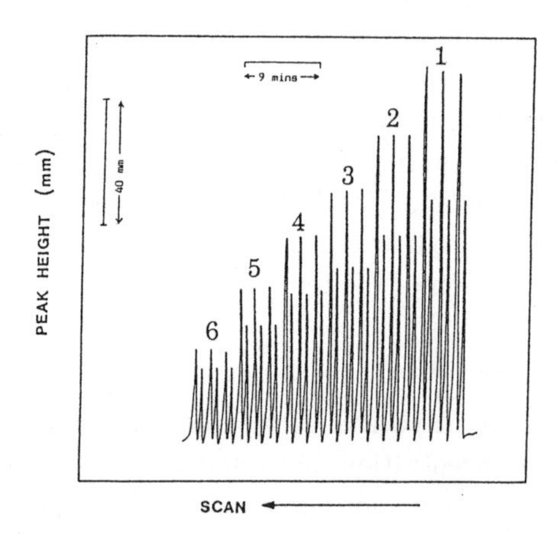

Fig. 13. Manifold of the FIA system used for the simultaneous spectrophotometric determination of iron(II) and total iron with Jones reductor microcolumn (A) and typical responses (B) [268]. S — sample injected (40 μl); a,b — pulse suppressors; GBR — glass beads reactors for stabilisation of base-line; W — waste. Concentrations of Fe(II) and Fe(III) in μM in B: (1) 90, 55; (2) 85, 45; (3) 80, 35; (4) 75, 25; (5) 70, 15; (6) 65, 5. The second peaks in each pair represent total iron. (*Reprinted by permission of copyright owner.*)

the FIA system for spectrophotometric simultaneous determination of iron(II) and total iron [268] (Fig. 13). Sparingly soluble cerium(IV) arsenite packed in a microcolumn was used as a strongly oxidizing agent in spectrophotometric determination of promethazine [269].

Several insoluble compounds packed into a flow-through reactor have been employed for indirect detection. Microcolumns packed with mercury(II) thiocyanate and silver thiocyanate were used for indirect spectrophotometric determination of chloride and bromide [270], while a reaction column packed with spherical beads of cellulose containing solid barium chloranilate particles was employed for spectrophotometric determination of sulphate ion [271]. In the latter system sulphite reacts with barium chloranilate to release chloranilate ion, which is monitored spectrophotometrically at 530 nm, and an additional on-line cation-exchange column is used to remove interfering cations. Several systems have been developed with AAS detection for indirect determinations. A cupric sulphide packed column was used in determination of cyanide [272]; a column containing glass beads and packed with silver chloride was used for AAS determination of ammonia, thiosulphate and cyanide [273], and copper(II) carbonate entrapped in a polyester during polymerisation was used for determination of glycine [274]. A solid lead sulphate column was employed for indirect potentiometric determination of sulphate with lead ion selective electrode, where a detection limit of 1 μM was reported [275].

Some applications were also reported for systems where the substrate of the reaction with the analyte was immobilised on an ion-exchange resin packed into the minicolumn. A fluorimetric determination of paracetamol was based on the oxidation of the analyte with hexacyanoferrate(III) immobilised on an anion-exchange resin [276]. A column packed with a strong anion-exchange resin in the tetrahydroborate form was employed in a flow injection AAS system for generation of arsine [277]. It was shown that such a system exhibits a potential for minimizing interferences of transition metal ions.

An open-tubular reactor with iodine as oxidant was made by impregnation of walls of poly(vinyl chloride) tubing with iodine. This was utilised for fluorimetric determination of adrenaline [278].

The immobilised reagents for the reaction with the analyte can be placed directly in the light path of the detector in the systems with fluorimetric and chemiluminescence detections. This allows the design of simple systems without merging streams of reagents. Bis(2,4,6-trichlorophenyl) oxalate in solid form packed into a bed reactor was used in the FIA system with chemiluminescence detection for the determination of hydrogen peroxide [279]. For the same purpose three other systems were developed. One of them involved a reactor packed with bis(2,4,6-trichlophenyl) oxalate and a flow cell containing immobilised fluorophores (8-hydroxyquinoline or Rhodamine B), located in front of the multiplier tube [280]. H_2O_2 was determined with luminol immobilised on controlled-pore glass [281]. A flow cell with an immobilised fluorophore (3-aminofluoranthene) on an acrylate polymer was placed as close as possible to the photomultiplier tube in the system based on the reaction between hydrogen peroxide and the chemiluminescent compound 1,1′-oxalyldiimidazole [282]. Fluorimetric determination of potassium and ammonium ions was based on the immobilisation of a suitable ionophore on CPG beads in the flow cell placed in the light path of the spectrofluorimeter [283]. The detection is based on the formation of a fluorescent ion pair between the complex cation and organic anion 8-anilinonaphthalene-1-sulphonate in the hydrophobic surface of CPG.

Most applications in FIA systems were developed for enzyme reactors, of which some aspects were already discussed in Chapter 4. In spite of numerous other methods used for the immobilisation of enzymes, such as directly on the active surface of electrochemical detectors and optodes or the use in soluble form, the immobilisation in paced bed reactors is commonly considered most favourable in regard of the stability of the activity of the biocatalyst in time, although incorporation of each reactor in the FIA system creates some additional dispersion of the sample segment in the system. Physical methods of enzyme entrapment in various materials do not offer sufficiently stable preparations for the use in flow systems. Chemical immobilisation by covalent bonding between enzyme and support has been the most common technique used in FIA [284]. The most common support for enzyme immobilisation is controlled-pore glass (Table 4) and the obtained preparations can be easily used in packed-bed reactors. The first step in immobilisation is usually functionalisation with aminosilane, which bears a terminal amino group for a coupling enzyme with glutaraldehyde in the following step. In the most-often-used procedure, as the one developed for instance by Masoom and Townshend [285], aminopropylated CPG is used for this purpose. A two-step immobilisation procedure is usually carried out by activating CPG with solution of glutaraldehyde and then suspended in enzyme solution in a phosphate buffer at 4°C. Several different procedures were also reported. Glutamate dehydrogenase was immobilised on succinate-CPG in the presence of 1-ethyl-3(3-dimethylaminopropyl) carbodiimide [286]. In order to prepare a rotating reactor, CPG was spread on one side of a double-sided Scotch tape, and with such configuration activation with glutaraldehyde and enzyme attachment was carried out [287]. Urease was immobilised on porous glass coated with poly(glycidyl methacrylate) in solution containing enzyme, EDTA and 2-mercaptoethanol [288]. In the immobilisation of penicillinase on the inner walls of the glass tube it was found that the maximum amount of the enzyme can be immobilised after glass silylation with (p-aminophenyl)trimethoxysilane [289]. Covalent immobilisation of enzymes on aminopropylated CPG can also be carried out by circulation of glutaraldehyde and enzyme solutions through a reactor packed with CPG [290–293]. Such a procedure was also applied for immobilisation of various enzymes on alkylamino-bonded silica LiChrosorb NH_2 [294].

Table 4. Enzyme reactors used in FIA systems.

Support	Immobilised enzyme	Reference
Conrolled-pore glass	Alcohol dehodrogenase	295,297
	Alcohol oxidase	296
	Aldehyde dehydrogenase	297
	Alkaline phosphatase	298
	L-amino acid oxidase	286
	Ascorbate oxidase	287
	Aspartate aminotransferase	316
	Catalase	299
	Cholesterol esterase	300,301
	Cholesterol oxidase	293,300,301
	Creatinine deiminase	286
	Ceratinine iminohydrolase	302–304
	Ethanol dehydrogenase	305
	Formate dehydrogenase	306
	β-galactosidase	307
	Glucose oxidase	285,293,299,307–311
	Glucose-6-phosphate dehydrogenase	311
	Glutamate dehydrogenase	302,303,312
	Glutamate oxidase	313
	Glutaminase	313,314
	Glycerol dehydrogenase	305
	Hexokinase	311
	Invertase	299
	Leucine dehydrogenase	286
	Luciferase	315
	Malate dehydrogenase	316
	Mutarotase	299,307
	Oxalate decarboxylase	306
	Penicillinase	289
	Peroxidase	317
	Phosphoglucose isomerase	311
	Sulphite oxidase	287,292,317
	Tyrosinase	290
	Urease	288,318
	Uricase	293
Alkylamino-bonded silica	b-galactosidase, catalase, glucose oxidase	294

continued

Table 4 (*continued*)

Support	Immobilised enzyme	Reference
Sepharose	Ascorbate oxidase	322,326
	Carbonic anhydrase	323
	Sulphite oxidase	324
Fractogel	Glucose oxidase, peroxidase	325
Amino-Cellulofine	Fructose dehydrogenase	326
Poly(vinyl alcohol)	Mannitol dehydrogenase	327
Nylon	Glucose oxidase	328
	Urease	329
PTFE	Urease	330

CPG was also used to immobilise antibodies in enzyme-linked flow injection immunoassays of theophyline and insulin [319] and also immunoglobulin G [320]. In order to avoid non-specific protein adsorption 1,1-carbonyldiimidazole-activated glycerol-coated CPG beads were used. In a liposome flow injection immunoassay of theophyline fused silica particles were used to prepare the antibody flow-through reactor [321].

Numerous other supports were also reported for preparation of enzyme reactors. Ascorbate oxidase [322, 326], carbonic anhydrase [323] and sulphite oxidase [324] were immobilised on cyanogen bromide activated Sepharose. Fractogel activated with a 2-fluoro-1-methylpyridinium salt was used for co-immobilisation of glucose oxidase and peroxidase for FIA spectrophotometric determination of glucose [325]. Mannitol dehydrogenase was covalently immobilised with glutaraldehyde on an aminated poly(vinyl alcohol) bead [327]. Enzymes were also immobilised on inner walls of a nylon tube [328, 329] and on a microporous PTFE membrane in a gas-diffusion reactor using the perfluoroalkyl immobilisation approach [330].

7. Precipitation

The preconcentration of the analyte and indirect determination in FIA systems can also be achieved by the use of on-line precipitation. This sample treatment is not that common as solvent extraction or solid-phase extraction, but several interesting developments have been reported in the literature, which were reviewed by several authors [165, 331, 332]. Contrary to the systems with turbidimetric detections (Chapter 1), with the use of precipitation for analyte

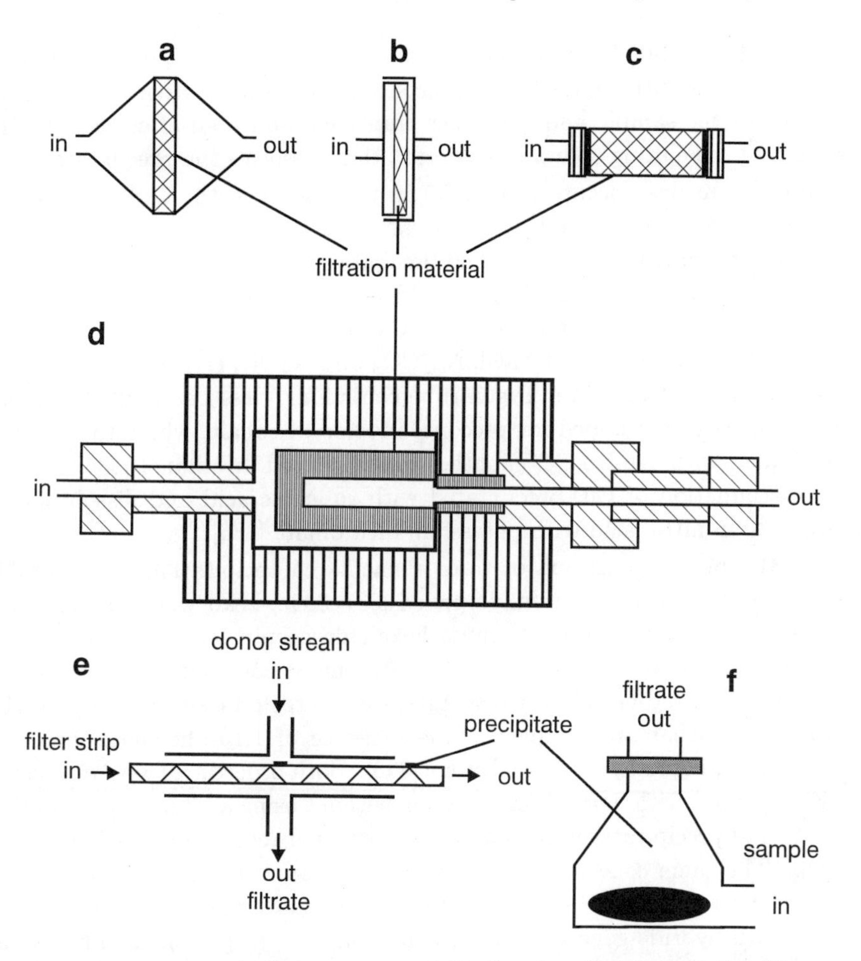

Fig. 14. Filtration devices used in FIA systems [332]: a,b — disposable in-line filters; c — packed column; d — HPLC filtration cartridge; e — filter with a filter strip; e — mixing chamber for solid/liquid reactions. (*Reprinted by permission of copyright owner.*)

preconcentration or indirect determinations the essential element of the system besides the precipitation coil is the filter. Various designs of filters used in FIA systems are shown in Fig. 14. For this purpose stainless steel filtration cartridges used in HPLC, disposable in-line filter cartridges, microcolumns packed with various beads or glass wool can be employed. The determination can be carried out with or without the precipitate dissolution. FIA systems with dissolution of accumulated precipitate are used mainly for the preconcentration of the analyte, while both options can be utilised for indirect determinations. In the systems with precipitation preconcentration most often a continuous pumping of the sample and reagent is employed, and then after washing the precipitate the eluting solution is injected that dissolves the precipitate.

In order to preconcentrate lead(II) in the system with AAS detection, a continuous aspiration of a 10–250 ml sample is used and precipitation with 1.5 M ammonia solution [333]. The precipitate collected on a stainless-steel filter is then dissolved in 2 M nitric acid which yields a concentration factor of up to 700. In a similar system Ag, Ca and Fe(III) were preconcentrated using as reagents solution of NaCl, Na_2CO_3 and NaOH, respectively, and precipitates were collected in a Tygon tube containing glass beads [334]. Similar procedures were developed for preconcentration of cobalt, where the precipitate formed with 1-nitroso-2-naphthol was dissolved in ethanol [335], and in the determination of Cu, precipitated with rubeanic acid and dissolution of precipitate in nitric acid with potassium dichromate [336].

For the preconcentration of trace elements coprecipitation with suitable carrier was also very successfully applied [337–339]. Lead was coprecipitated with an iron(II)-hexahydroazepinium hexahydroazepin-1-yl formate (hexamethyleneammonium hexmethylenedithiocarbamate) complex and dissolved in methyl isobutyl ketone [37], whereas Ag(I) was coprecipitated with the iron(II)-diethyldithiocarbamate complex in the presence of 1,10-phenanthroline and also dissolved in methyl isobutyl ketone [338]. Both procedures were developed for FIA systems with flame AAS detection without using a filter as quantitative collection of precipitates was obtained in a knotted reactor made of Microline tubing. The same collection of precipitate was found to be effective for preconcentration of Se(IV) by coprecipitation with lanthanum hydroxide in the system with hydride generation AAS detection [339] (Fig. 15). In this system each sample solution was spiked with lanthanum nitrate up to concentration 20 ppm and precipitation was carried out by merging with a 0.2 M ammonium buffer of pH 9.1.

The on-line precipitation of the analyte can not only be a convenient way of preconcentration of the analyte but it can also be used for the elimination of interferences present in complex matrices. The precipitation of calcium as oxalate and dissolution of precipitate in HCl allows the elimination of aluminium interference in calcium determination with AAS detection [340, 341]. Ag(I) can be separated from the sample matrix after precipitation with HCl and collection on nylon fibres [342] or glass beads [343], and then dissolved in ammonia or thiosulphate solution.

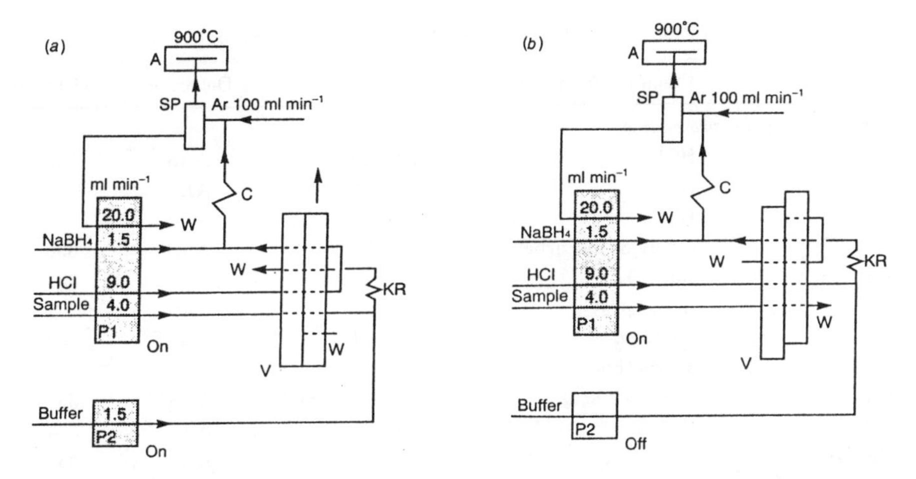

Fig. 15. Manifold of the FIA system with hydride generation AAS detection for determination of selenium(IV) with on-line preconcentration by coprecipitation with lanthanum hydroxide shown in the loading (precipitating) sequence (a) and in the elution stage (b) [339]. A — quartz cell; Ar — argon; C — reaction coil; SP — gas–liquid separator; P1, P2 — persitaltic pumps; V — injection valve; W — waste; KR — knotted reactor. (*Reprinted by permission of copyright owner.*)

The on-line precipitation can also be favourably employed for development of indirect determinations in FIA systems. The determination of chloride with AAS detection can be based on precipitation of AgCl and then dissolution of the precipitate in ammonia solution [343–346]. Chloride, bromide and iodide were precipitated as silver halides, collected on a micrcolumn packed with glass beads and then dissolution with cyanide, thiosulphate or ammonia allowed determination of each analyte [346]. Precipitation of calcium salts is used for indirect determination of oxalate [344] and carbonate [347]. Two different methods were developed for determination of sulphate [348, 349]. In both of them the excess of the precipitating cation was measured by flame AAS. Sulphide was determined in the system with on-line precipitation of CdS and a chelating sorbent column [350]. The cadmium sulphide passes the column and is detected by AAS as the first peak, then the excess of Cd(II) ions retained on the column is eluted with nitric acid, yielding a second peak. In determination of potassium the precipitation of sparingly soluble associates of a potassium-crown ether complex with tetraphenylborate ion was used [351]. The decrease of UV absorbance of tetraphenylborate was proportional to the content of potassium.

Table 5. Application of on-line precipitation for sample processing in FIA systems.

Analyte	Precipitating reagent	Eluent	Detection	Reference
Preconcentration procedures				
Ag	HCl	Ammonia	AAS	342
	HCl	Thiosulphate	AAS	334,343
	Fe(II)-diethyldithiocarbamate*	MIBK	AAS	338
Ca	Na_2CO_3	HCl	AAS	334
	$(NH_4)_2C_2O_4$	HCl	AAS	340,341
Co	1-nitroso-2-naphthol	Ethanol	AAS	335
Cu	Rubeanic acid	HNO_3,$K_2Cr_2O_7$	AAS	336
Fe(III)	NaOH	H_3PO_4	AAS	334
Pb	Ammonia	HNO_3	AAS	333
	HMA-HMDTC*	MIBK	AAS	337
Se(IV)	Ammonium buffer, $La(NO_3)_3^*$	HG-AAS	339	
Indirect determinations				
Chloride	$AgNO_3$	Ammonia	AAS	343–346
Chloride, bromide, iodide	$AgNO_3$	Cyanide, thiosulphate, ammonia	AAS	346
Oxalate	$CaCl_2$	HCl	AAS	344
Carbonate	$CaCl_2$	HCl	AAS	347
Sulphate	$BaCl_2$	–	AAS	348
	$Pb(NO_3)_2$ in 20% methanol	–	AAS	349
Sulphide	$Cd(NO_3)_2$	–	AAS	350
Ammonium	$FeCl_3$	–	AAS	344
K	18-crown-6, tetraphenylborate	–	UV	351
Lidocaine, tetracaine, procaine	Co(II)	–	AAS	352
Saccharin	Ag(I)	Ammonia	AAS	353
Sulphonamides	Cu(II)	–	AAS	354
Tannins	Cu(II)-acetate complex	–	AAS	355

*Coprecipitation of analyte

MIBK — methyl isobutyl ketone; HMA-HMDTC

As can be seen from Table 5, several FIA procedures with on-line precipitation have also been developed for various organic compounds.

8. Less Common Operations

A commonly used way to introduce reagents in the FIA system is merging of the streams of solutions, which causes their dilution, and when it is carried out after the injection valve the confluence of solutions is a source of additional dispersion of the sample zone. One of the methods to avoid this is the use of flow-through reactors. In another way the reagent can be introduced into the system through the membrane devices. This method with the use of Nylon 66 and Celgard 5511 membranes was applied to introduce glucose oxidase as a reagent for enzyme-catalysed determinations of sucrose and glucose, respectively [356, 357]. A porous PTFE membrane has been used to introduce color-forming reagent PAR in the determination of copper [358]. The Nafion tubing was used to introduce ammonium ion from ammonium hydroxide solution [359], whereas a silicon rubber membrane was employed for the introduction of bromine gas in the indirect determination of phenols [360]. The FIA system with the introduction of the two reagents for the spectrophotometric determination of phosphate has been developed using membrane reactors [361]. Hydrazine was introduced using Nafion cation exchange tubing while molybdate was introduced using a porous polypropylene membrane with 30% tri-n-butyl phosphate in heptane.

A membrane interface was also developed for hyphenation of a supercritical carbon dioxide extractor with a flow injection system with UV detection for measuring the solubility of two pharmaceuticals in supercritical CO_2 [362]. The developed phase separator with PTFE membrane was used to remove CO_2 from the extract sample in conjunction with a high-pressure sampling valve and a short capillary restrictor between the extractor and the phase separator.

Another membrane-based separation technique developed for on-line sample processing in FIA systems is pervaporation, combining continuous evaporation and gas diffusion [363, 364]. In this technique the sample does not contact the membrane but the volatile analyte or its volatile reaction product evaporates to a space between the donor solution and the membrane and diffuses through the membrane to the acceptor solution. This method was utilised for potentiometric determination of fluoride in various materials after its conversion to volatile trimethylfluorosilane. A system with an integrated pervaporation/detection unit was also developed for this purpose [364] (Fig. 16).

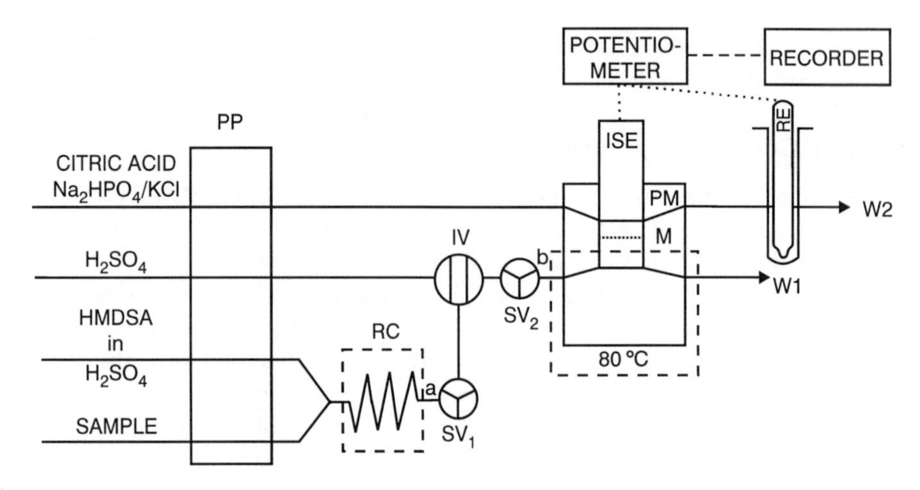

Fig. 16. Manifold of the FIA system with pervaporation/detection arrangement for the determination of fluoride in liquid samples [364]. PP — peristaltic pump; RC — reaction coil; N — injection valve; SV_1, SV_2 — switching valves to select between the injection and aspiration modes; PM — pervaporation module; ISE — ion-selective electrode; RE — reference electrode; M — membrane; W — waste. (*Reprinted by permission of copyright owner.*)

The chemical flow-through reactors used in FIA systems can be in certain cases favourably replaced by electrochemical reactors, for instance in the determination of nitrate that requires reduction of the analyte to nitrite when spectrophotometric detection based on diazotisation is employed. Usually for this purpose packed bed reactors with metallic reductors are used. They can be replaced by the electrochemical reduction with the use of a flow-through column electrode filled with glassy carbon grains coated with copper and cadmium [365, 366].

The use of the electroosmosis propulsion system in the FIA system [367] not only eliminates various drawbacks of the use of persitaltic and piston pumps or gas-pressurized delivery systems, but also allows the on-line preconcentration of the analyte based on electrostacking effects [368]. This can be carried out when conductance of the sample solution is significantly lower than that of the reagent carrier solution and it should be applicable for the preconcentration of virtually all ionic analytes.

9. References

1. L. Ilcheva and K. Cammann, *Fresenius Z. Anal. Chem.* **325** (1986) 11.
2. R. H. Atallah and D. A. Kalman, *Talanta* **38** (1991) 167.

3. R. T. Edwards, I. D. McKelvie, P. C. Ferret, B. T. Hart, J. B. Bapat and K. Koshy, *Anal. Chim. Acta* **261** (1992) 287.
4. D. Chen, A. Rios, M. D. Luque de Castro and M. Valcarcel, *Analyst* **116** (1991) 171.
5. D. Chen, A. Rios, M. D. Luque de Castro and M. Valcarcel, *Talanta* **38** (1991) 1227.
6. A. F. Danet and J. Martinez Calatayud, *Talanta* **41** (1994) 2147.
7. S. Motomizu and M. Sanada, *Anal. Chim. Acta* **308** (1995) 406.
8. L. Renmin, L. Daojie, S. Ailing and L. Guihua, *Talanta* **42** (1995) 437.
9. R.-M. Liu, D.-J. Liu and A.-L. Sun, *Analyst* **117** (1992) 1767.
10. M. de la Guardia, K. D. Khalaf, B. A. Hasan, A. Morales-Rubio and V. Carbonel, *Analyst* **120** (1995) 231.
11. M. Burguera, J. L. Burguera and O. M. Alarcon, *Anal. Chim. Acta* **214** (1988) 421.
12. J. L. Burguera, M. Burguera, P. Carrero, C. Rivas, M. Gallignani and M. R. Brunetto, *Anal. Chim. Acta* **308** (1995) 349.
13. M. Burguera, J. L. Burguera and O. M. Alarcon, *Anal. Chim. Acta* **179** (1986) 351.
14. B. Weltz, D. L. Tsalev and M. Sperling, *Anal. Chim. Acta* **261** (1992) 91.
15. A. Morales-Rubio, M. L. Mena and C. W. McLeod, *Anal. Chim. Acta* **308** (1995) 364.
16. S. Hinkamp and G. Schwedt, *Anal. Chim. Acta* **236** (1990) 345.
17. M. L. Balconi, M. Borgarello, R. Ferraroli and F. Realini, *Anal. Chim. Acta* **261** (1992) 295.
18. M. Valcarcel and M. D. Luque de Castro, *Non-chromatographic Continuous Separation Techniques* (Royal Society of Chemistry, London, 1991).
19. W. B. Furman, *Continuous Flow Analysis: Theory and Practice* (Dekker, New York, 1976).
20. E. H. Hansen and J. Ruzicka, *Anal. Chim. Acta* **87** (1976) 353.
21. J. F. van Staden, *Fresenius J. Anal. Chem.* **352** (1995) 271.
22. B. Bernhardson, E. Martins and G. Johansson, *Anal. Chim. Acta* **167** (1985) 111.
23. L. Risinger, G. Jahansson and T. Thorneman, *Anal. Chim. Acta* **224** (1989) 13.
24. S. D. Kolev and W. E. van der Linden, *Anal. Chim. Acta* **247** (1991) 51; **256** (1991) 301; **257** (1991) 317; **257** (1991) 331.
25. B. Olsson, H. Lundback and G. Johansson, *Anal. Chim. Acta* **167** (1985) 123.
26. S. A. Rosario, G. S. Cha, M. E. Meyerhoff and M. Trojanowicz, *Anal. Chem.* **62** (1990) 2418.
27. J. A. Koropchak and L. Allen, *Anal. Chem.* **61** (1989) 1410.
28. S. A. Rosario, M. E. Meyerhoff and M. Trojanowicz, *Anal. Chim. Acta* **258** (1992) 281.
29. N. Kasthurikrishnan and J. A. Koropchak, *Anal. Chem.* **65** (1993) 857.

30. B. O. Palsson, B. Q. Shen, M. E. Meyerhoff and M. Trojanowicz, *Analyst* **118** (1993) 1361.
31. L. Gorton and L. Ögren, *Anal. Chim. Acta* **130** (1981) 45.
32. M. Masson and A. Townshend, *Anal. Chim. Acta* **166** (1984) 111.
33. E. H. Hansen, J. Ruzicka and B. Rietz, *Anal. Chim. Acta* **89** (1977) 241.
34. W. D. Basson and J. F. van Staden, *Analyst* **104** (1979) 419.
35. A. N. Fudge, *Clin. Chem.* **30** (1984) 493.
36. J. Toffaletti and K. Krivan, *Clin. Chem.* **26** (1980) 1562.
37. J. F. van Staden and W. D. Basson, *Lab. Pract.* **29** (1980) 1279.
38. J. F. van Staden, *Water SA* **12** (1986) 43.
39. J. H. Aldstadt, D. F. King and H. D. Dewald, *Analyst* **119** (1994) 1813.
40. G. B. Martin and M. E. Meyerhoff, *Anal. Chim. Acta* **186** (1986) 71.
41. Q. Chang and M. E. Meyerhoff, *Anal. Chim. Acta* **186** (1986) 81.
42. A. N. Araujo, M. B. Etxebarria, J. L. F. C. Lima, M. C. B. S. M. Montenegro and R. P. Olmos, *Fresenius J. Anal. Chem.* **351** (1995) 614.
43. J. F. van Staden, *Fresenius J. Anal. Chem.* **346** (1993) 723.
44. W. E. van der Linden, *Anal. Chim. Acta* **151** (1983) 359.
45. S. D. Kolev and W. E. van der Linden, *Anal. Chim. Acta* **268** (1992) 7.
46. F. W. Nyasulu and H. A. Mottola, *J. Autom. Chem.* **9** (1987) 46.
47. W. Frenzel, *Anal. Chim. Acta* **291** (1994) 305.
48. D. C. Olsen, S. R. Bysoth, P. D. Dasgupta and V. Kuban, *Proc. Control Qual.* **5** (1994) 259.
49. P. J. Baxter, J. Ruzicka, G. D. Christian and D. C. Olsen, *Talanta* **41** (1994) 347.
50. D. Schepers, G. Schultze and W. Frenzel, *Anal. Chim. Acta* **308** (1995) 109.
51. L. N. Moskvin and J. Simon, *Talanta* **41** (1994) 1765.
52. J. S. Canham, G. Gordon and G. E. Pacey, *Anal. Chim. Acta* **209** (1988) 157.
53. R. Tryzell and B. Karlberg, *Anal. Chim. Acta* **308** (1995) 206.
54. H. Baadenhuijsen and H. E. H. Seuren-Jacobs, *Clin. Chem.* **25** (1979) 443.
55. G. Svensson and T. Anfält, *Clin. Chim. Acta* **119** (1982) 7.
56. G. Schultze, C. Y. Liu, M. Brodowski, O. Elsholtz, W. Frenzel and J. Möller, *Anal. Chim. Acta* **214** (1988) 121.
57. J. R. Clinch, P. J. Worsfold and F. W. Sweeting, *Anal. Chim. Acta* **214** (1988) 401.
58. A. Cerda, M. T. Oms, R. Forteza and V. Cerda, *Anal. Chim. Acta* **311** (1995) 165.
59. W. Frenzel, *Fresenius J. Anal. Chem.* **336** (1990) 21.
60. J. Möller and B. Winter, *Fresenius Z. Anal. Chem.* **320** (1985) 451.
61. J. Bartroli, M. Escalada, C. J. Jorquera and J. Alonso, *Anal. Chem.* **63** (1991) 2532.
62. M. Granados, S. Maspoch and M. Blanco, *Anal. Chim. Acta* **179** (1986) 445.
63. W. Frenzel, C. Y. Liu and J. Oleksy-Frenzel, *Anal. Chim. Acta* **233** (1990) 77.
64. E. B. Milosavljevic, L. Solujic and J. L. Hendrix, *Environ. Sci. Technol.* **29** (1995) 426.

65. Z. Zhu and Z. Fang, *Anal. Chim. Acta* **198** (1987) 25.
66. L. C. de Faria and C. Pasquini, *Anal. Chim. Acta* **245** (1991) 183.
67. T. Aoki and M. Wakabayashi, *Anal. Chim. Acta* **308** (1995) 308.
68. S. Motomizu and T. Yoden, *Anal. Chim. Acta* **261** (1992) 461.
69. P. Martorp, T. Anfält and L. Andersson, *Anal. Chim. Acta* **149** (1983) 281.
70. H. Ohura, T. Imato, Y. Asano, S. Yamasaki and N. Ishibashi, *Anal. Sci.* **6** (1990) 541.
71. M. R. Straka, G. Gordon and G. E. Pacey, *Anal. Chem.* **57** (1985) 1799.
72. J. F. Coetzee and C. Gunaratna, *Anal. Chem.* **58** (1986) 650.
73. J. R. Gord, G. Gordon and G. E. Pacey, *Anal. Chem.* **60** (1988) 2.
74. D. A. Hollowell, J. R. Gord, G. Gordon and G. E. Pacey, *Anal. Chem.* **58** (1986) 1524.
75. M. E. Meyerhoff and Y. M. Fraticelli, *Anal. Lett.* **24 (B6)** (1981) 415.
76. G. E. Pacey, M. R. Straka and J. R. Gord, *Anal. Chem.* **58** (1986) 502.
77. Z. Fang, Z. Zhu, S. Zhang, S. Xu, L. Guo and L. Sun, *Anal. Chim. Acta* **214** (1988) 41.
78. P. C. Hauser and Z. P. Zhang, *Fresenius J. Anal. Chem.* **355** (1996) 141.
79. A. Trojanek and P. Papoff, *Anal. Chim. Acta* **247** (1991) 73.
80. B. Karlber and S. Thelander, *Anal. Chim. Acta* **98** (1978) 1.
81. F. H. Bergamin, J. X. Medeiros, B. F. Reis and E. A. G. Zagatto, *Anal. Chim. Acta* **101** (1978) 9.
82. B. Karlberg, *Anal. Chim. Acta* **214** (1988) 29.
83. V. Kuban, *Crit. Rev. Anal. Chem.* **22** (1991) 477.
84. M. Trojanowicz and J. Szpunar-Lobinska, *Chem. Anal. (Warsaw)* **37** (1992) 517.
85. S. Motomizu and M. Oshima, *Analyst* **112** (1987) 295.
86. V. Kuban and F. Ingman, *Crit. Rev. Anal. Chem.* **22** (1991) 491.
87. F. F. Cantwell and J. A. Sweileh, *Anal. Chem.* **57** (1985) 329.
88. S. Motomizu and K. Korechika, *Japan Analyst* **38** (1989) T143.
89. A. Farran, J. de Pablo and S. Hernandez, *Anal. Chim. Acta* **212** (1988) 123.
90. D. C. Shelly, T. M. Rossi and I. M. Warner, *Anal. Chem.* **54** (1982) 87.
91. D. T. Burns, N. Chimpalee and M. Harriot, *Anal. Chim. Acta* **225** (1989) 123.
92. K. Bäckström and L.-G. Danielsson, *Anal. Chim. Acta* **232** (1990) 301.
93. V. Kuban and F. Ingman, *Anal. Chim. Acta* **245** (1991) 251.
94. T. M. Rossi, D. C. Shelly and I. M. Warner, *Anal. Chem.* **54** (1982) 2056.
95. L. Nord and B. Karlberg, *Anal. Chim. Acta* **164** (1984) 233.
96. B. Karlberg, *Fresenius Z. Anal. Chem.* **329** (1988) 660.
97. S. Lin and H. Hwang, *Talanta* **40** (1993) 1077.
98. K. Bäckström, L.-G. Danielsson and L. Nord, *Anal. Chim. Acta* **169** (1985) 43.
99. L. Nord and B. Karlberg, *Anal. Chim. Acta* **118** (1980) 285.
100. K. Bäckström, L.-G. Danielsson and L. Nord, *Anal. Chim. Acta* **187** (1986) 255.
101. T. Imaska, T. Harada and N. Ishibashi, *Anal. Chim. Acta* **129** (1981) 195.
102. S. Motomizu, M. Oshima and T. Kuroda, *Analyst* **113** (1988) 747.

103. V. Kuban, J. Komarek and D. Cajkova, *Coll. Czech. Chem. Communn.* **54** (1989) 2683.
104. Y. Sahleström and B. Karlbeg, *Anal. Chim. Acta* **185** (1986) 259.
105. J. L. Manzoori and A. Miyazaki, *Anal. Chem.* **62** (1990) 2457.
106. K. Ogata, K. Taguchi and T. Inamari, *Anal. Chem.* **54** (1982) 2127.
107. T. Sakai, Y. S. Chung, N. Ohno and S. Motomizu, *Anal. Chim. Acta* **276** (1993) 127.
108. J. Toei, *Talanta* **36** (1989) 691.
109. Y. Sahleström and B. Karlberg, *Anal. Chim. Acta* **179** (1986) 315.
110. M. H. Memon and P. J. Worsfold, *Anal. Chim. Acta* **201** (1987) 345.
111. M. H. Memon and P. J. Worsfold, *Analyst* **113** (1988) 769.
112. C. Thommen, A. Fromageat, P. Obergfell and H. M. Widmer, *Anal. Chim. Acta* **234** (1990) 141.
113. S. Motomizu and M. Kobayashi, *Anal. Chim. Acta* **261** (1992) 471.
114. F. Canete, A. Rios, M. D. Luque de Castro and M. Valcarcel, *Anal. Chem.* **60** (1988) 2354.
115. F. Canete, A. Rios, M. D. Luue de Castro and M. Valcarcel, *Anal. Chim. Acta* **224** (1989) 169.
116. J. A. G. Mesa, P. Linares, M. D. Luque de Castro and M. Valcarcel, *Anal. Chim. Acta* **235** (1990) 441.
117. I. Facchin and C. Paquini, *Anal. Chim. Acta* **308** (1995) 231.
118. C. A. Lucy and S. Varkey, *Anal. Chem.* **67** (1995) 3036.
119. Y. Hirai and K. Tomokuni, *Anal. Chim. Acta* **167** (1985) 409.
120. L. Nord, K. Backstrom, L.-G. Danielsson, F. Ingman and B. Karlberg, *Anal. Chim. Acta* **194** (1987) 221.
121. C. A. Lucy and F. F. Cantwell, *Anal. Chem.* **61** (1989) 101.
122. L. Fossey and F. F. Cantwell, *Anal. Chem.* **54** (1982) 1693.
123. R. A. Atallah, G. D. Christian and S. D. Hartenstein, *Analyst* **113** (1988) 463.
124. R. A. Atallah, J. Ruzicka and G. D. Christian, *Anal. Chem.* **59** (1987) 2909.
125. J. Toei, *Analyst* **113** (1988) 1861.
126. M. del Valle, J. Alonso, J. Bartroli and I. Marti, *Analyst* **113** (1988) 1677.
127. J. Kawase, A. Nakae and M. Yamanaka, *Anal. Chem.* **51** (1979) 1640.
128. S. Motomizu, Y. Hazaki, M. Oshima and K. Toei, *Anal. Sci.* **3** (1987) 265.
129. T. Sakai, N. Ohna, Y. S. Chung and H. Nishikawa, *Anal. Chim. Acta* **308** (1995) 329.
130. E. A. Novikov, L. K. Shpigun and Yu. A. Zolotov, *Anal. Chim. Acta* **230** (1990) 157.
131. Y. Sahleström, S. Twengström and B. Karlberg, *Anal. Chim. Acta* **187** (1986) 339.
132. J. Kawase, *Anal. Chem.* **52** (1980) 2124.
133. B. Karlberg, P. A. Johansson and S. Thelander, *Anal. Chim. Acta* **104** (1979) 21.
134. M. J. Whitaker, *Anal. Chim. Acta* **179** (1986) 459.
135. J. Möller and M. Martin, *Fresenius Z. Anal. Chem.* **329** (1988) 728.

136. J. Ruzicka and E. H. Hansen, *Anal. Chim. Acta* **99** (1978) 37.
137. S. Motomizu, M. Oshima, N. Yoneda and T. Iwachido, *Anal. Sci.* **6** (1990) 215.
138. J. L. Burguera and M. Burguera, *Anal. Chim. Acta* **152** (1983) 207.
139. M. Bengtson and G. Johnsson, *Anal. Chim. Acta* **158** (1984) 147.
140. M. Gallego and M. Valcarcel, *Anal. Chim. Acta* **169** (1985) 161.
141. J. Szpunar-Lobinska and M. Trojanowicz, *Anal. Sci.* **6** (1990) 415.
142. L. Nord and B. Karlberg, *Anal. Chim. Acta* **145** (1983) 151.
143. D. T. Burns, M. Chimpalee and M. Harriott, *Anal. Chim. Acta* **225** (1989) 241.
144. P. C. Rudner, J. M. C. Pavon, A. G. de Torres and F. S. Rojas, *Fresenius J. Anal. Chem.* **352** (1995) 615.
145. S. Motomizu, M. Onoda, M. Oshima and T. Iwachido, *Analyst* **113** (1988) 743.
146. T. Iwachido, M. Onoda and S. Motomizu, *Anal. Sci.* **2** (1986) 493.
147. R. Escobar, C. Lamoneda, F. de Pablos and A. Giuraum, *Analyst* **114** (1989) 533.
148. K. Kina, K. Shiraishi and N. Ishibashi, *Talanta* **25** (1978) 295.
149. K. Kimura, S. Itekami, H. Sakamoto and T. Shono, *Anal. Sci.* **4** (1988) 221.
150. Y. P. Wu and G. E. Pacey, *Anal. Chim. Acta* **162** (1984) 285.
151. D. T. Burns, N. Chimpalee, M. Harriott and G. M. McKillen, *Anal. Chim. Acta* **217** (1989) 183.
152. J. Szpunar-Lobinska, *Anal. Chim. Acta* **251** (1991) 275.
153. M. Silva, M. Gallego and M. Valcarcel, *Anal. Chim. Acta* **179** (1986) 341.
154. M. Gallego, M. Silva and M. Valcarcel, *Fresenius Z. Anal. Chem.* **323** (1986) 50.
155. K. Kimura, S. Iketani, H. Sakamoto and T. Shono, *Analyst* **115** (1990) 1251.
156. Z. Fang, Z. Zhu, Z. Zhang, S. Xu, L. Guo and L. Sun, *Anal. Chim. Acta* **214** (1988) 41.
157. M. Harriott and D. T. Burns, *Anal. Proc.* **26** (1989) 315.
158. K. Ogata, K. Taguchi and T. Imanari, *Japan Analyst* **31** (1982) 641.
159. B. Ya. Spivakov, T. A. Maryutkina, L. K. Shpigun, V. M. Shkinev, Yu. A. Zolotov, E. Ruseva and I. Havezov, *Talanta* **37** (1990) 889.
160. T. P. Lynch, A. F. Taylor and J. N. Wilson, *Analyst* **108** (1983) 470.
161. J. A. Sweileh and F. F. Cantwell, *Anal. Chem.* **57** (1985) 420.
162. S. Olsen, L. C. R. Pessenda, J. Ruzicka and E. H. Hansen, *Analyst* **108** (1983) 905.
163. Z. Fang, S. Xu and Z. Zhang, *Anal. Chim. Acta* **200** (1987) 35.
164. Z. Fang, *Spectrochim. Acta Rev.* **14** (1991) 235.
165. V. Carbonell, A. Salvador and M. de la Guardia, *Fresenius J. Anal. Chem.* **342** (1992) 529.
166. M. A. Marshall and H. A. Mottola, *Anal. Chem.* **57** (1985) 729.
167. Z. Fang and B. Weltz, *J. Anal. At. Spectrom.* **4** (1989) 543.
168. S. Xu, L. Sun and Z. Fang, *Anal. Chim. Acta* **245** (1991) 7.
169. Z. Fang, T. Guo and B. Welz, *Talanta* **6** (1991) 613.
170. S. R. Bysouth, J. F. Tyson and P. B. Stockwell, *Anal. Chim. Acta* **214** (1988) 329.

171. I. Sekerka and J. F. Lechner, *Anal. Chim. Acta* **234** (1990) 199.
172. J. Ruzicka and G. D. Christian, *Analyst* **115** (1990) 475.
173. Z. Fang, J. Ruzicka and Be. H. Hansen, *Anal. Chim. Acta* **164** (1984) 23.
174. Z. Fang, S. Xu, X. Wang and Z. Zhang, *Anal. Chim. Acta* **179** (1986) 325.
175. Yu. A. Zolotov, L. K. Shpigun, I. Ya. Kolorytkina, E. A. Novikov and O. V. Bazanova, *Anal. Chim. Acta* **200** (1987) 21.
176. S. Hirata and K. Honda, *Anal. Chim. Acta* **221** (1989) 65.
177. V. Kuban, J. Komarek and Z. Zdrahal, *Coll. Czech. Chem. Comm.* **54** (1989) 1785.
178. R. Purohit and S. Devi, *Anal. Chim. Acta* **259** (1992) 53.
179. R. Purohit and S. Devi, *Analyst* **120** (1995) 555.
180. A. Shah and S. Devi, *Anal. Chim. Acta* **236** (1990) 469.
181. T. Yamane, K. Watanabe and H. A. Mottola, *Anal. Chim. Acta* **207** (1988) 331.
182. L. Risinger, *Anal. Chim. Acta* **179** (1986) 509.
183. Y. Liu and J. D. Ingle Jr., *Anal. Chem.* **61** (1989) 520.
184. S. D. Hartenstein, J. Ruzicka and G. D. Christian, *Anal. Chem.* **57** (1985) 21.
185. T. Kumamaru, H. Matsuo, Y. Okamoto and M. Ikeda, *Anal. Chim. Acta* **181** (1986) 271.
186. S. Irata, Y. Umezaki and M. Ikeda, *Anal. Chem.* **58** (1986) 2602.
187. F. Malamas, M. Bengtsson and G. Johansson, *Anal. Chim. Acta* **160** (1984) 1.
188. A. A. Alwarthan, K. A. J. Habib and A. Townshend, *Fresenius J. Anal. Chem.* **337** (1990) 848.
189. J. E. Resing and M. J. Mottl, *Anal. Chem.* **64** (1992) 2682.
190. J. A. Resing and C. I. Measures, *Anal. Chem.* **66** (1994) 4105.
191. J. L. Nowicki, K. S. Johnson, K. H. Coale, V. A. Elrod and S. H. Lieberman, *Anal. Chem.* **66** (1994) 2732.
192. W. Jian and C. W. McLeod, *Talanta* **39** (1992) 1537.
193. M. M. G. Gomez and C. W. McLeod, *J. Anal. At. Spectrom.* **10** (1995) 89.
194. I. Ya. Kolorytkina, L. K. Shpigun, Yu. A. Zolotov and A. Malahoff, *Analyst* **120** (1995) 201.
195. E. Beinrohr, M. Cakrt, M. Rapta and P. Tarapei, *Fresenius Z. Anal. Chem.* **335** (1989) 1005.
196. M. A. Marshall and H. A. Mottola, *Anal. Chem.* **55** (1983) 2089.
197. W. M. Landing, C. Haraldsen and N. Paxeus, *Anal. Chem.* **58** (1986) 3031.
198. Z. Fang, M. Sperling and B. Welz, *J. Anal. At. Spectrom.* **5** (1990) 639.
199. S. Zhang, S. Xu and Z. Fang, *Quim. Anal.* **8** (1989) 191.
200. I. Ya. Kolotyrkina, L. K. Shpigun, Yu. A. Zolotov and G. I. Tsysin, *Analyst* **116** (1991) 707.
201. H. Bergamin F°, B. F. Reis, A. O. Jacintho and E. A. G. Zagatto, *Anal. Chim. Acta* **117** (1980) 81.
202. B. Patel, S. J. Haswell and R. Grzeskowiak, *J. Anal. At. Spectrom.* **4** (1989) 195.

203. J. R. Ferreira, E. A. G. Zagatto, M. A. Z. Arruda and S. M. B. Brienza, *Analyst* **115** (1990) 779.
204. K. K. Falkner and J. M. Edmond, *Anal. Chem.* **62** (1990) 1477.
205. M. R. P. Garcia, M. E. D. Garcia and A. Sanz-Medel, *J. Anal. At. Spectrom.* **2** (1987) 699.
206. S. Greenfield, T. M. Duwani, S. Kaya and J. F. Tyson, *Analyst* **115** (1990) 531.
207. P. P. Coetzee, I. Talkijaard and H. de Beer, *Fresenius J. Anal. Chem.* **336** (1990) 201.
208. I. G. Cook, C. W. McLeod and P. J. Worsfold, *Anal. Proc.* **23** (1986) 5.
209. A. G. Cox and C. W. McLeod, *Anal. Chim. Acta* **179** (1986) 487.
210. A. G. Cox, I. G. Cook and C. W. McLeod, *Analyst* **110** (1985) 331.
211. K. Pyrzynska, Z. Janiszewska, J. Szpunar-Lobinska and M. Trojanowicz, *Analyst* **119** (1994) 1553.
212. N. Furuta, K. R. Brushwyler and G. M. Hieftje, *Spectrochim. Acta* **B44** (1989) 349.
213. C. W. McLeod, I. G. Cook, P. J. Worsfold, J. E. Davies and J. Queay, *Spectrochim. Acta* **40** (1985) 47.
214. A. G. Cox, C. W. McLeod, D. L. Miles and J. M. Cook, *J. Anal. At. Spectrom.* **2** (1987) 553.
215. A. M. Naghmush, K. Pyrzyñska and M. Trojanowicz, *Anal. Chim. Acta* **288** (1994) 247.
216. A. M. Naghmush, K. Pyrzyñska and M. Trojanowicz, *Talanta* **42** (1995) 851.
217. P. Hernandez, L. Hernandez and J. Losada, *Fresenius Z. Anal. Chem.* **325** (1986) 300.
218. E. Olbrych-Sleszyñska, K. Brajter, W. Matuszewski, M. Trojanowicz and W. Frenzel, *Talanta* **39** (1992) 779.
219. A. M. Naghmush, M. Trojanowicz and E. Olbrych-Sleszyñska, *J. Anal. At. Spectrom.* **7** (1992) 323.
220. M. Trojanowicz and K. Pyrzyñska, *Anal. Chim. Acta* **287** (1994) 247.
221. S. Blain and P. Treguer, *Anal. Chim. Acta* **308** (1995) 425.
222. J. Ruzicka and A. Arndal, *Anal. Chim. Acta* **216** (1989) 243.
223. Z. Fang, M. Sperling and B. Welz, *J. Anal. At. Spectrom.* **5** (1990) 639.
224. S. Krekler, W. Frenzel and G. Schultze, *Anal. Chim. Acta* **296** (1994) 115.
225. V. Kuban, J. Havel and B. Patočkova, *Coll. Czech. Chem. Comm.* **54** (1989) 1777.
226. A. Astruc, R. Pinel and M. Astruc, *Anal. Chim. Acta* **228** (1990) 137.
227. M. R. Plantz, J. S. Fritz, F. G. Smith and R. S. Houk, *Anal. Chem.* **61** (1989) 149.
228. Y. P. de Peña, M. Gallego and M. Valcarcel, *Talanta* **42** (1995) 211.
229. M. Gallego, Y. P. de Peña and M. Valcarcel, *Anal. Chem.* **66** (1994) 4074.
230. Y. P. de Peña, M. Gallego and M. Valcarcel, *Anal. Chem.* **67** (1995) 2524.
231. Y. Daghbouche, S. Garrigues and M. de la Guardia, *Anal. Chim. Acta* **314** (1995) 203.
232. H. Chen, S. Xu and Z. Fang, *Anal. Chim. Acta* **298** (1994) 167.

233. S. P. Dolan, S. A. Sinex, S. G. Capar, A. Monaster and R. H. Clifford, *Anal. Chem.* **63** (1991) 2539.
234. A. Maquieira, H. A. M. Elmahadi and R. Puchades, *Anal. Chem.* **66** (1994) 1462.
235. A. Maquieira, H. A. M. Elmahadi and R. Puchades, *Anal. Chem.* **66** (1994) 3632.
236. K. Yoshimura, *Anal. Chem.* **59** (1987) 2922.
237. K. Yoshimura and H. Waki, *Talanta* **32** (1985) 345.
238. M. Valcarcel and M. D. Luque de Castro, *Analyst* **115** (1990) 699.
239. B. Fernandez, F. Lazaro, M. D. Luque de Castro and M. Valcarcel, *Anal. Chim. Acta* **229** (1990) 177.
240. K. Yosimura, *Bunseki Kagaku* **36** (1987) 656.
241. K. Yosimura, *Analyst* **113** (1988) 471.
242. F. Lazaro, M. D. Luque de Castro and M. Valcarcel, *Anal. Chim. Acta* **219** (1989) 231.
243. K. Yoshimura, S. Nawata and G. Kura, *Analyst* **115** (1990) 843.
244. N. Lacy, G. D. Christian and J. Ruzicka, *Anal. Chem.* **62** (1990) 1482.
245. L. F. Capitan-Vallvey, M. C. Valencia and G. Miron, *Anal. Chim. Acta* **289** (1994) 365.
246. P. Canizares and M. D. Luque de Castro, *Anal. Chim. Acta* **295** (1994) 59.
247. J. M. Fernandez-Romero and M. D. Luque de Castro, *Anal. Chim. Acta* **65** (1993) 3048.
248. L. C. R. Pessenda, A. O. Jacintho and E. A. G. Zagatto, *Anal. Chim. Acta* **214** (1988) 239.
249. K. Oguma, K. Nishiyama and R. Kuroda, *Anal. Sci.* **3** (1987) 251.
250. J. A. G. Neto, H. Bergamin F°, E. A. G. Zagatto and F. J. Krug, *Anal. Chim. Acta* **308** (1995) 439.
251. X. Yang, L. Risinger and G. Johansson, *Anal. Chim. Acta* **192** (1987) 1.
252. D. W. Bryce, A. Izquierdo and M. D. Luque de Castro, *Anal. Chim. Acta* **308** (1995) 96.
253. W. Frenzel, F. Titzenthaler and S. Elbel, *Talanta* **41** (1994) 1965.
254. O. Kondo, H. Miyata and K. Toei, *Anal. Chim. Acta* **134** (1982) 353.
255. H. Narasaki and M. Ikeda, *Anal. Chem.* **56** (1984) 2059.
256. P. Richter, J. M. Fernandez-Romero, M. D. Luque de Castro and M. Valcarcel, *Chromatographia* **34** (1992) 445.
257. J. Martinez Calatayud and J. V. G. Mateo, *Analyst* **116** (1991) 327.
258. J. V. G. Mateo and J. Martinez Calatayud, *Chem. Anal.* (Warsaw) **36** (1991) 1.
259. L. Anderson, *Anal. Chm. Acta* **110** (1979) 123.
260. J. F. van Staden, *Anal. Chim. Acta* **138** (1982) 403.
261. K. S. Johnson and R. L. Petty, *Limnol. Oceanogr.* **28** (1983) 1260.
262. S. Nakashima, M. Yagi, M. Zenki, A. Takahashi and K. Toei, *Fresenius Z. Anal. Chem.* **319** (1984) 506.
263. J. R. Clinch, P. J. Worsfold and H. Casey, *Anal. Chim. Acta* **200** (1987) 523.

264. A. Daniel. D. Birot, M. Lehaitre and J. Poncin, *Anal. Chim. Acta* **308** (1995) 413.
265. S. Motomizu, H. Mikasa and K. Toei, *Anal. Chim. Acta* **193** (1987) 343.
266. M. Trojanowicz, W. Matuszewski, B. Szostek and J. Michalowski, *Anal. Chim. Acta* **261** (1992) 391.
267. J. F. van Staden, A. E. Joubert and H. R. van Vliet, *Fresenius Z. Anal. Chem.* **325** (1986) 150.
268. A. T. Faizullah and A. Townshend, *Anal. Chim. Acta* **167** (1985) 225.
269. J. Martinez Calatayud and J. V. G. Mateo, *Anal. Chim. Acta* **264** (1992) 283.
270. A. M. Almuaibed and A. Townshend, *Anal. Chim. Acta* **245** (1991) 115.
271. K. Ueano, F. Sagara, K. Higashi, K. Yakata, I. Yoshida and D. Ishii, *Anal. Chim. Acta* **261** (1992) 241.
272. A. T. Haj-Hussein, G. D. Christian and J. Ruzicka, *Anal. Chem.* **58** (1986) 38.
273. F. T. Esmadi, M. Kharoaf and A. S. Attiyat, *Anal. Lett.* **23** (1990) 1069.
274. J. V. G. Mateo and J. Martinez Calatayud, *Anal. Chim. Acta* **274** (1993) 275.
275. T. C. Tang and H. J. Huang, *Anal. Chem.* **67** (1995) 2299.
276. J. Martinez Calatayud and C. G. Benito, *Anal. Chim. Acta* **231** (1990) 259.
277. S. Tesfalidet and K. Irgum, *Anal. Chem.* **61** (1989) 2079.
278. A. Kojlo and J. Martinez Calatayud, *Anal. Chim. Acta* **308** (1995) 334.
279. P. van Zoonen, D. A. Kamminga, C. Gooijer, N. H. Velthorst and R. W. Frei, *Anal. Chim. Acta* **167** (1985) 249.
280. X. Ding, P. Wang and G. Liu, *J. Chemiluminesc.* **40–41** (1988) 844.
281. K. Hool and T. A. Nieman, *Anal. Chem.* **59** (1987) 869.
282. M. Stigbrand, E. Ponten and K. Irgum, *Anal. Chem.* **66** (1994) 1766.
283. T. C. Werner, J. G. Cummings and W. R. Seitz, *Anal. Chem.* **61** (1989) 211.
284. J. Ruz, F. Lazaro and M. D. Luque de Castro, *J. Autom. Chem.* **10** (1988) 15.
285. M. Masoom and A. Townshend, *Anal. Chim. Acta* **166** (1984) 111, **179** (1986) 399.
286. C. S. Rui, Y. Kato and K. Sonomoto, *Biosens. Bioelectron.* **9** (1994) 429.
287. M. O. Rezende and H. A. Mottola, *Analyst* **119** (1994) 2093.
288. P. Solich, M. Polasek, R. Karlicek, O. Valentova and M. Marek, *Anal. Chim. Acta* **218** (1989) 151.
289. R. Gnanasekaran and H. A. Mottola, *Anal. Chem.* **57** (1985) 1005.
290. N. Kiba, M. Ogi and M. Furasawa, *Anal. Chim. Acta* **224** (1989) 133.
291. T. Yao and T. Wasa, *Electroanalysis* **5** (1993) 887.
292. T. Yao, M. Satomura and T. Nakahara, *Talanta* **41** (1994) 2113.
293. T. Yao, M. Satomura and T. Nakahara, *Electroanalysis* **7** (1995) 143.
294. T. Yao, R. Akasaka and T. Wasa, *Electroanalysis* **1** (1989) 413.
295. G. Maeder, J. L. Veuthey, M. Pelletier and W. Haerdi, *Anal. Chim. Acta* **231** (1990) 115.
296. W. Künnecke and R. D. Schmid, *Anal. Chim. Acta* **234** (1990) 213.
297. F. Lazaro, M. D. Luque de Castro and M. Valcarcel, *Anal. Chem.* **59** (1987) 1859.

298. M. Sanchez-Cabezudo, J. M. Fernandez-Romero and M. D. Luque de Castro, *Anal. Chim. Acta* **308** (1995) 159.
299. C. A. Koerner and T. A. Nieman, *Anal. Chem.* **58** (1986) 116.
300. A. Carpenter and W. C. Purdy, *Anal. Lett.* **23** (1990) 425.
301. A. Krug, R. Göbel and R. Kellner, *Anal. Chim. Acta* **287** (1994) 59.
302. F. Winquist, I. Lundström and B. Danielsson, *Anal. Chem.* **58** (1986) 145.
303. M. T. Jeppesen and E. H. Hansen, *Anal. Chim. Acta* **214** (1988) 147.
304. W. Matuszewski, M. Trojanowicz, M. E. Meyerhoff, A. Moszczynska and E. Lange-Moroz, *Electroanalysis* **5** (1993) 113.
305. I. L. Mattos, J. M. Fernandez-Romero, M. D. Luque de Castro and M. Valcarcel, *Analyst* **120** (1995) 179.
306. J. A. Infantes, M. D. Luque de Castro and M. Valcarcel, *Anal. Chim. Acta* **242** (1991) 179.
307. R. Puchades, A. Maquieira and L. Toro, *Analyst* **118** (1993) 855.
308. M. Akiba and S. Motomizu, *Anal. Chim. Acta* **214** (1988) 455.
309. B. A. Petersson, *Anal. Lett.* **22** (1989) 83.
310. W. Matuszewski, M. Trojanowicz and L. Ilcheva, *Electroanalysis* **2** (1990) 147.
311. M. Agudo, A. Rios and M. Valcarcel, *Anal. Chim. Acta* **308** (1995) 77.
312. J. M. Fernandez-Romero, M. D. Luque de Castro and R. Quiles-Zafra, *Anal. Chim. Acta* **308** (1995) 178.
313. Y. L. Huang, S. B. Khoo and M. G. S. Yap, *Anal. Lett.* **28** (1995) 593.
314. W. Matuszewski, S. A. Rosario and M. E. Meyerhoff, *Anal. Chem.* **63** (1991) 1906.
315. P. J. Worsfold and A. Nabi, *Anal. Chim. Acta* **179** (1986) 307.
316. G. C. Chemnitius and R. D. Schmid, *Anal. Lett.* **22** (1989) 2897.
317. M. D. Luque de Castro and J. M. Fernandez-Romero, *Anal. Chim. Acta* **311** (1995) 281.
318. T. Krawczynski vel Krawczyk, M. Trojanowicz and A. Lewenstam, *Talanta* **41** (1994) 1229.
319. I. H. Lee and M. E. Meyerhoff, *Mikrochem. Acta* **III** (1988) 207.
320. I. H. Lee and M. E. Meyerhoff, *Anal. Chim. Acta* **229** (1990) 47.
321. L. Locascio-Brown, A. L. Plant, V. Horvath and R. A. Durst, *Anal. Chem.* **62** (1990) 2587.
322. G. M. Greenway and P. Ongomo, *Analyst* **115** (1990) 1297.
323. K. Kashiwabara, T. Hobo, E. Kobayashi and S. Suzuki, *Anal. Chim. Acta* **178** (1985) 209.
324. K. Matsumoto, H. Matsubara, H. Ukeda and Y. Osajima, *Agric. Biol. Chem.* **53** (1989) 2347.
325. D. Narinesingh, V. A. Stoute, G. Davis, F. Shaama and T. T. Ngo, *Anal. Lett.* **24** (1991) 727.
326. K. Matsumoto, O. Hamada, H. Ukeda and Y. Osajima, *Anal. Chem.* **58** (1986) 2732.
327. N. Kiba, Y. Inoue and M. Furasawa, *Anal. Chim. Acta* **243** (1991) 183.

328. P. J. Worsfold, J. Farrelly and M. S. Matharu, *Anal. Chim. Acta* **164** (1984) 103.
329. M. Mascini and G. Palleschi, *Anal. Chim. Acta* **145** (1983) 213.
330. T. L. Spinks and G. E. Pacey, *Anal. Chim. Acta* **237** (1990) 503.
331. M. Valcarcel and M. Gallego, *Trends Anal. Chem.* **8** (1989) 34.
332. V. Kuban, *Fresenius J. Anal. Chem.* **346** (1993) 873.
333. P. Martinez-Jimenez, M. Gallego and M. Valcarcel, *Analyst* **112** (1987) 1233.
334. F. Esmadi, M. Kharaof and S. Attiyat, *Microchem. J.* **40** (1989) 277.
335. R. E. Santelli, M. Gallego and M. Valcarcel, *J. Anal. At. Spectrom.* **4** (1989) 547.
336. R. E. Santelli, M. Gallego and M. Valcarcel, *Anal. Chem.* **61** (1989) 1427.
337. Z. Fang, M. Sperling and B. Welz, *J. Anal. At. Spectrom.* **6** (1991) 301.
338. S. Pei and Z. Fang, *Anal. Chim. Acta* **294** (1994) 185.
339. G. Tao and E. H. Hansen, *Analyst* **119** (1994) 333.
340. C. E. Adeeyinwo and J. F. Tyson, *Anal. Proc.* **26** (1989) 58.
341. C. E. Adeeyinwo and J. F. Tyson, *Anal. Chim. Acta* **214** (1988) 339.
342. J. F. Tyson, S. R. Bysouth, E. A. Graszczyk and E. Debrah, *Anal. Chim. Acta* **261** (1992) 75.
343. F. F. Esmadi, I. M. Khaswneh, M. A. Kharoaf and A. S. Attiyat, *Anal. Lett.* **24** (1991) 1231.
344. P. Martinez-Jimenez, M. Gallego and M. Valcarcel, *Anal. Chem.* **59** (1987) 69.
345. P. Martinez-Jimenez, M. Gallego and M. Valcarcel, *J. Anal. At. Specrom.* **2** (1987) 211.
346. F. T. Esmadi, M. A. Kharoaf and A. S. Attiyat, *Analyst* **116** (1991) 353.
347. F. T. Esamdi, M. A. Kharoaf and A. S. Attiyat, *Talanta* **37** (1990) 1123.
348. M. Gallego and M. Valcarcel, *Mikrochim. Acta* **III** (1991) 163.
349. J. Zorro, M. Gallego and M. Valcarcel, *Mikrochem. J.* **39** (1989) 71.
350. B. A. Petterson, Z. Fang, J. Ruzicka and E. H. Hansen, *Anal. Chim. Acta* **184** (1986) 165.
351. S. Motomizu, K. Yoshida and K. Toei, *Anal. Chim. Acta* **261** (1992) 225.
352. R. Montero, M. Gallego and M. Valcarcel, *Anal. Chim. Acta* **215** (1988) 241.
353. M. C. Yebra, M. Gallego and M. Valcarcel, *Anal. Chim. Acta* **308** (1995) 275.
354. R. Montero, M. Gallego and M. Valcarcel, *J. Anal. At. Spectrom.* **3** (1988) 725.
355. M. C. Yebra, M. Gallego and M. Valcarcel, *Anal. Chim. Acta* **308** (1995) 357.
356. C. A. Koerner and T. A. Nieman, *Anal. Chem.* **58** (1986) 116.
357. D. Pilosof and T. A. Nieman, *Anal. Chem.* **54** (1982) 1698.
358. P. K. Dasgupta, R. S. Vithanage and K. Petersen, *Anal. Chim. Acta* **215** (1988) 277.
359. H. Hwang and P. K. Dasgupta, *Anal. Chem.* **59** (1987) 1356.
360. A. Trojanek and S. Bruckenstein, *Anal. Chem.* **58** (1986) 983.
361. S. J. Chalk and J. F. Tyson, *Talanta* **41** (1994) 1797.
362. J. D. Brewster, R. J. Maxwell and J. W. Hampson, *Anal. Chem.* **65** (1993) 2137.

363. I. Papaefsthathiou, M. T. Tena and M. D. Luque de Castro, *Anal. Chim. Acta* **308** (1995) 246.
364. I. Papaefsthathiou and M. D. Luque de Castro, *Anal. Chem.* **67** (1995) 3916.
365. R. Nakata, *Fresenius Z. Anal. Chem.* **317** (1984) 115.
366. R. Nakata, M. Terashita, A. Nitta and K. Ishikawa, *Analyst* **115** (1990) 425.
367. S. Liu and P. K. Dasgupta, *Anal. Chim. Acta* **268** (1992) 1.
368. P. K. Dasgupta and S. Liu, *Anal. Chem.* **66** (1994) 1792.

Chapter 7

Speciation Analysis Using Flow Injection Methodology

The importance of knowing that a given element, whether as a macro- or as a microcomponent, occurs in different physico-chemical forms in various natural matrices, has been widely recognised in the last 20 years in environmental science, biology and medicine. This is evident from the hundreds of original research papers published in scientific journals, and several books published in recent years [1–6]. The various chemical forms of an element of interest exhibit different reactivity, toxicity and bioavailability. The most difficult problem encountered in speciation is to develop a procedure which does not disturb the chemical equilibria between the forms of an element existing in a given matrix. It might then be concluded that the most satisfactory procedure should be based on the determination of the total amount of all elements in a given material, followed by the computer-aided calculation of the concentrations of particular species based on ionic and redox equilibrium constants. It is obvious, however, that the success of such a procedure is limited by the difficulties of taking into consideration all possible kinetic factors, adsorption processes, polymerisation reactions and heterogeneous processes. The availability of known stability constants is also limited; hence, in practice, the procedures for the analytical determination of particular species predominately involve the use of different detection methods, separation operations and the chemical conversion of elemental forms into detectable species.

IA offers several valuable advantages for speciation of both macro- and microelements. As its main advantage is a shorter analysis time than in conventional, manual procedures, in many cases the condition of not disturbing the equilibria in analysed samples is much better fulfilled. Also of great importance is the reduction of human participation in time-consuming operations such as sample conditioning, reagent manipulation and calibration of the measuring system. It is relatively simple to design a multidetector measuring system. The ability to use on-line preconcentration techniques and non-chromatographic separations, such as ion-exchange, dialysis, solvent or solid-phase extraction, is a great advantage.

Work on speciation using FIA has been reviewed in several papers [7–9]. Because of numerous ways that elements are distributed in natural matrices, FIA speciation studies have focused on three areas: the simultaneous determination of species in different oxidation states; the determination of the degree of complexation of metallic elements; and the determination of the content of a given complex compound. There have also been some attempts to use FIA for the speciation of organometallic compounds.

The literature concerning chemical speciation with FIA methods provides a large variety of designs for measuring systems utilising various detectors, as well as different examples of off-line and, especially, on-line sample manipulation. It is a relatively simple task to design a setup with multiple detectors, usually two, if each of them can selectively produce an analytical signal corresponding to the concentration of a particular chemical form of a given element. The detectors can be arranged in series [10, 11] if, in at least one of them, there is no substantial dispersion of the sample segment, or in parallel [12] with the sample split into two branches in the manifold. Simultaneous injection with two or more combined valves to various branches of the FIA manifold is also used.

In several cases, FIA speciation has been carried out using a detector which produces selective signals for individual species. This concept has been employed with a molecular emission cavity detector [15], a dual-electrode amperometric detector with different modification of electrodes [16] and a spectrophotometric diode-array detector [17].

Quite often, when single detector manifolds are employed, much greater creativity is required to design the whole measuring system successfully. The least complicated systems result when off-line sample pretreatment is involved [18–21]. Such systems utilise, to a limited extent, the advantages of the flow injection methodology. In most FIA systems designed for speciation, the determination of various species by the same detector is achieved through various methods of on-line sample manipulation. For each sample injection, a single analytical signal is produced in the detector, but a simple change of reagent solutions in the manifold between successive injections [22–25], or a simple and fast rearrangement of the manifold [26–36], gives (at different times) signals that correspond to different chemical forms of the element. Some of the most frequently used methods of on-line chemical conversion are reduction and oxidation, selective sorption on appropriate sorbents, or total retention followed by fractional elution as well as volatization. Less often, kinetic discrimination (i.e. differences in the reaction rate among various species) is used for this purpose [37]. The change of reagents in the manifold for the determination of various species is achieved by using selecting valves or by designing reversed FIA systems with the injection of the reagent into a continuously aspirated sample stream.

Different pretreatment of various aliquots of the same sample is achieved by splitting the sample zone between different branches of the FIA manifold. This different pretreatment, followed by the merging of sample segments into a single stream approaching the detector, can give two independent analytical signals for one injected sample solution [38–42]. Two signals from one injected sample can also be obtained with simultaneous injection from two synchronised valves into separate parts of the FIA manifold. The most common configuration in which signals are obtained for each of two different species is an FIA setup with selective on-line retention of one form or with retention of the sample followed by selective consecutive elution. A more detailed discussion of FIA manifolds used for speciation is given below, along with their chemical applications.

Recent years have also brought applications of sequential injection analysis for speciation [43–45].

1. Speciation of the Oxidation State of Elements

The majority of FIA systems have been developed for the speciation of different oxidation states of a target element. Two basically different concepts of measurement can be distinguished. In the first type of system, two consecutive measurements are made that differ in their on-line or off-line sample modification, so that firstly the total content of a given element is determined and then the content of particular species. In another type of FIA system, the selective determination of each different oxidation state is carried out.

The first of these two options, with the determination of the total content of the element, has been easier to develop. It has been most commonly used in the speciation of chromium, iron and arsenic, although FIA systems for the speciation of other elements have also been reported.

In the speciation of chromium, total chromium (Cr(III) plus Cr(VI)) is usually determined along with a separate determination of Cr(VI) using a spectrophotometric method with diphenylcarbazide and AAS detection. In the least complicated type of manifold, a serial arrangement of the spectrophotometric and AAS detectors is applied [10]. Several systems have been developed with a spectrophotometric detector only; these employ on-line oxidation of Cr(III) to Cr(VI) in order to determine the total chromium content [26, 39, 40, 46]. In several other systems, reversed FIA measurements have been employed with continuous sample aspiration and injection of reagents [39, 40]. In the system shown in Fig. 1 [40], the sample segment is split into two streams, in which Cr(VI) and total chromium are measured separately. The measurement of pH allows the calculation of the content of hydroxylated complexes of Cr(III) and of protonated and polymerised forms of Cr(VI). Other configurations that have been developed using double-beam detection with two spectrophotometric flow cells give the most reproducible results [46]. In determinations with AAS detection, on-line sorption of the diethyldithiocarbamate complex of Cr(VI) onto C18 has been adopted, with total chromium being determined after oxidation of Cr(III) with peroxydisulphate [47]. Another similar flow injection method with AAS detection and on-line proconcentration on a microcolumn packed with C18 bonded silica gel was based on the selective formation of diethyldithiocarbamate complex of chromium(VI) in the 1–2 pH range and chromium(III) in the 4–9 pH range in the presence of manganese(II) [48]. The procencentrated species are desorbed by methanol directly from the column to the nebuliser-burner system. The developed procedure was suitable for the determination of chromium species in various sea water samples. In the sequential injection procedure the reaction product of Cr(VI) with 1,5-diphenylcarbazide was ion-paired with perchlorate and extracted into an organic film wetting the inner wall of a Teflon tube [45]. After preconcentration the wetting film, with the extracted analyte, was eluted with acetonitrile and the analyte determined spectrophotometrically. For speciation Cr(III) was previously off-line oxidised by 5 min heating at 45°C with Ce(IV) solution. The developed method was employed in analysis of tap, lake and sea water. For chromium speciation also can be used a flow injection procedure with amperometric detection of Cr(VI), which is selective in the presence of Cr(III) and dissolved oxygen [49]. The use of phosphoric acid as the supporting electrolyte suppresses the interference from Fe(III). Chloride interferes in the determination at gold working electrodes but not at palladium working electrodes.

An FIA system with spectrophotometric and AAS detectors in series has also been reported for the speciation of iron with the determination of Fe(II) based on its reaction with 1,10-phenanthroline [10]. This reaction has also

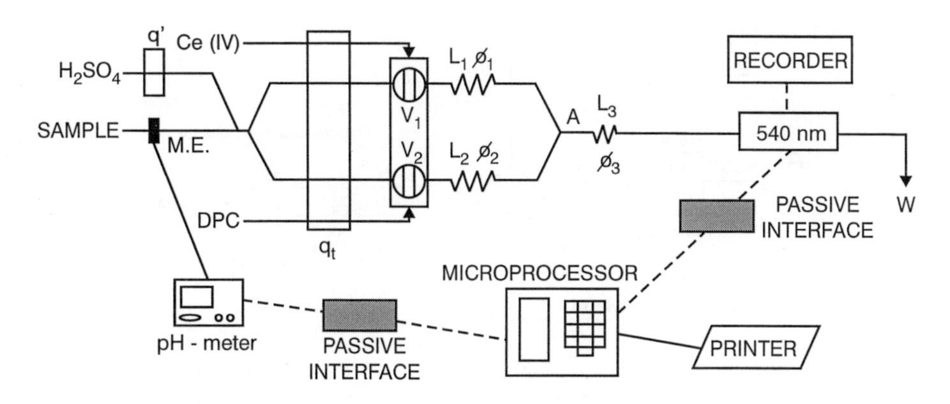

Fig. 1. Manifold of the reversed FIA system used for speciation of chromium with pH measurements of an aspirated sample [40]. q — peristaltic pumps; V — injection valves; L — reaction coils; M.E. — pH glass microelectrode. (*Reprinted by permission of copyright owner.*)

been utilised in other FIA systems with only spectrophotometric detection [37, 38, 50, 51]. In order to determine total iron, Fe(III) is reduced on-line on a column with Jones reductant [38]; alternatively, a complex of Fe(III) with 1,10-phenanthroline can be reduced photochemically [50]. A unique system for iron speciation using kinetic discrimination has also been developed (Fig. 2) [37]. It is based on the different kinetic-catalytic behaviour of Fe(II) and Fe(III) in the redox reaction between leucomalachite green and peroxodisulphate, with and without the presence of 1,10-phenanthroline as an activator. In a sandwich technique for the simultaneous determination of iron(II) and total iron, samples are inserted between zones of water and ascorbic acid solution, with subsequent addition of 1,10-phenanthroline [46]. The signal consists of a plateau region corresponding to Fe(II), followed by a peak corresponding to total iron.

A flow cell packed with an exchange resin was used in the system for speciation of iron based on integration of retention of the Fe(III)-SCN complex and spectrophotometric detection [52]. Inner-coupled injection valves enable discrimination between Fe(II) and Fe(III) by the use of a redox minicolumn housed in the loop of one of the valves. From the two absorbance measurements corresponding to a single injection, the Fe(III) and total iron can be determined. Speciation of iron can also be carried out in the flow injection system with spectrophotometric detection using a concept of relocation of the detector [53]. The detection is based on formation of an Fe(II) complex with 1,10-phenanthroline.

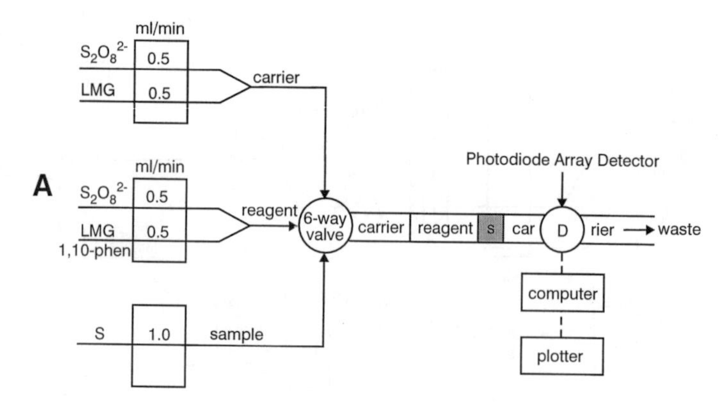

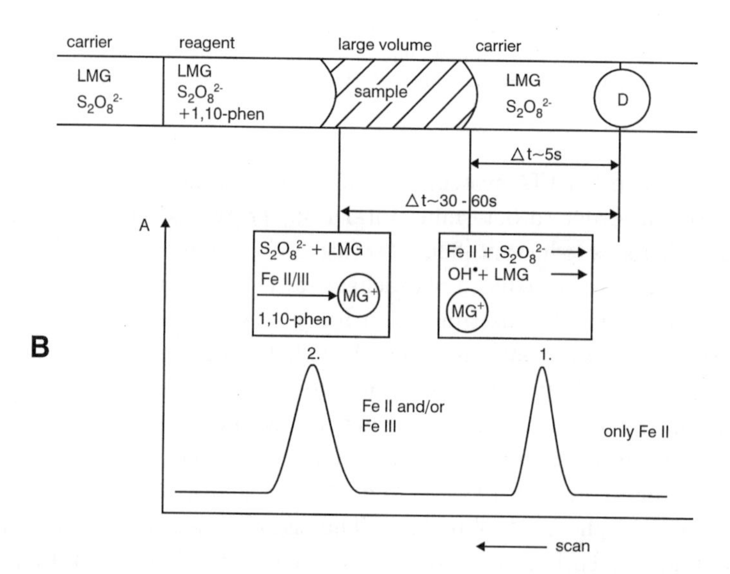

Fig. 2. Manifold of the FIA system used (A) and schematic diagram of the sample/reagent sequencing (B) in the speciation of iron utilising kinetic discrimination of analytes in the reaction between leucomalachite green (LMG) and $S_2O_8^{2-}$ with 1,10-phenanthroline (1,10-phen) as activator [37]. (*Reprinted by permission of copyright owner.*)

Detection of Fe(III) is based on its earlier reduction with ascorbic acid. For the same purpose of iron speciation a flow injection system with an opto-chemical integrated detector was developed [54]. Simultaneously Fe(III) was detected by measurement of the absorbance of the Fe(III)-sulphosalicylic acid complex, and Fe(II) by amperometric monitoring of the oxidation current of Fe(II) at a potential of +1.0 V vs. SCE at a platinum working electrode. The method applied to spiked tap water analysis was relatively free of interferences. For the speciation of iron a potential asynchronous merging zone technique was also examined [55], which is based on partial overlapping of the sample and reagent segments. The resulting detector signal consists of two consecutive peaks whose heights are proportional to the concentration of the detectable component present in the sample and another component which becomes detectable after the reagent has been added. A mathematical model was developed for the simulation of the detected signal.

Differences in on-line sample pretreatment have been utilised in the design of FIA systems for the speciation of As(III) and As(V) with spectrophotometric [22] and AAS [25,29] detection. The sequential determination of AsO_2^- and AsO_4^{3-} in mixtures is based on the formation of heteropolyacids with MoO_4^{2-} [22]. When the sample in the manifold merges with an iodate solution, arsenite is oxidised, and the signal obtained corresponds to the sum of the two arsenic anions. When the sample is not oxidised it gives a signal corresponding to arsenate alone. Using hydride generation AAS detection, total arsenic is determined using arsine generation in 6M HCl, while As(III) alone can be determined by the generation of arsine in a citrate buffer at pH 3 [25]. A more sophisticated AAS procedure is based on the generation of arsine in the FIA system and its subsequent trapping in a graphite furnace coated with appropriate absorbers [29]. The arsine is transferred to the graphite furnace through an electronically activated arm. As(III) is determined using the addition of NaOH to a water carrier stream, while potassium iodide is added when total arsenic is determined.

Various chlorine-containing species can be determined in flow injection systems with spectrophotometric detection. In the speciation of low concentrations of chlorite and chlorate ions, chlorite can be selectively determined using its reaction with iodide at pH 2, which liberates iodine; in 6 M HCl, both species produce iodine, which is measured spectrophotometrically [23]. The differentiation of residual free and combined chlorine is based on the different reaction of both species with 2,4-dinitrophenyldiazonium ion, depending on acidity, in a reversed flow injection system [56]. For the determination of chlorine and oxychlorine species automated selective iodometric methods have been developed based on iodometric measurements in a flow system with spectrophotometric detection [57]. Determinations of chlorine, chlorine dioxide and chlorite ion are based on various kinetics of reaction of these analytes and iodide at various pH. Additional improvement of selectivity can be gained by using suitable masking species, which allows the direct determination and speciation of these analytes in mixtures at a concentration well below the levels currently required for monitoring by the US EPA. In later studies it was found that flow injection methods, similarly to ion chromatography, were accurate and effective in synthetic solutions; however, the flow injection method was affected by chloramines and other oxidants in drinking water [19]. The determination of free and combined residual chlorine by flow injection spectrophotometry can be based on the oxidation of 4-nitrophenylhydrazine by free chlorine to the 4-nitrophenyldiazo cation and then its coupling with *N*-(1-naphthyl)ethylenediamine to give an azo dye [58]. It was found that only chlorine dioxide interferes in this procedure, but chloramines do not.

Iodide can be directly determined in a wide concentration range using potentiometry with an ion-selective membrane electrode, while iodine is determined after on-line reduction to iodide with sodium metabisulphite [24]. For iodine speciation in sea water an FIA system with spectrophotometric detection has been developed which is based on the catalytic, fading effect of either iodate or iodide on the indicator reaction of iron(III) thiocyanate and nitrite [59]. With and without an anion-exchange column in the flow system, one can determine iodate and total iodine, respectively. Iodide can be found by difference.

Total phosphorus and orthophosphate can be determined in waste water using an FIA system with two detectors [11]. Orthophosphate is determined as molybdovanadophosphoric acid using a spectrophotometric method with the solution from the flow cell of the spectrophotometer being directly introduced into an inductively coupled plasma. In another reported system with spectrophotometric detection, orthophosphate is determined along with the sum of orthophosphate and pyrophosphate, after the on-line separation and preconcentration of both analytes on an inert support modified with a diorganotin extractant [28]. The sum of the two anions is determined after the preliminary hydrolysis of pyrophosphate to orthophosphate using a soluble inorganic pyrophosphatase.

The speciation of inorganic selenium is carried out in FIA systems with cathodic stripping voltammetric detection using mercury film on a glassy carbon electrode [20], with anodic stripping voltammetry at a gold electrode [21], and with hydride generation AAS detection [30, 60], which allows Se(IV) determination. In all these cases, the sum of Se(IV) and Se(VI) is determined after the reduction of Se(VI) to Se(IV), carried out off-line by chemical [20, 21, 60] or on-line in a microwave unit [30]. In the system with anodic stripping voltammetry, interfering cations are removed on-line, using a Chelex-100 column [21].

Off-line reduction of Sb(V) is employed in the speciation of antimony by hydride generation AAS [18]. Cold-vapour AAS has been used in an FIA system for the speciation of inorganic and total mercury [27]. The determination of total mercury necessitates a sample digestion step, in which organic species are decomposed and all organic mercury is converted to the inorganic state.

It is a more complex and difficult problem to design FIA speciation systems with selective determination of each individual form of the element of interest. Several papers have reported manifolds of this type for the speciation of nitrate and nitrite, and also for chromium and iron speciation. In most samples nitrogen does not only occur as nitrate and nitrite, and so the methods for their determination are included in this type of FIA systems, although these two analytes are determined together and separately in most speciation systems. The simultaneous FIA determination of nitrate and nitrite has been carried out with spectrophotometric [12, 31, 42, 61–63], amperometric [41], AAS [32] and chemiluminescence [64,65] detection. In a measuring setup with a dual absorbance monitor connected to a dual-channel recorder, both species were determined using common methods in which nitrite is diazotised and the product is transformed into a coloured azo dye. The injected sample is split between two branches of the manifold and in one of them nitrate is reduced on-line to nitrite [29]. Two separate signals are measured for nitrate and the sum of nitrite and nitrate. The same method has been used by other authors in FIA systems with a single detector and two injection valves [31, 61]. In a flow injection system where simultaneously total nitrogen was determined (Fig. 3), the nitrogen-containing compounds (organic substances and nitrite and ammonium ions) were oxidised photochemically using a UV lamp and converted into nitrate, which was then reduced to nitrite and determined spectrophotometrically. The elimination of the refractive index effect caused by the heterogeneous flow and salinity variations in such a system can be achieved by simultaneous measurements of the signal at two different wavelengths [62]. One is carried out at a wavelength at which the product of the analyte reaction predominantly absorbs light and as a reference wavelength is used that one where no absorbance change due to the chemical reaction is observed.

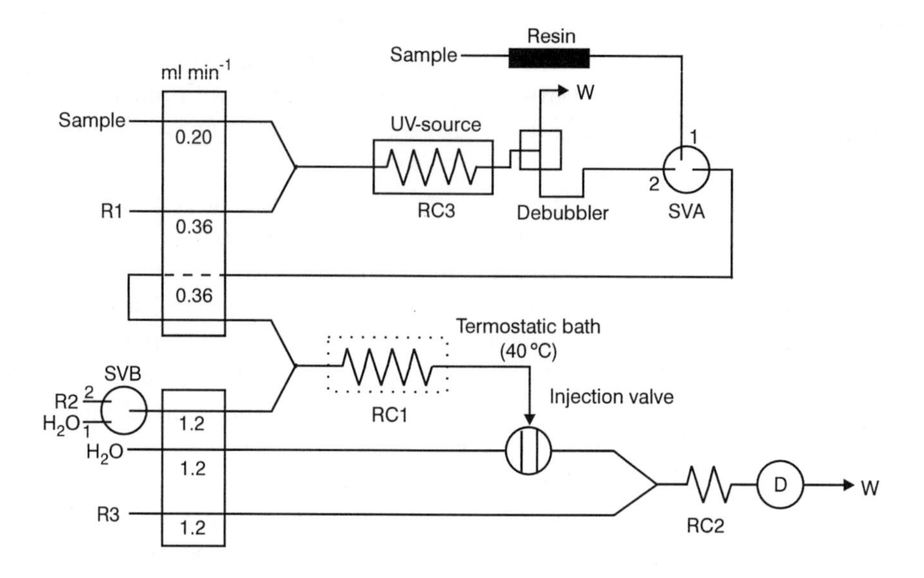

Fig. 3. Schematic diagram of the flow injection manifold for determination of nitrite, nitrate and total nitrogen [44]. R1 — persulphate alkaline solution; R2 — reducing solution; R3 — chromogenic reagent; resin, Amberlite XAD-7 non-ionic resin; UV-source, 15W lamp; SVA, SVB — selection valves; RC1, RC2 — reaction coils; RC3 — photo-oxidation coli; W — waste; D — detector. (*Reprinted by permission of copyright owner.*)

The on-line reduction of nitrate to nitrite is most often carried out on microcolumns with copperised metallic cadmium [12, 31, 41, 61, 62]; however, photoinduced reduction by UV radiation in the presence of complexones has been reported [42]. A simultaneous catalytic determination of nitrite and nitrate in an FIA system with spectrophotometric detection was based on the catalytic effect of nitrite on oxidation of naphthol green by potassium bromate in a phosphoric acid medium [63]. In the developed procedure a technique of double zone injection of the sample isolated by a cadmium-coated zinc reductor column was used with a single valve. The flow system produced two signals, one corresponding to nitrite and the other to the sum of nitrite and nitrate.

In the system which uses indirect biamperometric detection, the analytical signal is observed when nitrite oxidises iodide to iodine and a reversible couple is formed [41]. The injected sample is split into two segments in an FIA manifold. One of these is transported through a reducing minicolumn and a delay loop. For each injected sample, two peaks are obtained, of which the first corresponds to nitrite and the second to the sum of nitrite and nitrate. Sequential determinations of nitrate and nitrite with AAS detection, based on continuous liquid-liquid extraction, have been demonstrated [42]. Nitrate reacts with bis(2,9-dimethyl-1,10-phenanthrolinato) copper(I) to form an ion pair which is extracted into 4-methyl-2-pentanone in a flow injection manifold. In one aliquot of the sample, nitrite is oxidised by Ce(IV) and thus total nitrate is determined. In another, nitrite is converted to nitrogen with sulphamic acid, so that only the original nitrate is determined. Both species are then determined by measuring the AAS signal of copper in the organic phase. Nitrite and nitrate can also be determined in a reversed FIA system with the transfer of NO from the reduction of NO_2^- and/or NO_3^- to the gas phase through a PTFE membrane and chemiluminescence detection in the gas phase [64]. Nitric oxide produced by the injection of Ti(III) solution corresponds to the sum of nitrate and nitrite, whereas the injection of iodide results in the reduction of nitrite only. Detection is based on chemiluminescence from the reaction of NO with ozone. In the next work of the same group an FIA system has been further developed by the injection of hypochlorite as a reagent for simultaneous determination of ammonia [65]. A reversed system with continuous sample aspiration and sequential injection of appropriate reagents has also been used in the selective determination of nitrite with the formation of an azo dye and ammonia using the Nessler method [51]. As was already mentioned, the speciation of nitrogen can be carried out also in sequential system [43]. The determinations of nitrate and nitrite were based on the use of spectrophotometric detection with the commonly used Griess reaction with sulphanilamide and *N*-(1-naphthyl)ethylenediamine. Nitrate was previously reduced by hydrazine in an alkaline medium.

The speciation of chromium in FIA systems in which signals corresponding to Cr(III) and Cr(VI) are obtained for each injected sample has been reported with ICP-AES [67] and AAS [68] detection. In the former system, a microcolumn of activated alumina is used in the FIA manifold to separate Cr(VI) from Cr(III) and preconcentrate it. After the detection of Cr(III), an ammonia solution is injected to elute Cr(VI) from the column. In the system with AAS detection, two columns with differently functionalised cellulose sorbents were used for the selective sorption of Cr(III) and Cr(VI) from the continuously aspirated sample (Fig. 4) [68]. Sequential elution with appropriate solutions provides two peaks corresponding to the concentrations of Cr(III) and Cr(VI) in the sample.

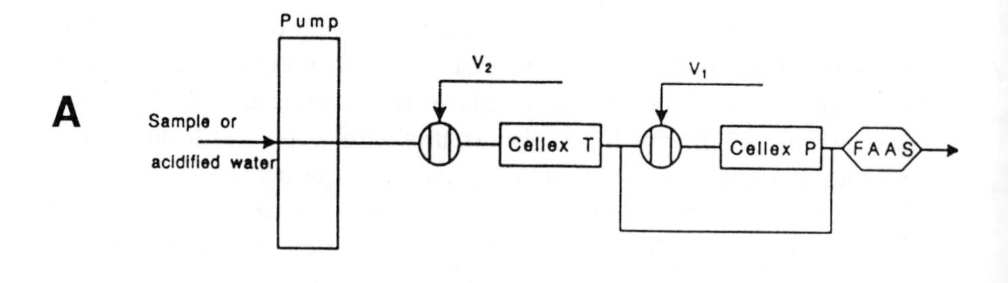

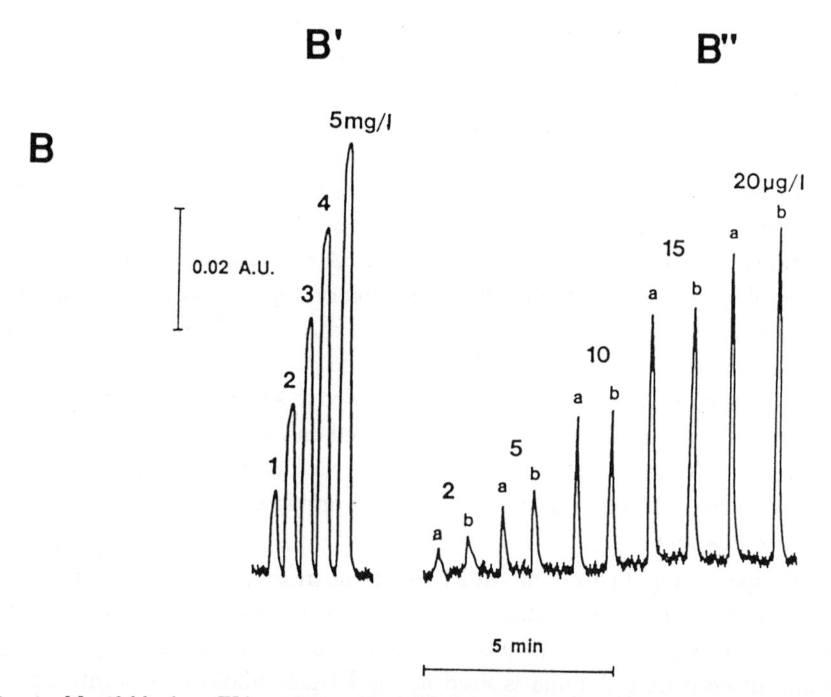

Fig. 4. Manifold of an FIA system with AAS detection and on-line preconcentration on cellulose sorbents, used for speciation of chromium (A) and a comparison of recordings (B) obtained with conventional aspiration (B′) and with an FI-AAS system (B″) with preconcentration of (a) Cr(III) and (b) Cr(VI). In B": aspirated sample volume 100 ml, flow rate 5.0 ml.min^{-1} [68]. (*Reprinted by permission of copyright owner.*)

In the speciation of iron, besides methods mentioned above, several different FIA methods giving individual signals for Fe(II) and Fe(III) have been developed. In a system with two spectrophotometric detectors and simultaneous sample introduction by two coupled injection valves, iron(II) is determined using 1,10-phenanthroline, while iron(III) is determined using thiocyanate [13]. In a system with AAS detection, a packed-bed reactor, with a Dowex 1 anion exchange resin, is used to retain chloride complexes of Fe(III), while Fe(II) species pass through the reactor to the detector [69]. This is followed by the separate elution of Fe(III) to the detector by a change in the carrier solution. On-line solid-phase extraction has also been employed in two other FIA systems for iron speciation with spectrophotometric [70] and AAS [71] detection. In the first system, Fe(II) is preconcentrated on a C_{18} column with immobilised Ferrozine and then eluted with methanol. The determination of Fe(III) requires on-line reduction with ascorbic acid [70]. In the second system, the injected sample is mixed with Ferrozine solution, and while Fe(III) is directly carried to the flame AAS detector, the Fe(II)-Ferrozine complex is temporarily retained on a C_{18} column and subsequently eluted with methanol. In an FIA system with amperometric detection, a dual-electrode assembly has been used with two glassy-carbon electrodes modified with different electrocatalysts, providing selective detection for Fe(II) and Fe(III) species [16]. The speciation of iron can also be carried out with a mixture of 1,10-phenanthroline and sulphosalicylic acid using a diode-array spectrophotometric detector and various methods of signal processing [17].

The speciation of vanadium has been carried out using flame AAS with an on-line procedure for the preconcentration of V(V) using a silica bonded ion-exchange resin [72]. During V(V) retention, the unretained V(IV) passes through to the AAS detector, after which the V(V) species are eluted with a plug of sodium hydroxide solution.

The simultaneous flow injection determination of sulphite, sulphide and sulphate has been reported for a system with a molecular emission cavity detector. This determination is based on the sequential appearance of S2 emission peaks corresponding to these species (Fig. 5) [15]. In another system with an atomic absorption spectrometer operating in the emission mode, 3 μl of sample is injected, and detection limits for sulphur anions in the range of 0.06–0.15 ng in injected samples have been obtained. A very complex multidetector FIA system has been developed for the determination of sulphur species in aqueous

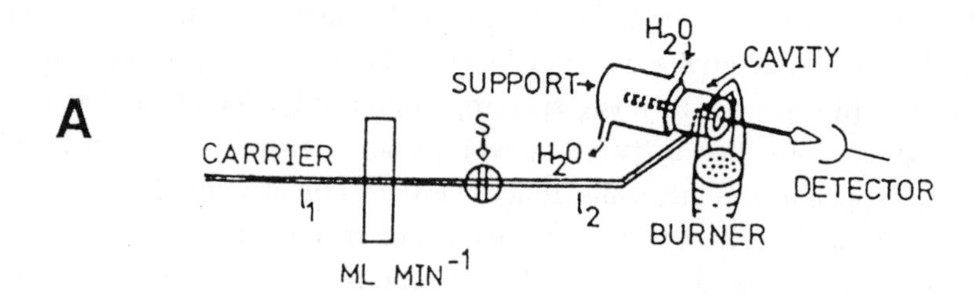

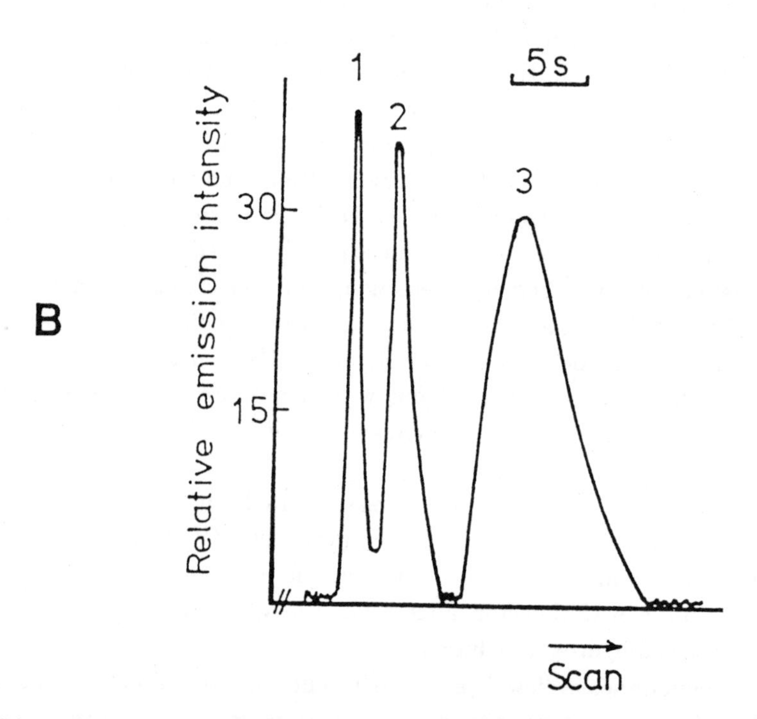

Fig. 5. Schematic diagram of an FIA system with molecular emission detection (A) used for speciation of sulphur anions and the response from a mixture of sulphur anions (B): 1 — sulphide; 2 — sulphite; 3 — sulphate [15]. (*Reprinted by permission of copyright owner.*)

samples of importance to the petroleum industry [14] (Fig. 6). Branch A of the manifold is used for the determination of sulphide by acidification of the sample to produce H_2S and gas diffusion into alkaline solution of $Na_2[Fe(CN)_5NO]$.

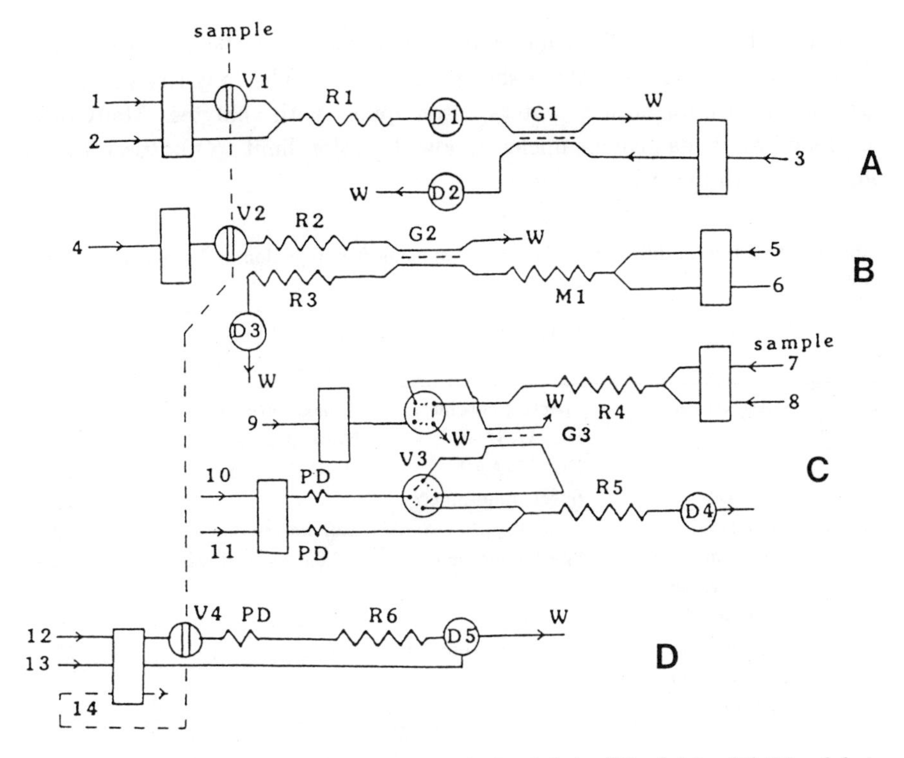

Fig. 6. Manifolds for determining: (A) sulphide/polysulphide; (B) sulphite; (C) thiosulphate; (D) sulphate [14]. V — valves; D — detectors; G — gas diffusion units; R — reaction coils; W — waste; 1–14 — reagent streams. (*Reprinted by permission of copyright owner.*)

In the same part of the manifold, sulphur produced upon the acidification of polysulphides is measured turbidimetrically. In branch B, sulphite is measured by acidification to produce SO_2, followed by gas diffusion and detection with pararosaniline. In branch C, thiosulphate is measured by dialysing the sample after on-line treatment with Zn(II) and formic acid, and then reducing $KMnO_4$ with the dialysate. Sulphate is measured by a turbidimetric $BaSO_4$ method in branch D.

Total and free sulphur dioxide can be determined in wine by an FIA method with gas diffusion, using *p*-aminobenzene as the colorimetric reagent [36]. Total SO_2 in wines is made up of free species (aqueous SO_2 and HSO_3^-) and species bound to aldehyde and ketone groups, especially acetaldehyde. The bound species are hydrolysed on-line in order to determine the total content.

A list of developed flow injection methodologies for the speciation of oxidation states of 12 elements is shown in Table 1. The majority of practical applications are for natural waters, often spiked with analytes. Many of the developed methods give a sufficiently low detection limit to find practical applications with natural samples.

Table 1. Speciation of different oxidation states using flow injection and sequential injection analysis.

Element	Species	Detection method	Analysed real samples	Reference
As	As(III),As(V)	Spectrophotometry	Spiked natural waters	22
		HG-AAS	Natural waters	25
		HG-ET-AAS	Natural waters	29
Cl	Chlorite, chlorate	Spectrophotometry	–	23
			Tap waters	19
	Residual chlorine, free and combined	Spectrophotometry	Spiked natural waters	56, 58
	Chlorine, chlorine dioxide, chlorite, chlorate	Spectrophotometry	Drinking water	57
Cr	Cr(III), Cr(VI)	Spectrophotometry	Leather treatment effluent	26
		Spectrophotometry	–	39,40,46
		Spectrophotometry	Natural waters	45
		AAS	Waters	47,68
		AAS	Plating effluents	48
		AAS + spectrophotometry	Corrosion test sea water	10
		ICP-AES	Reference waters	67
		Amperometry	–	49
Fe	Fe(II), Fe(III)	Amperometry	Spiked tap waters	16
		Spectrophotometry	–	13,17,37,38
		Spectrophotometry	Catalysts	50
			Ground waters	51
			Plant digests	53
			Reference sea water	70
			Wastes	52

continued

Table 1 (*continued*)

Element	Species	Detection method	Analysed real samples	Reference
		AAS	–	69,71
		AAS + spectrophotometry	Mineral process liquids	10
		Opto-electrochemistry	Spiked tap waters	54
Hg	Inorganic, total	Cold-vapour AAS	–	27
I	Iodide, iodine	Potentiometry	Pharmaceutical preparations	24
	Iodate, iodide	Spectrophotometry	Sea water, brines	59
N	Nitrate, nitrite	Spectrophotometry	–	12, 61
		Spectrophotometry	Natural waters	31,42,62,63
			Waste water, aerosols	43
		Amperometry	Natural waters	41
		AAS	Foodstuffs	32
		Chemiluminescence	River waters, waste water	64
	Nitrate, nitrite, total nitrogen	Spectrophotometry	Waste water	44
	Ammonia, nitrite	Spectrophotometry	Natural waters	66
	Ammonia, nitrate, nitrite	Chemiluminescence	–	65
P	Phosphate, total phosphorus	ICP-AES + spectrophotometry	Waste water	11
	Ortho-, pyrophosphate	Spectrophotometry	River water	28
S	Sulphide, sulphite, sulphate	Molecular emission	–	15
	Sulphide, polysulphide, sulphite, thiosulphate, sulfate	Spectrophotometry and turbidimetry	–	14
	SO_2, free and bound	Spectrophotometry	Wine	36

continued

Table 1 (*continued*)

Element	Species	Detection method	Analysed real samples	Reference
Sb	Sb(III), Sb(V)	HG-AAS	Sea water	18
Se	Se(IV), Se(VI)	Cathodic stripping voltammetry	Spiked natural waters	20
		Anodic stripping voltammetry	Spiked natural waters, lyophilised pig kidney	21
		HG-AAS	Reference waters	30
			Sea water	60
V	V(IV), V(V)	AAS	Yeast cells	72

2. Determination of the Degree of Complexation

In most natural matrices, elements occurring as cations in small fractions exist only as simple, hydrated cations and are mostly present in ion pairs or in complexes with inorganic and organic ligands. In natural waters, only a few percent of Al, Pb and Cr(III) cations are uncomplexed, while potassium and calcium have free ion concentrations above 80% [73].

The determination of the degree of complexation in FIA systems is not as widely developed as the determination of different oxidation states of elements (Table 2). The speciation capability of a simple FIA system with AAS detection and a microcolumn containing chelating resin with salicylate complexing groups has been investigated for a model mixture of Cu(II) with EDTA [34]. Different flow injection systems with potentiometric detection have been examined for speciation of fluoride using a fluoride ion-selective electrode [74]. The most suitable system with resampling was employed for a study of the effectiveness of releasing fluoride from complexes with aluminium, as well as for speciation of calcium with ion-selective electrode in ligand solutions and in natural waters.

Three different fractions of complexed species of Cd, Cu and Zn have been determined in an FIA system with AAS detection and two microcolumns with Chelex-100 and a strongly basic anion exchange resin (Fig. 7) [75]. Metal species such as hydrated free ions, labile metal complexes and (possibly) metals loosely associated with colloidal matter were retained on the Chelex-100. The strongly basic anion-exchanger retained negatively charged metal complexes and metal ions associated with negatively charged organic matter such as humic

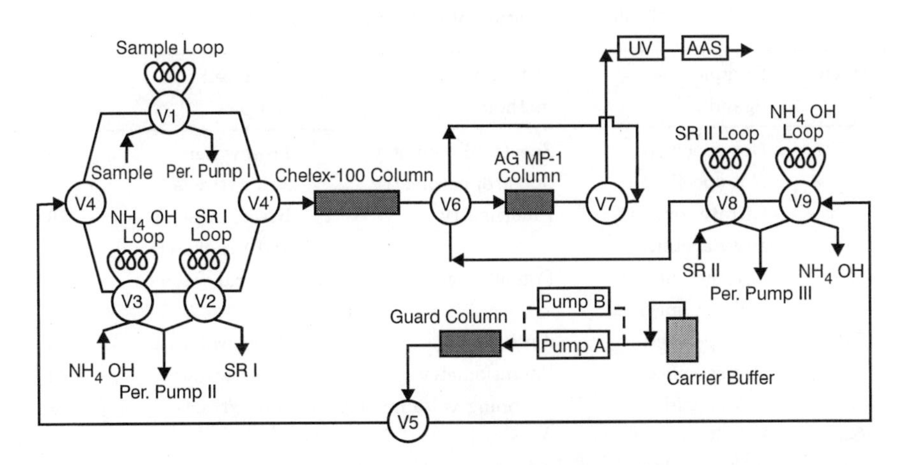

Fig. 7. Schematic diagram of an FIA system with AAS detection employing on-line sorption on Chelex- 100 and anion-exchange AG MP-1 columns used for the speciation of cadmium, copper and zinc [75]. (*Reprinted by permission of copyright owner.*)

material. The third, unretained fraction includes metals that are strongly associated with very large colloidal matter and do not dissociate in any of the columns used.

Two different FIA methods with spectrophotometric [35] and fluorimetric [76] detection have been developed for the speciation of aluminium complexes. In the setup with spectrophotometric detection, the determination was based on pyrocatechol violet chelation. The incorporation of on-line preconcentration on a cation-exchange column allows the measurement of the concentrations of non-labile monomeric forms of Al, while total reactive Al and total monomeric Al are differentiated by off-line sample pretreatment [35]. The fluorimetric detection method is based on the formation of a fluorescent Al-salicylaldehyde picolinohydrazone complex which is retained on a C18 sorbent packed in a flow cell located in a conventional spectrofluorimeter [76]. Three forms of Al (acid reactive Al, total monomeric Al and non-labile monomeric Al) can be determined in a similar way, as described for the system with spectrophotometric detection [35]. Two other forms, acid soluble and labile monomeric Al, can be evaluated by difference by injecting three sample aliquots into the continuous flow system and using on-line ion-exchange microcolumn. The flow injection system with spectrophotometric detection has also been applied to the determination of reactive aluminium in soil extracts [77]. Three different

Table 2. Speciation of complexes using FIA.

Element	Complexing ligand	Detection method	Analysed real samples	Reference
Al	Not specified	Spectrophotometry	Freshwater	35
	Not specified	Spectrophotometry	Soil extracts	77
	Citrate, oxalate, acetylacetone	Fluorimetry	River, mineral and tap waters	76
Ca	Glycine, citrate and not specified	Potentiometry	Natural waters	74
Cd	Not specified	AAS	River water	75
	Not specified	Potentiometry	River water	79
	Fulvic acid	Stripping voltammetry	Soil extracts	78
Cu	EDTA	AAS	–	34
	Acetate, glycine, iminodiacetate, NTA, EDTA	AAS	River water	75
F	Al(III) complexes	Potentiometry	–	74
Zn	Not specified	AAS	River water	75

used chromophores allow one to differentiate between the free Al^{3+}, aluminium bound by strongly complexing ligands and to weaker ligands. This gives the ability to order examined solid extracts or natural waters for aluminium toxicity.

The FIA method for the speciation of cadmium is based on the use of on-line Donnan dialysis and differential-pulse anodic stripping voltammetry for the determination of free cadmium concentrations [78]. The total cadmium concentration is determined with conventional electrothermal AAS.

The determination of free cadmium ions using the stopped-flow FIA system has also been reported for natural waters with a solid-state cadmium ion-selective electrode [79]. Interferences of trace amounts of Fe(III), Cu(II) and Pb(II) can be eliminated by reduction with hydroxylamine, complexation with Neocuproine and ion-exchange resin in sulphate form, respectively.

3. Speciation of Organometallic Compounds

Organometallic compounds with covalent metal-carbon bonds have chemical properties which are essentially different from those of coordination

Table 3. Speciation of organometallic compounds using FIA.

Element	Determined species	Detection method	Analysed real samples	Reference
As	As(III), As(V), monomethylarsonic dimethylarsinic acids	Hydride generation AAS	–	81
			Urine	82
			Natural waters	83
Hg	Hg(II), methylmercury	Cold vapour atomic fluorescence spectroscopy	Spiked tap and river waters	68
	Inorganic mercury, organomercury	Cold vapour AAS	Urine	85
Pb	Pb(II), tetraethyl lead	AAS	Spiked tap and surface waters	65
	Total lead, tetraethyl lead, tetramethyl lead	AAS	Gasoline	33

complexes or hydrated metal cations. They are much more volatile and lipophilic and are usually more toxic, although the degree of toxicity is strictly related to the nature of the monitored biosystem [80]. The known exceptions are the organometallic compounds of arsenic, which are less toxic than inorganic arsenic compounds.

Atomic spectroscopy detectors have been employed in all FIA systems developed for the speciation of organometallic compounds (Table 3). The speciation of arsenic has been carried out in a commercial setup with hydride generation AAS [81]. By carrying out hydride generation in selected acid media, determination of As(III) alone, the sum of mono- and dimethylated arsenic (with different sensitivities, which can be used for their differentiation), and all these species together with As(V), are possible. The conditions of hydride generation are adjusted in the FIA system by changing reagent solutions.

For arsenic speciation in urine also a flow injection system with hydride generation AAS detection was developed with ion-exchange separation of arsenic species and determination of total arsenic after microwave-assisted digestion [82]. In this system separation and sequential determination of inorganic As(III) and As(V) and monomethylarsonic and dimethylarsinic acids has been carried out. In the modified version of this procedure each arsine was cryogenically trapped in a PTFE coil, and then, based on their different boiling points, the arsine species were selectively liberated by using a heating cycle of microwave radiation, followed by AAS detection [83].

The sequential determination of inorganic mercury and methylmercury has been described for a system with cold vapour-atomic fluorescence spectrometry detection following the separation and preconcentration of methylmercury using a microcolumn of sulphydryl cotton with a relatively high affinity for methylmercury [84]. Inorganic mercury is not retained on the column and passes immediately to the detector. The retained methyl-mercury is eluted with hydrochloric acid.

In the FIA system with cold vapour AAS detection and on-line microwave sample pretreatment total and inorganic mercury in urine sample has been differentiated [85]. From the difference the content of the organomercury species in samples can be determined. The detection limit was 0.1 $\mu g\, l^{-1}$ Hg regardless of the mercury species under evaluation.

Two approaches, with flame AAS detection, have been reported so far for the FIA speciation of organolead compounds. In a simpler system oriented towards speciation in natural waters, both inorganic lead and alkyl lead species are separated and preconcentrated on-line on the cationic cellulose sorbent Cellex P with phosphonic acid functional groups [86]. Elution with nitric acid and then with ethanol gives two signals corresponding to the sum of inorganic lead and di- and trialkyllead compounds, and tetraethyllead, respectively. A more complex procedure has been developed for the FIA determination of total lead and the speciation of tetraethyl- and tetramethyllead in gasoline [33]. In this case, it is necessary to use on-line emulsification of the samples. This is achieved by the injection of the sample into a stream of commercial emulsogen solution. Demetallation using a solution of iodine in petroleum spirit has also been employed for both the speciation and determination of total lead. The speciation takes advantage of the different AAS sensitivities exhibited by the two tetraalkyllead species before and after demetallation.

The flow injection systems hyphenated to gas chromatography may also significantly simplify complex speciation analysis. In determination of methyl, ethyl and inorganic mercury, analytes are first preconcentrated on a dithiocarbamate resin packed in a 60 μl column in the FIA system, then eluted with thiourea solution and butylated with a Grignard reagent [87]. GC with a microwave-induced plasma atomic emission detector was employed for this

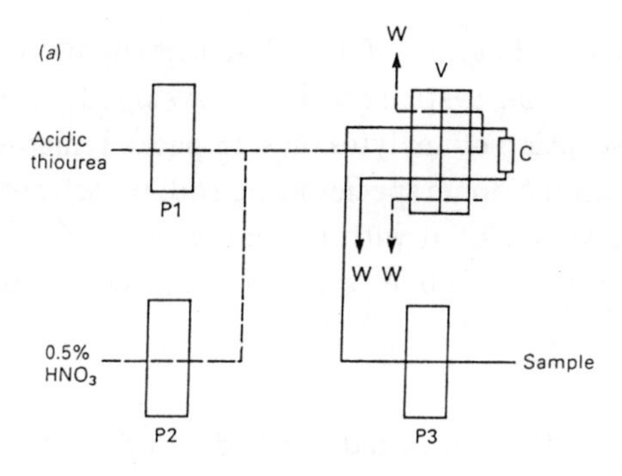

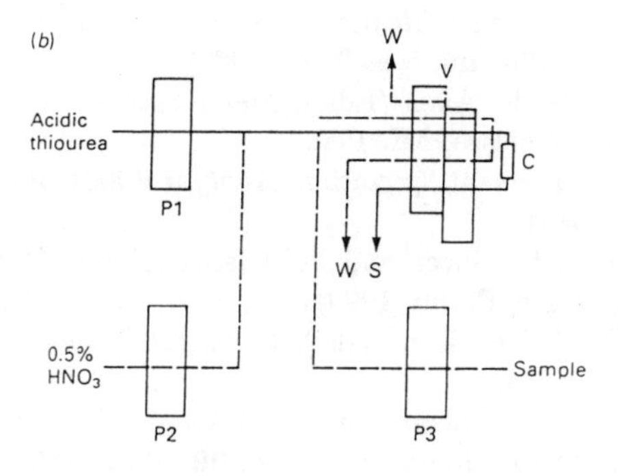

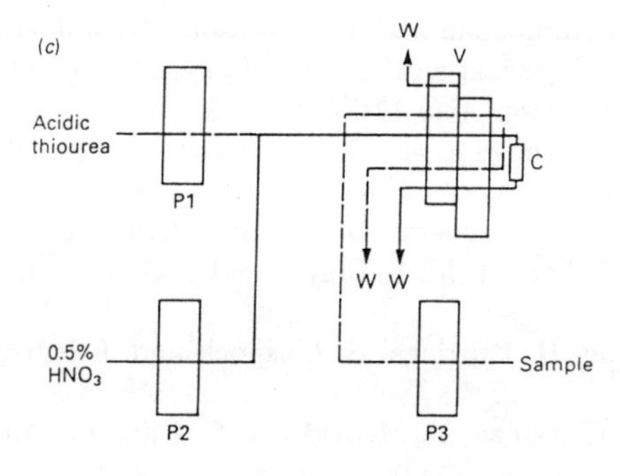

Fig. 8. Schematic diagram of the flow injection system used for speciation of mercury for preconcentration prior to a gas chromatography determination [87]. P1 — pump 1; P2 — pump 2; P3 — external pump; C — microcolumn; V — five-way valve, W —waste; and S — sample collection. Broken lines represent inactive parts in each sequence. (a) Sample enrichment operated in the fill mode of the valve with P3 pumping the sample through the column. (b) Sample elution operated automatically with P1 and the value in the injection mode. (c) Rinsing step performed immediately after sample elution with the valve in the iject mode. (*Reprinted by permission of copyright owner.*)

purpose. A schematic diagram of the FIA system used for this purpose is shown in Fig. 8. A similar system has been developed for organotin speciation analysis with the sorption of analytes on C18 packed microcolumns and *in situ* heterophase ethylation of ionic species using sodium tetraethylborate [88]. The derivatised species were eluted with methanol to the GC setup. For 10 to 50 samples a detection limit of about 0.1 ng l^{-1} has been reported.

4. References

1. M. Bernhard, F. E. Brinckman and P. J. Sadler (Eds.), *The Importance of Chemical Speciation in Environmental Processes* (Springer, Berlin, 1986).
2. G. G. Leppard (ED.), *Trace Element Speciation in Surface Waters and Its Ecological Implications* (Plenum, New York, 1983).
3. J. R. Kramer and H. E. Allen (Eds.), *Metal Speciation: Theory, Analysis and Application* (Lewis, Chelsea, MI, 1988).
4. G. E. Batley, *Trace Element Speciation: Analytical Methods and Problems* (CRC, Boca Raton, FL 1989).
5. J. A. C. Broekaert, S. Gücer and F. Adams (Eds.), *Metal Speciation in the Environment* (Springer, Berlin, 1990).
6. A. M. Ure and C. M. Davidson (Eds.), *Chemical Speciation in the Environment* (Blackie, London, 1995).
7. M. D. Luque de Castro, *Talanta* **33** (1986) 45.
8. M. D. Luque de Castro, *Mikrochim. Acta* **109** (1992) 165.
9. L. Campanella, M. Trojanowicz and K. Pyrzynska, *Talanta* **43** (1996) 825.
10. T. P. Lynch, N. J. Kernoghan and J. N. Wilson, *Analyst* **109** (1984) 839.
11. J. L. Manzoori, A. Miyuzaki and H. Tao, *Analyst* **115** (1990) 1055.
12. L. Anderson, *Anal. Chim. Acta* **110** (1979) 123.
13. T. P. Lynch, N. J. Kernoghan and J. N. Wilson, *Analyst* **109** (1984) 843.
14. K. Sonne and P. K. Dasgupta, *Anal. Chem.* **63** (1991) 427.
15. J. L. Burguera and M. Burguera, *Anal. Chim. Acta* **157** (1984) 177.
16. A. P. Doherty, R. J. Forster, M. R. Smyth and J. G. Vos, *Anal. Chem.* **64** (1992) 572.
17. M. Blanco, J. Gene, H. Iturriaga, S. Maspoch and J. Riba, *Talanta* **43** (1987) 987.
18. M. B. de la Calle Guntiñas, Y. Madrid and C. Camara, *Anal. Chim. Acta* **252** (1991) 161.
19. A. M. Dietrich, T. D. Ledder, D. L. Gallagher, M. N. Grabeel and R. C. Hoehn, *Anal. Chem.* **64** (1992) 496.
20. D. W. Bryce, A. Izquierdo and M. D. Luque de Castro, *Fresenius J. Anal. Chem.* **351** (1995) 433.
21. D. W. Bryce, A. Izquiredo and M. D. Luque de Castro, *Anal. Chim. Acta* **308** (1995) 96.
22. P. Linares, M. D. Luque de Castro, *Anal. Chem.* **58** (1986) 120.

23. D. G. Themelis, D. W. Wood and G. Gordon, *Anal. Chim. Acta* **225** (1989) 437.
24. D. E. Davey, D. E. Mulcahy and G. R. O'Connel, *Talanta* **37** (1990) 313.
25. R. Torralba, M. Bonilla, A. Palacios and C. Camara, *Analusis* **22** (1994) 478.
26. J. C. Andrade, J. C. Rocha and N. Baccan, *Analyst* **110** (1985) 197.
27. S. E. Birnie, *J. Autom. Chem.* **10** (1988) 140.
28. B. Ya. Spivakov, T. A. Maryutina, L. K. Shpigun, V. M. Shkinev, Yu. A. Zolotov, E. Ruseva and I. Havezov, *Talanta* **37** (1990) 889.
29. M. Burguera and J. L. Burguera, *J. Anal. At. Spectrom.* **8** (1993) 229.
30. L. Pitts, P. J. Worsfold and S. J. Hill, *Analyst* **119** (1994) 2785.
31. J. F. Van Staden, *Anal. Chim. Acta* **138** (1982) 403.
32. M. Silva, M. Gallego and M. Valcarcel, *Anal. Chim. Acta* **179** (1986) 341.
33. R. Borja, M. de la Guardia, A. Salvador, J. L. Burguera and M. Burguera, *Fresenius J. Anal. Chem.* **338** (1990) 9.
34. E. B. Milosavljevic, J. Ruzicka and E. H. Hansen, *Anal. Chim. Acta* **169** (1985) 321.
35. M. J. Quintela, M. Gallego and M. Valcarcel, *Analyst* **118** (1993) 1199.
36. J. Bartroli, M. Escalda, C. J. Jorquera and J. Alonso, *Anal. Chem.* **63** (1991) 2532.
37. H. Müller and V. Müler, *Anal. Chim. Acta* **230** (1990) 113.
38. A. T. Faizullah and A. Townshend, *Anal. Chim. Acta* **167** (1985) 225.
39. J. Ruz, A. Rios, M. D. Luque de Castro and M. Valcarcel, *Fresenius Z. Anal. Chem.* **322** (1985) 499.
40. J. Ruz, A. Torres, A. Rios, M. D. Luque de Castro and M. Valcarcel, *J. Autom. Chem.* **8** (1986) 70.
41. M. Trojanowicz, W. Matuszewski, B. Szostek and J. Michalowski, *Anal. Chim. Acta* **261** (1992) 391.
42. S. Motomizu and M. Sanda, *Anal. Chim. Acta* **308** (1995) 406.
43. M. T. Oms, A. Cerda and V. Cerda, *Anal. Chim. Acta* **315** (1995) 321.
44. A. Cerda, M. T. Oms, R. Forteza and V. Cerda, *Analyst* **121** (1996) 13.
45. Y. Luo, S. Nakano, D. A. Holman, J. Ruzicka and G. D. Christian, *Taalanta* **44** (1997) 1563.
46. J. Ruz, A. Rios, M. D. Luque de Castro and M. Valcarcel, *Anal. Chim. Acta* **186** (1986) 139.
47. M. Sperling, X. Yin and B. Weltz, *Analyst* **117** (1992) 629.
48. T. P. Rao, S. Karthikeyan, B. Vijayalekshmy and C. S. P. Iyer, *Anal. Chim. Acta* **369** (1998) 69.
49. K. W. Pratt and W. F. Koch, *Anal. Chem.* **58** (1986) 124.
50. R.-M. Liu, D.-J. Liu and A.-L. Sun, *Analyst* **117** (1992) 1767.
51. J. Alonso, J. Bartroli, M. del Valle and R. Barber, *Anal. Chim. Acta* **219** (1989) 345.
52. A. C. Lopes da Conceisao, M. T. Tena, M. M. Crreira dos Santos, M. L. Simoes Goncalves and M. D. Luque de Castro, *Anal. Chim. Acta* **343** (1997) 191.
53. E. A. G. Zagatto, H. Bergamin, S. M. B. Brienza, M. A. Z. Arruda, A. R. A. Nogueira and J. L. F. C. Lima, *Anal. Chim. Acta* **261** (1992) 59.

54. B. Haghigi and A. Safavi, *Anal. Chim. Acta* **354** (1997) 43.
55. M. Novic, Marjana Novic, J. Zupan, N. Zupan and B. Pihlar, *Anal. Chim. Acta* **348** (1997) 101.
56. A. Chaurasia and K. K. Verma, *Fresenius J. Anal. Chem.* **351** (1995) 335.
57. G. Gordon, K. Yoshino, D. G. Themelis, D. Wood and G. E. Pacey, *Anal. Chim. Acta* **224** (1989) 383.
58. K. K. Verma, A. Jain and A. Townshend, *Anal. Chim. Acta* **261** (1992) 233.
59. K. Oguma, K. Kitada and R. Kuroda, *Mikrochim. Acta* **110** (1993) 71.
60. M. G. C. Fernandez, M. A. Palacios and C. Camara, *Anal. Chim. Acta* **283** (1993) 386.
61. M. Novic, S. Tezak, B. Pihlar and V. Hudnik, *Fresenius J. Anal. Chem.* **350** (1994) 653.
62. A. Daniel, D. Birot, M. Lehaitre and J. Poncin, *Anal. Chim. Acta* **308** (1995) 413.
63. Z. Q. Zhang, L. J. Gao, H. Y. Zhan and Q. G. Liu, *Anal. Chim. Acta* **370** (1998) 59.
64. T. Aoki and M. Wakabayashi, *Anal. Chim. Acta* **308** (1995) 308.
65. T. Aoki, S. Fukuda, Y. Hosoi and H. Mukai, *Anal. Chim. Acta* **349** (1997) 11.
66. F. Canete, A. Rios, M. D. Luque de Castro and M. Valcarcel, *Analyst* **113** (1988) 739.
67. A. G. Cox, I. G. Cook and C. W. McLeod, *Analyst* **110** (1985) 331.
68. A. M. Naghmush, K. Pyrzyñska and M. Trojanowicz, *Anal. Chim. Acta* **288** (1994) 247.
69. G. E. Pacey and B. P. Bubnis, *Int. Lab. September* (1984) 30.
70. S. Blain and P. Treguer, *Anal. Chim. Acta* **308** (1995) 425.
71. S. Krekler, W. Frenzel and G. Schultze, *Anal. Chim. Acta* **296** (1994) 114.
72. B. Patel, S. J. Haswell and R. Grzeskowiak, *J. Anal. At. Spectrom.* **4** (1989) 195.
73. C. Steinberg, *Water Res.* **14** (1980) 1239.
74. M. Trojanowicz, P. W. Alexander and D. B. Hibbert, *Anal. Chim. Acta* **366** (1998) 23.
75. Y. Liu and J. D. Ingle Jr., *Anal. Chem.* **61** (1989) 525.
76. P. Canizares and M. D. Luque de Castro, *Anal. Chim. Acta* **295** (1994) 59.
77. D. J. Hawke and H. K. J. Powell, *Anal. Chim. Acta* **299** (1994) 257.
78. D. Berggren, *Int. J. Environ. Anal. Chem.* **41** (1990) 133.
79. M. Trojanowicz, P. W. Alexander and D. B. Hibbert, *Anal. Chim. Acta* **370** (1998) 267.
80. P. J. Craig (Ed.), *Organometallic Compounds in the Environment. Principles and Reactions* (Longman, Harlow, 1986).
81. T. R. Rude and H. Puchelt, *Fresenius J. Anal. Chem.* **350** (1994) 44.
82. J. L. Burguera, M. Burguera and C. Rivas, *Quim. Anal.* **16** (1977) 165.
83. J. L. Burguera, M. Burguera, C. Rivas and P. Carrero, *Talanta* **45** (1998) 531.
84. W. Jian and C. W. McLeod, *Talanta* **39** (1992) 1537.
85. M. Gallignani, H. Bahsas, M. R. Brunetto, M. Burguera, J. L. Burguera and Y. Petit de Pena, *Anal. Chim. Acta* **369** (1998) 57.

86. A. M. Naghmush, K. Pyrzyñska and M. Trojanowicz, *Talanta* **42** (1995) 851.
87. H. Emteborg, D. C. Bacter and W. Frech, *Analyst* **118** (1993) 1007.
88. J. Szpunar-Lobiñska, M. Ceulemans, R. Lobiñski and F. C. Adams, *Anal. Chim. Acta* **278** (1993) 99.

Chapter 8

Applications of Flow Injection Methods in Routine Analysis

1. Environmental Applications

The more than twenty years of intensive development of FIA have resulted in numerous applications in various areas of routine chemical analysis. The attractive simplicity of the mechanisation of sample treatment even with very complex matrices and the resulting significant shortening of the total analysis time compared to manual procedures have caused an increasing interest in this methodology not only in analytical research laboratories, but also in environmental, industrial, agriculture and clinical laboratories. It is greatly favoured by a broadening offer of complete commercial flow injection set-ups (see Chapter 9, as well as particular devices and accessories that can be assembled in FIA measuring systems in every analytical laboratory).

The largest number of applications was developed for environmental analysis, i.e. natural and treated waters, wastes, aerosols, dusts, and also discussed separately plant materials. Numerous methods are described especially for water analysis, and they have also been reviewed by several authors [1–7]. The developed methods include not only laboratory procedures but also field monitoring *in situ*, marine monitoring involving shipboard techniques and submersible applications [7]. A survey of over 200 applications of FIA in water analysis indicated that the overall quality of the application is acceptable [6]. RSD values were in general less than $\pm5\%$, usually between ±1 and $\pm3\%$. Among the analytical techniques involved, the most widely employed were photometry (75.1%) and potentiometry (11.6%).

In this presentation of methods reported in the literature, only such procedures are included, which were applied to natural samples, and their results were compared to those obtained with other methods, or were examined using standard reference materials, or at least they were tested using known standard additions to the natural samples.

1.1. *Inland and Tap Waters*

Inland waters (river, lake, underground) and tap water, because of the significant difference in the composition of matrices compared to sea water, will be discussed separately. The high salinity of sea water often requires particular pretreatment procedures, especially for determination of trace components. Because of the large number of methods developed for inland and tap waters, they are presented in separate sections devoted to macro-components, and inorganic and organic micro-components.

1.1.1. *Non-specific water quality parameters*

Flow injection methods have been applied not only to determination of individual components of waters but also to determination of non-specific water quality parameters such as chemical oxygen demand (COD), dissolved organic carbon (DOC) or dissolved inorganic carbon (DIC). Also, the pH of water should be included in these parameters, as it is a function of the presence of various different chemical species and is one of the most important hydrochemical parameters of natural waters.

COD is a widely used index of water and waste water quality which relates to the oxygen required for complete oxidation of the sample. It is an arbitrary measurement obtained by the oxidation of the sample by chromic acid. Such a procedure was adopted in several FIA methods [8–11], although methods using permanganate [12] or cerium(IV) sulphate [13] as the strong oxidising agent have been developed. In the systems employing dichromate oxidation samples are injected into a water stream which merges with an acidic dichromate carrier solution, but different designs of the reaction part of the manifold were proposed and, downstream, the absorbance of chromate is monitored at 445 nm. In the system with an HPLC pump, a 50-m-long 0.5 mm i.d. PTFE reaction coil was used in a thermostatted oil bath at 120°C and the reaction residence time necessary for the oxidation reaction was about 20 min [8]. By raising the temperature to 160°C a significantly higher degree of sample oxidation has been achieved using a 3 m reaction coil [9]. This resulted in a processing time of 3 min. Mercury(II) is added to the sample to prevent chloride interference and additionally a Ag(I) catalyst can be incorporated into the reagent. In the system based on the use of a 180 W microwave oven for supplying energy in the oxidation step (Fig. 1), a 10 m reaction coil was adopted [10]. In the system with amperometric detection the oxidation step was carried out off-line in a separate miniaturised vessel under standard COD determination conditions, and the excess of dichromate was determined in the FIA system by measuring the iodine released after adding potassium iodide [11].

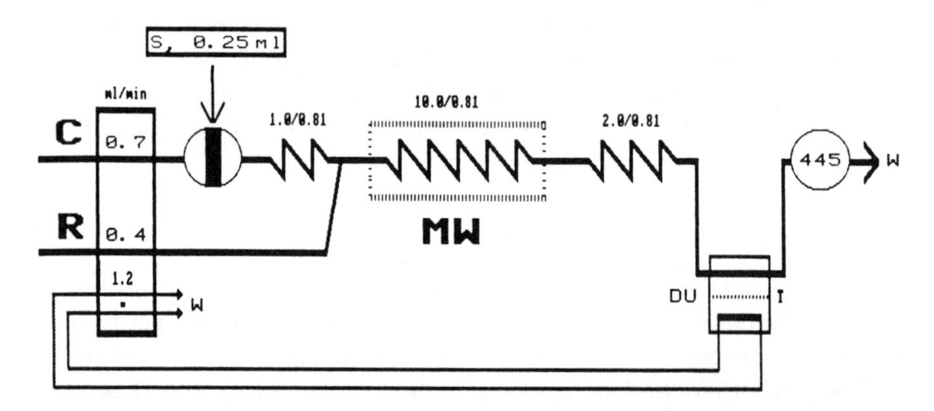

Fig. 1. Manifold for COD determination in the FIA system using the dichromate method with heat supplied by a microwave oven [10]. C — carrier stream of doubly distilled water; R — potassium dichromate-sulphuric acid reagent; S — injection of the sample. Reaction coils 1.0, 10.0 and 2.0 m, all with 0.81 mm i.d. MW, microwave oven; DU — degassing unit; T — PTFE membrane. (*Reprinted by permission of copyright owner.*)

Much milder operating conditions were reported for COD determination in the FIA system with cerium(IV) sulphate as oxidising solution and detection at 320 nm [13]. A 20 m reaction coil was thermostatted at 100°C and when 50 μl samples were injected at a frequency of 20 h^{-1} the determination range was 0.5–130 mg l^{-1} COD with high tolerance to chloride without any masking agents.

DOC is a parameter characterising organic pollution of water and wastes. It can be measured in the FIA system combining in-line UV photo-oxidation with indirect spectrophotometric monitoring of the CO_2 produced [14] (Fig. 6, Chapter 1). The injected sample is mixed with an alkaline peroxodisulphate reagent and irradiated with UV radiation in a simple Teflon photo-reactor. The CO_2 formed is passed through a Teflon membrane into a stream of phenolphthalein. The absorbance is monitored at 552 nm. The method is rapid (about 45 samples h^{-1}), yields a detection limit of 0.1 mg C l^{-1} and was successfully applied to river waters.

The content of DIC in surface waters is used as an indication of the degree of dissolution of chemical species such as silicates or carbonates. Although in official procedures titration is used to determine the total alkalinity, the alternative procedure can be based on determination of carbonate. The spectrophotometric flow injection method based on the on-line generation of carbon dioxide, followed by separation through a gas diffusion device inserted into the analytical path, has been proposed [15]. The system involved detection of carbon dioxide with cresol red solution with absorbance changes detected at 410 nm. Another reported spectrophotometric method is based on the ability of the hydrogencarbonate anion to catalyse the slow reaction of EDTA with Cr(III) of an aged aqueous solution [16]. The response of the FIA system is obtained from the difference of analytical signals generated before and after the catalytic action caused by DIC, which are detected at 540 nm.

Different approaches can be used for FIA determination of pH in natural waters, depending on the ionic strength of samples. For the determination in low ionic strength rainwater the optosensing was utilised that involves the use of a covalently bound reversible selective indicator on a cellulose substrate and reflectance spectroscopy [17]. The detection was carried out using an FIA microconduit manifold where a 75 μl sample volume was injected into a carrier of 0.5 mM acetate and measurements were carried out in the pH range of 3.5–5.0. In shipboard measurements of pH in sea water the absorbance of phenol red injected into a sea water stream was employed, where a working precision of ± 0.005 pH units was achieved with a sampling frequency of about 25 h^{-1} [18]. The measurement of pH together with another non-specific parameter conductivity can be carried out in the multicomponent FIA system including also spectrophotometric determination of residual chlorine, ammonium and nitrite ions [18a]. It operates in the reversed mode, so that the sample is continuously circulating through the system. The potentiometric glass-calomel microelectrode and conductimetric flow cell with platinum electrodes are located in the sample stream prior to the injection of any reagent.

1.1.2. *Inorganic macrocomponents*

As macrocomponents are considered elements which are present in natural waters at mg l^{-1} level (Na, K, Ca, Mg, Cl, S and Si), their content in various kinds of waters may differ essentially. Generally the analytical procedures for their determination are different than for trace elements, they usually do not require preconcentrations steps, and they are not subject to problems associated with contamination.

Sodium and potassium in waters are mostly determined, both conventionally and in FIA systems, using flame photometry or potentiometry with membrane ion-selective electrodes. The determination of sodium and potassium in the FIA system with flame photometry detection was reported in the setup with simultaneous determination of calcium and magnesium in the same injected sample using AAS detection with flame atomisation [19]. A 200 μl sample was injected into the stream of distilled water, then merged with a solution containing lanthanum nitrate and caesium and lithium chloride, which was split and transported to the flame photometer and AAS spectrometer. Simultaneous determination of sodium and potassium can be carried out in river and lake water with potentiometric detection using two ion-selective electrodes arranged in parallel in a flow injection system without a conventional reference electrode [20]. The sample is injected into the two different carrier streams alternately. Potassium was determined in the range of 0.6–22 mg l^{-1} and sodium in the range of 20–65 mg l^{-1}.

More frequently for the water quality control the determinations of calcium and magnesium are carried out. The spectrophotometric determinations of these two analytes can be carried out using different reagents. In direct determinations o-cresolphthalein complexone [21] and glyoxal bis(2-hydroxyanil) [22] were used as the colour-forming reagents. In indirect determinations the exchange reaction between the calcium and zinc complex of EGTA in the presence of 4-(2-pyridylazo) resorcinol (PAR) can be used [23]. The sensitivity and selectivity of spectrophotometric determination of calcium can be improved by the use of solvent extraction of an ion association complex formed by calcium bound with crown ether dicyclohexano-24-crown-8 and an anionic dye propyl orange [24]. The detection limit was estimated in this method as 0.2 μM. The calcium determinations with flame AAS detection were carried out in flow injection [19, 25] and sequential injection [26] systems. In all these procedures the lanthanum nitrate was employed for elimination of interferences. In the system with a dual channel spectrometer [19] and also in a sequential injection system [26], magnesium was determined. The developed FIA system was also successfully used for the determination of zinc in biological material on the concentration level of 20–40 mg kg^{-1} after mineralisation with nitric and perchloric acids [25].

For the determination of calcium and magnesium in waters in FIA systems also potentiometric detection with membrane ion-selective electrodes can be satisfactorily utilised. In spite of using in early works the arylphosphate ionophore of limited selectivity towards calcium, successful determinations in waters have been reported [21, 27]. In the FIA system where a tubular membrane electrode was used, due to negligible dispersion of the sample zone in the potentiometric detector, the system was additionally equipped in an AAS spectrometer that allows simultaneous determination of magnesium [27]. The same tubular electrode was used in the FIA system for simultaneous determination of calcium and nitrate in waters with sequential detection by two potentiometric sensors [28]. The calcium-selective electrodes with a neutral carrier and a photo-cured membrane formed on a metallic silver electrode in the FIA system were also used in the FIA system for water analysis [29]. The selective potentiometric determination of magnesium with a prototype magnesium-selective electrode in the FIA system required the use of the EGTA complexone in carrier solution for elimination of calcium interferences [30]. In the determination of magnesium in mineral waters at the concentration level of 10–19 mg l^{-1}, good agreement with AAS results was obtained.

The determination of chloride in waters in FIA systems is carried out most often using spectrophotometric or potentiometric detection. In order to extend the range of chloride concentration in water samples using the spectrophotometric mercury thiocyanate-Fe(III) method, a sample splitting and a complex forming reagent were used [31,32]. Another possible approach is on-line sample prevalve dilution [33]. The results of flow injection spectrophotometric determinations of chloride were compared with those obtained in the FIA system with turbidimetric determination based on the precipitation with silver ions [34]. For this purpose a complex flow system with a single spectrophotometric detector was employed and samples before injection were passed through a cationic resin column in protonated form. For most of analysed river water samples the chloride contents determined by both methods did not differ from each other at the 95% confidence level. The successful use of the potentiometric detection with various chloride electrodes was reported in FIA systems by different authors [35–37]. The analysed samples did not require any preliminary treatment, and as carrier solution or reagent solution in two line systems the potassium nitrate solution can be used.

The sulphate occurs in inland waters usually at the level of the millimole or a fraction of the millimole, hence it is included in this section. The determinations in flow injection measuring systems are usually carried out with turbidimetric detection, but they can also be performed with indirect spectrophotometric or potentiometric detection. With turbidimetric detection most of the developed systems were based on precipitation of barium sulphate [38–41]. In order to avoid accumulation of the precipitate in the flow system, several different precautions have been proposed, such as the addition of poly(vinyl alcohol) to barium chloride solution [38, 41], the alternate injection [39] or pumping [41] of a washing alkaline EDTA solution. In order to remove suspended solids and organic substances, which may interfere in spectrophotometric measurements at 420 nm, an active carbon filter between the sampler and the injection valve was successfully applied [40]. The developed procedures were used for sulphate determination in the range of 1–200 mg l^{-1}. For the same purpose a precipitation of colloidal lead sulphate in an ethanol water medium was proposed, with better stability of the lead sulphate suspension and a better detection limit than those obtained with the barium sulphate method [42]. Spectrophotometric determination of sulphate in waters can be carried out using various colour-forming reagents. The determinations in rain water at the 0.7–3.8 mg l^{-1} level were based on competitive reaction of sulphate and methylthymol blue with barium [43]. The determinations in river water samples based on the decoloration of the barium-dimethylsulphonazo-III complex required the sample to be passed through a cation exchange column in protonated form before being injected [44]. In another spectrophotometric method applied to sulphate determination in spring and rain waters, the sample plug in the carrier stream containing 50% ethanol passed through two columns in series. The first one was packed with a cation-exchange resin to remove interfering cations, whereas the second one was packed with cellulose beads containing solid barium chloranilate particles [45]. The detection was based on the monitoring of the released chloranilate ion at 530 nm. The direct spectrophotometric method for sulphate determination in river water was based on the monitoring of the absorbance of the $FeSO_4^+$ complex at 355 nm [46]. In two developed FIA systems with potentiometric detection for the determination of sulphate in waters with indirect measurements, a linearisation of potentiometric signal was utilised. In determinations with the use of a barium-selective electrode with a plasticised membrane, the sample after passing the column with cationic resin was mixed with barium chloride solution in 30% isopropanol (Fig. 2) [47]. In another system, where a commercial solid-state lead-sensitive electrode was applied, a solid lead sulphate reaction column was employed with a solvent containing 35% methanol, and a limit of detection

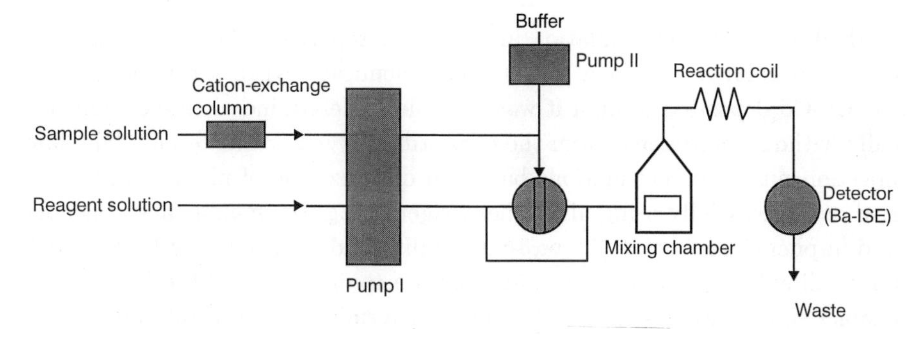

Fig. 2. Manifold for the determination of sulphate in waters based on flow injection titration with potentiometric detection using a barium ion-selective electrode [47]. As reagent solution a 50 μM $BaCl_2$ solution which contains 30% (v/v) isopropanol was used, and a 0.5 M lithium acetate solution was used as buffer solution. (*Reprinted by permission of copyright owner.*)

as low as 1 μM can be obtained [48]. In these last two cases the results of determination were compared with those obtained with ion chromatography.

1.1.3. *Inorganic microcomponents*

Natural waters are reservoir of numerous inorganic and organic compounds at submillimolar and trace level. The literature concerning methods of their determination using flow injection methods is very broad, and only certain parts of it are discussed. The separate section below is devoted to the determination of organic species. In this section the presentation of methods of determination of anionic species is followed by a discussion of determination of metallic trace elements.

For nitrogen-containing compounds a number of methods have been developed for the determination of nitrite, nitrate and ammonia, as each of these species has its essential environmental importance.

For the determination of nitrate various detection methods can be utilised. In spectrophotometric methods usually are first reduced on-line to nitrite in a copperised cadmium column, and then nitrite is diazotised and the product is coupled with amine to form a coloured azo dye. Most frequently diazotization is carried out with sulphanilamide and the product is coupled with N-(1-naphthyl)ethylenediammonium dichloride [49–52]. In the latter cited work nitrate was reduced on-line on a column electrode packed with co-electrodeposited Cu–Cd glassy carbon grains [52]. In the same work also a coulometric method of nitrate determination in water was reported, where by using a Ag deposited column electrode as an electrochemical oxygen scrubber prior to the Cu-Cd deposited column it was possible to determine nitrate coulometrically without interference from the coexisting oxygen or nitrite [52]. The flow injection nitrate determinations based on diazotization of nitrite can be carried out successfully using also other reagents, e.g. the *p*-aminoacetophenone and *m*-phenylenediamine [53] or 3-nitroaniline and N-(1-naphthyl) ethylenediamine dihydrochloride [55]. The appropriate configuration of the FIA system enables simultaneous spectrophotometric determination of nitrate and nitrite [54, 55]. Nitrate in rain water has also been determined indirectly utilising its UV light absorption at 210 nm [56]. Several luminescence methods have been reported for nitrate determination in flow injection measurements. In fluorimetric determination nitrate was first reduced to nitrite, which then reacted with 3-amino-1,5-naphthalenedisulphonic acid to form azoic acid [57]. This acid forms a fluorescent salt in an alkaline medium. The obtained detection limit of 10 nM for sample volume 160 μl is at least one order of magnitude lower than in spectrophotometric absorptive methods. Two different chemiluminescent methods have been reported for determination of nitrate in waters. One of them is based on formation of peroxonitrite during irradiation of an acidified nitrate solution with UV light [58]. Peroxonitrite is the oxidant of luminol. For determinations in natural waters an iminodiacetate ligand exchange resin column was used for on-line removal the interference of Fe(III), Cu(II), Co(II) and Ni(II). In another method applied in simultaneous flow injection determination of nitrate and nitrite in river waters, gas-phase chemiluminescence was utilised [59]. The chemiluminescence signal was produced by the reaction of nitric oxide (NO) with ozone. A microporous teflon membrane was used to transfer nitric oxide to the gas phase after reducing nitrate to nitrite with Ti (III) or nitrate alone by iodide in acidic media. The detection limits were 0.7 and 0.35 μg l^{-1} for nitrate and nitrite, respectively.

Besides the above-mentioned coulometric detection among various electrochemical methods, potentiometry and amperometry can also be used for determination of nitrate in waters in FIA systems. The detection of nitrate with a nitrate-selective electrode was used for the determination in waters in various two-component FIA systems. For instance, in the system for sequential determination of calcium and nitrate the interferences from coexisting chloride and hydrogen carbonate were removed using lead acetate solution [28]. The method was successfully applied for river water samples containing 1.5–41 mg l^{-1}

nitrate. A nitrate ion-selective electrode was also used in two-electrode systems where sensors were arranged in parallel with fluoride or potassium electrodes [20]. In these cases an acetic buffer containing silver sulphate to suppress interferences was used. The sample was injected into the two-carrier system alternately. Satisfactory results in simultaneous determination of nitrite and nitrate in water can also be obtained using flow injection biamperometry [60]. The detection in the system with two polarised platinum wire electrodes is based on the oxidation of iodide by nitrite. The injected sample is split in a FIA manifold (Fig. 3) into two segments and one of them is transported through a reductor column, where nitrate is reduced to nitrite. The obtained detection limits for nitrite and nitrate were 40 and 70 mg l^{-1}, respectively.

Other nitrogen-containing species present in natural waters are ammonia or ammonium ions. Most often, for determination of ammonia nitrogen in waters in FIA systems a spectrophotometric detection is used. For this purpose a sensitive method of determination of nitrite with diazotisation and coupling after oxidation of ammonia to nitrite can be used [61]. In the flow injection system the oxidant solution was obtained by on-line mixing of hypochlorite and potassium bromide, injected into a water stream, and then mixed with the sample stream. In such a reversed FIA system a sensitive detection of ammonia was reported in the range of 0.2–12 μM. Several procedures were developed for determinations with a Nessler reagent consisting of mercury (II) iodide and potassium iodide in alkaline solution [18, 62, 63]. The use of a resin column in the sample loop allows one to employ this detection at the $\mu g\ l^{-1}$ concentration level for determination of ammonia in rain water samples [63]. This method was also successfully used in the system mentioned above for simultaneous determination of pH, conductivity, residual chlorine, nitrite and ammonium in tap water and various kinds of inland waters [18]. A modified Bertelot reaction catalysed by sodium nitroprussiate was employed for spectrophotometric determination of ammonia in waters in a continuous flow system in which the sample was introduced into the carrier solution between two air bubbles with detection limit 5 $\mu g\ l^{-1}$ [64]. The air bubbles that limit dispersion of the sample segment in the flow system are removed with a permeation membrane before reaching the detector. The amperometric detection reported in the FIA system for determination of ammonia in waters was based on the use of a glassy carbon electrode modified with cupric hexacyanoferrate and coated with Nafion film [65]. Because of sensitivity to potassium and ammonium ions a gas-diffusion unit was used for the separation of ammonia.

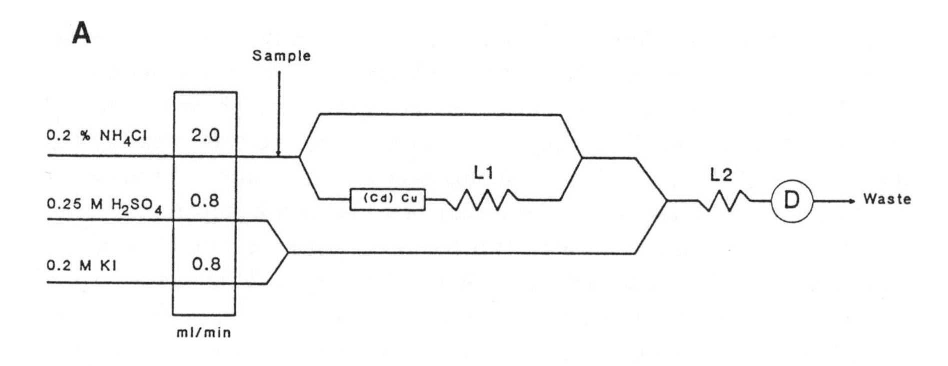

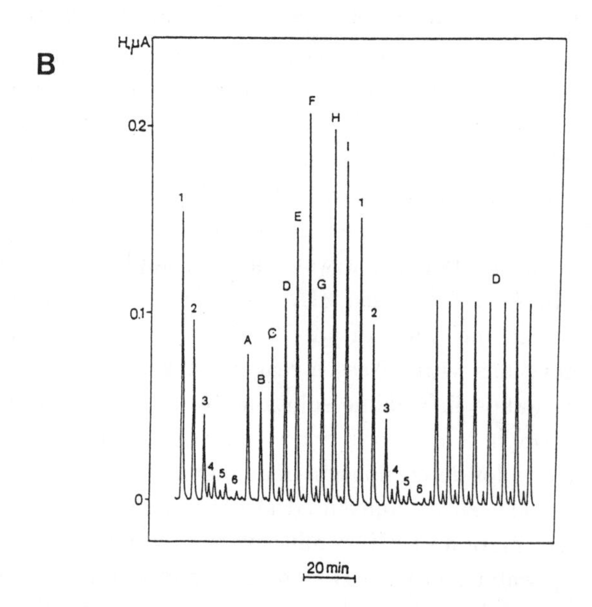

Fig. 3. Schematic diagram of the manifold used for simultaneous flow injection biamperometric determination of nitrate and nitrite in waters (A) and recording of peaks (B) for standard solutions (1–6) and neutral water samples (A–I) for injection volume 750 μl [60]. Concentration in standard solutions: 1 — 0.075; 2 — 0.050; 3 — 0,025 mM nitrate; and 4 — 7.5; 5 — 5.0; 6 — 2.5 μM nitrite. (*Reprinted by permission of copyright owner.*)

With a linear response of 2–40 μM the method was successfully applied to the determination of ammonium ions in lake water and rain water samples.

Chlorine occurs in natural waters mostly as chloride ion, whose determination in flow injection systems was discussed above. Because of using gaseous chlorine or chlorine dioxide for desinfection of tap water the quality control of tap water involves also determination of residual chlorine and other chlorine-containing oxyanions.

Among various colour-forming reagents that can be used for residual chlorine determination in waters, in flow injection systems have been employed N,N-diethyl-*p*-phenylenediamine [66], *o*-toluidine and methyl orange [18] and $3,3'$-dimethylnaphtidine [67]. The most sensitive response was obtained for the use of *o*-toluidine and $3,3'$-dimethylnaphtidine; however, the use of *o*-toluidine is largely restricted because of its carcinogenic properties. For the method based on the use of $3,3'$-dimethylnaphtidine the calibration plot was linear in the range of 0.1–1.0 mg l^{-1} of free residual chlorine; however, a selectivity study has shown some sensitivity to other forms of residual combined chlorine. A chemiluminescent method reported for the determination of free chlorine in tap water was based on light emission from the reaction between Rhodamine 6G and free chlorine (HOCl) with no interference from combined chlorine (chloramines) [68]. At sampling rate 240 samples h^{-1} and 20 μl injection reagent volume the limit of determination was estimated as 0.1 μM in the reversed FIA system.

For the same purpose electrochemical methods can be employed. The potentiometric detection of residual chloride can be carried out with various indicating electrodes. Using platinum and iodide-selective electrodes in the FIA system, 0.1–5.0 mg l^{-1} residual chlorine can be determined [69]. With the use of a gold-plated oxidation–reduction potential electrode and an Fe(III)/Fe(II) buffer the reduction of dissolved chlorine enables one to determine residual chlorine with the detection limit 3.5 μg l^{-1} [70]. Then if a lead (II) ion-selective electrode is used as an indicating electrode a sensitive response to residual chlorine is observed although the response mechanism is unclear [71]. The results obtained for tap water samples in the range of 0.1–1.0 mg l^{-1} were in good agreement with those obtained by photometric determination with *o*-toluidine. For flow injection determination of residual chlorine in waters, amperometric detection with two polarised platinum electrodes can also be used [72]. The detection is based on the oxidation of iodide and the limit of detection was estimated as 2.0 μg l^{-1}. The results obtained for determinations in swimming pool water were in good agreement with the batch spectrophotometric procedure using $3,3'$-dimethylnaphtidine.

Determinations of phosphorus-containing compounds in waters using flow injection techniques include determination of orthophosphates, pyrophosphates and dissolved organic phosphorus. The spectrophotometric determination of orthophosphate is based on formation of the heterocomplex between phosphate and molybdate and further reduction with different reducing species. With the use of tin (II) chloride stabilised with hydrazine a linear response was obtained in the range of 0.04–2.5 mg l^{-1} [73]. A significant improvement of the detection limit can be achieved by combining the flow injection technique with on-line solvent extraction. In a procedure where the ion associate formed between molybdophosphate and malachite green was extracted into a mixture of benzene and 4-methylpentan-2-one, the detection limit was 0.1 μg l^{-1} of phosphorus [74]. This method was successfully used for determinations in river waters. The molybdate method was applied also in the FIA system for the sequential determination of phosphate and silicate in river waters [75], as well as the sequential determination of ortho- and pyrophosphates in the same FIA system in river waters in the range of 10–30 μg l^{-1} [76]. The latter was carried out in the FIA system with an extraction-chromatographic mini-column, on which phosphate anions were separated and preconcentrated. The sum of the two phosphate anions was determined after enzymatic hydrolysis of pyrophosphate using inorganic pyrophosphatase. The determination of total organic phosphorus will be discussed in the next section.

Besides the above-mentioned sulphate in natural waters using FIA methods, trace levels of sulphite and sulphide can also be determined. Determination of sulphite in spiked waters of various origins was performed in the FIA system with an enzymatic system using immobilised sulphite oxidase [77]. Spectrophotometric detection of hydrogen peroxide produced in enzymatic reaction was carried out with a flow cell filled with a cation-exchanger, where both the derivatisation reaction and retention of the product took place. The detection limit obtained was 3 μg l^{-1}. The determination of sulphide content in waters can be carried out using potentiometric detection with an ion-selective electrode [37]. The reported determinations were carried out using a commercial antioxidant buffer in order to maintain suitable conditions for the ion-selective electrode functioning.

Determinations of fluoride in waters usually at the level of a fraction of mg l^{-1}, both conventional and flow injection ones, most frequently are carried out with potentiometric detection using a fluoride ion-selective electrode. In determinations carried out in rain water [78], tap water [79] and river water [20], buffer solutions were used that contain ligands able to release fluoride from complexes with aluminium and iron (III). It was shown in stopped-flow measurements, however, that decomplexation processes have to be carefully controlled [80]. The potentiometric determination of fluoride in waters can be performed in the FIA system simultaneously with determination of nitrate [20]. For fluoride determination in natural waters, optical detection methods can also be employed. Determinations with spectrophotometric detection were carried out in the reversed FIA setup with stopped flow [81]. In the flow system to the stream of sodium dodecyl sulphate the solution of lanthanum (III) with alizarin fluorine blue was injected, and then this stream was merged with the sample solution. The FIA determinations with fluorimetric detection were based on the ability of trace fluoride to increase the rate of formation of a fluorescent Al(III) - eriochrome red B complex in the presence of hexamethylenetetramine [82]. This procedure yields satisfactory results for tap and mineral waters. For indirect determinations of fluoride in waters the inductively coupled plasma atomic emission spectrometry was employed in the FIA system coupled with solvent extraction [83]. The method was based on the formation of lanthanum/alizarin complexone/fluoride complex and its extraction into hexanol containing N,N-diethylaniline, and then introduction of the organic layer into the plasma for lanthanum measurement. Such determinations exhibit satisfactory selectivity and the detection limit was estimated as 30 $\mu g\ l^{-1}$.

Iodide in rain waters was determined with spectrophotometric detection based on its catalytic effect on oxidation of arsenite by Ce (IV) [84].

Determination of boron can be carried out with spectrophotometric and fluorimetric detections. In the procedure based on formation of the boron complex with Azomethine-H the on-line preconcentration on N-methylglucamine functionalised resin with specificity for boron was employed [85]. For the sampling rate 10 samples h^{-1} the detection limit was estimated as 1 $\mu g\ l^{-1}$ and results obtained for natural waters in FIA system were in good agreement with ICP-AES determinations. The use of EDTA combined with the selectivity of resin resulted in an interference-free method for boron determination. A widely used reagent for boron determination chromotropic acid was also successfully applied in flow injection determinations with spectrophotometric [86] and fluorimetric [87] detections. The detection limits in these procedures were estimated as 8 $\mu g\ l^{-1}$ and 5 nM for spectrophotometric and fluorimetric detections, respectively. The spectrophotometric detection was employed for determination of boron in waters from the nuclear power industry, and the fluorimetric one for spiked river, lake and tap waters.

A sensitive spectrophotometric FIA procedure was developed for the determination of hydrogen peroxide in rain water [88]. The method was based on formation of the coloured product during condensation of N-ethyl-N-(sulphopropyl) aniline with 4-aminoantipyrene in the presence of hydrogen peroxide and peroxidase. In determinations at the concentration level of 0.15–23 μM in the rain water sample, no interference was observed from Mn (II) and Fe (III), which interfere in chemiluminescence detection. The detection limit was estimated as 0.14 μM.

An especially large number of flow injection methods have been developed for the determination of trace metal ions in natural waters. Among them are procedures for single component determinations with various detection methods, and procedures for speciation of trace elements and for multicomponent determinations. All these kinds of measuring systems will be presented below. The largest number of flow injection procedures was developed for single trace analyte determinations in waters.

Determination of aluminium in natural waters is carried out mostly with optical detections such as fluorimetry [89], AAS [90, 91], ICP-AES [81, 91] and ICP-MS [92]. In the system with fluorimetric detection of a Lumogallion complex of aluminium all Al species can be determined with the detection limit 3.7 nM except polymers. The example determinations were carried out in a lake water [89]. In determinations with AAS and ICP-AES detections an on-line preconcentration on cation-exchange [90], as well as on anion-exchange [91] resins in various chemical conditions, was applied. Using flame AAS detection aluminium at 20–40 mg l^{-1} was determined in tap water [90], whereas using ICP-AES detection determinations at 0.1–0.4 mg l^{-1} level were performed [91]. With sorption of an aluminium-8-hydroxyquinoline complex onto minicolumns containing non-polar Amberlite XAD-2, the fast reactive fractions in natural waters were determined with a limit of detection 1.8 μg l^{-1} for ICP-MS detection [92].

The selective determination of arsenite was carried out in an FIA system with spectrophotometric detection [93], while total arsenic was determined in the system with ICP-MS detection [94]. In spectrophotometric determinations using the molybdenum blue method the selectivity was achieved by on-line separation of the main interferents phosphate, arsenate and silicate using a strong anion-exchange microcolumn located in the aspiration line of the injection valve [93]. In the system with ICP-MS detection and hydride generator arsenic in the sample was on-line reduced with cysteine [94]. The detection limit in this method was estimated as 3 ng l^{-1}.

In trace determination of copper in FIA systems, the most often used detection method is atomic absorption spectrometry with flame atomisation [95–100]. An example of the use of potentiometric detection can also be found [101], and determinations in more sophisticated multicomponent measuring systems will be discussed separately in a further part of this section. In all single analyte systems the on-line copper preconcentration was employed. For this purpose a solid-phase extraction on silica immobilised 8-quinolinol [95, 96], pyrocatechol violet loaded XAD-2 [97] and Chelex 100 [101], adsorption preconcentration of Cu diethyldithiocarbamate chelate on the walls of a PTFE knotted reactor [98] and solvent extraction of copper chelates [99, 100] were reported. The lowest detection limits in the systems with AAS detection were on the level of a fraction of $\mu g\ l^{-1}$ [96, 98]. Determinations with potentiometric detection were carried out using a membrane ion-selective electrode, and their results were in good agreement with AAS determinations [101].

Gold at the concentration level of 13–120 $ng\ l^{-1}$ in river and lake waters has been determined in a flow injection system with ICP-MS detection and preconcentration on sulphydryl cotton fibre [102]. For preconcentrated sample volume 40 ml the detection limit was estimated as 0.19 $ng\ l^{-1}$.

For the determination of iron in tap, power plant, river and well waters several spectrophotometric flow injection procedures have been developed with various colour-forming reagents. The Fe(II) determination can be carried out with 2-nitroso-5-(N-propyl-N-sulphopropylamino) phenol [103] and 2-pyridyl-3′-sulphophenylmethanone 2-pyrimidylhydrazon [104]. The determination of the trace level of Fe(III) can be carried out by measuring the absorbance of the anion-exchange resin in a flow cell which retains an iron-thiocyanate complex formed in the FIA system [105]. In all three measured systems detection limits were on the level of several $\mu g\ l^{-1}$.

The determination of lead in natural waters in the flow injection system with flame AAS detection requires the on-line preconcentration in the system. For this purpose solid phase extraction on various sorbents [106, 107], Donnan dialysis [108], precipitation and accumulation of lead hydroxide [109] can be employed as well as ion-exchange followed by solvent extraction [110]. Preconcentration on solid sorbents can be carried out using activated alumina [106], functionalised cellulose sorbents [107] or a cation exchanger [110]. The lowest level of the detection limit 0.17 $\mu g\ l^{-1}$ for a 50 ml sample was obtained in preconcentration on cellulose sorbent [107]. By the use of different eluents a separate determination of tetra-alkyllead and sum of inorganic lead and organolead species of a smaller number of alkyl groups was demonstrated.

Manganese at mg l^{-1} concentration level in waters can be determined using flow injection spectrophotometry, where Mn (II) is on-line oxidised by solid lead (IV) dioxide suspended on silica beads [111]. The obtained permanganate ions are detected at 526 nm with detection limit 0.56 mg l^{-1}.

A detection limit lower by three orders of magnitude (0.7 μg l^{-1}) has been achieved in flow injection spectrophotometric determination of molybdenum [112]. The detection was based on catalytic action of molybdenum on the oxidation of iodide by hydrogen peroxide and was suitable for determination of molybdenum in tap and river waters.

The ultratrace amounts of selenium can be determined in waters by flow injection hydride generation atomic spectrometry with on-line preconcentration by coprecipitation with lanthanum hydroxide [113]. The precipitate is quantitatively collected on the inner walls of a knotted reactor, and then eluted with HCl into the hydride generation AAS system. For samples of tap and river water Se (IV) was determined at the level of 18–34 ng l^{-1}.

The spectrophotometric trace determination of vanadium in natural waters can be based in the flow injection system on the catalytic effect of vanadium (V) on the oxidative coupling reaction of 4-aminoantypyrine and N,N-dimethylaniline in the presence of bromate [114]. A schematic diagram of a manifold used and the typical signal recording is shown in Fig. 4. Although the method suffers from interferences (Cr (VI), Cu, Fe (III), Sn (II)), with Tiron as an effective activator it was successfully used for the determination of vanadium in river and lake waters at level 0.1–0.3 μg l^{-1}.

For flow injection determination of zinc in waters a spectrophotometric method with an enzymatic reactor has been developed [115]. With the reactor containing immobilised bovine carbonic anhydrase the method is based on the measurement of recovered esterase activity of the metal-free apoenzyme after taking up zinc. The spectrophotometric detection is based on monitoring of the product *p*-nitrophenol formed from substrate *p*-nitrophenyl acetate. The method was applied for determination in hot spring water at level 11 μg l^{-1}.

Due to properties of the detection method several different optimised flow injection systems with AAS detection can be employed for the determination of different elements by replacement of the hollow cathode lamp. Determinations of copper and zinc in natural waters were carried out in the system, where on-line preconcentration was carried out using columns with immobilised micro-organism *Penicillium notatum* [116]. In several papers the use of AAS detection with hydride generation was reported for determination of several elements in

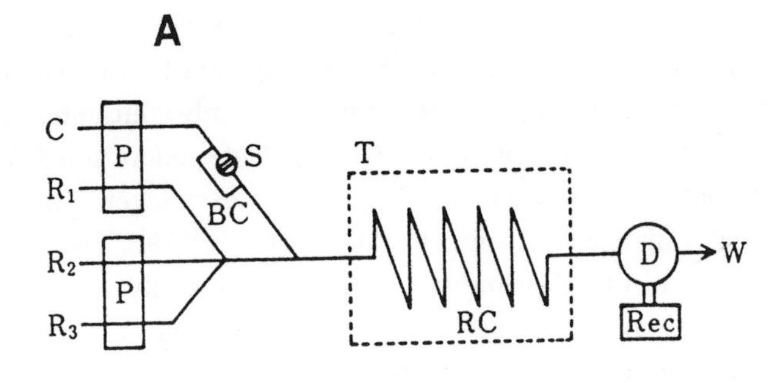

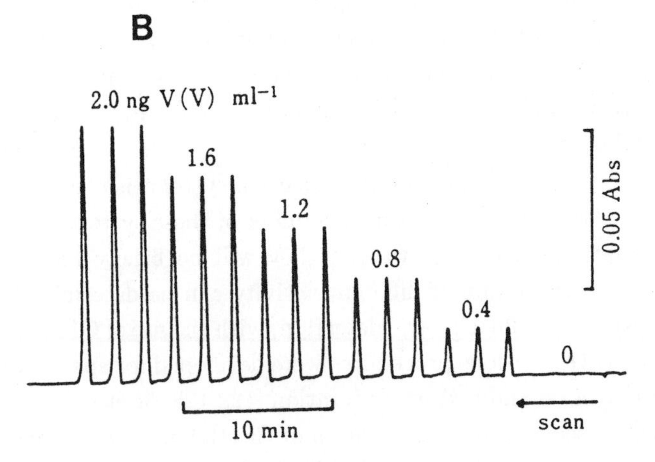

Fig. 4. Schematic diagram of a manifold for the catalytic determination of vanadium (A) and recorded flow injection signals (B) [114]. In A: C — carrier solution (0.1 M HCl); R_1, 4 mM 4 — aminoantipyrine with 30 mM N,N-dimethylaniline and 0.1 M Tiron; R_2 — 80 mM potassium bromate solution; R_3 — buffer solution (0.15 M sodium formate, NH_3, HCl); P — pump (0.8 ml min^{-1}); S — sample injector (200 μl); BC, bypass coil (3 m long, 0.25 m i.d.); RC, reaction coil (10 m long, 0.5 mm i.d.); T — thermostatted bath (55°C); D — spectrophotometer (555 nm); Rec — recorder; W — waste. (*Reprinted by permission of copyright owner.*)

waters in the same experimental conditions. In determination of As and Se in waters satisfactory results were obtained when samples were off-line passed through a Chelex 100 column in order to remove interferences [117]. It is also of great importance when determining the total content of elements that the entire sample be oxidized without filtration, since the elements are present in a variety of chemical forms. For removal of interfering copper and nickel in determination of selenium in the same system 1,10-phenanthroline can be satisfactorily employed [118]. In determination of bismuth and selenium at a concentration level of 0.01–0.56 $\mu g\ l^{-1}$ with hydride generation AAS detection analytes were preconcentrated on a strongly anion exchanger in the case of Se or on immobilised 8-hydroxyquinoline in the case of Bi [119]. The AAS detection with electrothermal atomisation can also be successfully combined with the flow injection system, where periodic preconcentration or removal of matrix with for example solid sorbents is carried out. A setup where interfacing the flow injection system to the graphite furnace was achieved simply by connecting the transfer capillary of the flow system to the sample introduction capillary of the auto sampler arm, was developed for determination of Cd, Cu, Pb and Ni in water samples [120]. The detection limits achieved were 0.8, 6.5, 17 and 36 ng l^{-1} for Cd, Pb, Cu and Ni, respectively, in the system where complexes of analytes with diethylammonium diethyldithiocarbamate were preconcentrated on octadecyl sorbent.

The speciation of elements in flow injection systems is discussed separately in Chapter 7; however, as an essential number of these systems were developed for determinations in natural waters, they will be briefly mentioned below. Three forms of aluminium of different activity can be determined in the flow injection system with fluorimetric detection, with the use of C18 sorbent packed in a flow cell [121]. Various procedures were developed for speciation of arsenic with AAS detection and hydride generation [122,123] or electrothermal atomisation [124]. A sequential determination of Cr (III) and Cr (VI) can be carried out in the flow injection system with ICP-AES detection and preconcentration of Cr (VI) on activated alumina [125]. Determination of Fe (II) and Fe (III) in natural waters was performed in the system with spectrophotometric detection with the use of 1,10-phenanthroline [126] and with amperometric detection using a dual electrode modified with electrocatalytic polymers [127]. A rapid sequential determination of inorganic mercury and methylmercury in tap and river waters at ng l^{-1} level was reported in the FIA system with cold vapour atomic fluorescence spectrometric detection [128]. Speciation of selenium was described in the system with hydride generation AAS detection and on-line microwave reduction [129]. The procedure of speciation of Cd, Cu and Zn in natural waters was developed in a two-column ion exchange system with AAS detection [130].

Another group of flow injection methods developed for trace determinations in natural waters are multicomponent procedures, based on the use of multicomponent detectors or appropriate design of the flow injection setup. The potentiometric stripping analysis employed in the FIA system offers a convenient way of exchange of matrix of the sample, which allows the optimisation of the resolution of analytical signals corresponding to different analytes. It can be successfully applied in determination of Cd, Cu and Pb in ground waters at a concentration level from 1 $\mu g\ l^{-1}$ up to a fraction of mg l^{-1} [131]. Typical multicomponent detection methods are inductively coupled plasma-atomic emission spectrometry (ICP-AES) and inductively coupled plasma-mass spectrometry (ICP-MS). The first one was applied in determination of 11 various components of natural waters after their preliminary on-line preconcentration on a Chelex 100 column [132], and in determination of Cd, Co, Cu and Pb after preconcentration using a column packed with iminodiacetic acid/ethylcellulose [133]. In the latter case the detection limits were evaluated at a level of 0.02–0.5 $\mu g\ l^{-1}$. Several times lower detection limits were reported for the system with ICP-MS detection, also involving an on-line preconcentration step using solid sorbent [134]. Satisfactory results were obtained for Mn, Co, Cu, Pb and U; however, for Ni positive errors were found due to spectral interference of CaO from coeluted calcium. The same detection combined with isotope dilution and preconcentration on an anion exchanger in the form of chlorocomplexes was employed for trace determination of Re and Pt in river waters [135].

For multicomponent determinations in natural waters, kinetic effects can also be exploited. In simultaneous determination of Hg (II) and Ag (I) the catalytic effect on the ligand substitution reaction rate between hexacyanoferrate (II) and α, α'-bipyridyl was utilised [136]. Based on absorbance measurements at different time intervals since sample injection 29 $\mu g\ l^{-1}$ mercury (II) and 48 $\mu g\ l^{-1}$ silver (I) was determined in tap water. In the reversed flow injection system with chemiluminescence detection total dissolved iron and total dissolved manganese were determined simultaneously in underground water [137]. In the method based on the measurement of the metal-catalysed light emission from luminol oxidation by potassium periodate, detection limits were evaluated as 3 ng l^{-1} for Fe (III) and 5 ng l^{-1} for Mn (II).

Using flow injection methodology, a spectrophotometric screening test for the presence of heavy metals in waters can also be carried out [138]. The developed procedure was based on the reaction of Cd, Co, Cu, Ni, Pb and Zn with dithizone and the extractive photometric determination in an FIA system. The results of tests carried out with samples of surface waters satisfactorily correlated with results obtained by atomic absorption spectrometry.

1.1.4. *Organic compounds*

For the determination of trace amounts of organic compounds in natural waters, high performance chromatographic methods or capillary electrophoresis are commonly employed. In certain cases, however, they can be replaced by flow injection methods, which allow the use of simpler instrumentation, and hence their more common use in routine water quality control.

Detergents are usually determined as a group of compounds, and not as single species; therefore in this case it is not necessary to employ sophisticated chromatographic procedures. With the use of flow injection systems anionic detergents can be determined in river water employing the extraction method with methylene blue. In a typical FIA system (Fig. 5) the extraction of the ion pair of the surfactant with methylene blue to chloroform and spectrophotometric detection after phase separation in a membrane separator is used [139]. This method was applied in determination of anionic detergents in river water at a concentration level of 0.08–6.4 mg l^{-1} of sodium dodecyl sulphate. Such determinations can also be carried out in the FIA system without phase separation, which was also employed for river water analysis [140].

Great attention is paid to development of flow injection methods for determination of pesticides. For this purpose chemical methods, procedures based on inhibition of the activity of appropriate enzymes and immunochemical methods, can be used. Only the last ones exhibit sufficient selectivity to be applied in quantitative determinations in natural waters. Two such methods have been developed for determination of atrazine in natural waters [141, 142]. In both of them an enzyme-linked immunoassay with fluorimetric detection was employed with detection limits 30 [141] and 75 [142] ng l^{-1}. Cross-reactivity studies have proved that the immunoassay is specific for atrazine while other triazine compounds are detected to a minor extent only [142]. Other chemical methods with spectrophotometric detection in the visible [143] or infrared [144] range proposed for determination of pesticides, and also methods based on inhibition of acetylcholinesterase [145–147], were examined only for natural water samples spiked with selected analytes. They can be used for semi-quantitative pre-screening of pesticides in water samples at $\mu g\ l^{-1}$ level.

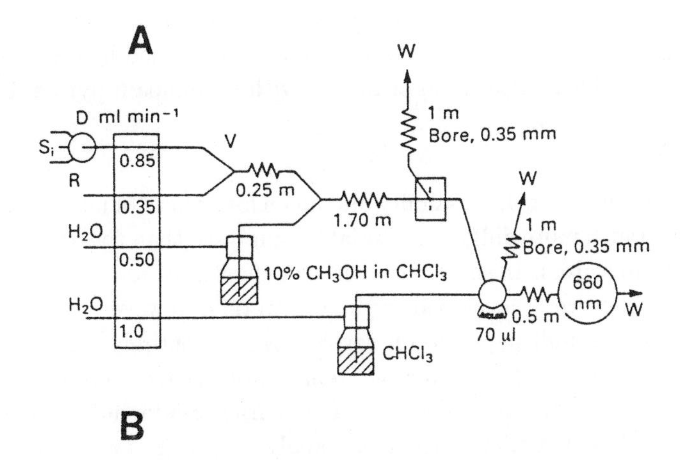

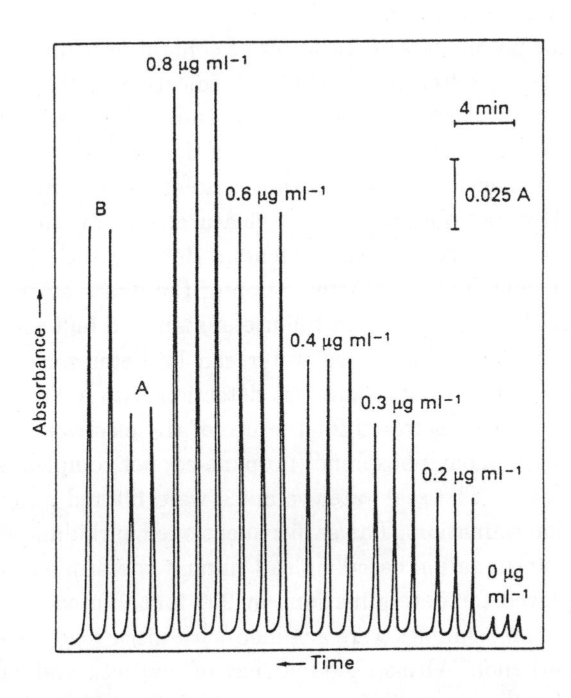

Fig. 5. Manifold for the determination of low levels of anionic surfactants in waters (A) and recorded flow injection signals for triplicate injections of 0–0.8 mg l^{-1} sodium dodecyl sulphate standards and two river water samples (B) [139]. In A: S_i — sample; R — methylene blue reagent; W — waste. All tubing is 0.5 mm i.d. except where indicated otherwise. (*Reprinted by permission of copyright owner.*)

Phenols in waters can be determined spectrophotometrically by the on-line extraction into chloroform of a coloured compound formed by the oxidation of the product of the condensation of phenol with 4-aminoantipyrine [148].

1.2. *Sea Water*

Sea water, due to its high salinity and much lower level of numerous trace elements, is a much more difficult material for analysis, than the above-discussed inland and tap waters. In some cases the same flow injection procedures can be employed, which were developed for inland waters; however, very often their modifications are indispensable or development of entirely different analytical procedures involving, for example, removal of matrix components or more efficient preconcentration of analytes. This paragraph includes also some applications of flow injection methods for analysis of brines and other strongly mineralised samples of various origins.

Non-specific parameters of sea water or content of main the components of sea water are rarely determined using flow injection methods. The shipboard spectrophotometric method for determination of pH of sea water was already mentioned above [18]. The pH value is one of the factors providing information about global changes of carbon dioxide content in the environment. It was shown that pH values obtained in FIA measurements are internally consistent with other carbon dioxide measurements. Because such pH measurements are calibrated though the characterisation of the dissociation constant of the indicator, this allows one to avoid the use of standard buffers during analyses.

The content of chloride in sea water can be determined in a flow injection system with direct potentiometric detection with a chloride ion-selective electrode. It was demonstrated for the use of the electrode containing in the membrane a bis(diphenylphosphino) propane-copper complex with a solid carbon contact [149]. The sea water samples were filtered and 50-fold diluted prior to the determination. The earlier-mentioned turbidimetric FIA method for sulphate determination based on lead sulphate determination was also successfully employed for determinations in 200-fold diluted sea water samples [42]. After dilution, samples were acidified with nitric acid to pH 3 and mixed 1:1 with 95% ethanol. With spiked matrices of synthetic and natural sea water a reliability of determination of a relatively high level of residual chlorine (10–17 mg l^{-1}) was examined using flow injection spectrophotometric methods with methyl orange and *o*-tolidine [150]. All results obtained with *o*-tolidine were elevated, which indicates for some side reactions of sample components with a colour-forming reagent. With methyl orange negative errors were observed for natural sea water, reflecting its chlorine demand.

Interferences caused by a sea water matrix were also found in flow injection determination of fluoride with spectrophotometric detection based on alizarin fluorine blue [81]. Satisfactory results were obtained by the standard addition method. Bromide was determined in sea water by flow injection spectrophotometry with chloramine-T and phenol red [151]. Because of the short reaction time the interferences of chloride and bicarbonate have been totally eliminated. In samples of coastal sea water, bromide was determined at concentration level 48–70 mg l^{-1} (after 10-fold dilution), and the accuracy of determination was examined by recovery tests. For determination of iodide in high saline waters a spectrophotometric flow injection method was developed based on the oxidation to iodine, which after permeation through a PTFE membrane is detected as tri-iodide at 350 nm [152]. The only interfering sulphide can be removed by preheating the samples after acidification. Iodide was determined with this method in marine pore water samples at concentration level 2–200 mg l^{-1}.

A common spectrophotometric method for determination of nitrite with diazotisation and coupling was employed for a submersible FIA measuring set-up for the determination of total oxidised nitrogen (nitrate plus nitrite) in coastal sea waters [153]. Using a 260 μl sample injection loop and a 20 mm path length flow cell, a detection range of 0.1–55 μM nitrate was obtained.

Several original FIA methods were developed for determination of orthophosphate in sea water. For determination of total phosphorus in the flow injection system with the molybdate method, an on-line capillary digestor for oxidation with potassium peroxodisulphate was used [154]. Interferences due to colour fading of malachite green caused by chloride were eliminated by adding a reducing agent sodium thiosulphate after the digestor. Total phosphorus was determined at a concentration level of 0.08–1.5 mg l^{-1}. The same molybdate method with malachite green can be used in so-called gel-phase absorptiometry, where an ion-association complex was adsorbed on Sephadex LH-20 beads located ion flow cell [155]. This method was successfully applied in determinations in sea water in the concentration range of 2–8 mg l^{-1} P. The same complex can be filtered through a membrane, which is then dissolved together with the membrane and the obtained solution is injected into the FIA system with spectrophotometric detection [156]. In such a procedure with a 10 ml sample the detection limit was estimated as 60 ng l^{-1} P. Determinations in sea water require at least two-fold dilution.

The reactive silicate can be determined in the reversed FIA system also with the molybdate method and spectrophotometric detection [157]. The method is based on formation of β-molybdosilicic acid in conditions where interferences from phosphate are suppressed and a wide range of salinity does not affect the determination. With the detection limit 0.1 μM Si determinations in sea waters were carried out at the concentration level of 0.5–70 μM, but results obtained were not verified with other methods.

Similarly to inland and tap waters, numerous flow injection methods have been developed for determination of the trace level of metal ions in sea water. One third of them only are the same as presented above for inland and tap waters.

For the determination of arsenic a flow injection method with ICP-MS detection mentioned above can be used, although at a large concentration of chloride the interference form $^{40}Ar^{35}Cl^{+}$ should be taken into account [94]. At a constant level of chloride satisfactory results can be obtained applying a correction factor evaluated during the optimisation of the procedure.

In differential determination of Cr (VI) and total chromium using on-line separation and preconcentration in the flow injection system with AAS detection, it was found that in sea water some difficulties may occur with complete oxidation of Cr (III) to Cr (VI), which is then procencentrated as a complex with diethyldithiocarbamate on C18 column [158]. It is a significant observation as practically in sea water Cr (III) only was detected. Quantitative oxidation was achieved by boiling the samples gently for about 25 min at pH 9 with potassium peroxydisulphate.

The determination of cobalt in sea water can be carried out in the flow injection system using sensitive catalytic methods with spectrophotometric detection [159, 160] or a procedure with chemiluminescence detection [161]. In the method based on the catalytic action of cobalt ion on the hydrogen peroxide oxidation of protocatechuic acid, the limit of detection 5 ng l^{-1} was obtained. In this method the preconcentration on the column with silica-immobilised 8-quinolinol and separation on a strongly acidic cation-exchange resin from Fe (III) and (II) was employed. The preconcentration is not needed in another catalytic flow injection method, based on the oxidation of N,N′-diethyl-*p*-phenylenediamine by hydrogen peroxide in the presence of Tiron as an activator [160]. The catalytic activity of Co (II) was enhanced by a sea water matrix, especially calcium ions. The limit of detection was in this case 1 ng l^{-1}. The preconcentration on an 8-hydroxyquinoline column and separation from alkaline earth metal ions was necessary in FIA determination of cobalt in the system with chemiluminescence detection [161]. The method was based on the Co-enhanced chemiluminescent oxidation of gallic acid in alkaline hydrogen peroxide. The limit of detection in this case was estimated as 8 pM.

An about three orders of magnitude (10 fM for 4 l sample) lower detection limit was reported for flow injection determination of gold in sea water using ICP-MS detection [162]. The method involves off-line preconcentration of gold by anion exchange as a cyanide complex $Au(CN)_2^-$ and a remarkable relative precision of 15% at the 100 fM level was reported. From the eluate additionally in the off-line procedure the organic matter and any present Si were removed.

Trace amounts of iron in sea water can be determined in FIA systems with a sensitive chemiluminescent procedure [163], or catalytic spectrophotometry [164]. The chemiluminescence detection is based on the reaction of brilliant sulphoflavin with hydrogen peroxide and iron (II) in a neutral medium. A cation-exchange column with 8-hydroxyquinoline immobilised on Fractogel placed in an injection loop of the valve was used to concentrate iron from sea water and to separate it from alkaline earth cations. When the sample was preconcentrated without addition ascorbic acid Fe (II) was determined, only, whereas when Fe (III) was off-line reduced in the sample with ascorbic acid the total dissolved iron could be determined. For a 4.4 ml sample volume a detection limit 0.45 nM was estimated [163]. Among developed catalytic procedures for the determination of iron with spectrophotometric detection the lowest detection limit 30 ng l^{-1} without separation of analyte was found for the method based on the selective catalytic effect of iron (III) on the oxidation of N,N′-dimethyl-*p*-phenylenediamnie by hydrogen peroxide [164]. These determinations were carried out in the reversed flow injection system.

A catalytic method with preconcentration on chelating sorbent can be used for determination of manganese (II) in sea water with spectrophotometric detection [165, 166]. In the method based on the catalytic effect of Mn(II) on the oxidation o N,N-diethylaniline by potassium periodate, for a 10 min aspiration of the sample for preconcentration the detection limit of 2 ng l^{-1} was obtained [165]. Manganese was determined in sea water samples at a level of 44–630 ng l^{-1}. In the second one, which also employed on-line preconcentration on chelating sorbent, the detection was based on the formation of the malachite green from the reaction of leucomalachite green and potassium periodate with Mn (II) acting as a catalyst [166]. When 15 ml of a sea water sample was used for preconcentration the limit of detection was 36 pM. Deep

Pacific ocean sea water containing about 270 pM Mn (II) was used to make standards. The use of a column with 8-hydroxyquinoline immobilised onto a vinyl polymer gel has eliminated any salinity effect. The method was used for determination in sea water samples containing 0.1–1.2 nM Mn (II).

The high sensitivity of catalytic methods was utilised also for determination of molybdenum with the spectrophotometric method mentioned earlier [112], based on the catalytic effect of the analyte on the reaction of iodide with hydrogen peroxide. In determinations in sea water some interferences were caused by large differences in the refractive index between carrier and sample solutions. They can be eliminated by the dilution of sea water 1:1 or by the addition of 3% sodium chloride to the carrier stream and all standards. The obtained detection limit (0.7 μg l^{-1}) is comparable to that found for the FIA system with ICP-AES detection involving on-line preconcentration on a mini-column with activated alumina (0.2 μg l^{-1}) [167]. The latter was used for determinations in sea water at the molybdenum level 10 μg l^{-1}, without interferences of a high concentration of dissolved sodium chloride.

In determination of selenium in the flow injection system with hydride generation AAS, one can determine both Se (IV) and Se (VI) when on-line reduction of Se (VI) in an acidified sample with HCl and heating to 140°C is employed [168]. Often observed formation of metal precipitates does not occur when hydrogen selenide is generated after the on-line reduction step. The detection for both species in the FIA system was estimated as 0.7 mg l^{-1} and the method was applied to determinations in sea water without and with spiking with selenium.

More serious difficulties can be encountered in determination of zinc in sea water in the flow injection system with fluorimetric detection based on the measurement of fluorescence of zinc complex with *p*-tosyl-8-aminoquinoline [169]. Although the detection limit was estimated for a 4.4 ml sample as 0.01 nM and interferences of alkali and alkaline earth ions are removed by analyte preconcentration on an 8-hydroxyquinoline column, the correction for cadmium interference in most cases must be done using direct determination of cadmium. Results of determinations in standard sea water solutions at levels 2.8 and 26 nM agreed with certified values without correction for Cd.

Several methods reported in the previous paragraph for multicomponent determinations in flow systems with ICP-AES [133] and ICP-MS [134, 135] detections, and also a method developed for several analytes with graphite furnace AAS detection [120], were also employed for determinations in sea water samples. The effect of a sea water matrix was found in determination of platinum; however, differences observed by various authors were not explained and eliminated [135]. In other procedures a sea water matrix did not affect results of determinations.

Determination of the trace level of metal ions in sea water was also optimised in several other systems with atomic spectroscopy detections. In determination of Cd, Cu and Pb with AAS detection and preconcentration of carabamate chelates on a C18 column, it was found that the applied preconcentration step increases tolerance against interferences for determination of Cd and Pb in sea water [170]. An especially marked effect was found for determination of cadmium in the presence of magnesium in a sea water matrix. In AAS determinations with electrothermal atomisation ammonium diethyldithiophosphate was applied as a complexing agent with citrate as a masking agent [171]. By sorption on a C18 column this allows one to remove effectively the sea water matrix in determination of Cd, Cu and Pb with detection limits 3, 50 and 40 ng l^{-1}, respectively. For determination of Ca, Mg, Al, Fe, P and Si in samples of high salt content in the flow injection system with ICP-AES detection also a computerised system was employed for matrix matching to level the sample and standard saline content [172].

1.3. *Wastes*

Among the most difficult and diverse natural matrices are wastes of various origins. More often in wastes than in natural waters are utilised determinations of cumulative, non-specific parameters corresponding to the content of certain groups of compounds.

For determination of chemical oxygen demand (COD) in waters, related to the amount of oxygen required for oxidation of the sample, methods reported above for water analysis are used [8–10, 13]. They are based on the use of dichromate [8–10] as oxidant or Ce (IV) [13]. These procedures were employed for determination of COD in wastes from the food industry [8–10, 13], chemical industry [8, 9], laboratories [8], textile industry [8, 9], municipal wastes [8] and gas-works effluents [9]. The range of determined COD values extended form 50 (food industry) to 37000 mg l^{-1} (gas-works effluents).

In a sewage treatment effluent a determination of dissolved organic carbon based on in-line UV photo-oxidation with spectrophotometric detection was used [14], although it was found that results of such determinations can exhibit positive errors for samples of elevated salinity, probably through the formation of chlorine, which oxidises the indicator phenolphthalein.

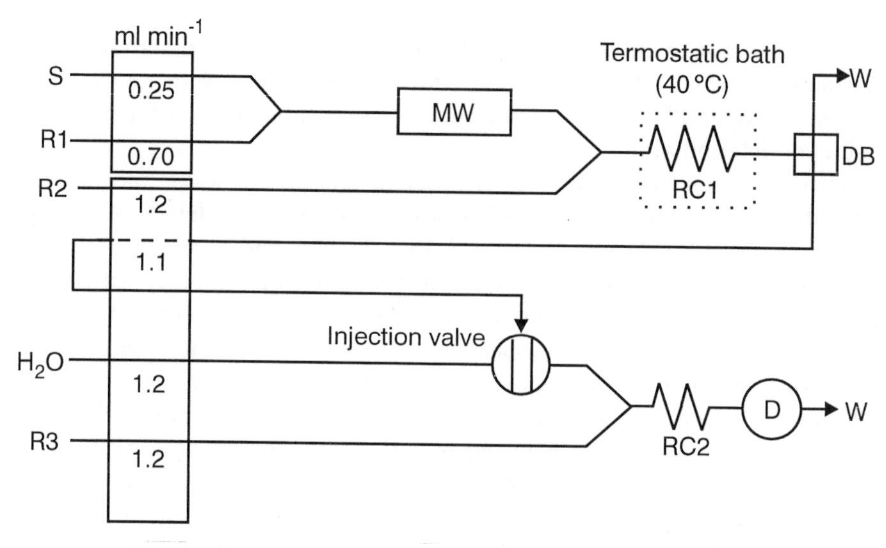

Fig. 6. Manifold for the determination of total nitrogen in waste water with spectrophotometric detection [173]: S — aspirated sample stream; R1 — peroxodisulphate solution; R2 — reducing solution of hydrazinium sulphate; R3 — chromogenic reagent consisting of sulphanilamide and N-(1-naphthyl) ethylene diamine; MW — microwave device; DB — debubbler; RC1 — reaction coil (2 m 0.8 mm i.d.); RC2 — reaction coil (1 m 0.8 mm i.d.); W — waste; D — detector. (*Reprinted by permission of copyright owner.*)

The FIA system with spectrophotometric detection (Fig. 6) was used for the determination of total nitrogen in municipal waste waters, collected at the inlet and outlet of different treatment plants [173]. The method is based on the oxidation of nitrogen-containing compounds to nitrate by peroxodisulphate using a microwave oven. Nitrate is determined spectrophotometrically after reduction to nitrite. Organic carbon may interfere only at a high concentration exceeding 200 mg l^{-1}. This method can be used in concentration range of 0.2–20 mg l^{-1} N and results of its use for determinations in wastes were in good agreement with those obtained by Kjeldahl digestion.

The determination of phosphorus compounds in waste waters is important for eutrophication control. Besides determination of orthophosphate with methods presented for water analysis, total phosphorus is also often determined [154, 174, 175]. These determinations can be carried out by on-line oxidation of phosphorus compounds to phosphate with potassium peroxodisulphate followed by spectrophotometric [154] or amperometric [174] detection.

In spectrophotometric determinations a two orders of magnitude lower detection limit was obtained (2 μg l^{-1} P) [154], than in amperometric ones. A simultaneous determination of orthophosphate and total phosphorus can be carried out in the system with spectrophotometric and ICP-AES detectors arranged in series [175]. Phosphate was determined specrophotometrically as molybdovanadophosphoric acid and the solution from the spectrophotometer was directly introduced into an inductively coupled plasma. The determination of total phosphorus was performed by measuring the emission intensity at 177.499 nm. Another determined parameter can be dissolved reactive phosphorus, which comprises orthophosphate and condensed phosphates. For determination of this parameter a spectrophotometric molybdate method in a reversed flow injection setup can be used [176]. Such a method was successfully employed for determinations in estuarine waters with salinity up to 3% with a detection limit of 2 μg l^{-1} P.

For simultaneous determination of orthophosphate and silicate a sequential injection analysis method based on the formation of molybdovanadophosphate in the system shown schematically in Fig. 7 can be used [177]. In the optimised system the reaction of one of the two species at each end of the sample zone takes place. The interferences were eliminated by adjusting the acid concentration by segmenting the sample zone by the addition of oxalic acid. This method was used for determinations in urban waste waters in the range of 1.3–17 mg l^{-1} P and of 12–90 mg l^{-1} Si.

Methods for determination of nitrate and nitrite in waters in various configurations of flow systems were also employed for determination of these analytes in wastes. With the use of a membrane ion-selective electrode nitrate was determined in wastes at the concentration level of 0.3–35 mg l^{-1} N [178]. Although in most cases a good agreement with the brucine method was found, at low concentrations positive errors can occur in the presence of chloride. The spectrophotometric method based on diazotisation and coupling was used in the merging zones system for determination of nitrite at the level of 9–43 μg l^{-1} N [179]. The same method in a more complex system with cadmium reductor can be used for simultaneous determination of nitrate and nitrite for monitoring an urban waste water treatment plant [180]. Similarly to other works, in this one some difficulties were reported for functioning of the reducing column, hence it can be advantageous to replace it by photoreduction proposed in the system with biamperometric detection [181]. Nitrate was effectively photoreduced by UV irradiation with 1 mM EDTA as an activator. Detection

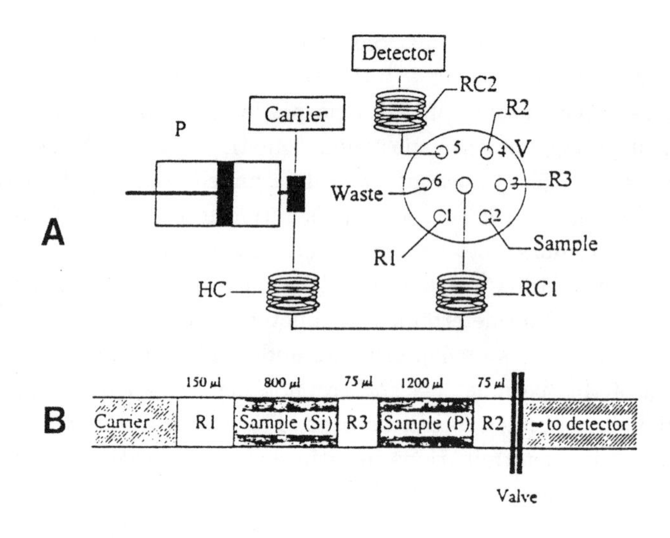

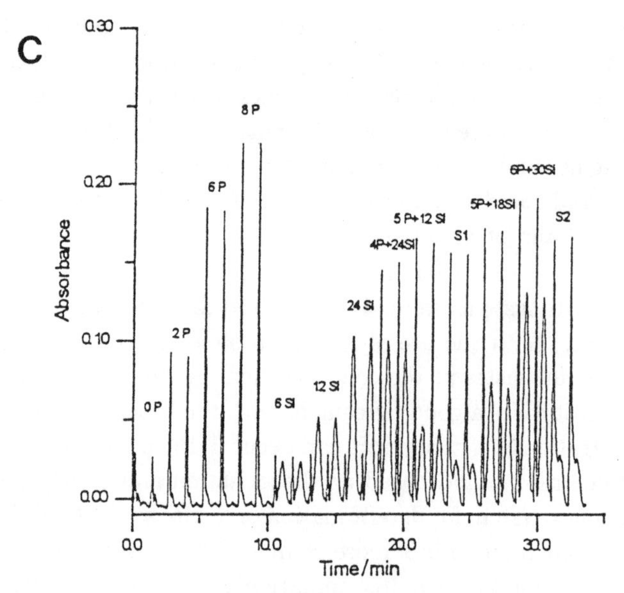

Fig. 7. Schematic diagram of the sequential injection analysis system used for the simultaneous determination of phosphate and silicate (A) and example recording of obtained signals with concentration expressed as mg l^{-1} (S1 and S2 correspond to waste water samples) (C) [177]. In A: P — titration burette 5 ml; V — six-port valve; HC — holding coil; RC1 — reaction coil 1; RC2 — reaction coil 2; R1-R3 — reagents; S — sample; C — carrier; D — detector. B: zone sequence. (*Reprinted by permission of copyright owner.*)

limits in this procedure were estimated as 25 and 50 μg l^{-1} for nitrite and nitrate, respectively. This method for determination in wastes was used at a concentration level about three orders of magnitude higher.

Fluoride in ceramic industry waste water at the concentration level of 6–9 mg l^{-1} was determined in the FIA system based on pervaporation of a volatile derivative and potentiometric detection after collection in an alkaline solution [182]. This method, however, is not free of interferences in the presence of Al(III) and Fe(III) at their larger excess compared to the analyte level.

In FIA systems developed for spectrophotometric determination of chloride in wastes, original dialytic systems were used in order to separate an analyte from the complex matrix of wastes. For simultaneous determination of chloride and calcium (also determined spectrophotometrically with Cresolphthalein Complexone) a double on-line dialyser was employed [183] in the system shown in Fig. 3, Chapter 6. The determined chloride content was in the range of 0.3–16.8 g l^{-1}, and calcium at the concentration level of 0.1–0.6 g l^{-1}. In order to increase the mass transfer through passive neutral membranes, one can employ electrodialysis. Such more efficient transport was demonstrated in a spectrophotometric FIA system for determination of chloride in industrial wastes [184].

In flow injection determination of cyanide in wastes with spectrophotometric detection it is advantageous to utilise detection of an unstable intermediate, which allows one to increase the speed of determination [185], as well as to improve its sensitivity [186]. In determination based on the reaction of cyanide with isonicotinic acid and pyrazolone, additionally a gas-diffusion separator was employed. In determination of total cyanide the detection limit 6 μg l^{-1} at sampling frequency 40 h^{-1} was achieved. In the presence of 1,10-phenanthroline all interferences except cobalt can be eliminated. A slightly worse detectability (20 μg l^{-1}) was reported in the reversed FIA system without separation step with detection of an intermediate product in the reaction of cyanide with a pyridine-barbituric acid reagent [186]. Although this method suffers from several interferences it was successfully applied in determination of free cyanide in waste water samples at the level of 0.3–2.8 mg l^{-1}.

A reversed FIA system with alternating injection of reagent and washing EDTA solution can be favourably used in turbidimetric on-line sulphate monitoring in effluent water streams [187]. Interferences from suspended solids, organic substances and colour were removed by an activated carbon filter incorporated into the conduit of the sampling line before the peristaltic pump.

The proposed system was used for determination of 30–200 mg l^{-1} sulphate in wastes.

The content of sulphide in wastes at a level of 0.2–15 mg l^{-1} can be determined with potentiometric detection employing a sulphide ion-selective electrode [188]. The use of an antioxidant buffer containing ascorbic acid allows the monitoring of residual sulphide in sewage effluents that have been treated with hydrogen peroxide.

For simultaneous determination of three different contaminants of wastes two different systems with spectrophotometric detection and continuous sample aspiration were proposed [189]. In one of them sulphide and cyanide were determined with continuous potentiometric control of pH. In the second one, also with pH monitoring, nitrite and cyanide were determined. A possibility of practical application of these systems was examined with synthetic samples simulating wastes.

Several FIA methods were also developed for determination of cationic components of wastes, especially heavy metal ions present at toxic level.

For the determination of cadmium in wastes from a galvanic industry at a level of 11–97 mg l^{-1}, a potentiometric detection with a cadmium ion-selective electrode was reported [190]. In the flow injection system on-line cadmium preconcentration as chlorocomplexes on an anion-exchange resin mini-column was employed together with removal of Cu (II) interferences on a mini-column with Chelex 100. The developed system allows Cd determination with a detection limit of 11 $\mu g\ l^{-1}$, with good agreement with results obtained by the AAS method.

For the determination of calcium in samples of back water from paper mills a spectrophotometric FIA method was proposed based on complexation with o-cresolphthalein complexone [191] and also potentiometric detection was applied. In both methods, however, negative errors were observed compared to results obtained with dc-plasma emission spectrometry, which was explained by determination of ionised calcium, only, with potentiometric detection, whereas with spectrophotometric detection does not measure a part of calcium which is bound in stronger complexes than that utilised in the detection.

The determination of iron in wastes is possible by reduction of Fe (III) by ascorbic acid and formation of coloured chelate of Fe (II) with ferrozine. For determination in wastes two FIA systems, normal and reversed, based on such a detection were compared [192]. More favourable was found to be a reversed system due to a wider range of responses, higher sensitivity and a lower detection limit. In analysis of municipal waste water samples at level 0.1 $\mu g\ g^{-1}$ a satisfactory agreement of results compared to the AAS method was reported.

Manganese (II) in wastes can be determined with the above-mentioned spectrophotometric method based on on-line oxidation in a solid phase on a MnO_2 reactor to permanganate [111]. In the concentration range of 37–116 mg l^{-1} a good agreement with the AAS method was reported; however, this method can exhibit erroneous results at an iron content above 250 mg l^{-1} and Mn (II) level below 5 mg l^{-1} as well as at a vanadium (V) content exceeding 25 mg l^{-1}.

A simplification of a digestion procedure in the FIA system with a cold vapour AAS detection of mercury was invented for determination of mercury in Solvay effluent samples [193] in the manufacture of sodium hydroxide and chlorine. This was enabled by a negligible level of organic interferences in analysed wastes. The effect of the presence of sulphide up to 50 mg l^{-1} was overcome by the digestion procedure and the interference caused by chloride up to 10 g l^{-1} was negligible. Mercury was determined at a concentration level of 1.2–2.2 $\mu g\ l^{-1}$.

For analysis of wastes several multicomponent flow injection methods can also be found. An earlier-presented stop-flow spectrophotometric kinetic method was employed for simultaneous determination of Hg (II) and Ag (I) in wastes [136]. Prior to the determination the sample was mineralised off-line in order to remove organic components. Determinations of two- and three-component mixtures of Fe, Co and Ni can be carried out utilising differences in spectra of their complexes with $2,2'$-dipyridyl keton picolinoylhydrazone. Using a partial least-squares calibration such a method was employed for analysis of a washing solution in an automotive factory and a waste acidic mine effluent [194]. The determined content of all three components in the sample of waste at levels 3.8, 38.6 and 554 mg l^{-1} for Fe, Co and Ni, respectively, agreed well with results of ICP-AES determination.

Simultaneous determination of several cationic components in wastes can also be performed by the design of more sophisticated FIA systems [195]. This can be illustrated by the system developed for determination of Cu, Fe and Al with spectrophotometric detection at 575 nm with continuous sample aspiration. In this system two selecting valves for delivery of a suitable buffer solution and for injection of an appropriate colour-forming reagent were used. For determination of copper Cuprizone was used; for iron, bathophenanthroline;

and for aluminium, Eriochrome Cyanine R. The developed FIA system was tested with artificial wastes containing analytes at a concentration level of 1–4 mg l^{-1}. An example of further extension of this system in order to determine additionally pH and nitrite is shown in Fig. 8.

Several different flow injection methods were critically compared for the determination of ammonia in waste water samples including those based on the Berthelot reaction and on gas diffusion [196]. It was concluded that gas diffusion methods with spectrophotometric or conductivity detection are reliable alternatives to the classical indophenol blue method, having good precision and sensitivity. It was also concluded that conductivity detection has important practical advantages over the spectrophotometric methods examined.

Flow injection methods were also developed for determination of several organic contaminants of wastes. In a simple single line FIA system without phase separation anionic surfactants were determined from 5 ml samples by extraction as the ion pair with ethyl violet into toluene [197]. The results of determinations in wastes at a concentration level of 0.9–15 mg l^{-1} were in satisfactory agreement with results obtained by the Japanese Industrial Standard method. The solvent extraction was carried out manually and an extract was injected into the toluene carrier stream. Trace amounts of formaldehyde in waste water can be determined based on its inhibition of the brilliant green-sulphite reaction at pH 7 [198]. In this kinetic spectrophotometric method some interference was observed in the presence of methanol, ethanol and acetone at level 1 g l^{-1} and sulphide interferences by inhibition at concentrations greater than 2 mg l^{-1}. A spectrophotometric enzymatic method was proposed for the determination of total phenols in industrial waste waters [199]. A crude extract of sweet potato roots was used as a source of polyphenol oxidase directly in the carrier of the flow system. The detection was based on enzymatic formation of *o*-quinones which can couple to each other producing melanin-like pigments with a strong absorption at 410 nm. The results presented for samples of industrial wastes on the total phenols level of 1.7–8.0 mg l^{-1} agreed well with the standard AOAC method.

The use of the flow injection method was also reported for the determination of selected radio-nuclides in reactor wastes with radiometric detection. Both for the determination of ^{90}Sr [200] and ^{99}Tc as pertechnetate [201] determinations were carried out with sequential injection techniques with a flow-through scintillation counter. In determination of ^{90}Sr the separation was achieved using a sorbent extraction column with a resin that binds selectively strontium as

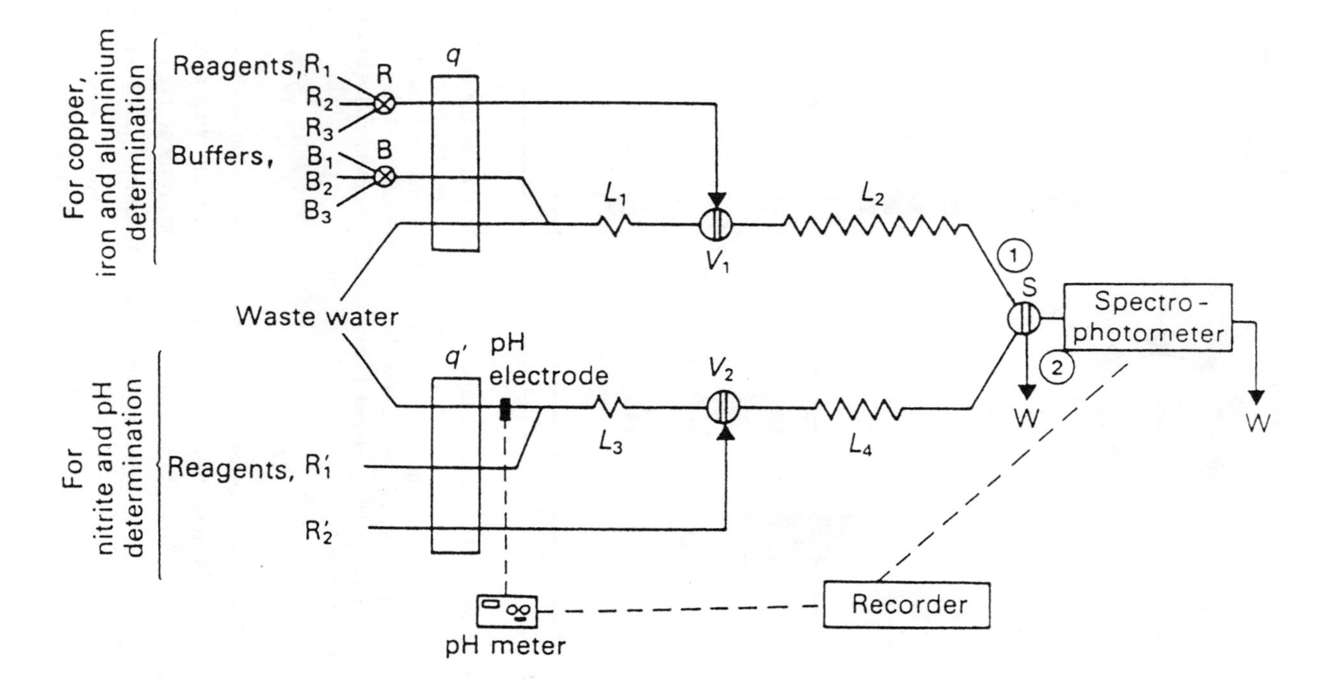

Fig. 8. Manifold for the simultaneous determination of pH, nitrite, copper, iron and aluminium in waste waters by reversed flow injection analysis [195]. (*Reprinted by permission of copyright owner.*)

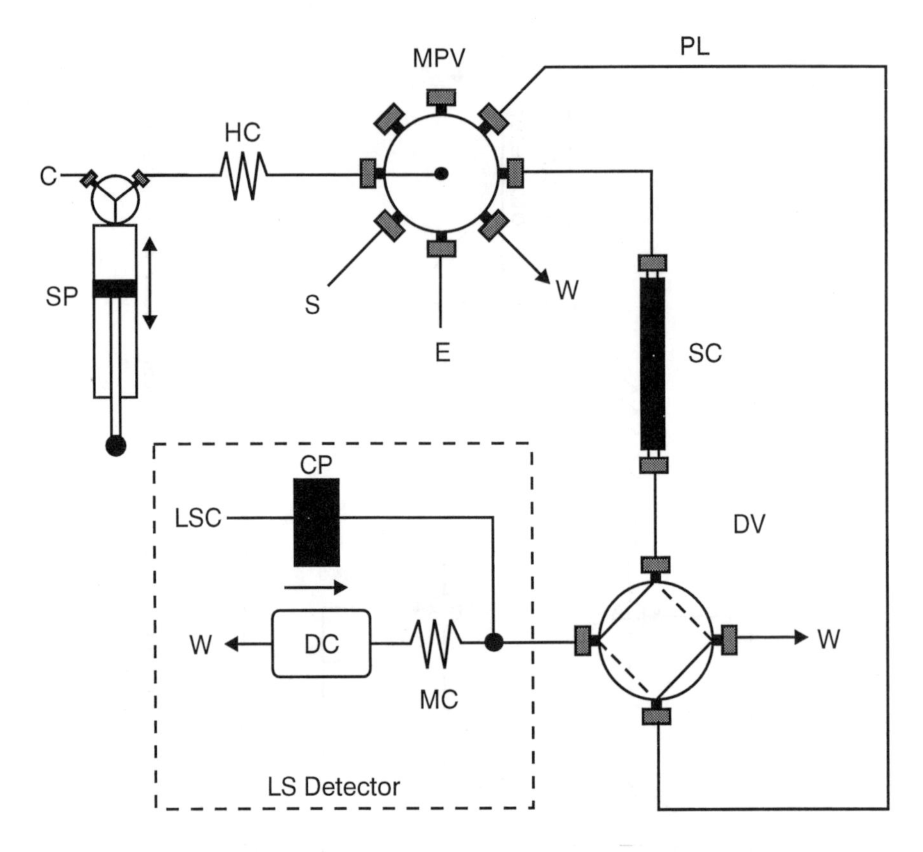

Fig. 9. Schematic diagram of a sequential injection analysis system for the determination of ^{99}Tc in nuclear waste [201]: C — carrier (water); SP — syringe pump; HC — holding coil; S — sample line; E — eluent lines; W — waste; MPV — multiposition valve; SC — resin column; PL — purge line; DV — two-way diverter valve; LSC — liquid scintillation cocktail; CP — cocktail pump; DC — detector flow cell; MC — mixing coil. (*Reprinted by permission of copyright owner.*)

a crown complex. Such a determination was employed for analysis of nuclear waste samples. In ^{99}Tc determinations (Fig. 9) a column with a resin that selectively retains pertechnetate and separates it from radioactive and stable interferences was used. In this case a stopped-flow technique has been used to improve the precision and detection limit of determination. Using a 15 min stopped-flow period a detection limit 2 ng (2 μg l^{-1}) was obtained for a 1 ml sample load.

1.4. *Sediments*

Flow injection methods were employed in analysis of sediments mainly for determination of heavy metals. The used methodologies are similar to or the same as those applied in analysis of water or wastes, and only sample pretreatment is obviously more complex.

The determination of arsenic with ICP-AES detection and hydride generation was applied to standard reference material river sediment [202]. Dried samples were off-line mineralised in the mixture of hydrochloric and nitric acids, and then heated with hydrochloric acid to remove any siliceous residue. They were introduced into the FIA system after addition of sodium iodide. Although some interferences were observed in the presence of Ni, Ag, Au, Bi, Te and Sn the obtained results were in good agreement with certified value 66 $\mu g\ g^{-1}$ As.

Mercury in sediments can be determined using flow injection methods with atomic spectroscopy detections. In the FIA procedure with atomic fluorescence spectrometric detection an on-line microwave digestion was developed [203]. The acidified slurried sample was injected into a carrier stream of hydrochloric acid and microwave digestion in 4 m Teflon coils permits in 50 s the quantitative extraction of mercury from the solid samples. In analysis of certified standard materials sewage sludge of domestic origin, sewage sludge amended soil and lake sediment at a concentration level of 0.2–2.1 $\mu g\ g^{-1}$ Hg, a good agreement was obtained with the certified content. For 400 μl sample volume at throughput 15 samples per hour the limit of detection was estimated as 0.09 ng g^{-1}. For the determination of mercury in a certified marine sediment sample a sequential injection system with a cold-vapour AAS detection was applied [204], in which sample reagent consumption is incomparably smaller, than in FIA systems. In this case the sample was 10 min off-line mineralised in the mixture of sulphuric and nitric acids in a microwave oven. The detection limit in such a system was 0.34 $\mu g\ l^{-1}$ at sampling rate 30 h^{-1}. None of these methods for mercury determination exhibited significant interferences.

In analysis of sediments several other flow injection methods with AAS detection were developed. In determination of arsenic and selenium in four various certified materials, the earlier-reported method with hydride generation AAS was used (Fig. 8, Chap. 2) [118]. Samples of sediments were mineralised for 3 h with a mixture of nitric, sulphuric and perchloric acids. Selenium (VI) in the digested solution was pre-reduced to Se (IV) in HCl solution, and arsenic (V) to As (III) by reacting with potassium iodide. In the FIA system

1,10-phenanthroline was used to mask interferences from copper and nickel in selenium determination. The detection limits for both arsenic and selenium were 75 ng g^{-1}.

For the determination of cadmium and copper in the same sediment reference material two FIA methods with AAS detection and trace metal enrichment on biological beds were reported. In one of them sorption was carried out on the yeast *Saccharomyces cerevisiae* [205], in another on cyanobacteria (*Spirulina platensis*) [206], both immobilised on controlled pore glass. The analysed sediment samples were digested off-line with a mixture of hydrochloric and nitric acids. Using both preconcentration methods the obtained results were in good agreement with the certified content, although the whole analytical process was essentially prolonged by the employed mineralisation procedure.

The earlier-presented FIA method with inductively coupled plasma-mass spectrometry detection and isotope dilution was utilised for simultaneous determination of Re, Pt and Ir in a partially oxidised organic-rich sediment from the Northwest Atlantic Ocean [135]. Sediment off-line digestion was accomplished with hydrofluoric and nitric acid dissolution aided by microwave heating, but before that were combusted overnight at 550–575°C in order to oxidise organic matter. Then with hydrochloric acid added HF and SiF_4 were removed. In order to preconcentrate analytes and separate from species giving isobaric interference in ICP-MS detection sediment solutions were passed through anion-exchange columns. As eluent 12 M nitric acid was used. 300–600 μl of the final solution was introduced into ICP-MS using flow injection. The mass detection limit estimated for this procedure was 5, 6 and 14 pg for Re, Ir and Pt, respectively. Improvement of detection limits even by a factor of ten may be realised, according to authors, by increasing the efficiency of the sample introduction system, for example, with ultrasonic nebulisation or electrothermal vaporisation.

The total selenium was determined in samples of several certified sediments in the flow injection system with fluorimetric detection, which allows also speciation of selenium [207]. Detection is based on the selective oxidation of the non-fluorescent 2-(α-pyridyl) thioquinaldinamide by Se (IV) to give a fluorescent oxidation product. Samples of sediments were mineralised off-line with a mixture of sulphuric and perchloric acids. Concentrated hydrochloric acid was added to ensure that all the selenium was present as Se (IV). The results of selenium determination were in satisfactory agreement with certified values.

1.5. *Air and Aerosols*

The methodology of FIA also finds applications in determination of gaseous analytes, mainly in the systems where they are first adsorbed in a suitable solution or diffuse from the gaseous phase into the liquid phase in the gas-diffusion unit. Several such systems were developed for environmental analysis of atmospheric air. Flow injection methods can be used for determination of volatile analytes, for trace analytes in cloudwater, aerosols or air particulates.

In one of the pioneering applications of potentiometric detection in FIA systems the nitrate ion-selective electrode was used for determination of particulate nitrate in air [178]. The aqueous extracts of particulate nitrate in air were analysed in the FIA manifold without any pretreatment. Collected in the City of Copenhagen gave extracts containing from 1.3 to 7.3 mg l^{-1} N. The obtained results were in good agreement with reference spectrophotometric determinations.

For environmental purposes chloride potentiometric detection in the flow injection system was employed for determination of hydrogen chloride formed in the combustion of chloroorganic compounds and for measurements of chloride emission from power plants [36]. In the latter case chloride was determined in a 3% hydrogen peroxide solution, which is routinely used for collection of gases emitted from power plants. The obtained results with chloride concentration from 1 to 59 mg l^{-1} were well correlated with results obtained by ion chromatography and the reference spectrophotometric method. An FIA system with another type of chloride membrane electrode was employed for the determination of chloride in atmospheric aerosol samples (sea spray), and also a good correlation with ion chromatography results was found [208].

Determination of sulphur dioxide in ambient air was carried out in an FIA system with spectrophotometric detection [209]. The passive-sampled SO_2 was oxidized by hydrogen peroxide and determined as sulphate ion in solution. The detection of sulphate was based on reaction with a barium-dimethylsulphonazo III complex in the FIA system. An in-line cation exchange column was used to eliminate the interference from triethanolamine, which was used as the sorbent in the passive sampler. This system was used for determination of SO_2 in air in the concentration range of 11–87 μl Nm^{-3}.

In another configuration of the flow injection system with spectrophotometric detection sulphur (IV) was determined in cloud water samples [210]. The cloud water samples were collected in the buffered absorber containing CDTA complexone to prevent interference from transition metal ions. The spectrophotometric detection was based on the reaction of sulphur (IV) with pararosaniline and formaldehyde. A single channel FIA system employed a passive cation-exchange membrane reactor for the introduction of NaOH into the carrier stream with sample segments and a pressurised porous membrane reactor for the introduction of acidic pararosaniline. For analysed samples the sulphur (IV) content was in the range of 0.34–0.43 μM. The limit of detection for a 200 ml injected sample was estimated as 0.16 mM S (IV).

In the flow injection system also formaldehyde was determined in fog water samples utilising for detection the fluorescence of 3,5-diacetyl-1,4-dihydrolutidine formed upon the reaction of formaldehyde with ammonium acetate and 2,4-pentanedione [211]. Interference from S (IV) was eliminated by addition of hydrogen peroxide. Response from other carbonyl compounds, which may be present in atmospheric water, and from methanol was found to be negligible. The detection limit of the method was found to be 0.1 μM and the detected level in analysed fog waters was in the range of 206–267 μM.

2. Food Analysis

Quality control of food is a large and important field of chemical analysis. Its results decide the distribution on the market, the way of processing of food articles and their storage. Among analytes which are determined in food products are those affecting the health value of the product, preservants, substances used in processing and storage, as well as components harmful to human and animal health. Continuous development of the food industry addresses increasing qualitative and quantitative requirements to analytical laboratories. A response to these trends is the increasing demand for mechanisation and automation of analysis, where flow injection methods play a significant role. The largest number of FIA methods was developed for analysis of liquid food products, where sample pretreatment steps prior to the detection are relatively simple (fruit juices, milk, wine, other beverages). An increasing number of methods are being developed also for analysis of solid food products. Especially in this case scientific and technical literature is dispersed among numerous specialised journals not easily available, and hence a review of developed methods cannot be complete.

2.1. *Fruit Juices and Soft Drinks*

Most often determined components of juices where flow injection methods can be used are sugars. Several different methods were reported for determination of sucrose, fructose and for simultaneous determination of several sugars.

Sucrose in juices can be determined with chemical methods as well as using biochemical procedures employing various enzymes. In spectrophotometric determination in the FIA system detection of sucrose is based on the reaction of periodate with fructose formed from sucrose inversion [212]. The consumption of periodate is measured as a transient lowering of the iodine concentration at 350 nm produced by the reaction of periodate with iodine. This method was applied to the determination of sucrose in sugar-cane juice samples with sampling frequency 30 h^{-1}. In another configuration of the flow injection system with spectrophotometric detection besides sucrose in the same injected sample solution total reducing sugars can be determined [213]. In this case detection was based on oxidation of reducing sugars by haxacyanoferrate (III) before and after hydrolysis of sucrose in the flow injection setup. The haxacyanoferrate (II) resulting from this reaction is measured spectrophotometrically after the addition of 1,10-phenenthroline and iron (III) to form Ferroin. Determinations also in this case carried out in sugar-cane juice samples were in good agreement with those of the standard volumetric procedure.

Various enzymatic procedures developed for FIA can be utilised successfully for determination of sucrose in juices. In amperometric determinations detection was based on oxygen consumption in enzymatic reaction of glucose catalysed by glucose oxidase. The juice samples were enzymatically off-line inverted for 5 min at 55°C to glucose and fructose and then injected into the flow injection system with circulating glucose oxidase solution [214]. In a system with closed-loop circulation of a reagent up to 700 samples h^{-1} can be determined, although the real effectiveness of sucrose determination was limited by off-line inversion. This method was employed for determination of sucrose in fruit juices and several other food products. An improvement in such a procedure can be achieved by the use of enzymes immobilised in flow-through reactors. This was carried out in the setup for determination of sucrose with chemiluminescence detection [215]. In the developed procedure the conversion of sucrose was performed in reactors with immobilised invertase and mutarotase, and then hydrogen peroxide formed in reaction of glucose catalysed by glucose oxidase reacts with luminol, giving chemiluminescence. The reaction with enzyme and reaction with luminol occur in a microporous membrane flow cell. This method was applied in analysis of soft drinks, which required on-line catalytic destruction of glucose in the sample prior to sucrose determination.

The working range for the developed method was from 5 μM to 1 mM sucrose. Another possibility of enzymatic determination of sucrose, also used for real samples of juices and soft drinks, is based on the application of sucrose phosphorylase [216]. In this case determination required the use of two other enzymatic steps and carried out fluorimetric detection was utilised for measurement of NADPH formed in the last step. Three used enzymes were co-immobilised in one flow-through CPG reactor. Such a setup allows the determination of sucrose in the range of 0.2–200 μM. An inhibiting effect of fructose was reduced by the use of a high (100 mM) concentration of phosphate in carrier solution.

Two different enzymatic systems can be employed for flow injection determination of fructose in fruit juices [217, 218]. In both of them the enzymes were immobilised in flow-through reactors. In the system with fructose 5-dehydrogenase as the redox acceptor hexacyanoferrate (III) was used and its reduced form was detected by amperometry with a platinum electrode. In fruit juice analysis interference of ascorbic acid was completely removed by using a pre-column of ascorbate oxidase [217]. Another enzyme used for this purpose, mannitol dehydrogenase, catalyses the reduction of D-fructose to mannitol with concomitant oxidation of reduced β-nicotinamide adenine dinucleotide (NADH). A decrease of NADH concentration is measured fluorimetrically [218]. The response is linearly related to fructose concentration in the range of 6–600μM, which was used for fructose determination in fruit juices, cola and wine with sampling rate 30 samples h^{-1}. No response to other sugars was observed at 10 mM level.

In a suitably designed FIA system fructose and glucose can be determined simultaneously [219, 220], or even three sugars, fructose, glucose and sucrose [221]. All these systems were used for analysis of fruit juices. In the system with fluorimetric detection multistep enzymatic reactions were employed with enzymes immobilised in flow-through reactors [219]. First, both sugars are enzymatically phosphorylated to glucose-6-phosphate and fructose-6-phosphate. The glucose-6 phosphate is oxidized with formation of the reduced form of the coenzyme NADPH, which is monitored fluorimetrically. Fructose-6-phosphate is first converted to glucose-6-phosphate with the aid of phosphoglucose and is then oxidised to form also NADPH. This method was used for determination of both sugars in juices with sampling frequency 30 h^{-1}. In another FIA system for the determination of glucose and fructose the spectrophotometric detection was employed with the standard addition procedure [220]. The sum of the two sugars was determined using the above-mentioned catalytic steps with detection of NADPH at 340 nm. Then glucose only was determined utilising glucose oxidase and hydrogen peroxide produced was determined spectrophotometrically with phenol, 4-aminoantipyrine and peroxidase at 500 nm. This method was employed for analysis of fruit juices, wines and beer with sampling rates from 15 to 2 h^{-1}, depending on the number of additions used.

Three sugars were determined simultaneously in juice samples in the system with three parallel enzyme reactors and with a multichannel amperometric detector [221]. Interference of ascorbic acid was removed by using an ascorbate oxidase reactor before the sample injection valve.

Several flow injection systems were developed also for determination of L-ascorbic acid in fruit and vegetable juices as vitamin C is an important micronutrient and plays many physiological roles. For its determination in juices can be used either chemical or enzymatic methods. A direct determination with UV detection at 245 nm was based on treatment of the sample with sodium hydroxide and as result of this a fraction of the ascorbic acid was decomposed into substances, which do not absorb in the UV region, and the decrease in signal was measured [219]. In the indirect method with UV detection samples with ascorbic acid are injected into a stream of a carrier solution containing triiodide ion or the triiodide-starch complex and decrease of signal is monitored at 350 or 580 nm [220]. Juice samples and jam for determination of ascorbic acid were diluted and stabilised with EDTA and formic acid. The flow injection determination of ascorbic acid with chemiluminescence detection was based on the reaction of the products of the photo-oxidation of ascorbic acid with visible light with lucigenin, sensitised by toluidine blue [221]. Interferences from some transition metal ions were removed on-line using a mini-column with a cation-exchange resin by passing the sample prior to injection into the flow system. The method was successfully employed for determination of ascorbic acid in juices at a concentration level of 60–900 mg l^{-1}, and also for determination in blood serum and pharmaceutical preparations.

Ascorbic acid in juices was also determined in various FIA systems with amperometric detection. This can be based on oxidation of ascorbic acid on a glassy carbon electrode, which occurs at an extremely low applied potential (+20 mV vs. Ag/AgCl) in the presence of Cu (II) in weakly acidic media [222]. This simple method when employed for determination in juices yielded results in good agreement with those of HPLC determinations. Also on direct oxidation of ascorbic acid at a platinum electrode was based a method of simultaneous determinations of ascorbic acid and glucose [223]. Direct measurement of a sample at +0.7 V vs. Ag/AgCl allowed one to obtain signal for ascorbic acid, whereas after passing the sample through an enzyme reactor with glucose oxidase a signal corresponding to the sum of ascorbic acid and glucose was measured. The results of determination in fruit juices validated with a recovery test were found satisfactory. With the use of a reactor with ascorbate oxidase an FIA system was designed, where at +0.6 V polarisation amperometric detection was carried out on a glassy carbon electrode [224]. On passage through the enzyme column a fraction of ascorbic acid is converted into non-electroactive dehydroascorbic acid and the decrease in signal is compared with the blank. Copper that has a catalytic effect on the decomposition of ascorbic acid was eliminated by the addition of 250 mg l^{-1} EDTA.

Sulphite is widely used as preservatives for food and beverages to prevent oxidation and bacterial growth. As it is also considered a hazard to human health, a rigorous control of its content is necessary in many products. For this purpose several methods have been developed, including the use of enzymatic biosensors. For determination of sulphite content in fruit juices two different FIA methods were employed. A spectrophotometric determination can be based on its inhibitory effect on the activity of polyphenol oxidase [225]. As a source of this enzyme a crude extract of a sweet potato root was used directly in the carrier of the FIA system and products of the enzymatic oxidation of catechol can be monitored at 410 nm. Ascorbic acid which produces strong interference can be removed by passing the sample through a column packed with ascorbate oxidase containing pieces of cucumber. Such a system based on natural plant materials allows one to obtain a detection limit for sulphite at level 2.2 μM with sample throughput 26 h^{-1}. The results of determination in fruit juices, wines and vinegar correlated well with results of the manual iodimetric method. Another method of determination of sulphite in juices in the FIA system was based on an on-line membrane separation of SO_2 released from a stream of sulphuric acid into a stream containing hydrogen peroxide in sulphuric acid [26]. The conductivity increase due to the oxidation of sulphur dioxide to H_2SO_4 is monitored by a bulk acoustic wave impedance sensor. At sampling rate 78 h^{-1} the limit of detection was estimated as 5.0 μM. This method has also been successfully used for determinations in juices, as well as in wines.

Using flow injection methods several other species can be determined in juices. Examples of these determinations are as follows. The spectrophotometric determination of citric acid can be based on reduction of Fe(III) by citric acid upon irradiation with visible or UV light to Fe(II), which reacts with 1,10-phenanthroline forming the reddish orange Ferroin [227]. In the stopped-flow measurements the limit of detection was estimated as 0.5 mg l^{-1}. In this case also ascorbic acid had to be removed from the samples using an enzymatic reactor containing ascorbate oxidase. Citric acid in juices is a main source of acidity, which can be measured conductimetrically, or spectrophotometrically [228]. In the first method a sample is injected into a stream of ammonia solution, and then the excess of the unreacted ammonia diffuses in a gas-diffusion cell through a PTFE membrane to acetic acid solution. In the second case a spectrophotometric titration is utilised, where the sample is mixed with sodium hydroxide in the presence of phenolophtalein. Both procedures have been successfully applied to the analysis of fruit juices having acidity at a level of about 1% (w/v) citric acid. Although results of the spectrophotometric procedure were more precise they had a negative bias due to the effect of sample colour.

Chloride in juices was determined in the flow injection setup with spectrophotometric detection based on formation of a thiocyanate Fe(III) complex as a result of the reaction of chloride with $Hg(SCN)_2$ [229]. In the measuring system the on-line dialysis of juices through polyetherimide-composed membranes was employed in order to remove interferences by suspended particles and colour organic compounds.

Thiourea is often added to citrus as a fungicide to assist in its conservation during cold storage. Because it is also toxic and has been labelled as having carcinogenic activity a flow injection fluorimetric method has been developed for its determination in fruit juices [230]. The method is based on the rapid oxidation of thiourea by thallium (III) with formation of fluorescent Tl (I) as $TlCL_3^{2-}$ anionic complex. In the optimised interference-free procedure a linear response was obtained over the range of 0.5–10 μM.

In samples of canned fruit juices also tin has been determined in the flow injection system with spectrophotometric detection [231]. The method was based on the formation of a complex between Sn(IV) and pyrocatechol violet in the flow system and its accumulation on Sephadex gel packed in a flow cell. In such a procedure the detection limit 0.3 μg l^{-1} and a linear range of 2–40 μg l^{-1} were reported. In order to carry out a determination the juice sample was heated until nearly dry, digested with nitric acid and followed by treatment with concentrate until evolution of NO_2, and diluted.

The obtained results on the level of 260–900 mg kg^{-1} were comparable to the AAS method.

2.2. *Milk and Dairy Products*

Milk and dairy products (yoghurts, cheese) can also be analysed using various flow injection methods. In most cases the analysed products have been processed prior to the injection into the measuring setup.

Great attention, for instance, is paid to determination of sugars, especially lactose, which occurs only in the milk of mammals. Its determination is routinely carried out in the dairy and related industries in order to ensure effective process and product control. In FIA systems lactose is determined using enzymatic methods. In the earlier-mentioned system with amperometric detection and continuous circulation of glucose oxidase also the content of lactose was determined in milk, ice cream and yoghurts at the concentration level of 4–16 g per 100 g of the analysed sample [214]. The samples were dissolved in an imidazole buffer and analysed after off-line addition of lactase. In a similar way also maltase activity was determined in modified infant milk after addition of maltose. In another flow injection system with amperometric detection β-galactosidase and glucose oxidase were immobilised in a flow-through reactor incorporated in series in the measuring system and the produced hydrogen peroxide was detected with the platinum electrode [232]. In order to remove glucose from the analysed samples, a pre-column packed with co-immobilised glucose oxidase and catalase was placed before two above-mentioned reactors. Milk samples prior to the injection were diluted 1:100 with a phosphate buffer. Determinations in human and cow's milk could be performed at a rate of 60 h^{-1}. The same enzymatic system was employed in the flow injection setup for determination of lactose with chemiluminescence detection [233]. The use of an on-line dialysis unit allows one to avoid the sample preparation step in determinations in milk and produces mechanised dilution of the sample. The sample throughput in these determinations was 20 h^{-1}.

In other dairy products FIA has been used for determination of glucose and fructose. Determination of glucose in ice cream was carried out in the system with amperometric detection, where glucose oxidase was immobilised on a nylon mesh and held over a platinum wire [234]. Ice cream samples were treated with 5 ml of 80 mM potassium hexacyanoferrate(III) solution and 5 ml 250 mM zinc sulphate solution in a phosphate buffer, diluted to 100 ml, filtered and then analysed for glucose content. A sequential determination of glucose and fructose in yoghurt and ice cream can be performed in the FIA system with fluorimetric detection based on monitoring of NADPH [219].

Another component of which the content is frequently controlled in milk is calcium. A simple method for its determination in the flow injection system was developed with the use of a photo-cured calcium selective electrode [235]. In order to determine the total calcium level in a full cream milk the sample was digested with a nitric and perchloric acid mixture to release all the complexed calcium. The obtained results correlated well with those obtained by atomic absorption spectrometry and labelled content. In a more complex, three-detector flow injection system (Fig. 10) a total and free calcium and chloride in milk can be determined simultaneously [236]. Free calcium is determined in dialysate with sepctrophotometric detection employing Cresolphthalein Complexone. The total calcium content is determined in the undialysed plug with flame AAS detection. Interferences from phosphates in milk are overcome by using a nitrous oxide-acetylene flame with the necessary suppression with potassium ions. A second dialyser in series is used to eliminate interferences, especially of casein, before the dialysed chloride is measured with a chloride selective electrode. The determinations with sampling rate 60 h^{-1} have not required any sample pretreatment.

The above-mentioned spectrophotometric method of determination of nitrite and nitrate was applied also to simultaneous determination of these analytes in cheese samples [55]. The samples were digested following the method recommended by the AOAC.

The content of lead in reference material, spiked skim milk powder was determined by atomic absorption spectrometry in the FIA system employing on-line solid phase preconcentration as pyrrolidinedithiocarbamate on an activated carbon column [237]. Prior to the determination the sample was mineralised.

The sum of acetoacetic acid and acetone, called oxidised ketone bodies, determined in milk is an indication of the appropriate energy balance in the diet of cows. For the determination of these species a spectrophotometric FIA method has been developed [238]. It was based on decarboxylation of acetoacetate to acetone at 100°C, then separation of acetone from the sample by gas-diffusion through a teflon membrane and spectrophotometric detection with detection limit 0.1 mM. A good agreement between flow injection and gas chromatography methods was demonstrated.

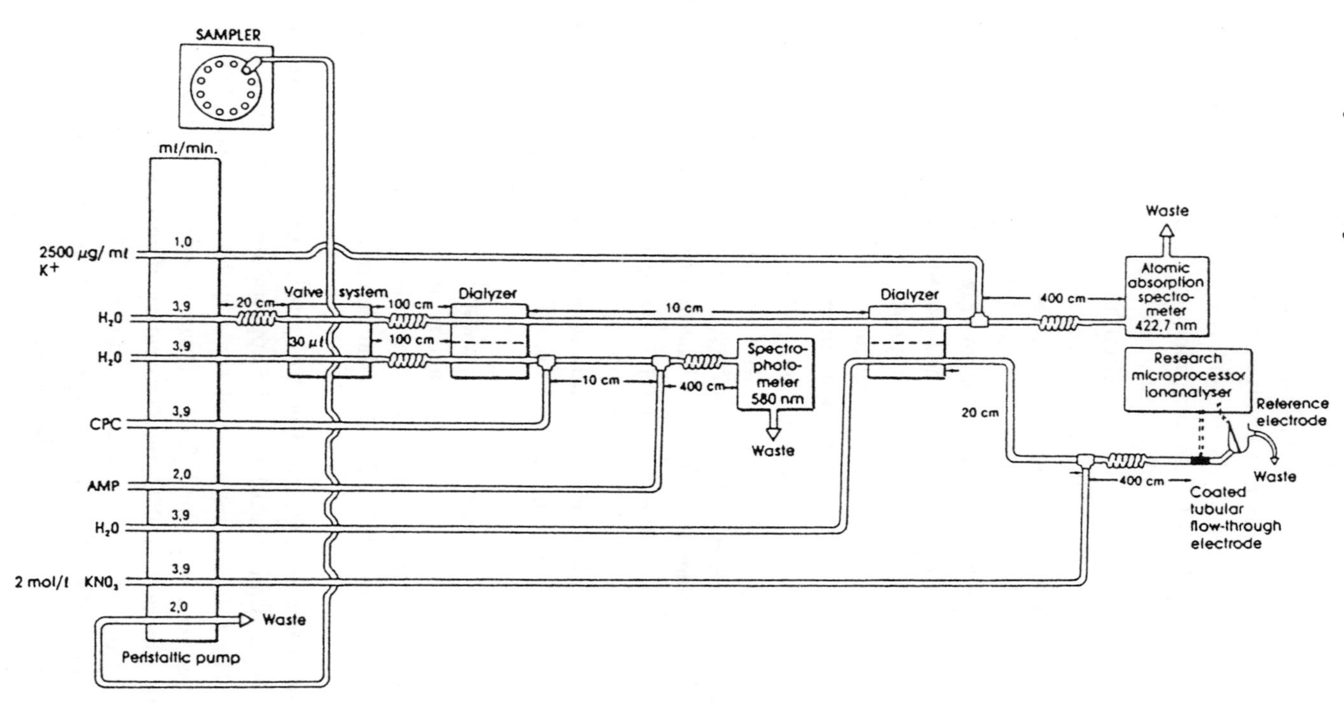

Fig. 10. Schematic diagram of the flow injection system for the simultaneous determination of free calcium, total calcium and chloride in milk [236]. Sample injection volume 30 ml. Internal tubing diameter 0.76 mm. (*Reprinted by permission of copyright owner.*)

2.3. *Wine*

Chemical quality control of wines includes determination of a large variety of compounds. Besides the content of ethanol and sugars, also glycerol, ascorbic acid, preservants, malic and lactic acids, tannins, main mineral components and trace heavy metals are determined. Wines do not belong to matrics that require time-consuming pretreatment, hence numerous flow injection methodolgies have been developed for their analysis.

The analytical literature yields many various flow injection methods for determination of ethanol, both chemical, relatively unspecific and enzymatic ones based on the use of alcohol oxidase or alcohol dehydrogenase. For determination of ethanol in undiluted samples of wine, beer, spirits and pharmaceutical preparations satisfactory application was reported for the flow injection procedure with amperometric detection and alcohol oxidase immobilised in a flow-through reactor [239]. A 10 μl sample is injected into the stream of phosphate buffer and transported to the gas-diffusion cell with a silicone-modified polypropylene membrane. An aliquot of the volatile components that passed through the membrane is transported with an acceptor stream of the enzyme reactor where the ethanol is converted to acetalaldehyde and hydrogen peroxide. The latter is detected at a platinum electrode. The sampling frequency was reported as 120–18 h^{-1} and there was a good correlation of obtained results with the standard distillation method. It was also demonstrated that the method is practically free of interferences. For practical applications it is especially advantageous to employ flow injection methods allowing simultaneous determination of ethanol and acetaldehyde [240] or ethanol and glycerol [241]. In both these methods flow-through enyzmatic reactors were used. In the system for determination of ethanol and acetaldehyde a dual injection valve was used, which inserts the sample into channels of different lengths, each of which includes an enzyme reactor [240]. In the reactor with alcohol dehydrogenase ethanol is oxidised by NAD^+ and semicarbazide is used as a trapping agent for acetaldehyde formed in the reaction. In the reactor with aldehyde dehydrogenase in the second line acetaldehyde is also oxidised by NAD^+. In both cases NADH formed is detected photometrically at 340 nm. In the other system for simultaneous determination of ethanol and glycerol NADH is monitored fluorimetrically [241]. Detection of ethanol is the same, whereas detection of glycerol is based on catalytic oxidation in the reactor with glycerol dehydrogenase. Both systems were used for analysis of in samples with sampling rates 46 [240] and 60 [241] h^{-1}. The samples were injected after suitable dilution to fit the concentration of the analytes within the linear calibration range.

Besides ethanol and glycerol during the fermentation process 2,3-butanediol is also formed. The control of the content of glycerol and 2,3-butanediol is important for the organoleptic characteristics of the final product. Determination of the sum of these two compounds can be performed by sequential injection analysis with potentiometric detection using a periodate ion selective electrode [242]. To minimize interferences caused by reducing anions present in the sample, a mini-column packed with the strong anion-exchange resin was incorporated in the system. With a sample throughput of 3 h^{-1} polyols were determined in a set of red and white wines.

Sulphite is commonly used as an antioxidant and antiseptic in winemaking; however, its potential toxicity makes its determination essential in quality control of wines. For determination of sulphite in flow injection systems chemical methods with suitable separation step are used as well as enzymatic methods, mainly with the use of sulphite oxidase. In methods using a gas-diffusion separation in an acceptor stream sulphite can be determined spectrophotometrically [243], amperometrcally [244], or using the above-mentioned detection with an acoustic wave impedance sensor [226]. In spectrophotometric detection the reaction between SO_2 and *p*-aminoazobenzene was employed and at sampling rate 50 h^{-1} in wine samples a good agreemnt was found with the routinely used method in a segmented-flow analyser. In the same above-cited paper also the determination of so-called total sulphur dioxide (free and bound to aldehyde and ketone groups) was proposed with additional use in the system of an on-line hydrolysis step to release all bound SO_2 [243]. In determinations with amperometric detection a working glassy carbon electrode was used polarised at +1.15 V with a detection limit observed at the 0.02 mg l^{-1} level. Wine samples were injected after 25-fold dilution, and determinations can be carried out at a sampling rate of 120 h^{-1}. In enzymatic determinations used for analysis of wines the amperometric detection was used with immobilised sulphite oxidase in a flow-through reactor. Hydrogen peroxide formed in enzymatic reaction is monitored by the use of a platinum electrode covered with a dialysis membrane in order to reduce interferences. For the same purpose and for protection of the immobilised enzyme against deactivation it was found to be successful to mix the wine sample with gelatin and filtrate on-line prior to the injection to the measuring setup. Satisfactory results were reported for white wines, whereas errors observed for red wine samples were attributed to large amounts of polyphenol compounds. The amperometric determinations with sulphite oxidase immobilised on a rotating disc were carried out in the presence of hexacynaoferrate (III) as the electron acceptor [246]. The analysed samples of white wines 5-fold diluted with phosphate buffer were prior to the detection passed through the column with immobilised ascorbate oxidase. Similarly, 5-fold diluted samples of wine were introduced into the FIA system, in which simultaneous determination of sulphite and phosphate was carried out [247]. Determination of sulphite was based on the use of sulphite oxidase, and of phosphate on a two-step reaction with purine nucleoside phosphorylase and xanthine oxidase. A significant improvement of detectability has been achieved by coating a working platinum electrode surface with poly(1,2-diaminobenzene). Determinations in wine samples were carried out at a sampling rate of 30 h^{-1}. For determination of sulphite in white wines also the above-mentioned method with spectrophotometric detection based on inhibition of polyphenol oxidase present in an extract of the sweet potato root was employed [225].

Determination of fructose and glucose in wines can be successfully carried out with enzymatic methods used for analysis of fruit juices with fluorimetric [218] and spectrophotometric [220] detections. For determination of the total content of reducing sugars a flow injection method with potentiometric detection can be used employing a picrate ion-selective electrode, based on a reaction between reducing sugars and picric acid [248]. The analysed samples were not pretreated prior to the injection. The measuremnent was based on the peak width at some fixed value of the analytical signal.

The quality of wine is also affected by the level of lactic and malic acids, as during the fermentation process malic acid is converted into lactic acid. Common analytical enzymatic procedures of their determination have been adapted for flow injection procedures for wine analysis. Both in the system for malic acid determination [249] and for simultaneous determination of lactic and malic acids [250] the oxidation of both analytes by NAD^+ is exploited in the presence of suitable dehydrogenase. Detection of produced NADH was carried out in both systems spectrophotometrically. In a single analyte system for determination of malic acid a kinetic procedure in the presence of hydrazine was used. In stopped-flow measurements used the elimination of the effect of interfering species was achieved [249]. Determinations in wine samples were performed after preliminary pH adjustment to the required value. Simultaneous determinations of lactic and malic acids were carried out in the reversed flow injection system with a dialyser [250]. Into the acceptor stream solutions of two enzymes are injected, yielding two peaks corresponding to the NADH formed for each determination. Such a procedure was free of interferences and it was successfully used for determinations in wine samples.

For determination of ascorbic acid content in wine the FIA method with fluorimetric detection can be used based on the oxidation of ascorbic acid catalysed by laccase to dehydroascorbic acid, which then reacts with *o*-pheneylenediamine to produce a fluorescent quinoxaline [251]. Prior to the injection the pH of the sample was adjusted, and EDTA was added in order to remove the interference from metal ions and laccase. The determination in the concentration range of 0.025–1.0 mg l^{-1} in the wine and beer samples was carried out by combining the merging zone principle with the stopped-flow technique.

The determination of tannins, which are partly responsible for the flavour of wines, can be carried out in the reversed FIA system by their precipitation using injection of the copper acetate complex into the sample stream [251]. The unprecipitated copper was determined by atomic absorption spectrometry. The analysed wine samples required 50- to 250-fold dilution to fit the tannin concentration within the linear range of the calibration graph (1–25 mg l^{-1} tannic acid).

For the simultaneous determination of potassium and calcium an FIA potentiometric system with ion-selective electrodes sensitive to potassium and calcium was used without any sample pretreatment [252]. The potassium was determined in the concentration range of 0.43–0.66 g l^{-1}, and calcium in the range of 34–45 mg l^{-1}. Then lead in port wine at a concentration level of 50–180 mg l^{-1} was determined by flow injection spectrophotometry with preconcentration on a mini-column with Chelex 100 resin [253]. The detection was based on the formation of a ternary complex between lead, malachite green and iodide. Interference from copper was eliminated by addition of picolinate to the buffer solution used in the FIA system. The analysed samples were injected after mineralisation. The obtained results were in good agreement with those obtained by AAS with electrothermal atomisation, although for 30% of the analysed samples the results differed by more than 10%. For the determination of iron (III) in wine samples at the concentration level of 1.5–3.0 mg l^{-1} a spectrophotometric FIA method was used based on the Fe(III)-thiocyanate complex by measuring the absorbance of the anion-exchange resin which retains the Fe (III) complex [105].

2.4. *Other Food Products*

Flow injection methods have been developed also for several other food products. The methods described below concern only processed food, whereas determinations in fruits, vegetables, meat or seafood are included in the next section dealing with biological matrices.

Several methods were developed for determination of various components in flour. The determination of starch can be successfully performed with amperometric detection with a tri-enzyme biosensor employing immobilised amyloglucosidase, mutarotase and glucose oxidase [254]. The analysed samples were first incubated with soluble α-amylase for 1 h. For this purpose samples were solubilised by incubation with dimethyl sulphoxide and hydrochloric acid. The determination in flour of various origins correlated well with spectrophotometric determination by the use of a Boehringer starch kit. The determination of glucose in flour samples can also be carried out amperometrcally using the enzymatic method reported above for determination in ice cream [234]. For the determination of trace levels of manganese and zinc in the FIA system a spectrophotometric detection was used. The manganese was determined based on its catalytic effect on oxidation of diphenylcarbazone in the presence of triethanolamine [255]. Analysed samples of flour and other products were mineralised with a nitric and perchloric acid mixture. In determination of zinc its reaction with PAR was employed following the preconcentration of the analyte as a thiocyanate complex in polyurethane foam contaning a mini-column [256]. Samples were mineralised with a mixture of nitric acid and hydrogen peroxide at microwave heating. Interferences from other transition metal ions were masked with citric acid. Mn (II) in rice flour was determined at the 40 mg g^{-1} level and Zn (II) at about the 20 mg g^{-1} level. For the determination of arsenic and selenium FIA systems with AAS hydride generation detection were used [117, 257]. In both cases flour samples were mineralised with nitric acid. In one procedure mineralisate after pH adjustment were off-line passed through the column with Chelex 100 in order to eliminate the interfering elements [117]. In both cases the obtained results on flour standard reference materials were in satisfactory agreement with certified valuse.

In samples of breakfest cereal and cake mix the sucrose was determined enzymatically with chemiluminescence detection in a three-enzyme system and detection of hydrogen peroxide via reaction with luminol [215]. The analysed samples were extraced with water.

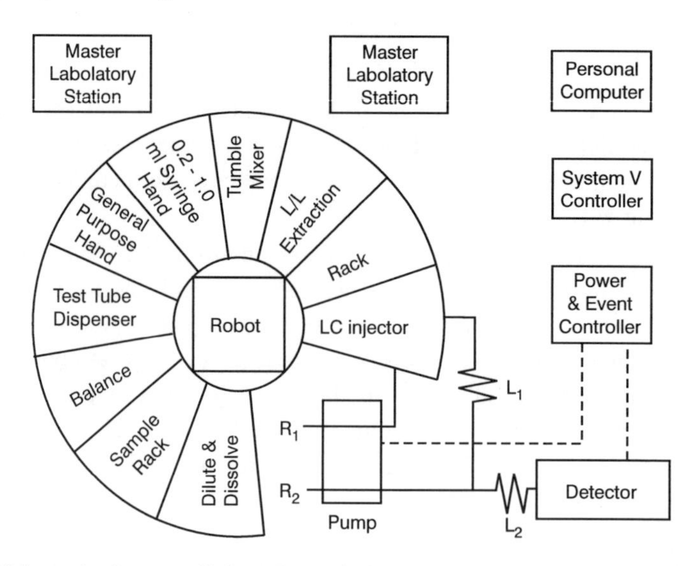

Fig. 11. Schematic diagram of the robotised flow injection system for spectrophotometric determination of polyphenols in olive oil [259]: (- - - -) active interface; (– – – –) pasive interface; R1 — Folin-Ciocalteu reagent, 5% (v/v); R2 — 0.1 M NaOH; L1 — 100 cm; L2 — 100 cm; sample loop 100 ml; flow rates $q_1 = q_2 = 0.6$ ml min^{-1}. (*Reprinted by permission of copyright owner.*)

A source of antioxidants in natural food such as polyphenols can be olive oil. A large amount of them is lost during the refining process; however, determination of their level is essential for estimation of the quality of oil. For this purpose two flow injection systems with spectrophotometric detection were developed which are based on the use of a Folin–Ciocalteu reagent. In both of them a solvent extraction of analytes to water or a water-methanol mixture was employed. In the flow injection extraction system without phase separation with iterative change of the flow direction is synchronised with the injection in such a way that the n-hexane solution of analysed oil never reaches the flow cell [258]. The analytes were determined with sampling rate 28 h^{-1} in the concentration range of 0.1–0.9 g l^{-1}. In another system extraction process was carried out in a robotic system coupled with detection in FIA arrangements [259] (Fig. 11). In this system the sampling rate was slightly lower (20 h^{-1}), but precision of determination expressed by relative standard deviation was much better (1.6%) than that obtained by the FIA method (4.5% [258]) or the manual method (10%). The FIA method was also developed for the determination of nickel in vegetable oil and chocolates using preconcentration on

8-hyroxyquinoline resin and spectrophotometric detection based on formation of the nickel-dimethylglyoxime complex [260]. The samples were dry-digested at 800°C and the residue was dissolved in hydrochloric acid. The nickel level in chocolate was found in the range of 0.5–8.0 mg g^{-1}, whereas in vegetable oil nickel was not detected.

Several substances contained in beer can also be determined using flow injection methods. The flow injection extraction method with UV spectrophoometric detection has been developed for determination of bittering compounds [261]. They have been identified as three pairs of diastereoisomeric compounds (isohumulone, isocohumulone and isoadhumulone). The FIA determination is based on on-line extraction and measuring absorbance at 275 nm. During the fermentation process in beer production the level of diacetyl should be controlled, as this compound contributes to the flavour of beer. For this purpose a flow injection-pervaporation procedure has been developed [262]. The diacetyl vapour in the air gap module diffuses through the hydrophobic membrane to an acceptor solution containing α-naphthol and creatine where a coloured product is formed and monitored photometrically. Determinations in ultrasonicated samples of beer with a sampling rate of 8 h^{-1} yield results which were in good agreement with the conventional method. The above-mentioned method for simultaneous determination of nitrite and nitrate with spectrophotometric detection was also used for analysis of beer samples [55].

The reversed FIA spectrophotometric method has been developed for simultaneous determination of creatinine and creatine and applied to the determination of these analytes in broth cube samples [263]. The determination of creatinine is based on Jaffe reaction with alkaline picrate, while the determination of creatine was based on reaction with 1-naphthol-biacetyl. The samples to be analysed were dissolved in water and filtered. The reported sampling rate was 42 h^{-1}. The determination of saccharin in mixtures of sweeteners was carried out in the system with continuous precipitation with silver ion [264]. The precipitate was retained on a filter, dissolved in ammonia and measured by AAS. The method cannot be applied to chloride-containing samples.

As a method developed for food analysis should be considered an FIA system for the determination of meat freshness, based on the simultaneous determination of polyamnines and hypoxanthine [265]. The developed system with amperometric detection was based on the use of putrescine oxidase and xanthine oxidase reactors, and the peroxidase electrode.

3. Biological Materials

Application of flow injection methods to analysis of biological materials of usually difficult matrices, requiring low limits of detection, is especially efficient when possibly a large part of the sample processing operation can be performed in the flow injection system. As most often solid samples are analysed, usually one time-consuming step of sample processing — transfer into solution — has to be performed off-line.

In the presentation of literature on this subject one may follow different groups of materials (e.g. leaves, vegetable, fruits, seeds, etc.) similarly to earlier sections of this chapter, but in this case it was much more difficult and it would required numerous repetitions of citations. Therefore in this section all reported developments are classified according to analytes, starting from inorganic anionic species, through cationic ones, and finally organic analytes. Similarly to earlier sections a literature survey includes both flow injection and sequential injection methods. When sufficient data are available in original papers methods of off-line dissolution of samples will also be reported.

3.1. *Plants*

For determination of boron in plant material several different flow injection procedures have been developed with application of azomethine-H and spectrophotometric detection [266–268]. In all these methods samples were prepared by ashing plant leaves at 550°C and dissolving in HCl solution. In the simplest system determinations were carried out with frequency 60 h^{-1} and interferences caused by the presence of Al, Cu and Fe were eliminated by the addition of EDTA to the buffer solution [266]. The application of the stopped-flow technique in an appropriate setup allows one to obtain a better sensitivity with the same analytical frequency [267]. The location of the detector in the loop of the injection valve gives a possibility of construction of measuring systems with so-called "multi-site detection", where as a result of a change of detector placement one can change the position of detection in the system. This concept resulted in the development of the system for determination of boron with the same spectrophotometric detection and sampling frequency of 120 h^{-1} [268]. The results of determinations at the concentration range of 28–145 mg kg^{-1} were in good agreement with those obtained by the ICP-AES method.

A flow injection system with potentiometric detection developed for liquid samples based on pervaporation can be directly adapted for determinations in solid biological samples [269], although after removal of the sample injection step, strictly speaking it is not a flow injection system any more. A dried and weighted sample of leaves was placed in a lower chamber of a pervaporation module and reagents necessary for the extraction of fluoride and its conversion into a volatile derivative were introduced.

In aqueous extracts of vegetables, after precipitation of organic matter with ammonium aluminium sulphate, the nitrate and nitrite were determined in the flow injection system with on-line solvent extraction and AAS detection [270]. The method is based on extraction of an ion pair of nitrate with Cu(I)-Neocuproine complex into 4-methyl-2-pentanone and measuring the AAS signal of copper in the organic phase. When nitrite is oxidised by Ce(IV), total nitrate is determined, whereas when it is converted with sulphamic acid to nitrogen, only the original nitrate is determined. The determinations were carried out in spiked samples, only. The total nitrogen was determined in the flow injection system in the plant digest obtained by sulphuric acid digestion [271]. A method was based on the use in the flow system of an isothermal distillation of ammonia and spectrophotometric detection of ammonia employing a Nessler reagent. At the sampling rate 100 h^{-1} the obtained results were in good agreement with those from the Indophenol method.

In homogenised samples of melons and raisin in imidazole/NaCl buffer phosphate was determined in the FIA system with amperometric detection [272]. The measuring system was constructed using immobilised nucleoside phosphorylase and xanthine oxidase and an amperometric platinum electrode polarised at 0.7 V. In the presence of ascorbic acid, glutathione or uric acid differential measurements can be performed using two identical electrodes: one with the enzymes and one without. Determined contents were in the range of 0.015–0.14 mg/100 mg of the product.

Two of the earlier-presented flow injection turbidimetric methods for determination of sulphate were successfully used for determinations in plant material [38, 41]. Samples were digested with nitric acid, and the excess of the nitric acid was driven off by perchloric acid [38]. These methods with precipitation, either barium sulphate or lead sulphate [273], were employed for determination of the total sulphur content in plant material. In the latter determinations were carried out in digests of powdered leaves in the range of 0.03–0.28% sulphur (w/w) in dry material. In this procedure an inducing of lead sulphate precipitation by on-line produced suspension of lead phosphate was applied.

Besides the above-mentioned distillation method for determination of the total content of nitrogen in plant material, conductometric determination of ammonia in Kjeldahl digest was also carried out with the gas diffusion step [274]. The vegetable tissues analysed included cotton, rice, orange, coffee and soy. Determinations were carried out with sampling rate 100 h^{-1}.

Determination of calcium in plant material can be performed also in the above-mentioned system with spectrophotometric detection with glyoxal bis(2-hydroxyanil) [22], and with AAS detection in the merging zones system [275]. In both cases plant digests were prepared by nitric-perchloric digestion. The spectrophotometric determinations can be carried out with sampling rate 180 h^{-1}, and those with AAS detection even with sampling rate 300 h^{-1}. Determinations of potassium were carried out in the systems with AAS [276] and flame emission spectrometry [275]. Also with AAS detection determinations of magnesium in the digests of the plant leaves were carried out [275], and in spinach digest with ion-selective potentiometry [30]. Similarly to the determinations of calcium, in both methods for magnesium determination similar methods of sample mineralisation were used.

Similarly to several other materials discussed earlier, numerous flow injection methods were also developed for determination of trace elements in mineralisates of biological material.

Among various trace elements, aluminium occurs in plant material at a relatively high level. In tea leaf digests it was found as 0.3–0.55 g l^{-1} in the FIA system with flame AAS detection and on-line preconcentration on a cation-exchange resin [90]. In coffee bean infusions, however, 7.3 mg l^{-1} was found, only, and in tea infusions from 4.4 to 12.6 mg l^{-1}. The determination of aluminium in plant digests simultaneously with iron was carried out in the FIA system with spectrophotometric detection using the above-mentioned concept of relocation of the detector in the system [277]. Aluminium was determined using Erichrome Cyanine R, and iron with 1,10-phenanthroline and ascorbic acid. The obtained results were in good agreement with those from the use of ICP-AES detection. The appropriate arrangement of this system allows also speciation of iron in terms of the oxidation state. Aluminium present in fruit juices containing smashed fruits was determined in the flow injection system with an automated slurry sample introduction and graphite furnace AAS detection [278]. The analysed samples were diluted with 0.2% nitric acid, and results of determinations were compared with those obtained where identical samples were pretreated by ashing and analysed by direct electrothermal atomic absorption spectrometry.

Arsenic in plant material can be easily determined in flow injection set-ups with hydride generation AAS detection [117, 118, 257]. Using these earlier-reported methods, arsenic and selenium were determined in standard reference material of orchard leaves [117, 118, 257], citrus leaves [118], wheat and rice flour [117, 257], and arsenic was determined in pine needles [117]. Using the same measuring system in orchard leaves were also determined bismuth, antimony and tellurium [257]. For the determination of arsenic in standard reference material orchard leaves in the flow injection system, also successfully employed was the anodic amperometric detection of As (III) with a platinum working electrode [279]. Samples were mineralised in a mixture of nitric, sulphuric and perchloric acid. Arsenic (V) is determined after reduction by hydrazine, and the excess of the hydrazine is removed on a cation-exchange column. The detection limit in this determination was estimated as 0.4 $\mu g\ l^{-1}$.

A spectrophotometric method developed for the determination of cadmium in fertilisers was also used for determinations in soybean leaves, digested with nitric acid at 155°C [280].

For the determination of cobalt in standard reference material oriental tabacco leaves, two methods were utilised with preconcentration of the analyte as a complex with 1-nitroso-2-naphthol-3,6-disulphonate (NRS) on alumina with spectrophotometric [281] and flame AAS [282] detections. The spectrophotometric determination was based on the detection of an NRS complex used for solid-phase extraction. In the procedure with atomic absorption detection at 20 min preconcentration the detection limit was estimated as 0.44 $\mu g\ l^{-1}$. For sample preparation a microwave-assisted digestion with a mixture of nitric and perchloric acid was employed.

The trace level of copper in various plant materials can be determined in flow injection systems with spectrophotometric detection and extraction preconcentration [100, 283], and in catalytic systems with fluorimetric detection [284]. An already-mentioned carbamate method with on-line solvent extraction was employed for determinations in carrot and spinach [100]. A similar method with the use of carbamate in the extraction system with an open-phase separator and the detection limit of 5 $\mu g\ l^{-1}$ was applied in determination in leaves of different plants, including citrus leaves standard reference material [283], with conventional mineralisation in a nitric and perchloric acid mixture.

For the determination of copper in the range of 0.2–300 $\mu g\ l^{-1}$, an FIA method was used, based on the catalytic effect of Cu on the 2,2′-dipyridyl ketone hydrazone/hydrogen peroxide reaction with fluorimetric detection [284]. The use of the stopped-flow measurement provided almost complete elimination of interferences. The analysed samples of rice, banana and pear were heated with a mixture of sulphuric-nitric acids with some addition of hydrogen peroxide. The results of determinations were in good correlation with those obtained by AAS.

Besides the above-reported spectrophotometric method [277], for the determination of iron in plant material the FIA system with flame AAS detection and slurry nebulisation into an air-acetylene flame was developed [285]. In such a system the detection limit was estimated as 0.6 mg l^{-1} at sampling frequency 120 h^{-1}. Preparation of samples of vegetables and fruits required grounding in a blender.

Depending on the required level of the detection limit for the determination of manganese in plant materials, flow injection systems with various spectrophotometric methods can be applied. In a method based on permanganate formation, the limit of detection was estimated as 0.3 mg l^{-1} [286], and in the above-reported catalytic method with diphenylcarbazone as 0.03 mg l^{-1} [255]. Both methods were employed for the determination of manganese in a wide variety of plant materials. The lowest contents (12-14 $\mu g\ g^{-1}$) were found in corn and bean seeds [255], and the highest (250–290 $\mu g\ g^{-1}$) in rice leaves [255] and sugar cane [286].

Due to participation in nitrogen fixation molybdenum is an essential trace element in plants. It can be determined with spectrophotometric flow injection methods based on various reagents and with different various separation methods. In the thiocyanate method a solvent extraction was used for determination in plant ash solutions down to 50 mg l^{-1} [287]. Two orders of magnitude better detectability was reported for the same materials using the catalytic method [288]. In the measuring system a cation-exchange resin column was employed to remove interfering ions. In different analysed materials (bean, soybean, sugar cane) the manganese content was form 0.3 to 2.0 $\mu g\ g^{-1}$ of the dry sample.

For the determination of zinc in plant materials flow injection methods with AAS [25, 285] and spectrophotometric [115, 289] detections were developed. In the so-called monosegmented system, due to restriction of dispersion in the system a sampling rate up to 400 h^{-1} was achieved for zinc determinations in plant digests containing from 20 to 38 mg kg^{-1} Zn [25]. An above-mentioned flow injection system with slurry nebulisation for determination of iron has also been used for determination of zinc in numerous vegetables and fruits [285]. A flow injection system based on measurement of recovered esterase activity with spectrophotometric detection was used for determination of zinc in tea Standard Reference Material, which was digested in sulphuric acid. In the used procedure it was needed to remove inhibiting metal ions from an enzymatic column by rinsing with dipicolinic acid [115]. In another method with spectrophotometric detection zinc from ashed samples was preconcentrated as chloro-complexes on an anion-exchange column, and after elution determined with the use of Zincon [289]. This method was employed for determination of zinc in plant leaves at the level of mg kg^{-1} of the dry analysed material.

Flow injection systems based on inhibition of enzyme activity can be applied for determination of selected pesticides in plant material. In the system with amperometric detection a determination of carbaryl in kiwi fruits was based on the inhibition of the acetylcholinesterase-catalysed hydrolysis of 4-aminophenylacetate [290]. Determinations of the pesticide residue were carried out in fruits treated earlier by immersion in aqueous solution of pesticide. Then the determination of trace amounts of atrazine in corn was based on the inhibition of the tyrosinase-catalysed oxidation of catechol in chloroform and spectrophotometric detection of *o*-quinone at 380 nm [291]. Blended samples of corn in a phosphate buffer were extracted with chloroform and injected into the FIA system. Satisfactory recoveries were obtained for spiked levels of atrazine in the 0.2–0.5 mg kg^{-1} range.

3.2. *Animal Tissues*

Not very many flow injection methods were developed with particular application for analysis of animal tissue material. As reported determinations, mostly of trace elements, are carried out in digests, some methods discussed earlier can be used also for this purpose. Some examples of such applications for various tissue samples are cited below.

Due to the low detection limit and common use the methods of arsenic determination in tissue samples are carried out most frequently in FIA systems with hydride generation AAS detection [117, 118]. In order to eliminate interferences in digests of oyster tissue Standard Reference Material it has been passed through a mini-column with Chelex 100 [117]. In determination of arsenic in SRMs lobster hepato and dogfish liver the detection limit was estimated as 75 ng g^{-1} of the solid sample [118].

Determination of Cd, Cu and Pb with graphite furnace AAS detection in the flow injection system was performed in four Standard Reference Materials: cod muscle, single cell protein, mussel tissue and plankton [171]. High-content heavy metal ions were removed from digests by extraction of analytes on a hydrophobic C18 bed with binding corresponding chelates with diethyldithiophosphate. Especially necessary was separation of iron and zinc in determination of lead. The results of determinations were in good agreement with certified values.

Two sensitive methods in flow injection systems were developed for determination of mercury in tissue material, testing their reliability with reference materials. In determinations with ICP-AES detection and vapour generation an on-line extraction preconcentration with phase separation on a membrane separator was reported [292]. After mixing the organic phase on-line with $SnCl_2$ in N,N-dimethylformamide, mercury vapour is generated directly from the organic phase. In such a system the detection limit was estimated as 4 mg l^{-1} and the method was employed for determinations in digests of pig kidney and dogfish muscle. Cu (II) and Ag (I) may interfere at a larger excess than 100. A system with ICP-MS detection was applied in determination of organomercury in biological reference materials [293]. The analyte was separated from samples as chloride extracted to toluene and back-extracted into aqueous solution of cysteine acetate. The detection limit was obtained at the same level, as with ICP-AES detection. It was found that 93% of mercury in dogfish muscle tissue and 39% in lobster hepatopancreas exists as methyl mercury. The results of determinations were in good agreement with GC results.

Determinations of trace contents of selenium were carried out with the hydride generation AAS detection mentioned above for arsenic determination [117, 118]. In this case it was needed to use 1,10-phenanthroline for masking of copper and nickel [118]. Determinations were performed in the same reference materials. For determinations of selenium in pig kidney reference material satisfactory results were also reported for the use of flow injection anodic stripping voltammetry at a gold electrode [294]. The use of a Chelex 100 column is indispensable for removing cationic interferences present in samples.

Tin was determined in animal tissues by flow injection hydride generation AAS in configuration with minimal manipulation of the samples with a great advantage of using a large volume of diluted digests [295]. A procedure with a detection limit of 2.5 $\mu g\ l^{-1}$ was applied to SRM bovine liver and to spiked different rat tissue digests.

For the determination of zinc in Standard Reference Materials fish flesh, fish tissue and tuna homogenate, a spectrophotometric flow injection system with preconcentration on polyurethane was employed [256].

4. Mineral Materials

4.1. *Soil*

Determination of soil components in its extracts is a wide field of practical application of FIA due to a common need of these determinations and the necessity of their performance with inexpensive procedures, not requiring highly trained personnel. The extraction procedures for soil samples are usually simpler than for food or biological matrices, and hence the use of FIA for this purpose is the obvious trend in agriculture analysis.

The application of the FIA system practically without dispersion (large sample volume, short volume of the system between injection valve and detector) allows convenient pH measurements in soil extracts. This was demonstrated for the FIA system with potentiometric detection with a PVC-based pH-sensitive membrane electrode where a sampling rate was obtained of about 110 h^{-1} [296]. A further miniaturisation of this system into an integrated micro-conduit resulted in obtaining a sampling rate of 200 h^{-1} [297].

Nitrogen occurs in soil in various chemical forms, both in inorganic and organic compounds. For the determination of nitrate in soil extracts flow injection systems with potentiometric detection with ion-selective electrode were successfully developed [178, 298, 299]. If the excess of chloride compared to nitrate is greater than 500 it should be masked or determined in order to make appropriate correction in determination of nitrate [299]. The potentiometric detection of nitrate together with the use of ion-selective electrodes for determination of potassium, calcium and chloride was utilised in the FIA system for the determination of these four components in soil extracts [300]. The results of potassium and calcium determination were validated with atomic absorption spectrometry, and those of chloride and nitrate with ion chromatography. For simultaneous determination of nitrate and nitrite in soil extracts the several times mentioned spectrophotometric flow injection method was also employed, based on the reaction of nitrite with 3-nitroaniline and then with *N*-(1-naphthyl)-ethylenediamine dihydrochloride [55]. The nitrate content was determined at the concentration level of 9–850 mg l^{-1}, and nitrite at 0.1–5.8 mg l^{-1}, depending on the intensity of the use of fertilisers.

Flow injection methods with spectrophotometric detection for determination of dissolved organic phosphorus in waters were also successfully applied for determination of total dissolved phosphorus in soil leachate samples [301]. The manifold contained a PTFE reaction coil for UV photo-oxidation of organic phosphorus compounds and a cation-exchange column for removal of interfering Fe (II). The detection was based on formation of molybdenum blue. The detection limit was estimated as 7 $\mu g\ l^{-1}$, and the content in analysed leachates was from 20 to 50 $\mu g\ l^{-1}$.

In determination of potassium in soil extracts with potentiometric detection a significant role is played by used extracting solution. Results with the best agreement compared to emission spectrometry were obtained after extraction of soil with 1 M sodium acetate [298]. For determination of calcium in soil extracts obtained with 1 M potassium chloride a flow injection system with spectrophotometric detection was used [22]. As colour-forming reagent glyoxal bis(2-hydroxyanil) was used and the obtained results correlated well with ICP-AES determinations [22].

For simultaneous determination of calcium, potassium and sodium in soil extracts obtained in 1 M ammonium acetate, a flow injection system with flame emission spectrometry was also used with a fast scanning monochromator [302]. The obtained results correlated well with the conventional flame photometric method.

Bromide can be determined in soil extracts in the FIA method with potentiometric detection, when its content exceeds 1 mg l^{-1} [302].

For the determination of boron in soil samples the direct introduction of a solid sample into the FIA system was proposed which was based on ultrasonic treatment in a sample cell of the manifold (Fig. 12) [303]. The detection was based on the reaction with azomethine-H and the sampling rate obtained was 25 h^{-1}. Leaching was carried for 0.5 min by circulating 0.1 M hydrochloric acid through the cell under ultrasonication at 80°C.

For speciation of Al in soil extracts flow injection systems with spectrophotometric detection employing various chromophores were used [304]. All chromophores differentiated between Al (III) bound in strong and weak complexes, and were suitable for ordering the soil solutions for relative toxicity. Kinetically labile and equilibrium-reactive Al (III) were determined with a detection limit of 60 nM.

Various methods were developed for flow injection determination of trace elements in soil. A flow injection system with on-line Donnan dialysis and

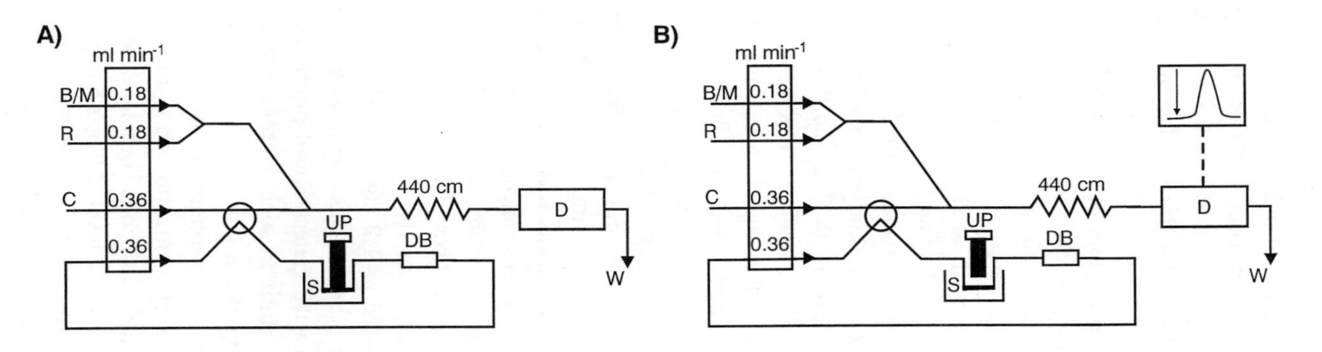

Fig. 12. Manifold of the flow injection system for the determination of boron in soils with spectrophotometric detection and direct introduction of solid samples in (A) the leaching position and (B) the injection position [303]. C — carrier solution; B/M — buffer/masking solution; R — reagent solution; S — sample cell; UP — ultrasonic probe; DB — debubbler; W — waste. (*Reprinted by permission of copyright owner.*)

voltammetric detection using differential pulse anodic stripping was developed for the determination of free cadmium concentration in soil extracts [305]. In a soil solution with a high concentration of dissolved organic carbon the inorganic fraction constituted 53% of the total Cd concentration. Chromium (VI) in concentration from 1 to 8.5 mg l^{-1} was determined in soil extracts in the FIA system with spectrophotometric detection based on reaction with diphenylcarbazide [306]. A sequential flow injection procedure was developed for simultaneous determination of cobalt and nickel in water and soil samples using spectrophotometric detection with PAR [307]. A different kinetics of dissociation of citrate complexes with analytes is used for simultaneous determination with detection limits 0.14 and 0.2 mg l^{-1} for nickel and cobalt, respectively. The masking of some cationic interferences is necessary. The above-mentioned method for the determination of manganese as permanganate formed by oxidation with periodate at 95°C in the flow injection system was also employed for analysis of soils extracts with a manganese level of 0.14–1.3 mg g^{-1} Mn in dry soil [286]. A flow injection enzymatic screening method for mercury in soil extracts was developed with fluorescence detection [308]. The method is based on inhibition of the activity of urease and fluorimetric detection of ammonia was carried out using reaction with *o*-phthalaldehyde. The detection limit 2 μg l^{-1} allowed one to carry out determination in soil samples in the range of 23–67 ng l^{-1}. For obtaining soil extracts a procedure with a complete breakdown of organomercurials was developed. For determination of molybdenum in soil extracts at the concentration level 0.3 mg l^{-1} in the analysed solution, a catalytic flow injection method was developed [309]. The method was based on the catalytic effect of Mo (VI) on the iodide oxidation by hydrogen peroxide with biamperometric detection of the iodine produced. The detection limit was estimated as 1.2 mg l^{-1} in the soil extract. Lead in soil extracts in the absence of other metals forming volatile hydrides can be determined in the FIA system with generation of plumbane and conductivity detection [311]. The detection limit in such a procedure was only 1 mg l^{-1}. An almost three orders of magnitude lower detection limit (5 mg l^{-1}) can be obtained in the system with spectrophotometric detection and on-line solvent extraction together with preconcentration on solid sorbent [310]. Such a method based on the formation of dithizone complexes was employed for determination of lead in soil extracts at a level of 0.5–2.5 mg l^{-1} in soil leachates. A method with hydride generation, atomic absorption spectrometric detection and preconcentration on a strongly basic anion-exchanger in the FIA system allows the obtaining of a particularly low detection limit 0.002 μg l^{-1} in determination of selenium [119]. Soil water extracts were prepared by refluxing air-dried soil with deionised water at boiling temperature for 2 h and collecting the filtrate. The interference of copper was overcome by adding thiourea to the hydrochloric acid eluent. Determination in soil extracts was carried out in the range of 0.04–0.56 mg l^{-1}.

Due to a need for sensitive methods for detection of explosives in soil and ground water, a flow injection immunosensing method was developed for the determination of 2,4,6-trinitrotoluene (TNT) [312]. A method is based on the use of anti-TNT antibody immobilised onto the inner walls of capillary and fluorimetric detection of the fluorophore-labelled antigen. Soil samples were extracted with acetone and after evaporation of solvent solid residues were dissolved in buffer solution used as carrier solution in the flow injection system. The detection limit was estimated as 1 ng l^{-1} TNT using a 100 μl injected sample volume.

4.2. *Ores, Minerals and Ceramic Materials*

This group of materials is predominated by samples of geological origin, such as ores and minerals. In flow injection measuring systems used for their analysis mostly spectrophotometric and atomic absorption spectrometry detection were applied. Some of them employed also for different materials were already reported.

For the determination of bismuth in silicate rocks an extraction spectrophotometric flow injection method was developed [313]. The procedure was based on solvent extraction of bismuth with lead tetramethylenedithiocarbamate and measurement of absorption of a Bi-carbamate complex. EDTA and cyanide were used as masking agents and a detection limit of 0.1 mg l^{-1} was obtained. The analysed samples were dissolved in nitric acid.

An already-mentioned spectrophotometric method for determination of cobalt based on complex formation with nitroso-R-salt preconcentrated in the FIA system on an alumina microcolumn was applied in determination of this element in standard reference pyrite ore [281]. For the preparation of samples, microwave-assisted digestion with a mixture of nitric and perchloric acid was used.

For determination of copper and gold in ores an FIA system with flame AAS detection was developed [314]. Determinations were based on solvent extraction of copper as its thiocyanate complex and gold as trichlorocomplex into methyl isobutyl ketone and then detection of extracted analytes by AAS. The detection limits were 1 $\mu g\ l^{-1}$ for copper and 1.8 $\mu g\ l^{-1}$ for gold. The reliability of the method was validated by determinations in standard ore samples with copper content from 5 to 177 $\mu g\ g^{-1}$, and gold content from 0.2 to 4.2 $\mu g\ g^{-1}$. A similar level of the detection limit for gold was reported in the flow injection method with AAS detection, where preconcentration of gold chlorocomplexes was carried out on micro-column with Amberlite XAD-8 resin [315]. Analysed samples of ores after heating at 570°C were dissolved in *aqua regia* and the level of determined gold was in the range of 0.8–8.8 mg kg^{-1}.

Iron was determined by flow injection spectrophotometry in silicate rocks [316, 317] and in geological core samples [318]. In the analysis of silicate rocks the samples were melted with a mixture of lithium carbonate and boric acid and the obtained product was dissolved in hydrochloric acid. Determination of iron was based on measurement of the absorbance of the chloro-complexes of iron at 335 nm [316]. In the sample aluminium was also determined with xylenol orange after reduction of Fe (III) to Fe (II) and elimination of other interferences with EDTA. Determination in natural samples was performed at a concentration level of Fe_2O_3 from 2 to 13% and Al_2O_3 from 14 to 17%. In another system also utilised for the determination in silicate rocks a simultaneous determination of iron (III) and titanium (IV) was carried out [317]. Determination was based on the formation of complexes with Tiron. In the system requiring the application of two spectrophotometers or a dual-channel spectrophotometer Fe (III) and Fe (II) were simultaneously determined in hydrochloric acid extracts of geological core samples [318]. In detection reactions with thiocyanate and 1,10-phenanthroline were utilised. Determinations in geological samples were carried out at the level of 0.17–1.5 g l^{-1} Fe.

In silicate rocks, prepared by melting as was mentioned above, manganese was also determined in the flow injection system with spectrophotometric detection with formaldoxime [319]. In the measuring system a mini-column with a cellulose phosphate were employed for removal of interfering polyvalent cations. Nickel was masked with 2-aminoethanethiol. In analysed standard rock samples manganese was at a concentration level of 0.018–0.22% as MnO_2.

Utilising their catalytic effects on the reaction of iodide with hydrogen peroxide, molybdenum and tungsten were simultaneously determined in the FIA system with spectrophotometric detection in mineral samples [320]. In the system with injection of two sample plugs into two carrier solutions a citric acid is used to differentiate a reaction rate in the presence of two analytes.

Determinations were carried out in standard reference minerals which melted with sodium peroxide and leached with water. Determined contents were in the range of 3–90 $\mu g\ g^{-1}$ Mo and 5–38 $\mu g\ g^{-1}$ W.

An above-mentioned flow injection AAS method with preconcentration as pyrrolidinedithiocarbamate on an activated carbon mini-column [273] was applied in determination of lead in Standard Reference Materials of lead vanadium and zinc concentrate. The samples were mineralised in a mixture of nitric and hydrochloric acids. The lead content in both analysed materials was 42.5 and 0.91%, respectively.

A very similar procedure was used in determination of silver in the FIA system with AAS detection [321]. The carbamate complex of silver was accumulated, not in the reactor with appropriate sorbent, but collected in a PTFE knotted reactor. The precipitate was dissolved in isobutyl methyl ketone and introduced directly into the nebuliser-burner system of the spectrometer. The method was verified using two standard geological materials with silver content 0.34 and 4.3 $\mu g\ g^{-1}$ Ag which were mineralised in two steps by a mixture of hydrofluoric, hydrochloric and perchloric acids and then by *aqua regia*.

A flow injection extraction system with spectrophotometric detection was reported for the determination of uranium at trace levels in ore leachates, which can be used for process control [322]. The leachate is extracted with solution of tributyl phosphate in heptane (Fig. 13). After separation of the organic phase the extract is mixed with a reagent stream containing 2-(5-bromo-2-pyridazolo)-5-diethylaminophenol and Zaphiramine, and the resulting ternary complex of U (VI) is measured spectrophotometrically. In determination in leachates with sampling rate 50 h^{-1} the detection limit was estimated as 0.1 mg l^{-1}.

The chemiluminescence determination of vanadium (V) was developed in the FIA system with luminol and hexacyanoferrate (II) immobilised on an anion-exchange resin column [323]. During the system operation reagents were eluted from the resin with phosphoric acid and mixed with the stream containing V (V). Interfering metal ions were removed on-line by a cation-exchange column placed upstream. With sampling rate 60 h^{-1} the method was employed for analysis of geochemical reference samples mineralised with a mixture of hydrofluoric, nitric and perchloric acids.

For effective simultaneous determination of 19 components of geological materials with ICP-AES detection it was hyphenated with the FIA system where calibration was carried out by means of a generalised standard addition

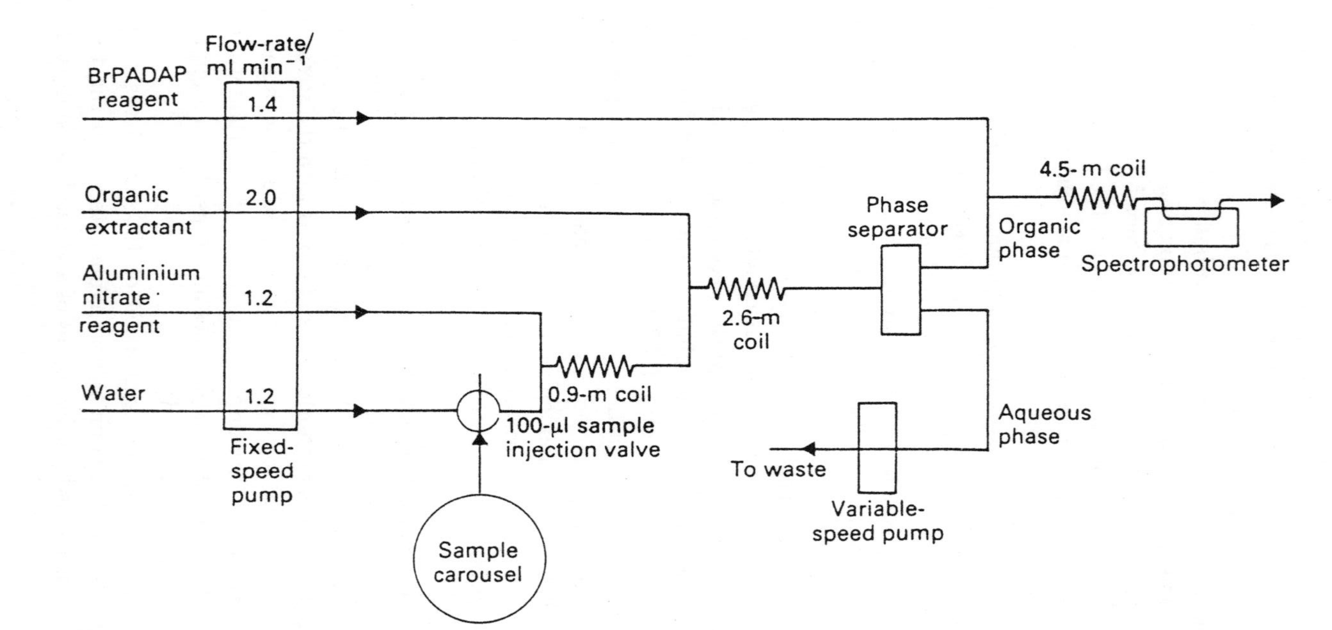

Fig. 13. Schematic diagram of the flow injection system for extraction spectrophotometric determination of trace levels of uranium in ore leachates based on reaction with 2-(5-bromo-2-pyridylazo)-5-di-ethyl-aminophenol [322].

method [324]. With the used configuration of merging zones and zone sampling only one standard solution is required per analyte. The efficiency of the developed method of calibration was verified by analysing 7 rock reference materials utilising this procedure with 120 additions per hour. Analysed materials were dissolved in a PTFE bomb with *aqua regia* and addition of hydrofluoric acid.

Besides rocks and ores flow injection determinations were employed for analysis of several other mineral materials. For the determination of sulphur in coal a flow injection system with fluorimetric detection based on the formation of a ternary complex between sulphate, zirconium and Calcein was elaborated [325]. The bomb combustion procedure in a fused-silica ignition capsule was applied to mineralisation of coal samples. Determination with sampling frequency 50 h^{-1} was examined using NIST coal reference samples containing 0.5 to 4.7% (w/w) sulphur.

Several components of ceramic materials were also determined in appropriate FIA systems. A system with AAS detection and a variable-volume injector was designed for determination of Fe, Ca, Mg, Na and K [326]. The last two components were determined with flame emission detection. Analysed samples were fused with lithium metaborate at 1000°C and dissolved in nitric acid. Fe, Mg, Na and K were determined in samples of porcelain, stoneware, kaolin, varnish and clay, and calcium in kaolin. Then for the determination of boron in various ceramic materials the FIA system with spectrophotometric detection based on pH changes produced by the reaction between boric acid and mannitol in the presence of an acid-base indicator was reported [327]. The developed system allowed one to achieve the detection limit 0.02 g l^{-1} B_2O_3. The analysed materials were dissolved in hydrochloric acid or fused with Na_2CO_3/ZnO at 900°C and leached with water. The detection limit is similar to that of the ICP-AES method and better than by AAS or potentiometry.

Determination of cobalt in glasses was reported with the use of the flow injection method with AAS detection and on-line preconcentration of the analyte on a Chelex 100 mini-column [328]. The use of chelating sorbent allows one simultaneously to separate the analyte from the main glass matrix elements (Na, Ca and Mg). Samples were dissolved in hydrofluoric acid with addition of nitric and perchloric acids. The determined cobalt was at the level of 20–30 $\mu g\ g^{-1}$.

Chromium (VI) was determined in workplace coal fly ash samples collected from a power plant using ultrasonic extraction and spectrophotometric detection with the diphenylcarbazide method [329]. Samples were treated with an ammonium buffer in an ultrasonic bath and then Cr (VI) was separated from Cr (III) on an anion-exchange column. In Standard Reference Material coal fly ash also As, Sb and Se were determined using hydride generation AAS [257].

4.3. *Fertilisers*

Flow injection methods can also be successfully used for quality control of fertilisers, both for determination of main plant nutrients (nitrogen, phosphate and potassium), and for determination of micro-elements.

A successful use of FIA with spectrophotometric and potentiometric detections for determination of main components of fertilisers was demonstrated in one of the pioneering works on FIA [330]. The determinations of potassium and nitrate were carried out using ion-selective electrodes, ammonium nitrogen was determined by the spectrophotometric Indophenol method, and phosphate was determined by the molybdenum spectrophotometric method with ascorbic acid. In the proposed measuring systems determinations were carried out with a sampling rate from 85 to 125 h^{-1} for various analytes and sufficient selectivity.

For the determination of total nitrogen content in fertilisers and animal feeds an above-mentioned flow injection method was based on determination of ammonia in Kjeldahl digests in the system with gas diffusion and conductivity detection [274]. The method was applied for determination of nitrogen content in fertilisers in the range of 2.1–5.5% N.

The content of cadmium in fertiliser digests in a range of 20–90 $\mu g\ l^{-1}$ was determined with the above-presented spectrophotometric method employing iodide and malachite green [280]. The detection limit was estimated as 2 $\mu g\ l^{-1}$ cadmium in injected samples. With the use of reaction with Nitroso-R-salt cobalt was determined in livestock mineral supplements [331].

4.4. *Alloys*

Various flow injection methods have also found applications in analysis of alloys. Due to special properties of analysed samples, rarely in this case can one find application of methods reported earlier for other materials.

Aluminium together with zinc in alloys were determined in a spectrophotometric system with simultaneous measurement at three wavelengths [332]. For this purpose a photometer with an integrated multidiode light source and computerised data acquisition and processing was developed. Determination was based on formation of complexes with xylenol orange. The analysed alloys were dissolved in hydrochloric and nitric acids, and the determined Al content was verified gravimetrically, and the Zn content by complexometric titrations. The determinations required chemical masking of iron (III), copper and cobalt.

A flow injection system producing a gradient of pH with carrier stream was used for spectrophotometric determination of bismuth and lead in a commercial anti-friction alloy [333]. Determination was based on formation of complexes with Arsenazo III at different pH. It was necessary to remove tin as stannic acid by precipitation and Fe(III) by reduction with ascorbic acid. The same two metals and also antimony and silver were determined in steels in FIA systems with flame AAS detection using the pulse-nebulisation technique, suitable for solutions with a high salt content [334]. Samples were dissolved in a mixture of hydrochloric and nitric acids. The detection limits in the solid steel were as follows: Pb 2.5, Bi 2.5, Sb 10, and Ag 0.3 $\mu g\ g^{-1}$.

Chromium in steel can be determined with the known addition procedure in the reversed FIA system with flame atomic absorption spectrometric detection [335]. Samples were dissolved similarly to the previously cited paper. In standard samples of steel the content of the determined chromium was in the range of 0.2–17%. In similar samples using sequential injection analysis chromium was determined together with iron [336]. Determination was based on the reaction of chromium with diphenylcarbazide and iron with salicylic acid. The procedure was employed in a single line setup with a set of valves for introduction of small aliquots of the sample and reagent solutions directly into the analytical path. The results of determinations were in good agreement with ICP-AES determinations. In the same work also another system for determination of nickel in steel with spectrophotometric detection based on reaction with dimethylglyoxime was developed.

Systems with spectrophotometric detection were developed also for flow injection determination of several other elements in alloys. Cobalt and manganese were determined in steel with the use of solvent extraction [337]. Detection was based on formation of cyanide complexes of Co (II) and on oxidation of manganese to permanganate, respectively. In both cases corresponding ion associates were extracted into chloroform. Then determinations of copper and nickel in copper-base alloys, including brasses, deoxidized copper, beryllium copper and German silver, were based on the measurement of the absorbance of aquo-complexes, after simple dissolution of the sample in a nitric acid-phosphoric acid mixture [338]. In the same alloys determination of zinc was based on formation of a complex with xylenol orange.

For the determination of lead in standard alloy samples containing from 0.7 to 1.4% of this element an above-reported spectrophotometric system was employed with formation of dithizone chelate [310].

Determination of As, Sb, Bi, Se and Te in steel can be performed in the system with AAS detection and hydride generation [257]. A sample of low-alloy steel was dissolved in nitric acid, then the acid was removed with sulphuric acid by evaporation to fumes of SO_3 and HCl was added.

A unique simplification of FIA determinations in alloys was achieved by incorporation into the measuring system on-line electrolytic dissolution of alloys. Such systems were developed for determination of molybdenum in steel with the use of the spectrophotometric thiocyanate method [338] and for multi-elemental analysis of stainless steel by ICP-AES detection [339]. The metallic sample in these systems acts as the anode in a flow-through electrolytic chamber and dissolved species in anodic oxidation process are directed towards the detector.

5. Clinical Analysis

Clinical analysis is, after environmental applications, the second-largest field of developed applications of FIA. Beside development of flow injection methodology itself, this can be connected with other trends observed in contemporary analytical chemistry such as great progress in biochemical methods of analysis (enzymatic and immunochemical biosensors) or common application of potentiometric ion-selective electrodes in determination of electrolytes in physiological fluids. In both these cases application of FIA techniques of measurement significantly simplifies measurements. The largest number of methods was developed for determinations in serum and plasma, although many of them were also extended for whole blood and urine. Flow injection methods are also being developed for control of dialysates, or content of various components in tissues or hair. Also in this case it has to be admitted that the present review is based on some arbitrary selection of references from the much larger amount published so far in various sources.

5.1. *Serum and Plasma*

Determinations in serum and plasma of human blood are the most frequent clinical applications of FIA. Due to the similar matrix composition and similar analytical problems to be solved this paragraph will also contain sporadically some examples of determinations which were also examined using the blood serum of animals. The developed procedures concern several groups of analytes, such as substrates of enzymatic reactions, proteins, drugs, electrolytes or trace metals.

Determination of the glucose level in blood for the identification and control of diabetes is the most often performed determination in clinical chemistry. An increasing number of procedures are being developed for determinations in whole blood and some of them employing flow injection analysis will be presented below, but most of them were developed for determination in blood serum and plasma. In all presented flow injection methods glucose is determined enzymatically, mostly with the use of glucose oxidase, but not exclusively. Enzymes are used in dissolved form in solution or immobilised on various supports. Various detection methods are used.

In flow injection determination of glucose with the use of glucose oxidase, detection is based most frequently on amperometric, spectrophotometric and chemiluminescence determinations of the hydrogen peroxide produced; less often on monitoring of consumed oxygen in enzymatic reaction or on changes in the acidity of the reaction solution resulting from the formation of gluconic acid. Amperometric detection was employed in the system with dissolved [340] or immobilised [341–344] enzyme. In the first case a closed flow system was reported where samples are injected into a continuously circulated enzyme solution in buffer and detection is based on amperometric monitoring of oxygen consumption (Fig. 14). Determinations in human blood serum can be carried out with sampling rate 700 h^{-1}. With the same portion of the enzyme solution more than 10.000 determinations can be carried out.

In amperometric systems with immobilised enzyme it was placed in a flow-through reactor [341, 342] or on the membrane at the working electrode surface on which hydrogen peroxide was monitored [343, 344]. In the system with a controlled pore glass (CPG) reactor determinations in blood serum required the use of an additional dialyser or incorporation a of column with copper(II) diethyldithiocarbamate [341]. In the systems with membrane biosensors discrimination of electroactive interferences was gained by the use of additional screening membranes [343, 344], as well as a layer of cation-exchanger Nafion [34]. In both cases satisfactory correlations were reported for results of glucose determinations in human blood serum with those obtained with the use of routine clinical analysers in the range of 3 to above 10 mM.

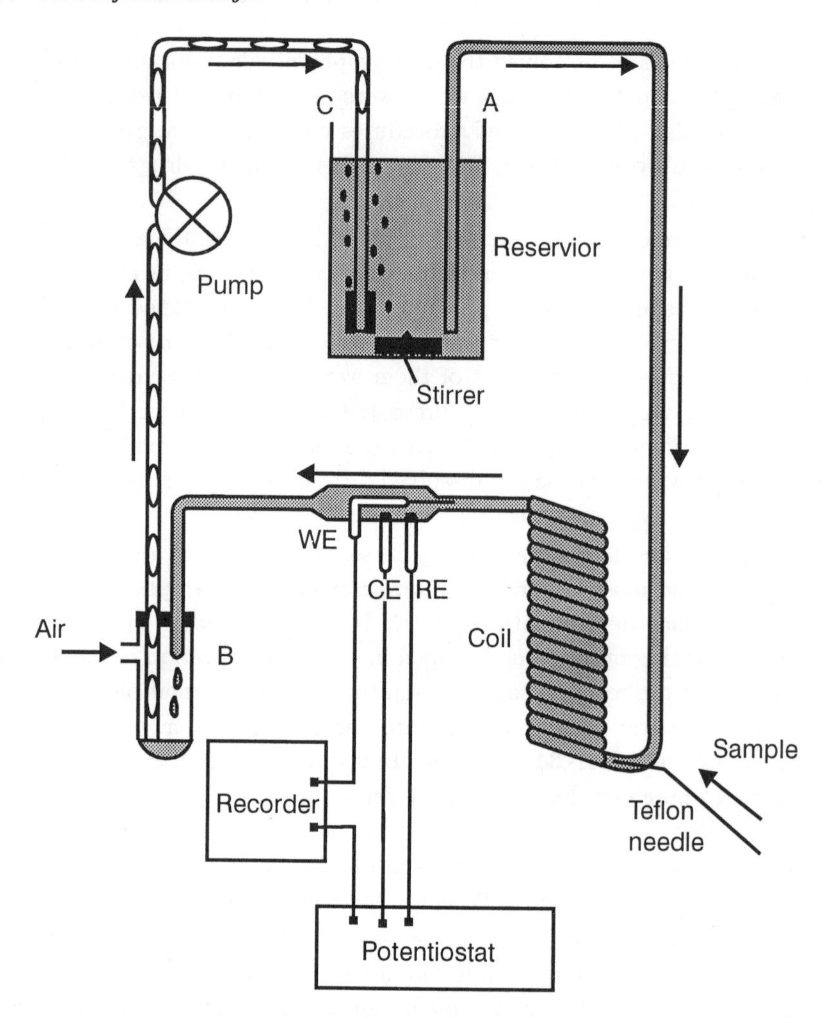

Fig. 14. Schematic diagram of a closed-loop flow injection system for the amperometric determination of glucose in serum [340]. WE — working electrode (Pt wire); CE — counter electrode; RE — reference SCE electrode. Arrows indicate the direction of flow. (*Reprinted by permission of copyright owner.*)

In FIA systems for enzymatic glucose determination with spectrophotometric detection various methods of detection of hydrogen peroxide can be employed. In a single line system with flow reversal samples are injected into GOD solution containing also peroxidase, phenol and 4-aminoantipyrine for detection of hydrogen peroxide [345]. Two-point measurement after flow reversal following the sample injection allows the measurement of the blank. In flow injection systems with glucose oxidase immobilised in a CPG reactor for detection of H_2O_2 reactions with 4-aminophenazone and N,N-dimethylaniline [346], and with Bindschedler's Green in the presence of Fe (II) as catalyst [347], were employed. In the first case the employed reaction requires the use of an additional reactor with peroxidase for performing colour reaction and a dialyser for removal of proteins interfering with detection. In the second case, in order to eliminate ascorbic acid interferences samples were 1000-fold diluted [347]. In another system a reactor with co-immobilised glucose oxidase and peroxidase was used in order to enhance the sensitivity of detection, whereas spectrophotometric detection was based on the reaction of hydrogen peroxidase with 3-methyl-2-benzothiazolinone hydrazone and 3-dimethylaminobenzoic acid [348]. A drawback of these reagents is their sensitivity to light. Yet in another system two reactors, one with glucose oxidase and the other with peroxidase, were employed. The second one, which served also as a spectrophotometric flow-through cuvette, contained a bed for sorption of the product of the reaction of hydrogen peroxide with 4-aminophenazone and 3-methyl-N-ethyl-N′(p-hydroxyethyl) aniline [349]. In both of the latter systems no interferences were found in determinations of glucose in human blood serum.

Chemiluminescence detection in flow injection systems for determination of glucose in serum is based on the reaction of hydrogen peroxide with luminol in the presence of a suitable catalyst. It was employed in the systems with dissolved glucose oxidase [350, 351], as well as with an enzyme reactor [352]. As a catalyst for reaction with luminol hexacyanoferrate (III) [350, 352] and a complex of copper with 1,10-phenanthroline was reported [351]. In one system only a dialyser unit was incorporated [352]. In the system with dissolved glucose oxidase satisfactory results were obtained for injected 6-fold diluted serum [351]. In the second system glucose oxidase was delivered to a flow injection system through the membrane under pressure where serum samples were assayed following off-line deprotonation, addition of luminol with a catalyst and overall 1000-fold dilution [351]. It was shown also that chemiluminescence detection in glucose determinations in serum yields results with better correlation to those obtained by electrochemical determination than spectrophotometric detection [352].

Glucose in blood serum and plasma in flow injection systems can be determined also using other enzymes than glucose oxidase. The glucose dehydrogenase was employed in a closed-loop FIA system for determination of glucose with spectrophotometric detection [353]. The NADH produced in enzymatic reaction is detected spectrophotometrically, and then glutamate dehydrogenase added to the circulating solution slowly oxidises NADH back to NAD+. This method was applied to human blood serum with sampling rate 120 h^{-1} and 300 determinations can be performed in the same 40 ml circulating solution.

Another enzyme pyranose oxidase catalyses a reaction of oxidation of D-glucose to D-glucosene with formation of hydrogen peroxide. It was used in immobilised form in a flow-through reactor for glucose determination in the FIA system with chemiluminescence detection [354]. Detection was based on the reaction of H_2O_2 with luminol in the presence of peroxidase immobilised on polymer beads packed into a transparent PTFE tube, which was spiralled to a coil functioning as a flow cell mounted in front of a photomultiplier tube. Although equimolar concentrations of Co (II), Cu (II) and Fe (II) enhanced the signal for glucose, their concentration in plasma is much smaller. Certain interferences were found in the presence of ascorbic acid, but results of determination in blood plasma gave correlation coefficient 0.998 compared to the routine method with the hexokinase.

Determination of glucose was incorporated besides urate and cholesterol into a three-component flow injection system for simultaneous assay in blood serum [355]. In all three cases immobilised enzymes glucose oxidase, cholesterol oxidase and uricase were used with amperometric detection. For determination of glucose and urate the Nafion film-coated platinum electrode was used, whereas for the cholesterol assay, the Nafion/cellulose acetate bilayer film-coated Pt electrode was applied in order to prevent dissolution of the Nafion layer by the detergent used in the carrier stream of the cholesterol channel. The system was tested using several control human serum.

Several flow injection systems with immobilised lactate oxidase and amperometric detection of hydrogen peroxide were designed for lactate determination in blood serum and plasma. The effect of the presence of electroactive interferences was not observed when a commercially available high-rejection lactate membrane was applied [347]. They can also be reduced by a Nafion layer deposited on a platinum electrode and appropriate dilution of real serum samples [356]. In other FIA systems the enzyme was immobilised in an electrodeposited film of poly-*o*-phenylenediamine [357–359]. A matrix of this conducting polymer itself was sufficient to minimise the effect of interferences in diluted serum for determination of lactate [357]. A better elimination of these interferences can be achieved by additional modification of the platinum working electrode with a bilayer of electrodeposited polypyrrole/polyphenol [358]. This allows one to determine lactate in undiluted serum. For the same purpose a flow injection system was constructed with a micro-dialysis fibre-based sampler [359]. This enables on-line dilution of the sample and removal of high molecular weight interferents. All these mentioned set-ups were examined for determination of lactate in human blood serum samples.

Another substrate especially often determined in blood serum and plasma is urea. Because urea is the final product of human protein and amino acid metabolism, the measurement of its blood level in an important factor of renal function. For analytical determination of urea the enzyme urease is commonly used that degrades substrate to CO_2 and ammonia. Due to acid-base properties of these substrates, especially potentiometric detection is widely applied in urea determination. Numerous such assays were developed in flow injection systems. In the system with pH detection using glass electrode urease was used dissolved in buffer solution [360]. The main interference in such a determination is the presence of bicarbonate in serum samples; thus, for accurate assay, the correction for blank values of individual serum samples must be made. The developed method was tested with control serum samples. In application of potentiometric detection with ammonium ion-selective electrode with ionophore nonactin-containing membrane a source of error can be endogenous ammonia and potassium ions. In the flow injection system with such a detection an on-line dialyser with an anion-exchange membrane was built in (Fig. 15) [361]. As the sample passes through the dialysis unit, neutral urea is transported through the membrane while the permeation of endogeneous ammonium ions or other cations in the sample is rejected. Urease was in this case immobilised on the surface of the nonactin membrane. Yet in another way this interference was eliminated in the system with enzymatic reactor with urease [362]. The nonactin-containing membrane was employed in the configuration of a gas-sensing membrane electrode with an outer hydrophobic gas-permeable membrane. In this case determinations were carried out injecting 200 μl of 10-fold diluted blood serum samples.

A flow injection system with a urease reactor was reported for measurements with spectrophotometric detection for determination of urea in blood serum and urine [363]. Enzymatically produced ammonia is converted to an indophenolate dye by oxidative coupling with hypochlorite and sodium

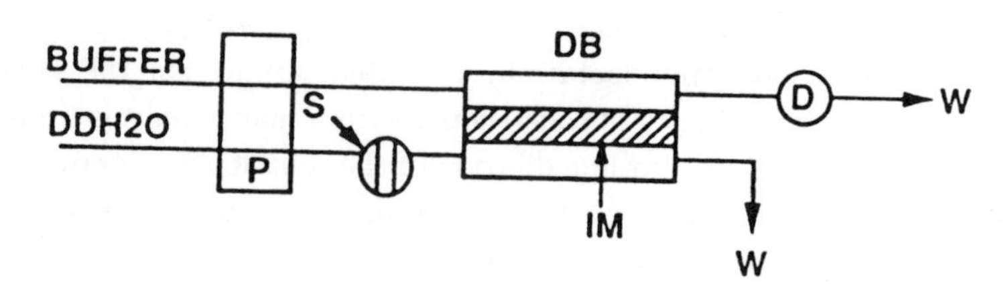

B

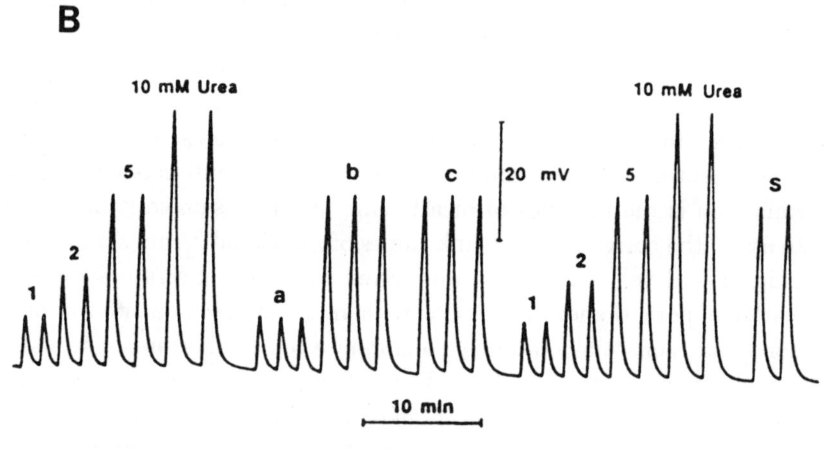

Fig. 15. Schematic diagram of a flow injection manifold for enzymatic determination of urea blood serum with potentiometric detection (A) and response recorded (B) for urea standard solutions, control serum (s), 1 mM urea +2 mM potassium (a), 5 mM urea +5 mM potassium (b), and 5 mM urea +1 mM ammonium (c) [361]. In A: P — peristaltic pump; S — sample injection; DB — dialysis blosk; D — enzyme-electrode detector; W — waste; IM — ionomer membrane. (*Reprinted by permission of copyright owner.*)

salicylate. Determinations required 1:50 dilution for serum samples and 1:1000 dilution for urine. Urea assays in serum can also be carried out in the FIA setup with enzymatic reactor and fluorimetric detection [364]. In this instance an additional gas-diffusion unit for separation of ammonia for further reaction with *o*-phthalaldehyde solution was used. This unit served to remove amino acids, amines and proteins from the sample which would otherwise have reacted with the *o*-phthaldehyde to produce fluorescence emission. The samples of serum were injected without dilution with sample frequency 25 h^{-1}. A very similar configuration of the flow injection system was reported for conductivity detection [365]. In this system, however, instead of immobilised purified urease a reactor filled with small pieces of bean that is a natural source of urease was incorporated. Ammonia produced in enzymatic reaction diffused in a gas-diffusion unit to the stream of deionised water in which conductivity was measured. Analysed samples were 1000-fold diluted prior to the injection.

The content of urea in blood serum is at the milimolar level, whereas that of another physiologically important substrate, creatinine, is in the range of 40–110 μM. Its content is an important indicator of the renal, muscular and thyroid functions of the organism. A low level of creatinine causes greater difficulties in determinations in terms of elimination of the effect of interferences present in the serum matrix. Determination of creatinine in serum can be carried out in FIA systems with various detections. With flow injection spectrophotometry creatinine was determined in undiluted serum samples [366]. The determined substrate was degraded to ammonia in a reactor with immobilised creatinine iminohydrolase, and then in the gas-diffusion unit ammonia was separated into a stream of solution of a pH-sensitive indicator. The resulting color change was monitored by reflectance measurement via optical fibres. The content of endogeneous ammonia was determined separately in the system prior to the enzymatic reaction. Determinations were carried out with sampling rate 60 h^{-1} at sample volume 30 μl. The same reaction of degradation of creatinine to ammonia was utilised in flow injection system where for the detection of ammonia an ammonia-sensitive oxide semiconductor structure was employed with a catalytic layer of iridium as a part of the gate of the structure [367]. A reactor with creatinine iminohydrolase was preceded in the flow system with a flow-through reactor with glutamate dehydrogenase to remove endogeneous ammonia. Determinations with sampling frequency 15 h^{-1} were carried out injecting 85 μl samples of plasma diluted 25-fold. In the FIA system with amperometric detection a three-step catalytic sequence was employed with detection of the produced hydrogen peroxide [368]. For the detection a miniaturised biosensor with a multilayer film structure containing a layer of poly (1,3-diaminobenzene), a multienzyme layer containing creatininase, creatinase and sarcosine oxidase and an outer layer of composite polymer containing Nafion was used. The detection limit 10–20 μM in the FIA system was sufficient to obtain satisfactory results for creatinine assays in human serum samples. The measurements were performed in a differential mode, i.e. the recorded signal was the difference between the creatinine sensor current and the creatine sensor current. The two sensors were simultaneously present in the detector.

For the determination of cholesterol in serum in the flow injection system usually a two-step enzymatic reaction is used with cholesterol esterase and cholesterol oxidase. In a setup with amperometric detection two enzymatic columns were used together with an anion-exchange resin column which effectively eliminated interferences of anodic detection of hydrogen peroxide before sample injection [369]. After elution from the anion-exchange column samples were 100-fold diluted prior to the flow injection determination. In a slightly more complex setup with sample splitting and amperometric detection two reactors can be utilised in such a way that in the same sample free and total (free plus esters) cholesterol can be determined in serum [370]. For the amperometric detection of hydrogen peroxide a peroxidase biosensor was used in the presence of hexacyanoferrate (II) in the carrier solution, which allows one to carry out detection at -30 mV vs. Ag/AgCl electrode. This allows the elimination of electroactive interferences present in serum. Determinations in such a system can be carried out with sampling rate 10 h^{-1} in 5 μl blood serum samples without preliminary dilution. Such a determination with the two mentioned enzymes can also be carried out with spectrophotometric detection [371]. For H_2O_2 detection the reaction with 2,2′-azinobis (3-ethylbenzothiazoline-6-sulphonate) catalysed by peroxidase was employed. Among possible interferents only triglyceride levels higher than 5 mM yielded a positive interference. Injected samples were 10 times diluted.

Among other substrates the FIA was employed for determinations of L-tyrosine in serum with the use of tyrosinase immobilised in a flow-through reactor and fluorimetric detection [372]. Tyrosine catalyses the oxidation of L-tyrosine to L-dopaquinone which rearranges to fluorescent dihydroxyindole. Analysed serum samples were deproteinised with sodium tungstate and sulphuric acid and filtered prior to the injection. The method is free from interferences when cysteine and ascorbic acid were on-line removed in the system using a C18 column. The detection limit for 50 μl injected samples was 20 nM.

Another wide application of FIA in analysis of serum and plasma is determination of electrolytes. In the determination of potassium in serum in FIA systems it was shown that results of determinations with potentiometric determination with potassium ion-selective electrode are in very good agreement with flame photometric determinations also carried out in the FIA system [298,

373]. This was demonstrated both for a conventional electrode with plasticised membrane and internal solution [298], as well as for potassium-selective ISFET [373]. Potassium was also determined in the FIA system with fluorimetric detection [374]. The detection was based on activation by the potassium ion of the reaction between phospho (enol) pyruvate and adenosino diphosphate catalysed by pyruvate diphosphate catalysed pyruvate kinase. The decrease in NADH concentration which participates in the reaction with pyruvate catalysed by lactate dehydrogenase is measured fluorimetrically. Interferences by sodium and ammonium are eliminated in the presence of lithium and Kriptofix-221 and by the use of glutamate dehydrogenase. Additionally the sensitivity of method allows the dilution of samples over 100 times, which decreases the effect of interferences. Pyruvate kinase and lactate dehydrogenase were immobilised in a flow-through reactor, similarly to glutamate dehydrogenase. Results obtained in serum were satisfactory, although because of the complexity of the system the sampling frequency was 6 h^{-1} only.

For the determination of lithium in blood serum an extraction-spectrophotometric flow injection procedure was established [375]. In a merging-zones setup parallel to a water carrier a 50 μl sample and to an organic phase stream a solution of 14-crown-4-dinitrophenol in chloroform was injected. The organic phase was separated in a flow-through phase separator and transported to the spectrophotometer. It was demonstrated that the determination of lithium in serum samples was possible with sampling rate 100 h^{-1}. Calcium was determined in serum in the FIA system with potentiometric detection using a membrane ion-selective electrode [21]. For determination of total calcium a standard addition method was used and the obtained results correlated well with spectrophotometric determinations. Ionised calcium was determined from direct measurement at minimum dispersion of the sample zone. For determination of total magnesium in blood serum a flow injection method was developed with polarographic detection [376]. The detection was based on measurement of the adsorptive reduction current peak of the Eriochrome Black T-magnesium complex using rapid scan square-wave voltammetry. Results of determination in serum samples were in satisfactory agreement with AAS determinations. Determinations were carried out in 200-fold diluted samples with sampling rate up to 35 h^{-1}.

All the above-mentioned electrolytes (potassium, lithium, calcium and magnesium) and also sodium, copper, iron and zinc were determined simultaneously in serum in the FIA system with ICP-AES detection [377]. For injections of undiluted serum samples matrix interferences were observed which were minimised at low injection volumes or by using a high RF power. Such effects were not observed for diluted samples. Besides this multicomponent assay for determination of level of micro-elements in blood serum several other flow injection methods were developed with various detections. Copper was determined catalytically with fluorimetric detection [284], as well as with amperometric [378], AAS [379] and spectrophotometric [380–382] detections. The already-mentioned earlier fluorimetric determination was based on the catalytic effect of cupric ion on the 2,2′-dipyridylketone hydrazone reaction with hydrogen peroxide. Serum samples for this determination were deproteinised and 250-fold diluted. Determination was performed with the stopped-flow technique [284]. Also, determination with biamperometric detection was based on catalytic effect [378]. Copper(II) catalyses the oxidation of thiosulphate by Fe(III), and formation of Fe (II) in the presence of excess of Fe(III) is a source of current signal in the detector with two polarised platinum electrodes. For results obtained for deproteinised blood plasma samples a good agreement was found with flame AAS non-flow determinations at a concentration level of 0.4–1.2 mg l^{-1} Cu (II). Flame AAS detection was also used in the FIA system for copper determination in serum [379]. In the system with two independent pumps and hydrodynamic sample injection (Fig. 16) determinations were carried out for 1:1 diluted serum samples with sampling frequency 100 h^{-1}.

Determinations of copper in serum were also carried out in flow injection systems with spectrophotometric detection in two-component systems together with determination of zinc [380] and iron [381, 382]. In simultaneous determinations of copper and zinc based on reaction with Zincon two sample plugs were sequentially introduced into the carrier stream. With appropriate setup design in one sample segment zinc is determined, only, and in another one a sum of zinc and copper [380]. Several interfering cations were masked by introducing into the carrier stream citrate, triphosphate and 1,10-phenanthroline. The analysed samples were slightly diluted during deproteinisation. Simultaneous determination of copper and iron was conducted in FIA systems with a diode-array detector. In one of the developed systems a mixture of two chromogenic agents was employed and determination was based on measurements at 10 or 11 wavelengths [381], whereas in the second one only one chromogenic agent was used and measurements were made at two wavelengths [382]. The zinc interferences in the second method were eliminated by the addition of nitrilotriacetate to a chromogenic reagent solution. In both cases satisfactory

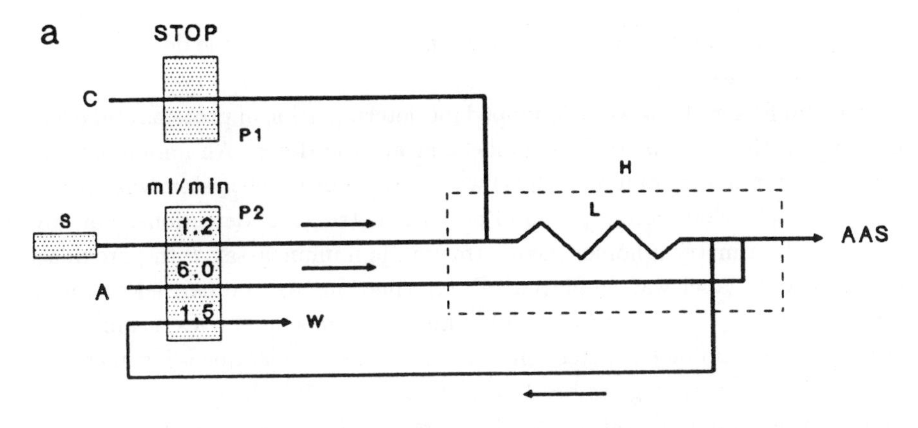

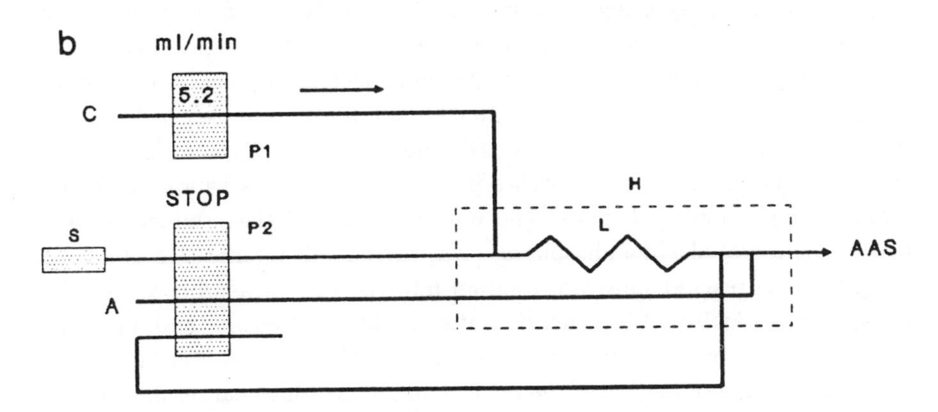

Fig. 16. Schematic diagram of the flow injection system with hydrodynamic injection in loading (a) and injection (b) configuration developed for the determination of copper in blood serum samples with flame AAS detection [379]. P1, P2 — pumps; S — auto sampler; H — hydrodynamic injector; C — water carrier; A — augmenting water stream; L — sampling loop; W — waste stream. (*Reprinted by permission of copyright owner.*)

results were obtained in deproteinised standard blood sera. For the determination of the trace amount of selenium in blood plasma and serum a flow injection system with hydride generation AAS detection has been developed [383]. Determinations in clinical samples required digestion in a mixture of nitric, sulphuric and perchloric acids which was combined with 12.5-fold dilution of raw samples. For injected sample volume 330 μl the detection limit was estimated as 1.2 μg l^{-1}.

Several FIA systems were developed for determination of proteins and drugs in serum with predominated immunochemical procedures. An immunochemical assay of immunoglobulin G (IgG) was carried out in the system with fluorimetric [384], turbidimetric [385] and potentiometric [386] detections. For flow injection fluorimetry a homogeneous enzymatic immunoassay with peroxidase as an enzyme label was employed. Determinations can be carried out with sampling rate up to 60 h^{-1} in 700-fold diluted serum samples. Determinations of IgG with turbidimetric detection were conducted in a stopped-flow merging-zones manifold with a glass-bead-packed column [385]. In this case 800-fold dilution was employed and due to the merging-zones technique the consumption of expensive antiserum was less than 1 μl per assay. In a setup with potentiometric assay with ammonium ion-selective electrode a sandwich-type enzyme immunoassay using an immobilised antibody reactor and adenosine-antibody conjugates was reported [386]. A single protein assay takes about 12 min, including regeneration of the reactor. Determinations in the control serum sample were carried out in a 10.000-fold diluted sample. A flow injection immunoassay of α-fetoprotein was developed with amperometric detection [387]. This heterogeneous sandwich-type enzyme immunoassay was conducted off-line using a multiwell plate, and only detection of the product of enzymatic reaction of alkaline phosphatase was carried out by injection of a 20 μl sample of the reacted substrate solution into the FIA measuring setup. Determinations required 2 h incubation after addition of serum, and another 3 h after addition of antibody-enzyme conjugate. This method was successfully employed for determination of the analyte in human maternal serum samples. In the used amperometric detection of 4-aminophenol the detection limit was estimated as 24 nM. The same kind of detection was reported also in the system with a continuously circulated reagent mixture for determination of the activity of amylase in human blood serum samples [214]. An indicating system was the same one as reported for enzymatic determination of glucose, and the activity of amylase was determined by its catalytic effect on the hydrolysis of starch to glucose.

Among drugs determined by flow injection procedures in serum, several different immunochemical methods can be found for determination of respiratory stimulant theophyline with potentiometric [388], amperometric [389] and fluorimetric [390] detections. In theophyline determination the adenosine deaminase was used as the labelling enzyme with detection by means of an ammonium ion-selective electrode [388]. The same immunoassay concept was also utilised for determination of insulin with peroxidase as the label and spectrophotometric detection [388]. In flow injection amperometry the enzyme label was alkaline phosphatase and *p*-aminophenol was detected [389]. Fluoroimmunoassays of theophyline and valproic acid were flow injection adaptations of commercial kits available for therapeutic drug assays. In another fluorimetric method used in FIA configuration determination was based on inhibition by theophyline of the catalytic effect of alkaline phospahatase on the hydrolysis of 4-methylumbelliferone [391]. A competitive heterogeneous enzyme immunoassay in the flow injection system was developed also for determination of digoxin [392]. Again amperometric detection was based on the use of alkaline phosphatase conjugate to digoxin. In enzyme immunoassays of phenytoin where as a result of reaction of the enzyme used as label NADH is formed the 2,6-dichloroindophenol was employed as a redox coupling agent of which form reduced by NADH is monitored amperometrically [393]. Such a system does not require pre-columns or analytical columns for isolation of the NADH response due to the low applied potential for detection. A non-competitive flow injection immunoassay with fluorimetric detection was also developed for determination of α-(difluoromethyl) ornithine in blood plasma [394].

In a complex matrix of blood serum or plasma flow injection methods can be used for determination of several analytes which cannot be classified into the above-mentioned groups of compounds. They are both organic and inorganic species. The above-mentioned in food analysis section chemiluminescence method of determination of ascorbic acid can also be employed for determinations in blood serum [221]. The serum samples were treated with perchloric acid to separate proteins and after centrifugation adjusted to pH 3.0. The flow injection method with the same detection was also applied in determination of choline-containing phospholipids in human serum [395]. The detection is based on peroxyoxalate chemiluminescence detection of hydrogen peroxide enzymatically formed from choline-containing phospholipids in a reactor with immobilised phospholipase D and choline oxidase. The hyphenation of the FIA system with preparative HPLC was applied for the determination of individual choline-containing phospholipids in human serum.

For the determination of ammonia in blood plasma an FIA method was developed with potentiometric detection with ammonium ion-selective electrode

[396]. In a setup containing a gas-diffusion unit determinations were carried out with sampling rate 30 h^{-1} in spiked serum samples. The enzymatic method of ammonia determination in plasma with fluorimetric detection was based on the use of immobilised glutamate dehydrogenase [397]. The concentration of NADH consumed in the reaction is measured fluorimetrically. No interferences were found in determinations with sampling frequency 40 h^{-1}. The same procedure of indirect detection was also utilised in determinations of bicarbonate in serum in the flow injection system [398]. Determination was based on reaction of the analyte with phospho (enol) pyruvate catalysed by carboxylase. The analysed reaction was coupled with a derivatisation reaction in which NADH was consumed. In determinations carried out with the stopped-flow technique the sampling frequency 20 h^{-1} was reported. Also, determinations of phosphate in blood serum can be carried out in FIA systems with enzymatic methods [272, 399]. In both cases reactions catalysed by nucleoside phosphorylase and xanthine oxidase that produce uric acid and hydrogen peroxide were utilised. In flow injection systems both these products can be monitored amperometrically. [272], or one can apply an additional enzymatic step catalysed by peroxidase and a product of this reaction can be measured fluorimetrically [395]. In both systems the main enzymes were immobilised on the membrane [272] or in the flow-through reactor [399]. Serum samples were diluted 100- [272] or 250-fold [399]. Sampling rates were reported as 10 [272] and 30 [399] h^{-1}.

5.2. *Whole Blood*

For various reasons, for clinical diagnostic purposes determinations in whole blood are more advantageous than in blood serum or plasma. First of all the time of determination is shorter as the whole procedure is usually simpler. It requires less personnel, reduces the possibility of contamination in case of trace determinations, does not require in most cases a well-equipped clinical laboratory and allows bed-site testing. Hence one can find in the existing literature several examples of development of flow injection systems that allow FIA determinations in injected whole blood samples. They were developed mostly for determination of the substrate of enzymatic reactions and trace element determinations.

For the determination of glucose in whole blood flow injection systems were developed with amperometric [400, 401] and chemiluminescence [402] detections. In the system with amperometric detection and immobilised glucose dehydrogenase an on-line dialysis and segmented sample injection was used in order to decrease the red cell volume fraction dependence [400]. Due to modification of the graphite electrode with phenaxozinium ion the detection of NADH can be carried out at potential 0 mV vs. Ag/AgCl reference electrode. Interference-free determinations can be performed with sampling frequency 40 h^{-1} with sample volume 50 μl. In another amperometric FIA system a glucose biosensor was employed with glucose oxidase immobilised on the surface of a carbon paste-tetrathiofulvalene electrode protected from the solution with a polycarbonate membrane [401]. The results obtained in whole blood samples with this electrode were satisfactory; however, results in the FIA system were elevated in comparison to the reference routine method. Also on the use of glucose oxidase was based the chemiluminescence method with detection of hydrogen peroxide [402]. Satisfactory functioning of the system with dialyser was reported with the use of an open tubular enzyme reactor; however, it was pointed out that developed system was too complex for practical applications and too slow (35 h^{-1}) to be attractive compared to already existing static systems.

Determination of urea in whole blood can be carried out in flow injection systems with potentiometric [403] or spectrophotometric [404] detections. In potentiometric ones the nonactin-based ion-selective electrode was employed with a polyester membrane with a covalently immobilised urease. Variations in the hematocrit level in blood samples had no effect on the measurements. The concentration of potassium in the blood samples was measured separately, and mathematical corrections were made, A sampling rate 40 h^{-1} was reported. In a setup with spectrophotometric detection of ammonia with the use of the color change of an acid-base indicator a flow cell combining gas diffusion and optosensing was utilised [404]. On one membrane of an integrated gas diffusion/detection unit urease was immobilised. The measuring system was not affected by variations in pH of samples, the buffering capacity or the hematocrit level. Also no interference due to endogeneous ammonia was encountered. Determinations were carried out with frequency 30 samples h^{-1}.

For the determination of creatinine an ammonia-sensitive semiconductor was used [367], and samples were 25-fold diluted prior to the injection.

In the systems for determination of ethanol in whole blood the catalytic oxidation in the presence of alcohol dehydrogenase and measurement of the absorbance of NADH at 340 nm were reported [405, 406]. In one of them the enzyme was used in dissolved form [405], whereas in another one it was immobilised in a flow-through reactor [406]. In order to eliminate pretreatment steps in whole blood samples, absorbance measurements can be carried out kinetically at stopped flow [405], or with the use of an on-line dialyser [406]. Stopped-flow determinations were carried out with sampling rate 70–80 h^{-1}, and in the system with dialyser 20 h^{-1}. In both cases results of determinations in whole blood samples were in good agreement with gas chromatography results.

The level of ammonia in blood is also an object of interest in clinical medicine for diagnostics of several diseases. In an FIA system developed both for the ammonia determination in plasma and whole blood a diffusion of ammonia into a stream of a pH-sensitive indicator was applied [407]. The color changes are monitored spectrophotometrically. It was found that erythrocytes contain far more ammonia than plasma, and hence its level in hemolysed whole blood was much higher than in untreated blood samples. With small changes in designed systems whole blood, plasma and precipitated samples can be analysed. For 90 μl samples the time of a single assay was 1 min.

The possibility of trace determination of lead in whole blood was investigated in two different FIA systems. In the system with flame AAS detection lead was preconcentrated on-line by coprecipitation with an organic collector and separation of a precipitate in a knotted reactor made of ethyl vinyl acetate tubing [408]. An enrichment factor of 20 was obtained for a 30 s coprecipitation. This method was applied in determination in reference blood samples, but the obtained results were elevated about 10% in comparison to preliminary recommended values. For 30-fold diluted whole blood samples a system with potentiometric stripping detection was examined with a dialysis membrane coated mercury film electrode [409]. Lead was determined in certified whole blood samples.

Determinations of cobalt were carried out in a highly mechanised FIA system with electrothermal AAS detection and on-line microwave-assisted mineralisation of samples [410]. The accuracy of the procedure was investigated using reference materials of whole blood containing a level of Co from 5 to 10 μg^{-1}. A similar system but with flame atomisation AAS was developed for determinations of copper and zinc [411]. In determinations of these elements and also of calcium and magnesium with flame AAS detection, it was found that the sensitivity of detection can be improved after addition of non-ionic surfactants to the samples [415].

5.3. *Urine*

Several components determined for diagnostic purposes in blood are also often determined in urine. Some procedures are quite similar to those developed for blood plasma or serum; however, in some cases it is necessary to design different FIA systems to carry out determinations in the urine matrix.

Urea can be determined in urine with the same enzymatic FIA method as was developed for blood serum with immobilised urease [363]. Because of the much higher level of urea in urine the analysed samples were 1000-fold diluted. As also the level of ammonia is much higher it had to be determined separately in the FIA system without enzymatic detector and the determined amount was subtracted from the results obtained with the urease reactor. The urea level in the analysed sample was in the range of 0.17–0.66 M.

One of the most often determined substrates in urine is creatinine, which is an indicator of the renal, muscular and thyroid functions. It is commonly determined in clinical laboratories with the spectrophotometric method based on the Jaffe reaction with sodium picrate and the same method was also adapted in the flow injection system [412]. The analysed samples were 10 times diluted with a phosphate buffer containing EDTA. In determination with sampling frequency 102 h^{-1} no interferences were observed. In electrochemical enzymatic determination of creatinine the endogeneous ammonia even more strongly interferes than in the case of urea determination. In the flow system with potentiometric detection employing a nonactin-based ammonia electrode and immobilised creatinine iminohydrolase a splitting of the sample into two branches of the flow system was used [413]. With the concomitant addition of alkali metal ions to the carrier solution, this configuration allows for simple correction of ammonia and alkali metal ion content. The analysed samples did not require any pretreatment. In the developed system with amperometric detection determination was based on the use of three immobilised enzymes, creatinine iminohydrolase, leucine dehydrogenase and L-amino acid oxidase, and monitoring of oxygen consumed in the last enzymatic reaction [414]. In order to consider the presence of endogeneous ammonia a branched configuration of the system was designed with differential measurement, or ammonia was eliminated in a separate enzyme reactor with glutamate dehydrogenase. Depending on the configuration of the used system samples were diluted prior to the injection from 5 to 20 times. Without any other sample preparations determinations were carried out in the concentration range of 2–30 mM.

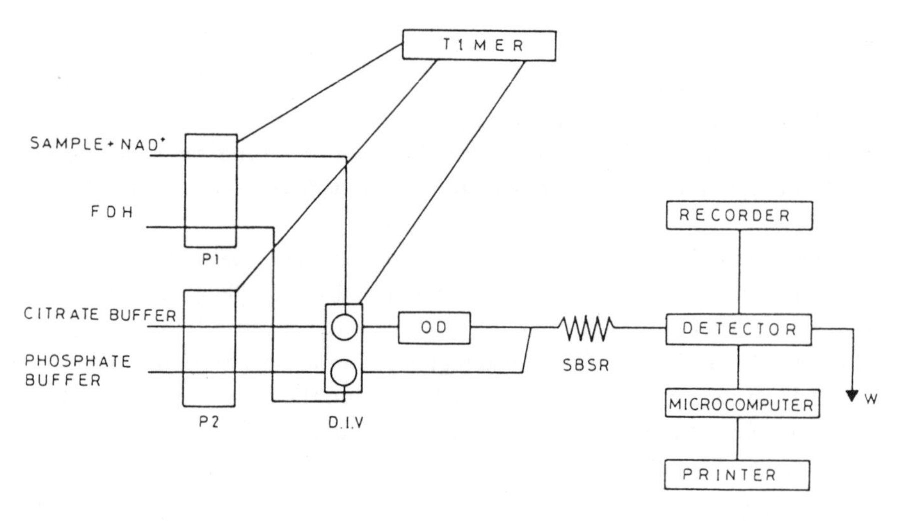

Fig. 17. Schematic diagram of the flow injection system for kinetic-enzymatic determination of oxalate in urine with spectrophotometric detection [416]. P1, P2 — peristaltic pumps; FHD — formate dehydrogenase solution; OD — oxalate decarboxylase reactor; SBSR — single-bed string reactor; DIV — dual injection valve; w — waste. (*Reprinted by permission of copyright owner.*)

In enzymatic flow injection systems oxalate in urine can also be determined. In the system with chemiluminescence detection the immobilised oxalate oxidase was employed and the generated hydrogen peroxide was detected via reaction with luminol and hexacyanoferrate (III). The developed method has a detection limit 34 μM and effectiveness 55 samples h^{-1}. In spectrophotometric determinations two enzymes were employed, oxalate decarboxylase and formate dehydrogenase, and detection was based on measurement of the reduced form of the coenzyme NADH [416]. Measurements were carried with stopped flow, and samples prior to the injection were treated with hydrochloric acid, filtered and neutralised. In the FIA system parallel injections of the sample with NAD+ and solution of formate dehydrogenase to two streams of carrier solution were made (Fig. 17). The first stop of the flow is made when the sample plug reaches the reactor with oxalate decarboxylase, the second when the sample plug reaches the detector cell. Determinations can be carried out with sampling rate 12 samples h^{-1}.

The already-mentioned enzymatic determination of ascorbic acid with laccase added to the samples and fluorimetric detection can also be used for urine analysis [251]. Samples with added EDTA were 50 times diluted and measured with sampling rate 30 h^{-1}. Great diagnostic importance has a level of kynurenic acid in urine, one of the tryptophan metabolites. For its assay flow injection fluorimetry was employed [417]. The detection was based on fluorescence of the analyte on irradiation with UV light in the presence of hydrogen peroxide. For preliminary separation of the analyte urine samples were passed through C18 columns, and then eluate containing eluted analyte was injected into the FIA system.

Flow injection systems were also developed for determination of several drugs in urine. Phenothiazine derivatives are used as psychotropic drugs, and their detection can be based on photochemically induced fluorescence [418]. The developed system was applied to detection of unsubstituted phenothiazine, thionine, Azure A and methylene blue with detection limits from 13 to 35 $\mu g\ l^{-1}$. For these determinations urine was diluted at least 500 times. Benzodiiazepines are most frequently prescribed as sedative and hypnotic drugs. A flow injection setup was developed for screening these drugs in urine [419]. It is based on the diazotization of the analytes in pre-hydrolysed samples and formation of dyes with 1-naphthol. The dyes are separated on a XAD-2 minicolumn, eluted and detected spectrophotometrically.

Several inorganic constituents of urine are also determined in flow injection systems. Potassium can be determined potentiometrically with potassium ISFET in 4 times diluted samples with 0.1 M NaCl [373]. Determination of phosphate can be carried out with amperometric detection based on reduction of 12-molybdophosphoric acid [420]. In determinations in 10 times diluted samples with sampling rate 70 samples h^{-1} the detection limit was estimated as 20 nM. Iodide was determined with chemiluminescent detection using head-space separation of iodine in a closed vessel in the presence of potassium dichromate [421]. The released iodine is carried out with a nitrogen stream to iodide trapping solution. The detection of iodine is based on reaction with luminol in the presence of cobalt (II). This method with detection limit 10 $\mu g\ l^{-1}$ was used for urine analysis with the cobalt concentration range of 61–125 $\mu g\ l^{-1}$.

In FIA of trace elements in urine atomic spectrometric techniques predominate as detection methods. A procedure of arsenic speciation was developed with hydride generation AAS detection [422]. Speciation was based on elution with various eluents of arsenic species retained on a cation-exchange column, while total arsenic was determined after microwave-assisted mineralisation/oxidation in a parallel and on-line module of the system. In the FIA setup for determination of chromium (III) in the elevated samples of urine the analyte was preconcentrated on activated alumina and with ICP-AES detection a detection limit 50 ng l^{-1} was reported for 10 ml samples [423]. The same sorbent was also exploited for preconcentration of cadmium in the FIA system with flame AAS detection [424]. Determinations in urine samples were carried out with a slotted quartz tube mounted on a burner, which increased the sensitivity about 3 times. Copper determination was carried out in the flow system with amperometric detection with platinum microelectrodes [425]. The sample cleanup and preparation can be conveniently achieved by the use of a C18 column off-line and formation of a copper dithiocarbamate complex. For the detection an anodic oxidation of this complex on the platinum electrode is used. For detection of total mercury in urine the FIA system with off-line oxidation of organomercury species and cold vapour AAS detection was developed [426]. In the optimised system the detection limit was achieved as 0.23 μg l^{-1}. Because the on-line oxidation of organomercury species was not complete, an off-line rapid potassium permanganate/sulphuric acid digestion procedure had to be employed. With a similar detectability the same detection system can be applied in differentiation of inorganic and total mercury in urine [427]. Total mercury is determined after an on-line oxidation with persulphate with microwave-assisted heating. The determined contents in urine were in the range of 1.8–6.7 μg l^{-1}. An especially low detection limit of 1.5 ng l^{-1} was reported for determination of uranium in the FIA system with ICP-MS detection [428]. Determinations with sampling rate 60 h^{-1} were carried out without preconcentration or off-line pretreatment of urine samples, except for acidification with a few drops of nitric acid. Results obtained for urine samples at μg l^{-1} concentration level were validated by a laser-induced fluorescence method. Other authors with the same detection determined 9 trace elements in SRM freeze-dried urine, in most cases reporting a very good agreement with certified or suggested levels [429]. Multielement determinations in urine were also applied in the FIA system with ICP-AES detection and on-line preconcentration on a microcolumn with chelating resin [133]. The analysed samples were diluted 1:1 and acidified with nitric acid. Simultaneous determinations of Cd, Co, Cu and Pb can be carried out with sampling rate 12 h^{-1}.

5.4. *Other Specimens*

Besides these basic objects of clinical analysis discussed in previous paragraphs, flow injection procedures have also found several applications in determination of various analytes in tissues of various organs, hair, and also in dialytic fluids for the control of the dialysis process.

The example applications to tissue analysis concern determination of trace heavy metals with AAS detection. Determination of zinc and cadmium in samples of human kidney and liver tissue was carried out in the FIA system with flame AAS detection [430]. The samples of analysed materials (1.0–1.2 g) were mineralised in nitric acid in a microwave oven, and then the digest was aspirated to the flow injection system. 20 μl samples were injected into Zn determinations and 100 μl for Cd determinations. The SRM bovine liver and oyster tissue were used for standardisation of the system. The detection limits were reported as 0.5 and 0.1 $\mu g\ g^{-1}$ of the initial sample for Zn and Cd, respectively. For determination of antimony (III) and total content of this element in liver tissue microwave-assisted mineralisation was also employed for determinations in the FIA system with hydride generation AAS detection [431]. The response for Sb (III) was selectively obtained from samples in a 1 M acetic medium, and for total antimony after sample treatment with sulphuric acid and potassium iodide. The detection limits obtained for digests were 0.15 and 0.10 $\mu g\ l^{-1}$ for Sb(III) and total antimony, respectively.

Flow injection methods were applied also for determination of heavy metals in hair. For the determination of cobalt a system was developed with flame AAS detection, on-line preconcentration and separation of the analyte from the matrix as an ion pair of the cobalt-nitroso-R-salt complex with tetrabutylammonium cation on C18 sorbent [432]. The interference from iron was masked with citrate. Samples were mineralised in a mixture of nitric and perchloric acids. Co was determined in hair samples at content level 0.12 $\mu g\ g^{-1}$. Copper and zinc were simultaneously determined by flow injection spectrophotometry with Zincon as the colour-forming reagent [433]. The analysed samples were dry-ashed at 450°C, the residue was dissolved in nitric acid and masking solution (citrate, triphosphate and 1,10-phenanthroline) was added. The determination was based on the use of the pH gradient technique. Determined contents in hair samples were in the range of 50–130 $\mu g\ g^{-1}$ Zn and of 9–14 $\mu g\ g^{-1}$ Cu. The spectrophotometric detection based on reaction with Ferrozine was employed in determination of iron [192]. Hair samples were mineralised in a mixture of nitric and sulphuric acids and hydrogen peroxide, and the determined contents were at a level of 4 $\mu g\ g^{-1}$. Chemiluminescence detection in the FIA system developed for determination of vanadium in geological samples was also used for determinations in hair [323]. Samples were mineralised in nitric and perchloric acids and the content of determined vanadium was from 0.12 to 0.16 μg g^{-1}.

In kidney dialysates flow injection methods were used for determination of urea and aluminium. Urea was determined using the spectrophotometric biosensor method [434]. The urease was cross-linked into a cellulose pad with an acid-base indicator dye covalently bound to the surface of the cellulose. The sensor was placed within a flow injection optosensing system in an integrated conduit. The same determinations were carried out with urea fluorescent optodes where urease was co-immobilised with fluorescent dye-labelled bovine serum albumin on the tip of the fibre [521]. The dialysis buffer containing between 3 and 10 mM urea in the gradient mixing system was diluted continuously with buffer to obtain optimal analysis conditions. Determinations of aluminium were carried out in FIA systems with atomic spectrometric detections. In the configuration with ICP-AES detection an on-line preconcentration of aluminium on a mini-column with an anion-exchanger was applied [91]. Although it was found that anions forming complexes with Al interfere in the determination, the developed system was successfully employed for determination in dialysates at a level from 60 to 110 μg l^{-1}. Samples were mixed with buffer prior to the injection. For the same purpose also the FIA system with flame AAS detection can be used with Al separation on a Chelex 100 column [435].

6. Pharmaceutical Applications

The quality control of pharmaceutical preparations is more and more challenging nowadays for chemical analysts and requires fast, mechanised or automated analytical procedures. It concerns either the determination of the content of the main therapeutic component or studies on the kinetics of its release from the preparation. The possibility of replacement in some cases of the most common, but usually time-consuming and expensive, chromatographic methods with selective and mechanised flow injection methods for determination of a given component of interest is always attractive. Hence, the number of such procedures developed for pharmaceutical applications is constantly increasing. Because most of these analytes are organic compounds the most frequently molecular spectroscopy at various wavelengths and luminescence methods are exploited for detection in these systems. It is more difficult to achieve sufficient selectivity with electrochemical detection methods, although several examples of flow injection systems with such detections will also be presented.

Molecular spectroscopy is used for detection in flow injection systems for pharmaceutical determinations in various ranges of wavelength. The acetaminophen (paracetamol), a widely used analgesic and antipyretic drug, can be determined in both the visible [435] and the infrared [437] range. In the first case determination is based on the reaction of its oxidation by 2-iodylbenzoate and detection of the product at 445 nm. The interface of the FIA system with the dissolution apparatus has been found useful in automating the drug dissolution test. The sampling rate achieved in this system was as high as 360 h^{-1}. In the second case for the detection was used a Fourier transform infrared spectrometer with steel flow cell equipped with a zinc selenide, internal reflection element. The detection is based on the hydrolysis of the analyte to *p*-aminophenol and its oxidation reaction with potassium hexacyanoferrate (III). For the detection three different bands of the final reaction product are used. Both methods were successfully employed for commercial preparations.

Various flow injection procedures were developed for spectrophotometric determinations of ascorbic acid in different preparations. The above-mentioned method based on direct absorbance measurement at 245 nm with partial decomposition of the analyte in the presence of NaOH to products non-absorbing UV radiation was employed for this purpose [219]. Ascorbic acid may be determined spectrophotometrically at 360 nm based on reduction of vanadotungstophosphoric acid in the FIA system [438]. In this method the reagent can be immobilised in a flow-through reactor [439]. Ascorbic acid can also react with immobilised cupric orthophosphate, and the released cuprous ion reacts with Bathocuproine to form a coloured complex. In the developed system the detection limit was estimated as 0.3 μM for measurements with sampling rate 80 h^{-1}. Then in the sequential analysis procedure ascorbic acid reduces iron (III) and formed Fe (II) gives a coloured chelate with 1,10-phenanthroline [440]. All these methods were successfully used for analysis of commercial preparations.

Aromatic primary amines, of which sulphonamides are widely used in pharmaceutical preparations, in oxidative coupling with 4-N-methylaminophenol form a coloured product utilised for detection in flow measurements [441]. The FIA procedure allows one to minimise errors caused by the instability of the coupling intermediate and oxidation of the amine in comparison to manual operations. Such a method was employed for determination of sulphacetamide and sulphamethoxazole.

L-carnitine is an essential nutrient for early infant development, and hence of great importance is its determination in infant formulae. For this purpose a flow injection enzymatic method with spectrophotometric detection and immobilised carnitine acetyltransferaseis is used. In order to reduce the coenzyme consumption the merging zones technique was used in this determination. The necessary sample preparation included the precipitation of proteins and oxidation of trace levels of endogeneous thiol artefacts by hydrogen peroxide. The excess of H_2O_2 was removed with catalase. Determinations can be carried out at a sampling frequency up to 50 h^{-1}, and the obtained results were verified with an enzymatic commercial UV test kit.

Clotiazepam and triazolam belong to benzodiazepines used therapeutically as tranquillisers. In FIA systems they can be determined by direct UV spectrophotometry at 260 and 228 nm, respectively [443]. They can also be detected amperometrically at potentials -0.95 and -1.125 V with the use of a sessile mercury drop working electrode. Both methods were employed for determination in commercial preparations [443].

Antibiotic doxycycline can be determined by indirect spectrophotometry in the flow injection system at 395 nm using a flow-through reactor with copper carbonate entrapped in a polymeric material [444]. Detection was based on measurement of the absorbance of the complex with the analyte formed by Cu (II) eluted from a polyester bed in the reactor.

Alkaloid ergonovine has an important role as a direct stimulant on smooth muscle. In the developed FIA spectrophotometric method the on-line derivatisation of ergonovine maleate with *p*-dimethylaminobenzaldehyde is applied [445]. Reaction is carried out in the presence of Fe (III) and the reaction rate is improved by UV irradiation. In reported determinations the sampling rate 34 h^{-1} was achieved and they are more selective than determinations based on reaction of the analyte with sodium periodate and hydrogen peroxide [446].

The ethylenediamine in pharmaceutical preparations is associated with the anti-asthmatic drug aminophylline. Its flow injection spectrophotometric determination can be based on reaction with Cu (I) and pyridine-2 carbaldehyde with formation of an orange chelate [447]. The addition of EDTA allows one to mask successfully several interfering cations. Also, the tolerated levels of primary, secondary and tertiary amines, quaternary ammonium salts and selected alkaloids were examined. Determinations were carried out at 475 nm with sampling rate up to 55 h^{-1}.

Reserpine is used in preparations with antihypertensive properties. In the developed FIA system it is oxidised by manganese dioxide embedded on the walls of a Tygon tubular reactor and the product of oxidation is monitored at 385 nm [448]. Reaction is catalysed by the presence of Mn (II) in solution. Reported determinations were carried out in commercial preparations and at US Pharmacopoeia standards.

Terbutaline is used in anti-asthmatic drugs and its flow injection determination is based on the reaction with 4-aminoantipyrine and potassium hexacyanoferrate (III) [449].

An appropriate chemometric method in data processing in spectrophotometric measurements may be utilised for mulicomponent determinations of several components in pharmaceutical preparations. A simultaneous stopped-flow method was developed for determination of paracetamol, acetylsalicylic acid and caffeine with FTIR detection [450]. For partial least-squares data treatment measurements of 14 characteristic bands were acquired. The analytes were leached from the pharmaceutical preparations with ethanol solution in dichloromethane in an ultrasonic bath. For calibration eight standard mixtures were used and the accuracy error was lower than 5%. Using a photodiode array detector in the FIA system simultaneously etafedrine, phenylephrine, doxylamine and theophyline were determined by measuring the third-derivative spectra in the range of 214–290 nm [451]. Processing of multicomponent data was carried out by software provided by the manufacturer of a diode-array spectrophotometer.

Numerous applications in FIA of pharmaceutical preparations were reported with fluorimetric detection. In the procedure developed for adrenaline determination it is oxidised by iodine to adenochrome, which in turn is converted by alkali into fluorescent adrenolutine [452]. A solid-phase tubular reactor where iodine was impregnated on the inner wall of the PVC tubing was developed for this determination.

Derivatives of acridine are commonly used as disinfectants, as they inactivate or inhibit the action of some viruses. 9-amino acridine can be monitored directly in the FIA system based on its own fluorescence or colour [453]. In optimised conditions with fluorimetric detection the sampling rate 160 h^{-1} and detection limit 60 ng l^{-1} were reported for such determinations. Fluorimetric detection was also used for the determination of ascorbic acid in the system mentioned above [251]. In a merging zones system using stopped-flow techniques the oxidation of ascorbic acid catalysed by laccase to dehydroascorbic acid was employed, which then reacts with *o*-pheneylenediamine to form a fluorescent product. Alkaloid berberine occurring in an oriental herbal pharmaceutical preparation was determined fluorimetrically in the FIA system with extraction of its ion pair with perchlorate into 1,2-dichloromethane [454]. The antihypertensive drug captopril was determined also in the flow injection fluorimetric system based on oxidation of the drug by a Ce (IV) solution and the fluorimetric monitoring of the formed Ce (III) [455]. Coumarin derivatives scopoletine and umbelliferone, which occur simultaneously in many medicinal plants, were determined in a double-injection single-line setup with fluorimetric detection [456]. The method was based on the reversible shift of the fluorescence excitation spectrum induced by pH changes. The method was successfully employed for analysis of several synthetic mixtures of both analytes in various ratios. Alkaloid emetine can also be determined by FIA fluorimetry with photoderivatisation in the presence of barium peroxide [457]. In determinations of paracetamol the on-line oxidation with potassium hexacyanoferrate (III) immobilised on an anion-exchange resin in a flow-through reactor was utilised [458]. The oxidation product additionally is merged with N,N′-dimethylformamide to enhance the fluorescence. Besides the mentioned spectrophotometric system for the determination of antihypertensive reserpine for its determination also fluorimentic detection in the FIA system was developed [459]. Performed on-line photolytic oxidation in acetic acid leads to a strong fluorescent product, which is detected. The described system is simpler than that with spectrophotometric detection and a packed-bed reactor. The setup for thiamine (Vitamin B_1) determination is based on the oxidation of the analyte by mercury (II) to fluorescent thiochrome [460]. In this system the detection limit was estimated as 34 nM and the sampling rate 22 h^{-1}. A more complex flow system is required for the determination of vitamin K3 (menadione) with fluorimetric detection [462]. First, hydrolysis of the sulphite derivative of the menadione was carried out, and then reduction in the zinc reactor to fluorescent 2-methyl-1,4-dihydronaphthalene. The detection limit was estimated as 1 μg l^{-1} and the maximum sampling rate as 70 h^{-1}. Then determination of cardiac glycoside digoxin in tablets can be carried out in a single-line system where tablet extracts are injected into a stream of alcoholic hydrochloric acid solution containing benzoyl peroxide and ascorbic acid [461]. An on-line heating at 75°C facilitates the rapid formation of fluorescent species which are detected. The system is an automable method for dissolution testing, uniformity content and bulk assay of a tablet formulation.

The flow injection systems with chemiluminescence detection were developed for determination of ascorbic acid [221], penicillamine [463] and steroids [464] in various preparations. The first one, based on reaction of lucigenin with the products from the photooxidation of the analyte sensitised by toluidine blue, was already mentioned earlier [221]. It provides a very sensitive response (detection limit 0.2 nM), and interferences from Co (II), Mn (II) and Fe (II) on the chemiluminescence reaction can be eliminated by incorporating a mini-column with a cation exchanger into the system. Penicillamine is used for the treatment of a number of diseases, including arthritis and heavy-metal poisoning. In the FIA system for its determination a sample is injected into an inert carrier stream, which then joins the reagent stream of fluorescer quinine and Ce (IV) [463]. No serious interferences have been observed from the classical excipients tested. Then determination of steroids is based on the sensitising effect on the cerium (IV) - sulphite reaction [464]. The method was employed for determination of cortisone, hydrocortisone, dexamethasone, prednisolone, methylprednisolone, progesterone, corticosterone, testosterone and betamethasone with different sensitivity and in different concentration ranges.

The turbidimetric detection has found an application for determinations of drugs exhibiting cationic properties whose injection into the solution of an appropriate anionic dye can form precipitating ion associates. This concept was applied in determination of amitriptyline with bromocresol purple [465] and diphenhydramine with bromophenol blue [466].

Several drugs were determined indirectly with AAS detection with flame atomisation. The determination of glycine was based on a release of Cu (II) from a copper (II) carbonate precipitate packed into a column, and the complex formed was fed to an AAS detector [467]. The procedure was examined in terms of tolerance to the presence of chloride, acetylsalicylic acid, ascorbic acid and sorbitol. A bacteriostatic drug, isoniazid, was determined by oxidation with MnO_2 incorporated in polyether resin beads [468]. The measurements are based on the determination of Mn (II) released on oxidation. An antihelmintic agent, levamisole, is determined in the FIA system by precipitation with HgI_4^{2-}, dissolution of the precipitate in ethanol and the measurement of dissolved mercury ions by AAS [469]. An analgesic and antipyretic drug, metamizol, can be determined in the flow injection system with a flow-through reactor with PbO_2 in a polyester resin and AAS detection of released Pb(II) [470]. Then sulphonamides can be determined indirectly based on continuous precipitation with copper or silver ions in the FIA system [471]. Determinations were conducted in reversed FIA systems with constant sample aspiration where the analyte concentration was directly related to the decrease in absorbance of the precipitating cation. The copper method was more selective.

More limited applications in determination of drugs can be found for electrochemical detections. Acetylsalicylic acid can be determined from the direct measurement with a membrane electrode sensitive towards salicylate [472]. Determinations of penicillins in pharmaceutical samples were carried out in an enzymatic system with potentiometric detection employing a glass pH electrode (Fig. 18) [473]. The developed setup contains a flow-through reactor with immobilised penicillinase, and penicillins G and V can be selectively determined in the range of 0.05–0.50 mM at sampling rate 150 h^{-1}.

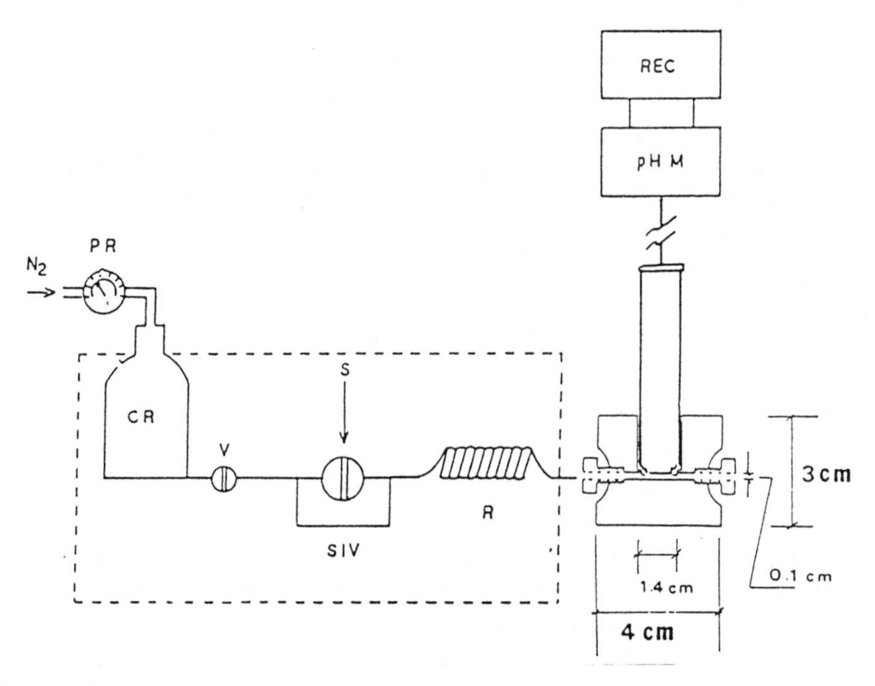

Fig. 18. Experimental setup for the determination of penicillins in pharmaceutical samples using a flow injection enzymatic system with potentiometric detection using a glass pH electrode [473]. PR — pressure regulator; CR — carrier solution reservoir; V — on/off valve; SIV — sample injection valve; R — urease reactor; pHM — pH meter; REC — recorder. Units inside the dashed rectangle were thermostatted. (*Reprinted by permission of copyright owner.*)

For determination of impurities of penicilloic acid in various penicillin preparations, another flow injection method was reported with polarographic detection at a dropping mercury electrode [474]. With the sampled direct current mode the penicilloate gives a linear calibration graph for 0.2 μM to 0.1 mM. The results of determination in the presence of a large excess of penicillin were in good agreement with titrimetric determinations. Amperometry with solid electrodes in a typical three-electrode configuration was employed for determination of warfarin [475] and ethanol [239]. Warfarin was determined directly from the current corresponding to oxidation of the analyte on a glassy carbon electrode at +1.05 V. Determination of ethanol in medicine was carried out in an enzymatic system with immobilised alcohol oxidase and a gas-diffusion module presented earlier [239]. The achieved sampling rate was 120–180 h^{-1} and the operational half-life of the immobilised enzyme was 800 injections. Biamperometric detection with two polarised platinum electrodes and indicating system Fe (III)/Fe (II) was applied in flow injection determination of phenothiazine derivatives promazine and thioridazine [476] and also epinephrine [477].

FIA may also find efficient applications in determinations of inorganic constituents of pharmaceutical preparations with various detection methods. The determination of bismuth was reported with the use of solvent extraction of the tetraiodobismuthate (III) anion into chloromethane as an ion pair with tetramethylenebis (triphenylphosphonium) cation [337] or into propylene carbonate as hydrogen tetraiodobismuthate (III) [478]. In both cases spectrophotometric detection was based on the measurement of the absorbance of the tetraiodobismuthate(III) complex. Various samples of pharmaceutical preparations were mineralised in Kjeldahl flasks with a mixture of hydrogen peroxide and perchloric acid [478]. The spectrophotometric determination of bromide employing reaction with chloramine T and phenol red was developed for bromide salts determination in drugs [151]. This was then exploited for content uniformity tests and dissolution studies of drug formulation. The setup with potentiometric iodide detection was used for determinations of iodide in the range of 1 μM to 0.1 M [479]. When sodium metabisulphite is used for on-line reduction in the same system, both iodide and iodine can be determined in some preparations. Determinations of iron were carried out in the FIA system with amperometric detection [480] and in the SIA system with spectrophotometric detection [480]. In the FIA system detection was conducted with a glassy carbon electrode with a redox polymer with osmium redox sites. The

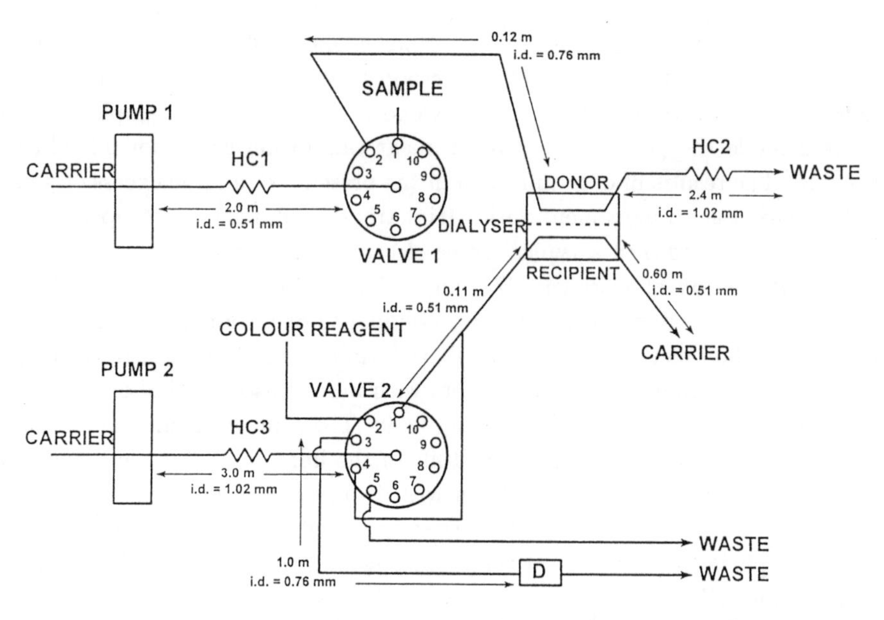

Fig. 19. Schematic diagram of a sequential injection flow analysis system for determination of iron (III) in pharmaceutical samples with spectrophotometric detection based on formation of a complex with Tiron [481]. HC1, HC2, HC3 — holding coils; D — detector. (*Reprinted by permission of copyright owner.*)

modified electrode behaves as an efficient electrocatalyst for the reduction of Fe (III) at -0.05 V vs. SCE with the limit of detection in the FIA system 3 μM. The analysed samples were dissolved in sulphuric acid and Fe (II) was oxidised with hydrogen peroxide. In the sequential injection system (Fig. 19) Fe (III) was separated from a sample matrix by dialysis, then complexed with Tiron, and the resulting complex was monitored spectrophotometrically [481]. The sample rate 8 h^{-1} and the limit of detection 45 mg l^{-1} were reported for this procedure. The essential advantage of the SIA system is a very limited reagent consumption.

7. Biotechnology

For the control of biotechnological processes of fermentation, cultivation of various micro-organisms or mammalian cells, besides measurements of dissolved oxygen, pH or temperature, increasing attention is paid to the monitoring of the concentration of chemical species providing information about occurring biochemical processes. Depending on the kind of process, such monitoring may involve typical nutrients (glucose, glutamine), metabolites (acetate, lactate, ammonia), gene regulators (phosphate, copper, indoleacrylic acid) or expected products (antibiotics, recombinants proteins, etc.). For this purpose various flow injection procedures were designed, mostly enzymatic ones with different detection methods.

The possibility of application of this methodology was examined especially for basic nutrients. For the glucose determination the enzymatic system with glucose oxidase biosensor was employed with a linear response range from 0.03 to 30 mM [482]. In the developed system also simultaneous determination of L-glutamine was carried out using one of the enzymatic methods presented below. For the determination of glucose in fermentation broth samples a sequential injection analysis system with a tubular glucose oxidase reactor and chemiluminescence detection with luminol was examined [483]. Determinations were carried out in the linear range of response from 0.03 to 0.6 mM, with sampling rate 54 h^{-1} and a 30 s stopped-flow in the enzyme reactor. In comparison to the analogous FIA procedure, the SIA method exhibits better sensitivity, smaller reagent consumption and a larger sampling rate.

Several various flow systems were designed for glutamine determination in bioreactor media. Those with potentiometric detection utilise immobilised glutaminase and ammonium sensitive membrane electrode with nonactin. In all proposed configurations of the FIA systems it was necessary to eliminate the interference from endogeneous ammonia and potassium ions. In set with enzyme immobilised on the outer surface of the ammonium electrode the on-line dialyser was added to the setup with the anion-exchange membrane [361]. As the sample passes through the dialysis unit, anionic analyte molecules move through the membrane while permeation of ammonium or potassium ions is retarded. In another configuration an on-line tubular cation-exchange unit was placed between the injection valve and the detector with ammonium ISE [482, 484, 485]. Interferent cation species within the sample plug are exchanged for other cations (e.g. Li^+) contained within a reservoir solution surrounding the ion-exchange tubing. Such a design was also employed in two-component FIA systems, where besides L-glutamine glucose was determined [482] or ammonia nitrogen [485] in bioreactor media from the process of cultivation of mammalian cells. Yet in another design of measuring flow setup a high level of background interfering ions provides a convenient means for compensation of background interferent ions in the injected samples [486]. In the latter system detection was also made with ammonium ISE while glutaminase was immobilised in a CPG reactor. In the bioreactor glutamine was determined in the concentration range of 0.2–4 mM. For the same purpose also amperometric detection can be employed in the system where glutaminase and glutamate oxidase are co-immobilised in a flow-through reactor [487]. An additional reactor with glutamate dehydrogenase was needed to remove endogeneous glutamate present in the culture medium. A linear response was observed in the range of 0.01–0.2 g l^{-1}.

L-lactate is one of the most important metabolites in numerous bioprocesses. The FIA system successfully applied for its determination in a fermentation medium was based on amperometric detection with a bienzyme modified carbon paste electrode with lactate oxidase and peroxidase [488]. Due to the low potential of detection (0 mV vs. Ag/AgCl) interferences of oxidizable substances can be prevented. Determinations in fermentation media were carried out in the range of 1–10 mM lactate.

As products of biotechnological processes in off-line FIA systems glutamic acid and β-lactams were determined. For the glutamic acid determination a potentiometric biosensor CO_2 with immobilised bacteria *Escherichia coli* was employed [489]. Determinations in fermentation broth were carried out in the range of 180–580 mg l^{-1}. The penicillin V in fermentation samples was determined by flow injection spectrophotometry [490]. The analyte was on-line hydrolysed in the presence of dissolved penicillinase and the formed penicilloic acid in the presence of $HgCl_2$ reduced molybdoarsenic acid to form a molybdenum blue, detected photometrically. For determination of penicillins and cephalosporins in samples from the fermentation process an FIA system with an enzyme thermistor detector was also developed [491]. Enzymes reactions were carried out in a small solid bed reactor, and for this purpose penicillinase, bactopenicillinase and cephalosporinase were used. In both systems with spectrophotometric and enthalpimetric detection similar detection limits were reported for penicillin V 1.7 and 1.0 μg l^{-1}. The sample frequency, however, was twice as large as for the spectrophotometric detection, than for the enthalpimetric one.

Besides the several systems mentioned above, a larger number of flow injection measuring systems were developed for on-line monitoring of the biotechnological processes. They will be discussed separately below, in the section on process analysis. They predominate over all developed applications of flow injection methods in process analysis.

8. Process Analysis

For many years process analysis has been an indispensable element of the control of different industrial processes, biotechnological systems, food processing, or production of pharmaceuticals. Process analysis can also be treated as part of environmental monitoring, such as control of waste treatment. It includes such functions as monitoring of the safety of industrial installations and the safety of their functioning from the environmental point of view. Already for many years measured physical quantities are not the only parameters of process control, and chemical process instrumentation is widely developed, including semiconductor gas detectors, numerous spectrophotometric analysers of dissolved species, up to process chromatographs or X-ray analysers. Is there in this long list of methods and devices some room for flow injection systems? The answer to this question has been discussed since the middle of the eighties [492–495].

Instrumentation and analytical methods to be employed in process analysis have to fulfil several specific requirements, contrary to scientific instrumentation or typical laboratory instruments. From the instrumental point of view, process instrumentation must be as much as possible maintenance-free, resistant to aggressive chemicals, should work with maximum independence from the human operator, and should be equipped in signal transmission devices. An analytical method for process analysis, besides the necessary sensitivity for a given purpose, should be precise and fast, with possibly short delay from sampling to obtaining result. These last factors allow one to evaluate the value of specific process parameter *measurability*, which determines the usefulness of a given method with given instrumentation [496]. Carrying out the process analysis with a measurability value of less than 0.8 is simply impractical. Based on the very broad literature it seems that many flow injection methods exhibit good precision, a satisfactory sampling rate and a short time delay between sampling and obtaining result; however, in one case only [497] was the measurability value determined.

It seems that each of basic flow injection systems shown in Fig. 20 in various conditions can fulfil the above-mentioned criteria for the process analysis method when an appropriate detection method is employed. It is much more difficult, however, to indicate instrumental possibilities. As is shown in Chapter 10, there is quite a large offer of commercial flow injection instrumentation, but most of these systems, with very few exceptions, are typical laboratory instruments.

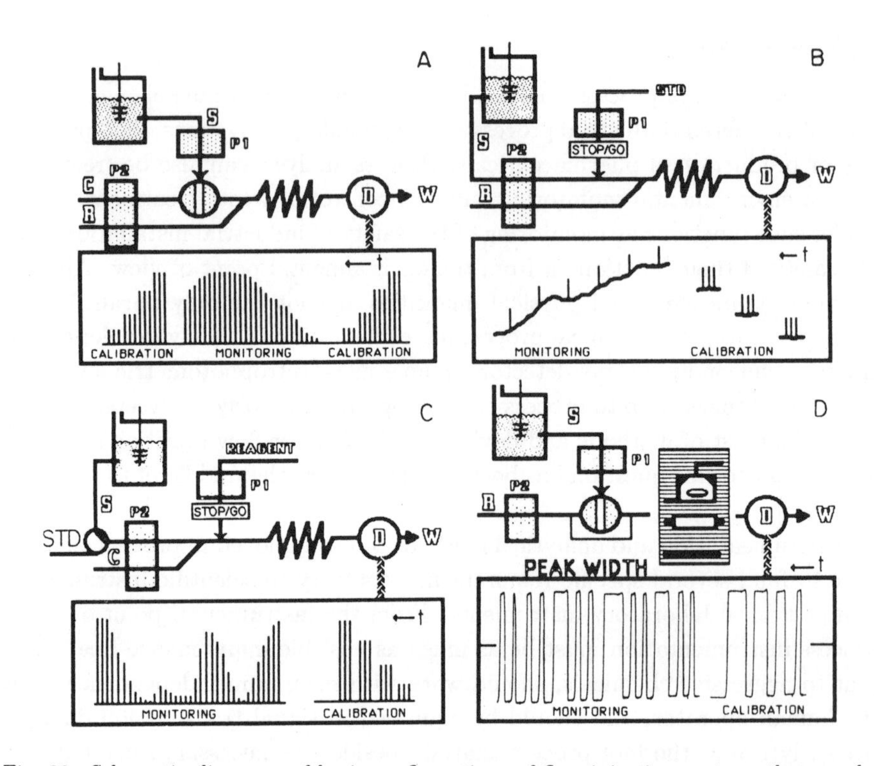

Fig. 20. Schematic diagrams of basic configurations of flow injection systems that can be potentially applied in continuous monitoring [494]. A — system with sample or standard injection; B — system with sample aspiration and standard injection; C — system with sample or standard aspiration and reagent injection (reversed FIA system); D — system with sample or standard injection for flow injection titrations. (*Reprinted by permission of copyright owner.*)

The example process applications presented below are collected entirely from the scientific literature. It is quite probable that the number of practical applications is much larger, but for technological and commercial reasons data about such applications are not published. With some exceptions, there are no reported publications where, based on laboratory experiments, predictions were made of possible process applications.

8.1. *Industrial Applications*

The reported applications of flow injection measurements were developed for various industrial processes; however, in most of them only laboratory instrumentation was employed.

In the system developed for control of the thermal conditioning cycle of power plants, in very diluted solutions pH, ammonia and hydrazine were determined [497]. All determinations were carried out with potentiometric detection using a glass pH electrode, an ammonium ion-selective electrode with plasticised membrane, and a free chlorine sensor for indirect determination of hydrazine. The conditions of determination were optimised separately for each analyte, and then the values of measurability were determined for each procedure taking into account the rate constant of the process, dynamic parameters and precision of individual determinations. This approach has shown the usefulness of procedures for determination of ammonia and hydrazine for process control. Determinations of pH exhibited too low precision and too small a sampling rate to be used for this purpose.

For the monitoring of trace HCN in gas streams from the petroleum-refining process, a flow injection analyser was constructed with a gas permeation unit and spectrophotometric detection based on reaction with chloramine T and mixture of isonicotinic acid and 3-methyl-1-phenyl-2-pyrazolin-5-one [498]. HCN through the walls of the silicone tubing is collected into caustic, then oxidised by chloramine and reacts with color-forming reagents. The analyser was tested on-line on both thermal and catalytic process units with gas streams containing besides HCN also H_2S, CO_2, NH_3 and hydrocarbons. Placement of the gas permeation unit in the sampling loop of the injection valve allows one to cover a broad concentration range, from a few ppbv to at least 10 ppmv HCN.

A fully automated FIA system was developed for on-line monitoring of ammonium thiosulphate production [499]. Schematically the process and places of application of analytical process control are shown in Fig. 21. For the control three parameters have to be monitored: the concentration of ammonium sulphite in the feed stream, the concentration of the product ammonium thiosulphate and the concentration of ammonium sulphite at the reactor outlet. For the monitoring of large concentrations of substrate and product a spectrophotometric detection based on reduction iodine was employed, whereas small concentrations of sulphite were detected with pararosaniline. Although the final configuration of the system sampling rate was 8 h^{-1}, only the described on-line monitoring was successfully employed in the pilot plant.

Several FIA systems were also developed for off-line process applications. Such procedures can be potentially used also in the on-line regime. The flow injection setup was designed for HCl determination in concentrated solution

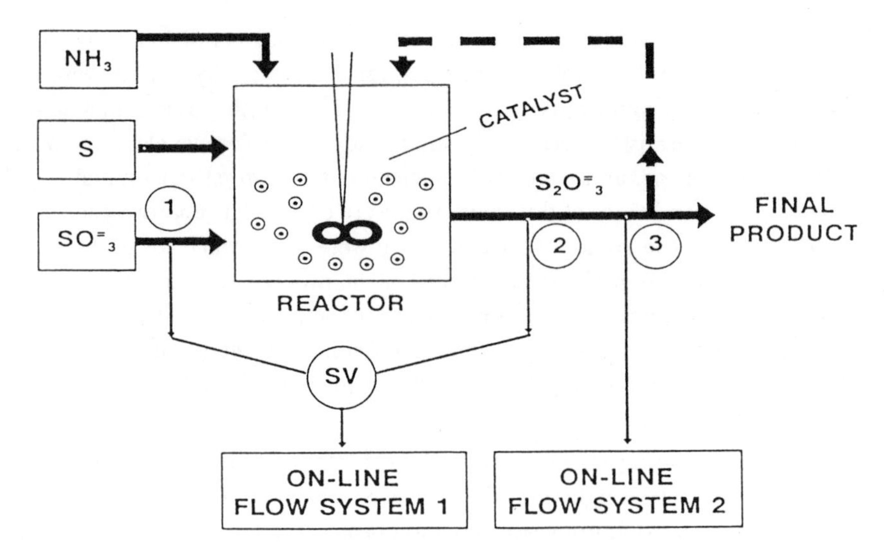

Fig. 21. Schematic diagram of the pilot plant and on-line monitoring in production of ammonium thiosulphate using FIA systems with spectrophotometric detection [499]. 1 to 3 — points for sampling; SV — switching valve. (*Reprinted by permission of copyright owner.*)

from production plants [500]. It was based on potentiometric detection with a chloride-selective electrode and prevalve dilution of samples. For the determination of morphine in process stream samples from extraction of alkaloids, an FIA system with chemiluminescence detection was developed [501]. The detection was based on the reaction of morphine and acidic permanganate solution in the presence of tetraphosphoric acid. Results of determinations in process liquors were in good agreement with HPLC results. For the determination of sulphuric acid in metallurgical process stream an FIA system for photometric titration using a bromophenol blue indicator was developed [502]. Determinations can be carried out with frequency 110 h^{-1}. For a wide range of metallurgical samples containing from 22 to 904 g l^{-1} H_2SO_4 results of determinations correlated well with the hydroxide batch titration method. Example curves for such titrations are shown in Fig. 22.

The on-line determinations of sulphite in brine were carried out in the flow injection system with spectrophotometric detection with disulphide 5,5′-dithiobis(2-nitrobenzoic acid) [503]. For process applications a reversed system was designed in order to eliminate refractive index effects and to reduce reagent consumption.

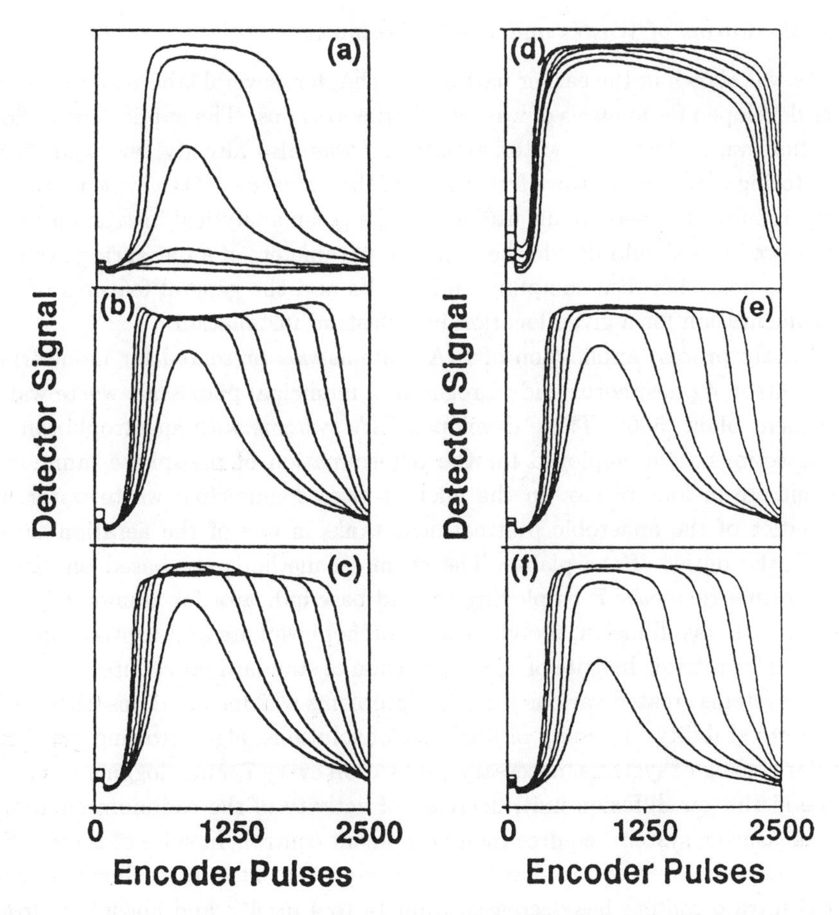

Fig. 22. The titration curves recorded for photometric titration of sulphuric acid in the presence of 0.1 M concentration of various metals in the FIA system using a bromophenol blue indicator [502]. a — As (V); b — Cr (III); c — Cu (II); d — Fe (III); e — Pb (II); f — Zn (II). Successive curves indicate increasing concentration of sulphuric acid from 0.0487 to 0.1455 M. (*Reprinted by permission of copyright owner.*)

For particular process application also two systems were developed for determination of total alkali and total phenol in alkaline phenoxine liquors as it occurs in industrial process streams [504]. Monitoring of alkali is based on a peak width measurement with conductometric detection, while total phenol is measured by reaction with hypochlorite and ammonia, where indophenol blue is monitored photometrically.

8.2. *Monitoring of Wastes and Waste Treatment*

As was shown in the earlier part of this chapter, several laboratory systems were developed for analysis of wastes of various origins. The application of flow injection methodology to water monitoring was also already reviewed [505]. Monitoring of wastes, or the effectiveness of their treatment, is a typical process analysis subject, where about usefulness of a given analytical instrumentation and procedures should decide the dynamic parameters of a measuring system, the time delay between sampling and results and the general features of the instrumentation for a given location in industrial installation.

Strictly process application of FIA systems was presented for monitoring and control of phosphorus and nitrogen in a municipal pilot scale waste water treatment plant [506]. Three developed FIA systems with spectrophotometric detections were employed for the determination of phosphate, ammonia and nitrate in four places: in the inlet of filtered municipal waste water, in the outlet of the anaerobic pretreatment tank, in one of the aeration tanks and in the outlet of the plant. The chemical methods are based on classical calorimetric methods employing an acid-base indicator for ammonia in the system with gas diffusion, molybdate for orthophosphate and azotisation and coupling for nitrite. In spite of the application of standard procedures the measuring systems created various technical problems which caused unsatisfactory long-term stability (necessity of filtration of solutions, algae growing, need for regular rinsing of systems, necessary calibration every 7 min, clogging of membrane in the gas-diffusion unit, decrease of activity of the cadmium column). The measuring system requires then permanent control. In spite of these difficulties the employed control has led to an improvement in the effluent quality as the nitrate content has decreased from 14 to 4 mg l^{-1} and phosphate from 6 to 3 mg l^{-1}.

Another spectrophotometric flow injection system was designed for the evaluation of a pilot plant decontamination of waste waters with a high content of chromium (VI) from leather factories [507]. A sample segment was introduced between solutions of various concentrations of diphenylcarbazide that allows one injection to obtain two signals corresponding to two different working ranges of concentration. In the system a sampling rate 45 h^{-1} was achieved. In the used process of waste treatment, Cr(VI) decontamination was based on the photoelimination in the presence of CdS as photocatalyst. The application of the developed procedure allows Cr(VI) monitoring in the concentration range of 0.01–4.9 mg l^{-1}.

The flow injection system fulfilling the industry specifications was developed for on-line determination of residual aluminium in potable and treated waters [508]. Aluminium was determined spectrophotometrically based on reaction with pyrocatechol violet. The portable monitor operates on a 30 min analytical cycle (48 results per day), which includes an autocalibration step. Over the course of the three-month field test, the reliability of the monitor was greater than 90%.

8.3. *On-Line Monitoring in Biotechnology*

The majority of developed authentic process applications of FIA were reported for biotechnological processes of different scales. Partly this can be attributed to the fact that many of these processes are carried out in conditions which are similar to those used in laboratory FIA measurements. On the other hand, it is overlapped with very intensive progress in the development of biosensors and their application in flow measurements has certain advantages. Flow injection methodology enables operating with small volumes of solutions taken from the reactor, and allows its easy dilution and on-line pretreatment if necessary. A short time of interaction of the sample with biosensor creates less serious problems in clogging of the detector surface. Flow conditions provide better possibilities of repetitive rinsing and conditioning of the biosensor. In very few cases other detections were used than enzymatic or immunochemical methods. Almost always systems for continuous control of biotechnological processes are equipped in a sample filtration device [509].

The largest number of systems for biotechnological applications were developed for the determination of glucose, a basic nutrient in many processes. The amperometric detections predominate in these designs [510–514], although also chemiluminescent [515], potentiometric [516] and enthalpimetric [517] detections were employed. An amperometric system with an oxygen electrode and glucose oxidase immobilised in a flow-through reactor was used for on-line process analysis and control of the production of alkaline protease, penicillin V and β-galactosidase/EcorI fusion protein [510]. The optimised enzyme cartridges with various enzymes, including glucose oxidase and pyranose oxidase [511], were utilised for on-line flow injection monitoring of glucose during the cultivation of *Saccharomyces cerevisiae* also with the use of an oxygen electrode for the detection [512]. Reactors with both enzymes can be used for 5000 assays with sodium azide as a supplement in the carrier buffer. For the monitoring of glucose in the production of L-lactic acid during fermentation in an aqueous two-phase system, an FIA step using micro-dialysis sampling and amperometric detection was developed [513]. As biosensor in flow cell a carbon paste electrode containing glucose oxidase and peroxidase was employed. In the same flow-through detector an analogous biosensor with lactate oxidase and peroxidase for monitoring of the lactate level was also used.

Glucose was also determined amperometrically in the FIA system for three-component monitoring in bioreactor fermentation broth (together with ethanol and glutamate) [514]. In this system three enzyme reactors, a six-position switching valve and a flow-through hydrogen peroxide electrode covered by a cellulose dialysis membrane were used (Fig. 23). Since a high salt content in the monitored fermentation broth affected the measurement, an additional ion-exchange column was inserted between the injector and the switching valve to remove the interfering chloride ions. Determinations were carried out with injected sample volume 2 μl.

A system with chemiluminescence detection of glucose was developed for monitoring lactic acid fermentations [515]. Simultaneously in the system, lactic acid, protein content and cell density were monitored. The detection of glucose was based on the reaction of hydrogen peroxide with luminol in the presence of Fe (III). Cells in the injected sample were retained by a membrane module, which also diluted the sample. In such a system the most important fermentation variables were measured every 10 min with a delay of less then 5 min. For monitoring the glucose concentration during cultivation of *Escherichia coli*, a flow injection setup with potentiometric detection with a pH-sensitive FET with glucose oxidase and catalase was employed [516]. The application of the detector was limited to three weeks because of membrane detachment. For process monitoring also a four-channel enzyme thermistor system in the FIA setup was applied [517]. For the glucose determination glucose oxidase was immobilised in a flow-through reactor and the system was employed for monitoring the cultivation of *Cephalosporium acremonium* for production of cephalosporin C and for production of protease.

Similarly to glucose, for monitoring of lactate very often amperometric systems are used [510–513, 518], mostly with immobilised lactate oxidase [510, 513, 518], but also with L-lactate-2-monooxigenase [512, 513]. These systems were developed for monitoring the production of penicillin V [510] and lactic acid [512, 513], but also for on-line monitoring of lactate formation during mammalian cell cultivation [518]. In the latter case lactate oxidase was immobilised on the membrane attached to the surface of a platinum working

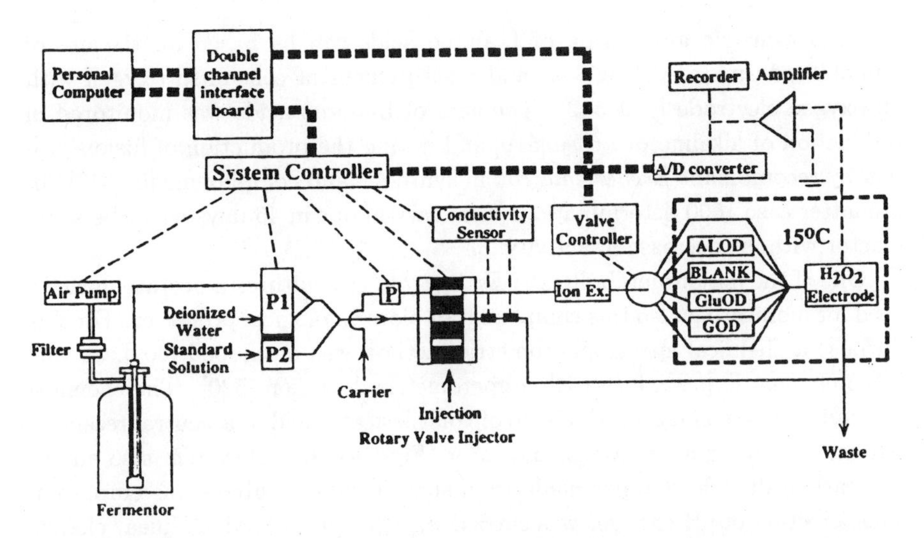

Fig. 23. Schematic diagram of the FIA system for sequential enzymatic monitoring of glucose, ethanol and glutamate in bioreactor fermentation broth containing a high salt concentration [514]. P, P1, P2 — pumps; Ion Ex. — ion-exchange column; GOD — reactor with glucose oxidase; ALOD — reactor with alcohol oxidase; GluOD — reactor with glutamate oxidase. (*Reprinted by permission of copyright owner.*)

electrode. The enzyme reaction was coupled with a water-soluble mediator 1,2-dimethylferricium-cyclodextrin inclusion complex. A dialysis membrane was placed over the lactate oxidase membrane to extend the linear detection range up to 40 mM. Mentioned already is the application of chemiluminescent system for detection of lactate [515]. For 13 h monitoring of lactate during a kefir fermentation a fibre-optic lactic acid biosensor in the flow injection system was employed [519]. The biosensor was based on an oxygen optode with immobilised lactate oxidase and consumption of oxygen was determined via dynamic quenching of the fluorescence of decacyclene by molecular oxygen. Using the zone sampling technique the linear range was extended in this system up to 60 mM.

Flow injection amperometry was also reported for the monitoring of ethanol with immobilised alcohol oxidase [511, 512, 514]. Ethanol was monitored during cultivation of *Saccharomyces cerevisiae* [512], and in fermentation broth containing a high salt content [514]. In the latter case simultaneously glutamate was measured (Fig. 23).

Amperometric monitoring of L-amino acids can be based on the use of immobilised amino acid oxidase and measurement of consumed oxygen with an oxygen electrode [510–512]. The sum of L-amino acids was monitored in production of alkaline protease [510] and during the production of fusion protein by recombinant *Escherichia coli* in synthetic and complex media [512]. In the latter case 1500 determinations were carried out in 46 days with the same reactor without decrease of activity.

One of the major metabolites of *Escherichia coli* cultures is acetate, hence a need for monitoring also this component in biotechnological processes. For this purpose a chemical spectrophotometric method was developed [509], as well as an enzymatic procedure with amperometric detection [520]. The chemical method was based on a gas diffusion of volatile acetic acid to a neutral receiving stream containing an acid-base indicator (Fig. 24). In the system also an on-line carbon dioxide stripper made from silicone-rubber tubing was employed. The detection of pH changes was carried out with phenol red. A linear change of the absorbance was observed up to 80 mM acetate in injected samples. In a 32 h fermentation process acetate were determined every 10 min. The enzymatic determination of acetic acid was based on the oxidation of sarcosine in the presence of sarcosine oxidase, which is competitively inhibited by acetate.

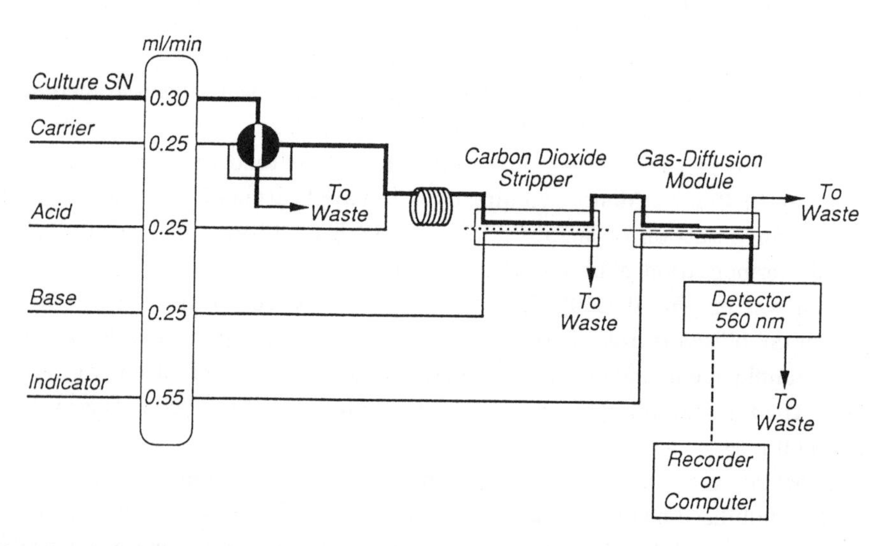

Fig. 24. Schematic diagram of the flow injection system for the determination of acetic acid with spectrophotometric detection in the fermentation process [509]. (*Reprinted by permission of copyright owner.*)

The measurement was carried out with an oxygen electrode and enzyme was immobilised in a flow-through reactor. The determined content of acetate during the fermentation process was up to 1 g l^{-1}.

The monitoring of phosphate in *Escherichia coli* fermentations was carried out in the FIA system with spectrophotometric detection using the molybdate method [509]. Phosphate is an essential nutrient and it can also be used to control the expression of some recombinant proteins. Due to the frequently occurring colour of fermentation media, measurements were carried out at 800 nm, where the background medium absorbance was much lower than at 660 nm, usually used for measurement of molybdenum blue. It was also necessary to detergent to diluent solution in order to prevent precipitation of proteins. In the fermentation process the concentration of phosphate was monitored by means of the FIA system up to 16 mM, and it was found to be more convenient than ion chromatography due to the larger sample throughput, simplicity, lower cost and better sensitivity in determination of low concentrations.

During some fermentation processes it is necessary to monitor the ammonium nitrogen concentration. In production of alkaline protease and penicillin for this purpose a flow injection potentiometry with an ammonia gas sensing electrode was used [510].

The urea in the cephalosporin cultivation medium was monitored in the range of 0.1–2 g l^{-1} using the FIA system with an enzyme-modified bio-field-effect transistor [616]. The gate of the commercial pH-sensitive ISFET was covered with a urease-containing layer. The cultivation medium was diluted 1:1 with potassium chloride solution of pH 10 to reduce the buffer capacity and to adjust the sample pH to 6. The operating time of the sensor was 4–5 weeks.

The same configuration of a detector with immobilised penicillin G amidase instead of urease was employed for determination of penicillin G in the medium of a *Penicillium chrysogenum* cultivation [516]. In the same process also the concentration of penicillin V was monitored with the use of a single fibre fluorescent optode with immobilised penicillinase [521]. In this case the penicillin V cultivations were performed in a synthetic medium with phenoxyacetic acid as a precursor. As label for fluorimetric detection fluorescein isothiocynatae was used. As the signal from an enzyme optode was extremely sensitive to changes of the pH value of the sample, a pH optode and a penicillin V optode were used simultaneously to lower the effect of pH. For measurements of penicillin V during penicillin fermentation an interesting comparison of the functional properties of the two on-line FIA enzymatic systems shown in Fig. 25 with

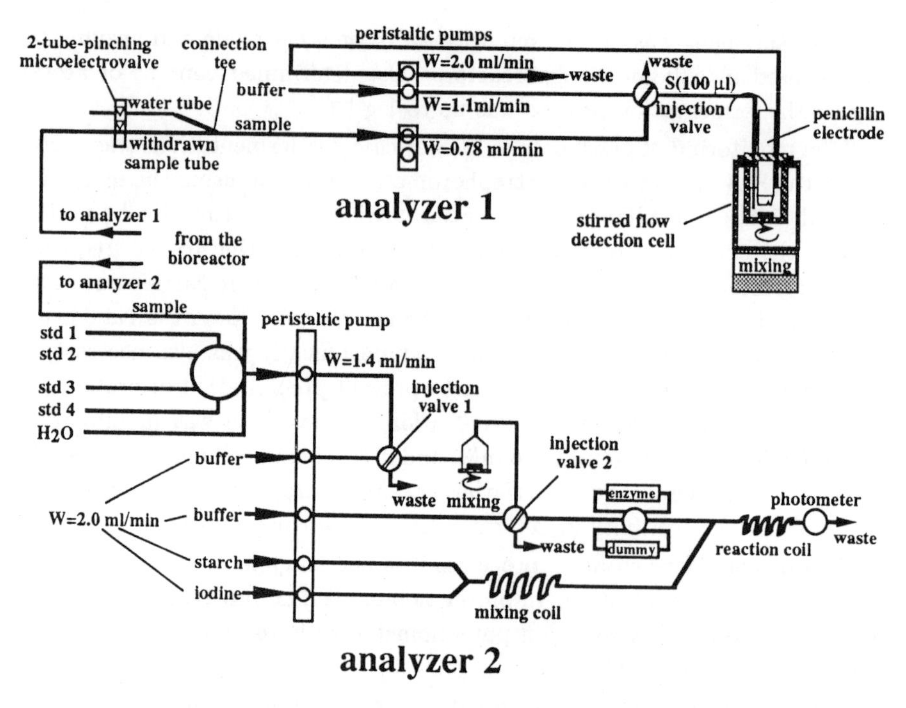

Fig. 25. Schematic diagrams of flow injection systems for on-line monitoring of penicillin V during penicillin fermentation with potentiometric detection using a penicillin biosensor (analyser 1) and with spectrophotometric detection based on the reaction of penicilloic acid with an iodine-starch complex (analyser 2) [522]. (*Reprinted by permission of copyright owner.*)

different methods of detection was reported [522]. In this case also the fermentation was carried out using a high yielding stain of *Penicillium chrysogenum* and both analysers were used for on-line monitoring for 200 h. In the potentiometric analyses the enzyme electrode consists of β-lactamase immobilised onto a pH glass electrode. In the other one with spectrophotometric detection the enzyme is immobilised in the reactor, and the product of enzymatic reaction penicilloic acid is detected by decolorisation of an iodine-starch complex. The configuration of the second setup allows detection of both penicillin and penicilloic acid. The potentiometric flow injection setup was evaluated as being more advantageous because of its simplicity and easy handling. It allows one to carry out measurements in the concentration range of 0.5–30 g l^{-1} potassium penicillin V with sampling rate 3 h^{-1}. The enzyme electrode needed to be rinsed every third day. In this system also the reagent consumption was much smaller.

In biotechnological processes also protein concentration changes are monitored. For determination of proteins as a source of nitrogen in lactic acid fermentation a flow injection system employing a biuret method was developed as it allows one to measure the total amount of peptide bonds [515]. The change of the colour of the biuret reagent was monitored photometrically. The system contained an on-line membrane module to separate cells from the sample before the mixing with the biuret reagent. For 50 μl injection volume the linear range of response was 0.2–8 g l^{-1} protein. The concentration of proteins, and also glucose, lactic acid and cell density, was measured during the process in-line every 10 min over many hours. The control of the level of specific proteins during the fermentation process requires application of immunochemical methods. For measurements of monoclonal antibodies produced in fermentations of mouse-mouse hybridoma cells and pullulanase isoenzymes produced in a fermentation by a *Clostridium* species, FIA systems with spectrophotometric detection were developed [523]. The monitoring is based on measurements of the turbidity caused by aggregates formed between the target proteins and their antibodies. The assays were optimised to cover concentration ranges of 1–1000 mg l^{-1} and were employed to monitor processes that took between 240 and 450 h by automatic activation three times a day for a triple assay. For future application in bioprocess monitoring also a flow injection setup with a calorimetric immunoassay for the determination of various IgG was developed [521]. This sandwich type assay was based on the use of an enzyme thermistor sensitive to glucose and on the use of protein A immobilised on a Sepharose support. A sandwich was formed by binding of IgG on immobilised protein A and subsequent binding of protein A labelled with β-galactosidase.

Besides the mentioned simple sugars, in some biotechnological processes it is necessary to detect changes of various disaccharides such as sucrose, maltose or lactose [510, 517]. In employed enzymatic determinations they are converted into simple sugars, and then glucose is determined enzymatically. The conversion of maltose is carried out in the presence of α-glucosidase, lactose with β-galactosidase and sucrose with invertase and mutarotase. The β-D-glucose produced in each of these reactions is detected amperometrically based on consumed oxygen [510]. These enzymatic reactions were also used in detection with a four-channel thermistor system [517]. In the systems with series of enzymatic reactors glucose present in the sample can be removed in a separate reactor by a glucose dehydrogenase. More practical and less expensive is on-line dilution of the sample after the reactor with glucose oxidase such that the level of glucose will be negligible compared to that from the conversion of disaccharides [510]. A continuous flow injection monitoring of maltose was reported in control of alkaline protease production [510], and lactose in production of penicillin V [510]. In protease production also glucose, sucrose and maltose were monitored using the mentioned thermistor system [517]. The monitoring of maltose was employed also to the control of a starch hydrolysis by immobilised thermostable enzymes with α-amylase and pullulanase activity [517]. The thermistor system was successfully employed for process control over periods of up to 300 h as a stand-alone version, controlled by a computer.

For bioprocess monitoring also a flow injection setup was developed for rapid determination of micro-organisms with amperometric detection [524]. Detection was based on reduction of suitable mediator potassium hexacyanoferrate (III) by microbial metabolism of fungi or bacteria. This determination was employed for on-line bioprocess control during an *E. coli* batch fermentation with sampling frequency up to 10 h^{-1}. The basic advantage of such a method is that it indicates the metabolic state of the microbial cells, which is a more important parameter than the increase in biomass.

Although so many numerous flow injection systems, based mainly on biodetection, have been reported for on-line monitoring of medium components in bioprocesses, suitable attention should be paid to possible faults and to the necessity of validation of the used method [525]. Serious errors in determinations can be caused by on-line aseptic sampling and by inappropriate preconditioning (dilution, pH and buffer capacity changes, precipitation of proteins in the manifold tubings). Especially in the initial stage of the method development, the concentration of monitored components should be measured with two independent instruments.

9. References

1. H. Casey and S. Smith, *Trends Anal. Chem.* **4** (1985) 256.
2. F. L. Boza, M. D. Luque de Castro and M. Valcarcel Cases, *Analusis* **13** (1985) 147.
3. J. F. van Staden, *Water SA* **13** (1987) 197.
4. J. F. van Staden, *Water SA* **15** (1989) 153.
5. M. Trojanowicz, R. L. Benson and P. J. Worsfold, *Trends Anal. Chem.* **10** (1991) 11.
6. M. Agudo, A. Rios and M. Valcarcel, *Trends Anal. Chem.* **13** (1994) 409.

7. K. N. Andrew, N. J. Blundell, D. Price and P. J. Worsfold, *Anal. Chem.* **66** (1994) 917A.
8. T. Korenaga and H. Ikatsu, *Anal. Chim. Acta* **141** (1982) 301.
9. J. M. H. Appleton, J. F. Tyson and R. P. Mounce, *Anal. Chim. Acta* **179** (1986) 269.
10. M. L. Balconi, M. Borgarello, R. Ferraroli and F. Realini, *Anal. Chim. Acta* **261** (1992) 295.
11. M. Novic, B. Pihlar and M. Duler, *Fresenius Z. Anal. Chem.* **332** (1988) 750.
12. T. Korenaga and H. Ikatsu, *Analyst* **106** (1981) 301.
13. T. Korenaga, X. Zhou, K. Okada and T. Moriwake, *Anal. Chim. Acta* **272** (1993) 237.
14. R. T. Edwards, I. D. McKelvie, P. C. Ferret, B. T. Hart, J. B. Bapat and K. Koshy, *Anal. Chim. Acta* **261** (1992) 287.
15. S. Motomizu, K. Toei, T. Kuwaki and M. Oshima, *Anal. Chem.* **59** (1987) 2930.
16. N. Maniasso, S. Sato, M. F. Gine and A. O. Jacintho, *Analyst* **121** (1996) 1617.
17. B. A. Woods, J. Ruzicka, G. D. Christian, N. J. Rose and R. J. Charlson, *Analyst* **113** (1988) 301.
18. R. G. J. Bellerby, D. R. Turner, G. E. Millward and P. J. Worsfold, *Anal. Chim. Acta* **309** (1995) 259.
18a. F. Canete, A. Rios, M. D. Luque de Castro and M. Valcarcel, *Analyst* **113** (1988) 739.
19. W. D. Basson and J. F. van Staden, *Fresenius Z. Anal. Chem.* **302** (1980) 370.
20. R.-M. Liu, D.-J. Liu and A.-L. Sun, *Analyst* **117** (1992) 1335.
21. E. H. Hansen, J. Ruzicka and A. K. Ghose, *Anal. Chim. Acta* **100** (1978) 151.
22. A. O. Jacintho, E. A. G. Zagatto, B. F. Reis, L. C. R. Pessenda and F. J. King, *Anal. Chim. Acta* **130** (1981) 361.
23. G. Nakagawa, H. Wada and C. Wei, *Anal. Chim. Acta* **145** (1983) 135.
24. S. Motomizu, M. Oshima, N. Yoneda and T. Iwachido, *Anal. Sci.* **6** (1990) 215.
25. B. F. Reis, M. A. Z. Arruda, E. A. G. Zagatto and J. R. Ferreira, *Anal. Chim. Acta* **206** (1988) 253.
26. A. N. Araujo, R. C. C. Costa, J. L. F. C. Lima and B. F. Reis, *Anal. Chim. Acta* **358** (1998) 111.
27. J. Alonso, J. Bartroli, J. L. F. C. Lima and A. A. S. C. Machado, *Anal. Chim. Acta* **179** (1986) 503.
28. J. Alonso-Chamarro, J. Bartroli, S. Jun, J. L. F. C. Lima and M. C. B. S. M. Montenegro, *Analyst* **118** (1993) 1527.
29. P. W. Alexander, T. Dimitriakopoulos and D. B. Hibbert, *Talanta* **44** (1997) 1397.
30. N. A. Chaniotakis, J. T. Tsagatakis, E. A. Moschou, S. J. West and X. Wen, *Anal. Chim. Acta* **356** (1997) 105.
31. J. Ruzicka, J. W. B. Stewart and E. A. G. Zagatto, *Anal. Chim. Acta* **81** (1976) 387.
32. W. D. Bason and J. F. van Staden, *Water Res.* **15** (1981) 333.
33. J. F. van Staden, *Fresenius Z. Anal. Chem.* **322** (1985) 36.

34. C. C. Oliveira, R. P. Sartini, E. A. G. Zagatto and J. L. F. C. Lima, *Anal. Chim. Acta* **350** (1997) 31.
35. M. Trojanowicz and W. Matuszewski, *Anal. Chim. Acta* **151** (1983) 77.
36. W. Frenzel, *Fresenius Z. Anal. Chim.* **335** (1989) 931.
37. J. F. L. C. Lima and L. S. M. Rocha, *Intern. J. Environ. Anal. Chem.* **38** (1990) 127.
38. F. J. Krug, H. Bergamin Filho, E. A. G. Zagatto and S. S. Jørgensen, *Analyst* **102** (1977) 503.
39. J. F. van Staden, *Fresenius Z. Anal. Chem.* **310** (1982) 239.
40. J. F. van Staden, *Fresenius Z. Anal. Chem.* **312** (1982) 438.
41. F. J. Krug, E. A. G. Zagatto, B. F. Reis, O. Bahia, A. O. Jacintho and S. S. Jørgensen, *Anal. Chim. Acta* **145** (1983) 179.
42. R. E. Santelli, P. R. S. Lopes, R. C. L. Santelli and A. L. R. Wagner, *Anal. Chim. Acta* **300** (1995) 149.
43. B. C. Madsen and R. J. Murphy, *Anal. Chem.* **53** (1981) 1924.
44. O. Kondo, H. Miyata and K. Toei, *Anal. Chim. Acta* **134** (1982) 353.
45. K. Ueno, F. Sagara, K. Higashi, K. Yakata, I. Yoshida and D. Ishii, *Anal. Chim. Acta* **261** (1992) 241.
46. A. Kojlo, J. Michalowski and M. Trojanowicz, *Anal. Chim. Acta* **228** (1990) 287.
47. O. Lutze, B. Ross and K. Cammann, *Fresenius J. Anal. Chem.* **350** (1994) 630.
48. T.-C. Tang and H. J. Huang, *Anal. Chem.* **67** (1995) 2299.
49. J. F. van Staden, A. E. Joubert and H. R. van Vliet, *Fresenius Z. Anal. Chem.* **325** (1986) 150.
50. J. R. Clinch, P. J. Worsfold and H. Casey, *Anal. Chim. Acta* **200** (1987) 523.
51. H. Casey, R. T. Clarke, S. M. Smith, J. R. Clinch and P. J. Worsfold, *Anal. Chim. Acta* **227** (1989) 379.
52. R. Nakata, M. Terashita, A. Nitta and K. Ishikawa, *Analyst* **115** (1990) 425.
53. S. Nakashima, M. Yagi, M. Zenki, A. Takahashi and K. Toei, *Fresenius Z. Anal. Chem.* **319** (1984) 506.
54. J. F. van Staden, *Anal. Chim. Acta* **138** (1982) 403.
55. M. J. Ahmed, C. D. Stalikas, S. M. Tzouwara-Karayanni and M. I. Karayannis, *Talanta* **43** (1996) 1009.
56. J. Slanina, F. Bakker, A. G. M. Brujin-Hes and J. J. Möls, *Fresenius Z. Anal. Chem.* **289** (1978) 38.
57. S. Motomizu, H. Mikasa and K. Toei, *Anal. Chim. Acta* **193** (1987) 343.
58. L. Renmin, L. Daojie, S. Ailing and L. Guihua, *Talanta* **42** (1995) 437.
59. T. Aoki and M. Wakabayashi, *Anal. Chim. Acta* **308** (1995) 308.
60. M. Trojanowicz, W. Matuszewski, B. Szostek and J. Michalowski, *Anal. Chim. Acta* **261** (1992) 391.
61. R. Liu, H. Wang, A. Sun and D. Liu, *Talanta* **45** (1997) 405.
62. F. J. Krug, J. Ruzicka and E. H. Hansen, *Analyst* **104** (1979) 47.
63. H. Bergamin, B. F. Reis, A. O. Jacintho and E. A. G. Zagatto, *Anal. Chim. Acta* **117** (1980) 81.

64. C. Pasquini and W. A. de Oliveira, *Anal. Chem.* **57** (1985) 2575.
65. R. Liu, B. Sun, D. Liu and A. Sun, *Talanta* **43** (1996) 1049.
66. D. J. Leggett, N. H. Chen and D. S. Mahadevappa, *Fresenius Z. Anal. Chem.* **315** (1983) 47.
67. E. Pobozy, K. Pyrzyñska, B. Szostek and M. Trojanowicz, *Microchem. J.* **51** (1995) 379.
68. T. Nakagama, M. Yamada and T. Hobo, *Analyst* **114** (1989) 1275.
69. M. Trojanowicz, W. Matuszewski and A. Hulanicki, *Anal. Chim. Acta* **136** (1982) 85.
70. N. Ishibashi, T. Imato, H. Ohura and S. Yamasaki, *Anal. Chim. Acta* **214** (1988) 349.
71. A. Sakai, A. Hemmi, H. Hachiya, F. Kobayashi, S. Ito, Y. Asano, T. Imato, Y. Fushinuki and I. Taniguchi, *Talanta* **45** (1998) 575.
72. W. Matuszewski and M. Trojanowicz, *Anal. Chim. Acta* **207** (1988) 59.
73. T. A. H. M. Janse, P. F. A. van der Wiel and G. Kateman, *Anal. Chim. Acta* **155** (1983) 89.
74. S. Motomizu and M. Oshima, *Analyst* **112** (1987) 295.
75. A. O. Jacintho, E. A. M. Kronka, E. A. G. Zagatto, M. A. Z. Arruda and J. R. Ferreira, *J. Flow Injection Anal.* **6** (1989) 19.
76. B. Y. Spivakov, T. A. Maryutina, L. K. Shpigun, V. M. Shkinev, Y. A. Zolotov, E. Ruseva and I. Havezov, *Talanta* **37** (1990) 889.
77. M. D. Luque de Castro and J. M. Fernandez-Romero, *Anal. Chim. Acta* **311** (1995) 281.
78. J. Slanina, W. A. Lingerar and F. Bakker, *Anal. Chim. Acta* **117** (1980) 91.
79. W. Frenzel and P. Brätter, *Anal. Chim. Acta* **188** (1986) 151.
80. M. Trojanowicz, P. W. Alexander and D. B. Hibbert, *Anal. Chim. Acta* **366** (1998) 23.
81. M. E. Leon-Gonzalez, M. J. Santos-Delgado and L. M. Polo-Diez, *Anal. Chim. Acta* **219** (1989) 329.
82. V. Marco, F. Carillo, C. Perez-Conde and C. Camara, *Anal. Chim. Acta* **283** (1993) 489.
83. J. L. Manzoori and A. Miyazaki, *Anal. Chem.* **62** (1990) 2457.
84. T. Deguchi, A. Tanaka, T. Sanemasa and H. Nogai, *Bunseki Kagaku* **32** (1983) 23.
85. I. Sekerka and J. F. Lechner, *Anal. Chim. Acta* **234** (1990) 199.
86. T. Lussier, R. Gilbert and J. Hubert, *Anal. Chem.* **64** (1992) 2201.
87. S. Motomizu, M. Oshima and Z. Jun, *Anal. Chim. Acta* **251** (1991) 269.
88. B. C. Madsen and M. S. Kromis, *Anal. Chem.* **56** (1984) 2849.
89. S. H. Sutheimer and S. E. Cabaniss, *Anal. Chim. Acta* **303** (1995) 211.
90. H. Salaciñski, P. G. Riby and S. J. Haswell, *Anal. Chim. Acta* **269** (1992) 1.
91. M. R. Pereiro Garcia, M. E. Diaz Garcia and A. Sanz Medel, *J. Anal. At. Spectrom.* **2** (1987) 699.
92. B. Fairman, A. Sanz Medel and P. Jones, *J. Anal. At. Spectrom.* **10** (1995) 281.
93. W. Frenzel, F. Titzenthaler and S. Elbel, *Talanta* **41** (1994) 1965.

94. M. Huang, S. Jiang and C. Hwang, *J. Anal. At. Spectrom.* **10** (1995) 31.
95. M. A. Marshall and H. A. Mottola, *Anal. Chem.* **57** (1985) 729.
96. V. Kuban, J. Komarek and Z. Zdrahal, *Coll. Czech. Chem. Commun.* **54** (1989) 1785.
97. A. M. Naghmush, M. Trojanowicz and E. Olbrych-Seleszyñska, *J. Anal. At. Spectrom.* **7** (1992) 323.
98. H. Chen, S. Xu and Z. Fang, *Anal. Chim. Acta* **298** (1994) 167.
99. V. Kuban, J. Komarek and D. E. Cajkova, *Coll. Czech. Chem. Commun.* **54** (1989) 2683.
100. J. Szpunar-Lobiñska and M. Trojanowicz, *Anal. Sci.* **6** (1990) 415.
101. Y. A. Zolotov, L. K. Shpigun, Y. A. Kolotyrkina, E. A. Novikov and O. V. Bazanova, *Anal. Chim. Acta* **200** (1987) 21.
102. M. M. Gomez Gomez and C. W. McLeod, *J. Anal. At. Spectrom.* **10** (1995) 89.
103. N. Ohno and T. Sakai, *Analyst* **112** (1987) 1127.
104. H. Ishii, M. Aoki, T. Aita and T. Odashima, *Anal. Sci.* **2** (1986) 125.
105. F. Lazaro, M. D. Luque de Castro and M. Valcarcel, *Anal. Chim. Acta* **219** (1989) 231.
106. Y. Zhang, P. Riby, A. G. Cox, C. W. McLeod, A. R. Date and Y. Y. Cheung, *Analyst* **113** (1988) 125.
107. A. M. Naghmush, K. Pyrzyñska and M. Trojanowicz, *Talanta* **42** (1995) 851.
108. J. A. Koropchak and L. Allen, *Anal. Chem.* **61** (1989) 1410.
109. P. Martinez-Jimenez, M. Gallego and M. Valcrcel, *Analyst* **112** (1987) 1233.
110. G. Tao and Z. Fang, *Fresenius J. Anal. Chem.* **360** (1998) 156.
111. J. F. van Staden and L. G. Kluever, *Anal. Chim. Acta* **350** (1997) 15.
112. Z. Fang and S. Xu, *Anal. Chim. Acta* **145** (1983) 143.
113. G. Tao and E. H. Hansen, *Analyst* **119** (1994) 333.
114. S. Nakano, M. Tago and T. Kawashima, *Anal. Sci.* **5** (1989) 69.
115. K. Kashiwabara, T. Hobo, E. Kobayashi and S. Suzuki, *Anal. Chim. Acta* **178** (1985) 209.
116. A. Maquierra, H. A. M. Elmahadi and R. Puchades, *Analyst* **121** (1996) 1633.
117. H. Narasaki and M. Ikeda, *Anal. Chem.* **56** (1984) 2059.
118. C. C. Y. Chan and R. S. Sadana, *Anal. Chim. Acta* **270** (1992) 231.
119. S. Zhang, S. Xu and Z. Fang, *Quim. Anal.* **8** (1989) 191.
120. M. Sperling, X. Yin and B. Welz, *J. Anal. At. Spectrom.* **6** (1991) 295.
121. P. Canizares and M. D. Luque de Castro, *Anal. Chim. Acta* **295** (1994) 59.
122. R. Torralba, M. Bonilla, A. Palacios and C. Camara, *Analusis* **22** (1994) 478.
123. J. L. Burguera, M. Burguera, C. Rivas and P. Carrero, *Talanta* **45** (1998) 531.
124. M. Burguera and J. L. Burguera, *J. Anal. At. Spectrom.* **8** (1993) 229.
125. A. G. Cox, I. G. Cook and C. W. McLeod, *Analyst* **110** (1985) 331.
126. J. Alonso, J. Bartroli, M. del Valle and R. Barber, *Anal. Chim. Acta* **219** (1989) 345.
127. A. P. Doherty, R. J. Forster, M. R. Smyth and J. G. Vos, *Anal. Chem.* **64** (1992) 572.
128. W. Jian and C. W. McLeod, *Talanta* **39** (1992) 1537.

129. L. Pitts, P. J. Worsfold and S. J. Hill, *Analyst* **119** (1994) 2785.
130. Y. Liu and J. D. Ingle Jr., *Anal. Chem.* **61** (1989) 525.
131. A. Hu, R. E. Dessy ande A. Graneli, *Anal. Chem.* **55** (1983) 320.
132. S. D. Hartenstein, G. D. Christian and J. Ruzicka, *Can. J. Spectrosc.* **30** (1985) 144.
133. S. Caroli, A. Alimonti, F. Petrucci and Zs. Horvath, *Anal. Chim. Acta* **248** (1991) 241.
134. D. Beauchemin and S. S. Berman, *Anal. Chem.* **61** (1989) 1857.
135. D. C. Colodner, E. A. Boyle and J. M. Edmond, *Anal. Chem.* **65** (1993) 1419.
136. J. Wang and R. He, *Anal. Chim. Acta* **294** (1994) 195.
137. Y. Zhou and G. Zhu, *Talanta* **44** (1997) 2041.
138. B. Romberg and H. Müller, *Anal. Chim. Acta* **353** (1997) 165.
139. M. del Valle, J. Alonso, J. Bartroli and I. Marti, *Analyst* **113** (1988) 1677.
140. F. Canete, A. Rios, M. D. Luque de Castro and M. Valcarcel, *Anal. Chem.* **60** (1988) 2354.
141. Ch. Wittmann and R. D. Schmid, *Sens. Actuators B* **15–16** (1993) 119.
142. J. Gaascon, A. Oubina, B. Ballesteros, D. Barcelo, F. Camps, M. P. Marco, M. A. Gonzales-Martinez, S. Morais, R. Puchades and A. Maquieira, *Anal. Chim. Acta* **347** (1997) 149.
143. M. de la Guardia, K. D. Khalaf, V. Carbonell and A. Morales-Rubio, *Anal. Chim. Acta* **308** (1995) 462.
144. Y. Daghbouche, S. Garrigues and M. de la Guardia, *Anal. Chim. Acta* **314** (1995) 203.
145. J. L. Marty, N. Mionetto, S. Lacorte and D. Barcelo, *Anal. Chim. Acta* **311** (1995) 265.
146. A. Günther and U. Bilitewski, *Anal. Chim. Acta* **300** (1995) 117.
147. J. J. Rippeth, T. D. Gibson, J. P. Hart, I. C. Hartley and G. Nelson, *Analyst* **122** (1997) 1425.
148. J. Müller and M. Martin, *Fresenius Z. Anal. Chem.* **329** (1988) 728.
149. E. Wang and S. Kamata, *Anal. Chim. Acta* **261** (1992) 399.
150. D. J. Legett, N. H. Chen and D. S. Mahadevappa, *Analyst* **107** (1982) 433.
151. P. I. Anagnostopoulou and M. A. Koupparis, *Anal. Chem.* **58** (1986) 322.
152. J. T. Hakedal and P. K. Egeberg, *Analyst* **122** (1997) 1235.
153. A. R. J. David, T. McCormack, A. W. Morris and P. J. Worsfold, *Anal. Chim. Acta* **361** (1998) 63.
154. M. Aoyagi, Y. Yasumasa and A. Nishida, *Anal. Chim. Acta* **214** (1988) 229.
155. K. Yoshimura, S. Nawata and G. Kura, *Analyst* **115** (1990) 843.
156. J. P. Sasanto, M. Oshima, S. Motomizu, H. Mikasa and Y. Hori, *Analyst* **120** (1995) 187.
157. J. Thomsen, K. S. Johnson and R. L. Petty, *Anal. Chem.* **55** (1983) 2378.
158. M. Sperling, X. Yin and B. Welz, *Analyst* **117** (1992) 629.
159. T. Yamane, K. Watnabe and H. A. Mottola, *Anal. Chim. Acta* **207** (1988) 331.
160. A. Malahoff, I. Y. Kolotyrkina and L. K. Shpigun, *Analyst* **121** (1996) 1037.
161. C. M. Sakamoto-Arnold and K. S. Johnson, *Anal. Chem.* **59** (1987) 1789.

162. K. K. Falkner and J. M. Edmond, *Anal. Chem.* **62** (1990) 1477.
163. V. A. Elrod, K. S. Johnson and K. H. Coale, *Anal. Chem.* **63** (1991) 893.
164. I. Y. Kolotyrkina, L. K. Shpigun, Y. A. Zolotov and A. Malahoff, *Analyst* 120 (1995) 201.
165. I. Y. Kolotyrkina, L. K. Shpigun, Y. A. Zolotov and G. I. Tsysin, *Analyst* **116** (1991) 707.
166. J. A. Resing and M. J. Mottl, *Anal. Chem.* **64** (1992) 2682.
167. N. Furuta, K. R. Brushwyler and G. M. Hieftje, *Spectrochim. Acta* **B44** (1989) 349.
168. M. G. C. Fernandez, M. A. Palacios and C. Camara, *Anal. Chim. Acta* **283** (1993) 386.
169. J. L. Nowicki, K. S. Johnson, K. H. Coale, V. A. Elroad and S. H. Lieberman, *Anal. Chem.* **66** (1994) 2732.
170. Z. Fang, T. Guo and B. Welz, *Talanta* **38** (1991) 613.
171. R. Ma, W. Van Mol and F. Adams, *Anal. Chim. Acta* **293** (1994) 251.
172. M. F. Gine, H. Brgamin, B. F. Reis and R. L. Tuon, *Anal. Chim. Acta* **234** (1990) 207.
173. A. Cerda, M. T. Oms, R. Forteza and V. Cerda, *Anal. Chim. Acta* **351** (1997) 273.
174. S. Hinkamp and G. Schwedt, *Anal. Chim. Acta* **236** (1990) 345.
175. J. L. Manzoori, A. Miyazaki and H. Tao, *Analyst* **115** (1990) 1055.
176. S. Auflitsch, D. M. W. Peat, I. D. McKelvie and P. J. Worsfold, *Analyst* **122** (1997) 1477.
177. F. Mas-Torres, A. Munoz, J. M. Estela and V. Cerda, *Analyst* **122** (1997) 1033.
178. E. H. Hansen, A. K. Ghose and J. Ruzicka, *Analyst* **102** (1977) 705.
179. E. A. G. Zagatto, A. O. Jacintho, J. Mortatti and H. Bergamin, *Anal. Chim. Acta* **120** (1980) 399.
180. D. Gabriel, J. Baeza, F. Valero and J. Lafuente, *Anal. Chim. Acta* 359 (1998) 173.
181. I. Gil Torro, J. V. Garcia Mateo and J. Martinez Calatayud, *Anal. Chim. Acta* **366** (1998) 241.
182. I. Papaefstathiou, M. T. Tena and M. D. Luque de Castro, *Anal. Chim. Acta* **308** (1995) 246.
183. J. F. van Staden, *Fresenius J. Anal. Chem.* **346** (1993) 723.
184. J. F. van Staden and C. J. Hattingh, *Talanta* **45** (1998) 485.
185. Z. Zhu and Z. Fang, *Anal. Chim. Acta* **198** (1987) 25.
186. H. Ma and J. Liu, *Anal. Chim. Acta* **261** (1992) 247.
187. J. F. van Staden, *Fresenius Z. Anal. Chem.* **326** (1987) 754.
188. M. G. Gleister, G. J. Moody and J. D. R. Thomas, *Analyst* **110** (1985) 113.
189. A. Rios, M. D. Luque de Castro and M. Valcarcel, *Analyst* **109** (1984) 1487.
190. C. M. C. M. Couto, J. L. F. C. Lima, M. C. B. S. M. Montenegro, B. F. Reis and E. A. G. Zagatto, *Anal. Chim. Acta* **366** (1998) 155.
191. J. Nyman and A. Ivaska, *Talanta* **40** (1993) 95.

192. M. I. Pascual-Reguera, I. Ortego-Carmona and A. Molina-Diaz, *Talanta* **44** (1997) 1793.
193. C. Pasquini, W. F. Jardim and L. C. de Faria, *J. Autom. Chem.* **10** (1988) 188.
194. C. Moreno, M. P. Manuel-Vez, I. Gomez and M. Garcia-Vargas, *Analyst* **121** (1996) 1609.
195. A. Rios, M. D. Luque de Castro and M. Valcarcel, *Analyst* **110** (1985) 277.
196. A. Cerda, M. T. Oms, R. Forteza and V. Cerda, *Anal. Chim. Acta* **311** (1995) 165.
197. Y. Hirai and K. Tomokuni, *Anal. Chim. Acta* **167** (1985) 409.
198. A. Safavi and A. A. Ensafi, *Anal. Chim. Acta* **252** (1991) 167.
199. I. da Cruz Vieira and O. Fatibello-Filho, *Anal. Chim. Acta* **366** (1998) 111.
200. J. W. Grate, R. Strebin, J. Janata, O. Egorov and J. Ruzicka, *Anal. Chim. Acta* **68** (1996) 333.
201. O. Egorov, M. J. O'Hara, J. Ruzicka and J. W. Grate, *Anal. Chem.* **70** (1998) 977.
202. R. R. Liversage, J. C. Van Loon and J. C. de Andrade, *Anal. Chim. Acta* **161** (1984) 275.
203. A. Morales-Rubio, M. L. Mena and C. W. McLeod, *Anal. Chim. Acta* **308** (1995) 364.
204. F. M. Bauza de Mirabo, A. Ch. Thomas, E. Rubi, R. Forteza and V. Cerda, *Anal. Chim. Acta* **355** (1997) 203.
205. A. Maquieira, H. A. M. Elmahadi and R. Puchades, *Anal. Chem.* **66** (1994) 1462.
206. A. Maquieira, H. A. M. Elmahadi and R. Puchades, *Anal. Chem.* **66** (1994) 3632.
207. M. J. Ahmed, C. D. Stalikas, P. G. Veltsistas, S. M. Tzouwara-Karayanni and M. I. Karayannis, *Analyst* **122** (1977) 221.
208. P. C. Hauser, *Anal. Chim. Acta* **278** (1993) 227.
209. Y. Yang, X. X. Zhang, T. Korenaga and K. Higuchi, *Talanta* **45** (1997) 445.
210. P. K. Dasgupta and V. K. Gupta, *Environ. Sci. Technol.* **20** (1986) 524.
211. S. Dong and P. K. Dasgupta, *Environ. Sci. Technol.* **21** (1987) 581.
212. E. A. G. Zagatto, I. L. Mattos and A. O. Jacintho, *Anal. Chim. Acta* **204** (1988) 259.
213. I. L. Mathas, E. A. G. Zagatto and A. O. Jacintho, *Anal. Chim. Acta* **214** (1988) 247.
214. D. P. Nikolelis and H. A. Mottola, *Anal. Chem.* **50** (1978) 1665.
215. C. A. Koerner and T. A. Nieman, *Anal. Chem.* **58** (1986) 116.
216. M. Kogure, H. Mori, H. Ariki, C. Kojima and H. Yamamoto, *Anal. Chim. Acta* **337** (1997) 107.
217. K. Matsumoto, O. Hamada, H. Ukeda and Y. Osajima, *Anal. Chem.* **58** (1986) 2732.
218. N. Kiba, Y. Inoue and M. Furusawa, *Anal. Chim. Acta* **243** (1991) 183.
219. A. Jain, A. Chaurasia and K. K. Verma, *Talanta* **42** (1995) 779.

220. J. Hernandez-Mendez, A. Mateos, M. J. A. Parra and G. G. de Maria, *Anal. Chim. Acta* **184** (1986) 243.
221. T. Perez-Ruiz, C. Martinez-Lozano and A. Sauz, *Anal. Chim. Acta* **308** (1995) 299.
222. A. Sano, T. Kuwayama, M. Furukawa, S. Takitani and M. Nakamura, *Anal. Sci.* **11** (1995) 405.
223. W. Matuszewski, M. Trojanowicz and L. Ilcheva, *Electroanalysis* **2** (1990) 147.
224. G. M. Greenway and P. Ongomo, *Analyst* **115** (1990) 1297.
225. O. Fatibello-Filho and I. da Cruz Vieira, *Anal. Chim. Acta* **354** (1997) 51.
226. X. Su, W. Wei, L. Nie and S. Yao, *Analyst* **123** (1998) 221.
227. E. Luque-Perez, A. Rios and M. Valcarcel, *Anal. Chim. Acta* **366** (1998) 231.
228. K. Grudpan, P. Sritharathikhun and J. Jakimunee, *Anal. Chim. Acta* **363** (1998) 199.
229. S. Morais, M. I. Alcaina-Miranda, F. Lazaro, M. Planta, A. Maquieira and R. Puchades, *Anal. Chim. Acta* **353** (1997) 245.
230. T. Perez-Ruiz, C. Martinez-Lozano, V. Tomas and R. Cadajus, *Talanta* **42** (1995) 391.
231. L. F. Capitan-Vallvey, M. C. Valencia and G. Mirou, *Anal. Chim. Acta* **289** (1994) 365.
232. T. Yao, R. Akasaka and T. Wasa, *Electroanalysis* **1** (1989) 413.
233. R. Puchades, A. Maquieira and L. Torro, *Analyst* **118** (1993) 855.
234. G. J. Moody, G. S. Sanghera and J. D. R. Thomas, *Analyst* **111** (1986) 605.
235. L. T. Di Benedetto, T. Dimitrakopoulos, J. R. Farrell and P. J. Iles, *Talanta* **44** (1997) 349.
236. J. F. Van Staden and A. van Rensburg, *Fresenius J. Anal. Chem.* **337** (1990) 393.
237. Y. P. de Pena, M. Gallego and M. Valcarcel, *Talanta* **42** (1995) 211.
238. P. Marstorp, T. Anfält and L. Andersson, *Anal. Chim. Acta* **149** (1983) 281.
239. W. Künnecke and R. D. Schmid, *Anal. Chim. Acta* **234** (1990) 213.
240. F. Lazaro, M. D. Luque de Castro and M. Valcarcel, *Anal. Chem.* **59** (1987) 1859.
241. I. L. Mattos, J. M. Fernandez-Romero, M. D. Luque de Castro and M. Valcarcel, *Analyst* **120** (1995) 179.
242. G. C. Luca, B. F. Reis, E. A. G. Zagatto, M. C. B. S. M. Montenegro, A. N. Araujo and J. C. F. C. Lima, *Anal. Chim. Acta* **366** (1998) 193.
243. J. Bartroli, M. Escalda, C. J. Jorquera and J. Alonso, *Anal. Chem.* **63** (1991) 2532.
244. M. Granados, S. Maspoch and M. Blanco, *Anal. Chim. Acta* **179** (1986) 445.
245. K. Matsumoto, H. Matsubara, H. Ukeda and Y. Osajima, *Agric. Biol. Chem.* **53** (1989) 2347.
246. M. O. Rezende and H. A. Mottola, *Analyst* **119** (1994) 2093.
247. T. Yao, M. Satomura and T. Nakahara, *Talanta* **41** (1994) 2113.
248. T. I. M. S. Lopes, A. O. S. S. Rangel, J. L. F. C. Lima and M. C. B. S. M. Montenegro, *Anal. Chim. Acta* **308** (1995) 122.

249. C. G. De Maria, T. M. Munoz, A. A. Mateous and L. G. De Maria, *Anal. Chim. Acta* **247** (1991) 61.
250. J. L. F. C. Lima, T. I. M. S. Lopes and A. O. S. S. Rangel, *Anal. Chim Acta* **366** (1998) 187.
251. M. C. Yebra, M. Gallego and M. Valcarcel, *Anal. Chim. Acta* **308** (1995) 357.
252. L. Ilcheva, R. Yanakiev, V. Vasileva and N. Ibekwe, *Food Chem.* **38** (1990) 105.
253. T. I. M. S. Lopes, A. O. S. S. Rangel, R. P. Sartini and E. A. G. Zagatto, *Analyst* **121** (1996) 1047.
254. J. A. Hamid, G. J. Moody and J. D. R. Thomas, *Analyst* **115** (1990) 1289.
255. N. Maniasso and E. A. G. Zagatto, *Anal. Chim. Acta* **366** (1998) 87.
256. D. S. de Jesus, R. J. Casella, S. L. C. Ferreira, A. C. S. Costa, M. S. de Carvalho and R. E. Santelli, *Anal. Chim. Acta* **366** (1998) 263.
257. M. Yamamoto, M. Yasuda and Y. Yamamoto, *Anal. Chem.* **57** (1985) 1382.
258. J. A. G. Mesa, P. Linares, M. D. Luque de Castro and M. Valcarcel, *Anal. Chim. Acta* **235** (1990) 441.
259. J. A. Garcia-Mesa, M. D. Luque de Castro and M. Valcarcel, *Anal. Chem.* **65** (1993) 3540.
260. R. Purohit and S. Devi, *Analyst* **120** (1995) 555.
261. Y. Sahleström, S. Twengström and B. Karlberg, *Anal. Chim. Acta* **187** (1986) 339.
262. J. M. Izquierdo-Ferrero, J. M. Fernandez-Romero and M. D. Luque de Castro, *Analyst* **122** (1997) 119.
263. G. del Campo, A. Irastorza and J. A. Casado, *Fresenius J. Anal. Chem.* **352** (1995) 557.
264. M. C. Yebra, M. Gallego and M. Valcarcel, *Anal. Chim. Acta* **308** (1995) 275.
265. T. Yao, M. Satomura and T. Wasa, *Anal. Chim. Acta* **261** (1992) 161.
266. F. J. Krug, J. Mortatti, L. C. R. Pessenda, E. A. G. Zagatto and H. Bergamin, *Anal. Chim. Acta* **125** (1981) 29.
267. M. A. Z. Arruda and E. A. G. Zagatto, *Anal. Chim. Acta* **199** (1987) 137.
268. A. R. A. Nogueira, S. M. B. Brienza, E. A. G. Zagatto, J. L. F. C. Lima and A. N. Araujo, *Anal. Chim. Acta* **276** (1993) 121.
269. I. Papaefstathiou and M. D. Luque de Castro, *Anal. Chem.* **67** (1995) 3916.
270. M. Silva, M. Gallego and M. Valcarcel, *Anal. Chim. Acta* **179** (1986) 341.
271. E. A. G. Zagatto, B. F. Reis, H. Bergamin and F. J. Krug, *Anal. Chim. Acta* **109** (1979) 45.
272. K. B. Male and J. H. T. Luong, *Biosensors Bioelectron.* **6** (1991) 581.
273. S. M. B. Brienza, R. P. Sartini, J. A. G. Neto and E. A. G. Zagatto, *Anal. Chim. Acta* **308** (1995) 269.
274. C. Pasquini and L. C. de Faria, *Anal. Chim. Acta* **193** (1987) 19.
275. E. A. G. Zagatto, F. J. Krug, H. Bergamin, S. S. Jørgensen and B. F. Reis, *Anal. Chim. Acta* **104** (1979) 279.
276. B. F. Reis, A. O. Jacintho, J. Mortatti, F. J. Krug, E. A. G. Zagatto, H. Bergamin and L. C. R. Pessenda, *Anal. Chim. Acta* **123** (1981) 221.

277. E. A. G. Zagatto, H. Bergamin, S. M. B. Brienza, M. A. Z. Arruda, A. R. A. Nogueira and J. L. F. C. Lima, *Anal. Chim. Acta* **261** (1992) 59.
278. M. A. Z. Arruda, M. Gallego and M. Valcarcel, *Anal. Chem.* **65** (1993) 3331.
279. J. A. Lown and D. C. Johnson, *Anal. Chim. Acta* **116** (1980) 41.
280. J. A. G. Neto, H. Bergamin, E. G. Zagatto and F. J. Krug, *Anal. Chim. Acta* **308** (1995) 439.
281. K. Pyrzyñska, Z. Janiszewska, J. Szpunar-Lobiñska and M. Trojanowicz, *Analyst* **119** (1994) 1553.
282. M. Trojanowicz and K. Pyrzyñska, *Anal. Chim. Acta* **287** (1994) 247.
283. T. Blanco, N. Manisasso, M. F. Gine and A. O. Jacintho, *Analyst* **123** (1998) 191.
284. F. Lazaro, M. D. Luque de Castro and M. Valcarcel, *Anal. Chim. Acta* **165** (1984) 177.
285. J. C. de Andrade, F. C. Strong III and N. J. Martin, *Talanta* **37** (1990) 711.
286. M. Mesquita, A. O. Jacintho, E. A. G. Zagatto and R. F. Antonia, *J. Braz. Chem. Soc.* **1** (1990) 28.
287. H. Bergamin, J. X. Medeiros, B. F. Reis and E. A. G. Zagatto, *Anal. Chim. Acta* **101** (1978) 9.
288. L. C. R. Pessenda, A. O. Jacintho and E. A. G. Zagatto, *Anal. Chim. Acta* **214** (1988) 239.
289. J. R. Ferreira, E. A. G. Zagatto, M. A. Z. Arruda and S. M. B. Brienza, *Analyst* **115** (1990) 779.
290. C. La Rosa, F. Pariente, L. Hernandez and E. Lorenzo, *Anal. Chim. Acta* **308** (1995) 129.
291. A. Hipolito-Moreno, M. E. Leon-Gonzalez, L. V. Perez-Arribus and L. M. Polo-Diaz, *Anal. Chim. Acta* **362** (1998) 187.
292. P. C. Rudner, J. M. C. Pavon, A. G. de Torres and F. S. Rojos, *Fresenius J. Anal. Chem.* **352** (1995) 615.
293. D. Beauchemin, K. W. M. Siu and S. S. Berman, *Anal. Chem.* **60** (1988) 2587.
294. D. W. Bryce, A. Izquierdo and M. D. Luque de Castro, *Anal. Chim. Acta* **308** (1995) 96.
295. M. Burguera, J. L. Burguera, C. Rivas, P. Carrero, R. Brunetto and M. Gallignani, *Anal. Chim. Acta* **308** (1995) 339.
296. C. Hongbo, E. H. Hansen and J. Ruzicka, *Anal. Chim. Acta* **169** (1985) 209.
297. J. Ruzicka and E. H. Hansen, *Anal. Chim. Acta* **161** (1984) 1.
298. J. Ruzicka, E. H. Hansen and E. A. G. Zagatto, *Anal. Chim. Acta* **88** (1977) 1.
299. E. H. Hansen, J. Ruzicka and A. K. Ghose, *Soil Nitrogen as Fertilizer or Pollutant.* IAEA, Vienna 1980, p. 77.
300. T. J. Cardwell, R. W. Cattrall, P. C. Hauser and I. C. Hamilton, *Anal. Chim. Acta* **214** (1988) 359.
301. D. M. W. Peat, I. D. McKelvie, G. P. Matthews, P. M. Haygarth and P. J. Worsfold, *Talanta* **45** (1997) 47.
302. J. F. van Staden, *Analyst* **112** (1987) 595.

303. D. Chen, F. Lazaro, M. D. Luque de Castro and M. Valcarcel, *Anal. Chim. Acta* **226** (1989) 221.
304. D. J. Hawke and H. K. J. Powell, *Anal. Chim. Acta* **299** (1994) 257.
305. D. Berggren, *Intern. J. Environ. Anal. Chem.* **41** (1990) 133.
306. S. S. Jørgensen and M. A. B. Regitano, *Analyst* **105** (1980) 292.
307. R. E. Taljaard and J. F. van Staden, *Anal. Chim. Acta* **366** (1998) 177.
308. D. Narinesingh, R. Mungal and T. T. Ngo, *Anal. Chim. Acta* **292** (1994) 185.
309. M. Trojanowicz, A. Hulanicki, W. Matuszewski, M. Palys, A. Fuksiewicz, T. Hulanicka-Michalak, S. Raszewski, J. Szyller and W. Augustyniak, *Anal. Chim. Acta* **188** (1986) 165.
310. E. A. Novikov, L. K. Shpigun and Y. A. Zalotov, *Anal. Chim. Acta* **230** (1990) 157.
311. P. C. Huaser and Z. P. Zhang, *Fresenius J. Anal. Chem.* **355** (1996) 141.
312. U. Narang, P. R. Gauger and F. S. Ligler, *Anal. Chem.* **69** (1997) 2779.
313. J. Szpunar-Lobiñska, *Anal. Chim. Acta* **251** (1991) 275.
314. S. Lin and H. Hwang, *Talanta* **40** (1993) 1077.
315. S. Xu, L. Sun and Z. Fang, *Anal. Chim. Acta* **245** (1991) 7.
316. T. Mochizuki, Y. Toda and R. Kuroda, *Talanta* **29** (1982) 659.
317. S. Kozuka, K. Saito, K. Oguma and R. Kuroda, *Analyst* **115** (1990) 431.
318. T. P. Lynch, N. J. Kernoghan and J. N. Wilson, *Analyst* **109** (1984) 843.
319. K. Oguma, K. Nishiyama and R. Kuroda, *Anal. Sci.* **3** (1987) 251.
320. R. Liu, D. Liu, A. Sun and G. Liu, *Analyst* **120** (1995) 565.
321. S. Pei and Z. Fang, *Anal. Chim. Acta* **294** (1994) 185.
322. T. P. Lynch, A. F. Taylor and J. N. Wilson, *Analyst* **108** (1983) 470.
323. W. Qin, Z. Zgang and C. Zhang, *Analyst* **122** (1997) 685.
324. M. F. Gine, F. J. Krug, H. Bergamin, B. F. Reis, E. A. G. Zagatto and R. E. Bruns, *J. Anal. At. Spectrom.* **3** (1988) 673.
325. N. Chimpalee, D. Chimpalee, S. Suparuknari, B. Boonyanitchayakul and D. T. Burns, *Anal. Chim. Acta* **298** (1994) 401.
326. M. de la Guardia, A. Morales-Rubio, V. Carbonell, A. Salvador, J. L. Burguera and M. Burguera, *Fresenius J. Anal. Chem.* **345** (1993) 579.
327. S. Sanchez-Ramos, M. J. Medino-Hernandez and S. Sagrado, *Talanta* **45** (1998) 835.
328. M. C. Valdes-Hevia y Temprano, J. P. Parajon, M. E. D. Garcia and A. Sanz-Medel, *Analyst* **116** (1991) 1141.
329. J. Wang, K. Ashley, E. R. Kennedy and Ch. Neumeister, *Analyst* **122** (1997) 1307.
330. E. H. Hansen, F. J. Krug, A. K. Ghose and J. Ruzicka, *Analyst* **102** (1977) 714.
331. M. Martinelli, H. Bergamin, M. A. Z. Arruda and E. A. G. Zagatto, *Quim. Anal.* **8** (1989) 129.
332. M. Trojanowicz and J. Szpunar-Lobiñska, *Anal. Chim. Acta* **230** (1990) 125.
333. J. Martinez Calatayud, R. M. Albert and P. Camplco, *Anal. Lett.* **20** (1987) 1379.
334. N. Zhou, W. Frech and E. Lundberg, *Anal. Chim. Acta* **153** (1983) 23.

335. J. F. Tyson and A. B. Idris, *Analyst* **109** (1984) 23.
336. P. B. Martelli, B. F. Reis, E. A. M. Kronka, H. Bergamin, M. Korn, E. A. G. Zagatto, J. L. F. C. Lima and A. N. Araujo, *Anal. Chim. Acta* **308** (1995) 397.
337. M. Harriot and D. T. Burns, *Anal. Proc.* **26** (1989) 315.
338. H. Bergamin, F. J. Krug, B. F. Reis, J. A. Nobrega, M. Mesquita and I. G. Souza, *Anal. Chim. Acta* **214** (1988) 397.
339. I. G. Souza, H. Bergamin, F. J. Krug, J. A. Nobrega, P. V. Oliveira, B. F. Reis and M. F. Gine, *Anal. Chim. Acta* **245** (1991) 211.
340. C. M. Wolff and H. A. Mottola, *Anal. Chem.* **50** (1978) 94.
341. M. Masoom and A. Townshend, *Anal. Chim. Acta* **166** (1984) 111.
342. W. V. de Alwis, B. S. Hill, B. I. Meiklejohn and G. S. Wilson, *Anal. Chem.* **59** (1987) 2688.
343. B. A. Petersson, *Anal. Chim. Acta* **209** (1988) 231.
344. W. Matuszewski, M. Trojanowicz and A. Lewenstam, *Electroanalysis* **2** (1990) 607.
345. J. Toei, *Analyst* **113** (1988) 475.
346. L. Gorton and L. Ögren, *Anal. Chim. Acta* **130** (1981) 45.
347. M. Akiba and S. Motomizu, *Anal. Chim. Acta* **214** (1988) 455.
348. D. Narinesingh, V. A. Stoute, G. Davis, F. Shaama and T. T. Ngo, *Anal. Lett.* **24** (1991) 727.
349. J. M. Fernandez-Romero and M. D. Luque de Castro, *Anal. Chem.* **65** (1993) 3048.
350. C. Ridder, E. H. Hansen and J. Ruzicka, *Anal. Lett.* **15** (1982) 1751.
351. D. Pilosof and T. A. Nieman, *Anal. Chem.* **54** (1982) 1698.
352. P. J. Worsfold, J. Farrelly and M. S. Matharu, *Anal. Chim. Acta* **164** (1984) 103.
353. P. Roehring, C. M. Wolff and J. P. Schwing, *Anal. Chim. Acta* **153** (1983) 181.
354. N. Kiba, A. Itagaki, S. Fukumara, K. Saegusa and M. Furusawa, *Anal. Chim. Acta* **354** (1997) 205.
355. T. Yao, M. Satomura and T. Nakahara, *Electroanalysis* **7** (1995) 143.
356. W. Matuszewski and M. Trojanowicz, *Pol. J. Chem.* **69** (1995) 1257.
357. F. Palmisano, D. Centonze and P. G. Zambonin, *Biosensors Bioelectron.* **9** (1994) 471.
358. T. Krawczyñski vel Krawczyk, M. Trojanowicz, A. Lewenstam and A. Moszczyñska, *Biosensors Bioelectron.* **11** (1996) 1155.
359. F. Palmisano, D. Centonze, M. Quinto and P. G. Zambonin, *Biosensors Bioelectron.* **11** (1996) 419.
360. J. Ruzicka, E. H. Hansen, A. K. Ghose and H. A. Mottola, *Anal. Chem.* **51** (1979) 199.
361. S. A. Rosario, G. S. Cha, M. E. Meyerhoff and M. Trojanowicz, *Anal. Chem.* **62** (1990) 2418.
362. T. Krawczyñski vel Krawczyk, M. Trojanowicz and A. Lewenstam, *Talanta* **41** (1994) 1229.

363. P. Solich, M. Polasek, R. Karlicek, O. Valentova and M. Marek, *Anal. Chim. Acta* **218** (1989) 151.
364. D. Narinesingh, R. Mungal and T. T. Ngo, *Anal. Biochem.* **188** (1990) 325.
365. L. C. de Faria, C. Pasquini and G. de Oliveira Neto, *Analyst* **116** (1991) 357.
366. M. T. Jeppsen and E. H. Hansen, *Anal. Chim. Acta* **214** (1988) 147.
367. F. Winquist, I. Lundström and B. Danielsson, *Anal. Chem.* **58** (1986) 145.
368. M. B. Madaras and R. P. Buck, *Anal. Chem.* **68** (1996) 3832.
369. A. Carpenter and W. C. Purdy, *Anal. Lett.* **23** (1990) 425.
370. T. Yao and T. Wasa, *Anal. Chim. Acta* **207** (1988) 319.
371. A. Krug, R. Göbel and R. Kellner, *Anal. Chim. Acta* **287** (1994) 59.
372. N. Kiba, M. Ogi and M. Furusawa, *Anal. Chim. Acta* **224** (1989) 133.
373. P. D. van der Wal, E. J. R. Sudhölter and D. N. Reinhoundt, *Anal. Chim. Acta* **245** (1991) 159.
374. J. M. Fernandez-Romero, M. D. Luque de Castro and R. Quiles-Zafra, *Anal. Chim. Acta* **308** (1995) 178.
375. K. Kimura, S. Iketani, H. Sakamoto and T. Shono, *Analyst* **115** (1990) 1251.
376. R. Goldik, C. Yarnitzky and M. Ariel, *Anal. Chim. Acta* **234** (1990) 161.
377. C. W. McLeod, P. J. Worsfold and A. G. Cox, *Analyst* **109** (1984) 327.
378. J. Michalowski and M. Trojanowicz, *Anal. Chim. Acta* **281** (1993) 299.
379. S. Xu and Z. Fang, *Microchem. J.* **51** (1995) 360.
380. R. Liu, D. Liu and A. Sun, *Talanta* **40** (1993) 511.
381. V. Kuban, D. B. Gladilovich, L. Sommer and P. Popov, *Talanta* **36** (1989) 463.
382. H. Wada, T. Murakawa and G. Nakagawa, *Anal. Chim. Acta* **200** (1987) 515.
383. K. McLaughlin, D. Dadgar, M. R. Smyth and D. McMaster, *Analyst* **115** (1990) 275.
384. T. A. Kelly and G. D. Christian, *Talanta* **29** (1982) 1303.
385. P. J. Worsfold, A. Hughes and D. J. Mowthorpe, *Analyst* **110** (1985) 1303.
386. T. H. Lee and M. E. Meyerhoff, *Anal. Chim. Acta* **229** (1990) 47.
387. Y. Xu, H. B. Halsall and W. R. Heineman, *Clin. Chem.* **36** (1990) 1941.
388. I. H. Lee and M. E. Meyerhoff, *Mikrochim. Acta* **III** (1988) 207.
389. E. P. Gil, H. T. Tang, H. B. Halsall, W. R. Heineman and A. S. Misiego,. *Clin. Chem.* **36** (1990) 662.
390. P. Allain, T. Turcant and A. Premel-Cobic, *Clin. Chem.* **35** (1989) 469.
391. M. Sanchez-Cabezudo, J. M. Fernandez-Romero and M. D. Luque de Castro, *Anal. Chim. Acta* **308** (1995) 159.
392. K. R. Wehmeyer, H. B. Halsall, W. R. Heineman, C. P. Volle and I. W. Chen, *Anal. Chem.* **58** (1986) 135.
393. H. T. Tang, H. B. Halsall and W. R. Heineman, *Clin. Chem.* **37** (1991) 245.
394. P. C. Gunaratna and G. C. Wilson, *Anal. Chem.* **65** (1993) 1152.
395. M. Wada, K. Nakashima, N. Kroda, S. Akiyama and K. Imai, *J. Chromatogr. B* **678** (1996) 129.
396. M. E. Meyerhoff and Y. M. Fraticelli, *Anal. Lett.* **24** (1981) 415.
397. R. Quiles, J. M. Fernandez-Romero, E. Fernandez and M. D. Luque de Castro, *Anal. Chim. Acta* **294** (1994) 43.

398. R. Quiles, J. M. Fernandez-Romero and M. D. Luque de Castro, *Clin. Chim. Acta* **235** (1995) 169.
399. M. D. Luque de Castro, R. Quiles, J. M. Fernandez-Romero and E. Fernandez, *Clin. Chem.* **41** (1995) 99.
400. T. Buch-Rasmussen, *Anal. Chim. Acta* **237** (1990) 405.
401. H. Gunasingham and C. H. Tan, *Analyst* **115** (1990) 35.
402. B. A. Petersson, *Anal. Lett.* **22** (1989) 83.
403. B. A. Petersson, *Anal. Chim. Acta* **209** (1988) 239.
404. B. A. Petersson, H. B. Andersen and E. H. Hansen, *Anal. Lett.* **20** (1987) 1977.
405. P. J. Worsfold, J. Ruzicka and E. H. Hansen, *Analyst* **106** (1981) 1309.
406. G. Maeder, J. L. Veuthey, M. Pelletier and W. Haerdi, *Anal. Chim. Acta* **231** (1990) 115.
407. G. Svensson and T. Anfält, *Clin. Chim. Acta* **119** (1982) 7.
408. Z. Fang, M. Sperling and B. Welz, *J. Anal. At. Spectrom.* **6** (1991) 301.
409. J. H. Aldstadt, D. F. King and H. D. Dewald, *Analyst* **119** (1994) 1813.
410. M. Burguera, J. L. Burguera, C. Rondon, C. Rivas, P. Carrero, M. Gallignani and M. R. Brunetto, *J. Anal. At. Spectrom.* **10** (1995) 343.
411. J. L. Burguera, M. Burguera and M. R. Brunetto, *At. Spectrosc.* **14** (1993) 90.
412. T. Sakai, H. Ohta, N. Ohno and J. Imai, *Anal. Chim. Acta* **308** (1995) 446.
413. W. Matuszewski, M. Trojanowicz , M. E. Meyerhoff, A. Moszczyñska and E. Lange-Moroz, *Electroanalysis* **5** (1993) 113.
414. C. S. Rui, Y. Kato and K. Sonomoto, *Biosensors Bioelectron.* **9** (1994) 429.
415. E. H. Hansen, S. K. Winther and M. Gundstrup, *Anal. Lett.* **27** (1994) 1239.
416. J. A. Infantes, M. D. Luque de Castro and M. Valcarcel, *Anal. Chim. Acta* **242** (1991) 179.
417. K. Mawatari, F. Iinuma and M. Watanabe, *Anal. Biochem.* **190** (1990) 88.
418. B. Laassis, J. J. Aaron and M. C. Mahadero, *Talanta* **41** (1994) 1985.
419. D. Gambart, S. Cardenas, M. Gallego and M. Valcarcel, *Anal. Chim. Acta* **366** (1998) 93.
420. S. M. Harden and W. K. Nonidez, *Anal. Chem.* **56** (1984) 2218.
421. J. L. Burguera, M. D. Brunetto, Y. Contreras, M. Burguera, M. Gallignani and P. Carrero, *Talanta* **43** (1996) 839.
422. J. L. Burguera, M. Burguera and C. Rivas, *Quim. Anal.* **16** (1997) 165.
423. A. G. Cox and C. W. McLeod, *Anal. Chim. Acta* **179** (1986) 487.
424. A. Karakaya and A. Taylor, *J. Anal. At. Spectrom.* **4** (1989) 261.
425. D. L. Luscombe, A. M. Bond, D. E. Davey and J. W. Bixler, *Anal. Chem.* **62** (1990) 27.
426. C. P. Hanna, J. F. Tyson and S. McIntosh, *Anal. Chem.* **65** (1993) 653.
427. M. Gallignani, H. Bahsas, M. R. Brunetto, M. Burguera, J. L. Burguera and Y. P. de Pena, *Anal. Chim. Acta* **369** (1998) 57.
428. A. Lorber, Z. Karpas and L. Halicz, *Anal. Chim. Acta* **334** (1996) 295.
429. D. R. Wiederin, R. E. Smyczek and R. S. Houk, *Anal. Chem.* **63** (1991) 1626.
430. M. Burguera, J. L. Burguera and O. M. Alarcon, *Anal. Chim. Acta* **214** (1988) 421.

431. C. Rondon, J. L. Burguera, M. Burguera, M. R. Brunetto, M. Gallignani and Y. Petiddepena, *Fresenius. Anal. Chem.* **353** (1995) 133.
432. X. Liu and Z. Fang, *Anal. Chim. Acta* **316** (1995) 329.
433. R. Liu, D. Liu, A. Sun and G. Liu, *Analyst* **120** (1995) 569.
434. T. D. Yerian, G. D. Christian and J. Ruzicka, *Anal. Chim. Acta* **204** (1988) 7.
435. M. R. Pereiro Garcia, A. L. Garcia, M. E. D. Garcia and A. Sanz-Medel, *J. Anal. At. Spectrom.* **2** (1987) 699.
436. K. K. Verma, A. Jain and K. K. Stewart, *Anal. Chim. Acta* **261** (1992) 261.
437. M. L. Ramos, J. F. Tyson and D. J. Curran, *Anal. Chim. Acta* **364** (1998) 107.
438. D. T. Burns, N. Chimpalee and S. Rattanariderom, *Anal. Chim. Acta* **243** (1991) 187.
439. A. V. Pereira and O. Fatibello-Filho, *Anal. Chim. Acta* **366** (1998) 55.
440. S. M. Sultan and N. I. Desai, *Talanta* **45** (1998) 1061.
441. K. K. Verma and K. K. Stewart, *Anal. Chim. Acta* **214** (1988) 207.
442. I. M. P. L. V. O. Ferreira, M. N. Macedo and M. A. Ferreira, *Analyst* **122** (1997) 1539.
443. R. M. Alonso, R. M. Jimenez, A. Carvajal, J. Garcia, F. Vicente and L. Hernandez, *Talanta* **36** (1989) 761.
444. J. L. Lopez Paz and J. Martinez Calatayud, *J. Pharm. Biomed. Anal.* **11** (1993) 1093.
445. A. M. Romero, G. G. Benito and J. Martinez Calatayud, *Anal. Chim. Acta* **282** (1993) 95.
446. J. Martinze Calatayud and S. S. Vives, *Pharmazie* **44** (1989) 614.
447. M. Milla, R. M. de Castro, M. Garcia-Vargas and J. A. Munoz-Leyva, *Anal. Chim. Acta* **179** (1986) 289.
448. S. R. Varma, J. Martinez Calatayud and H. A. Mottola, *Anal. Chim. Acta* **233** (1990) 235.
449. M. Strandberg and S. Thelander, *Anal. Chim. Acta* **145** (1983) 219.
450. Z. Bouhsain, S. Garrigues and M. de la Guardia, *Analyst* **121** (1996) 1935.
451. M. Blanco, J. Gene, H. Iturriaga and S. Maspoch, *Analyst* **112** (1987) 619.
452. A. Kojlo and J. Martinez Calatayud, *Anal. Chim. Acta* **308** (1995) 334.
453. J. Martinez Calatayud, A. S. Sampedro, P. V. Civera and C. G. Benito, *Anal. Lett.* **23** (1990) 2315.
454. T. Sakai, N. Ohno, Y. S. Chung and H. Nishikawa, *Anal. Chim. Acta* **308** (1995) 329.
455. R. S. Guerrero, S. S. Vives and J. Martinez Calatayud, *Microchem. J.* **43** (1991) 176.
456. P. Solich, M. Polasek and R. Karlicek, *Anal. Chim. Acta* **308** (1995) 293.
457. C. G. Benito. T. G. Sancho and J. Martinez Calatayud, *Anal. Chim. Acta* **279** (1993) 293.
458. J. Martinez Calatayud and C. G. Benito, *Anal. Chim. Acta* **231** (1990) 259.
459. J. Martinez Calatayud and C. G. Benito, *Anal. Chim. Acta* **245** (1991) 101.
460. C. Martinez-Lozano, T. Perez-Ruiz, V. Tomas and C. Abellan, *Analyst* **115** (1990) 217.

461. M. S. Bloomfield, S. Matchett and K. A. Prebble, *Analyst* **121** (1996) 1613.
462. I. G. Torro, J. V. G. Mateo and J. Martinez Calatayud, *Analyst* **122** (1997) 139.
463. Z. D. Zhang, W. R. G. Baeyens, X. R. Zhang and G. Van der Weken, *Analyst* **121** (1996) 1569.
464. N. T. Defteros and A. C. Calokerinos, *Anal. Chim. Acta* **290** (1994) 190.
465. J. Martinez Calatayud and C. M. Pastor, *Anal. Lett.* **23** (1990) 1371.
466. J. Martinez Calatayud, A. S. Sampedro and S. N. Sarrion, *Analyst* **115** (1990) 855.
467. J. Martinez Calatayud and J. V. G. Mateo, *Analyst* **116** (1991) 327.
468. L. L. Zamora, J. V. G. Mateo and J. Martinez Calatayud, *Anal. Chim. Acta* **265** (1992) 81.
469. S. L. Ortiz, J. G. Mateo and J. Martinez Calatayud, *Microchem. J.* **48** (1993) 112.
470. L. L. Zamora and J. Martinez Calatayud, *Talanta* **40** (1993) 1067.
471. R. Monero, M. Gallego and M. Valcarcel, *J. Anal. At. Spectrom.* **3** (1988) 725.
472. L. Rover, C. A. B. Garcia, G. de Oliveira Neto, L. T. Kubota and F. Galembeck, *Anal. Chim. Acta* **366** (1998) 103.
473. R. Gnanasekaran and H. A. Mottola, *Anal. Chem.* **57** (1985) 1005.
474. U. Forsman and A. Karlsson, *Anal. Chim. Acta* **139** (1982) 133.
475. T. Belal and J. L. Anderson, *Mikrochim. Acta* **II** (1985) 145.
476. J. Michalowski, A. Kojlo, B. Magnuszewska and M. Trojanowicz, *Anal. Chim. Acta* **289** (1994) 339.
477. J. V. G. Mateo and A. Kojlo, *J. Pharm. Biomed. Anal.* **15** (1997) 1821.
478. D. T. Burns and C. D. P. Dangolle, *Anal. Chim. Acta* **337** (1997) 113.
479. D. E. Davey, D. E. Mulcahy and G. R. O'Connell, *Talanta* **37** (1990) 313.
480. A. P. Doherty, M. A. Stanley, G. Arana, C. E. Koning, R. H. G. Brinkhuis and J. G. Vos, *Electroanalysis* **7** (1995) 333.
481. J. F. van Staden, H. du Plessis and R. E. Taljaard, *Anal. Chim. Acta* **357** (1997) 141.
482. M. E. Meyerhoff, M. Trojanowicz and B. O. Palsson, *Biotechn. Bioeng.* **41** (1993) 964.
483. X. Liu and E. H. Hansen, *Anal. Chim. Acta* **326** (1996) 1.
484. S. A. Rosario, M. E. Meyerhoff and M. Trojanowicz, *Anal. Chim. Acta* **258** (1992) 281.
485. B. O. Palsson, B. Q. Shen, M. E. Meyerhoff and M. Trojanowicz, *Analyst* **118** (1993) 1361.
486. W. Matuszewski, S. A. Rosario and M. E. Meyerhoff, *Anal. Chem.* **63** (1991) 1906.
487. Y. L. Huang, S. B. Khoo and M. G. S. Yap, *Anal. Lett.* **28** (1995) 593.
488. U. Spohn, D. Narasaiah, L. Gorton and D. Pfeiffer, *Anal. Chim. Acta* **319** (1996) 79.
489. M. Hikuma, H. Obana, T. Yasuda, I. Karube and S. Suzuki, *Anal. Chim. Acta* **116** (1980) 61.

490. I. Schneider, *Anal. Chim. Acta* **166** (1984) 293.
491. G. Decristoforo and B. Danielsson, *Anal. Chem.* **56** (1984) 263.
492. W. E. van der Linden, *Anal. Chim. Acta* **179** (1986) 91.
493. M. J. Whitaker, *Am. Lab. March* (1986) 154.
494. J. Ruzicka *Anal. Chim. Acta* **190** (1986) 155.
495. M. Gisin and C. Thommen, *Trends Anal. Chem.* **8** (1989) 62.
496. P. M. E. M. van der Grinten, *Control Eng.* **10** (1963) 51, 87; *J. Instrum. Soc. Am.* **12** (1965) 87; **13** (1966) 58.
497. M. L. Balconi, F. Sigon, R. Ferraroli and F. Realini, *Anal. Chim. Acta* **214** (1988) 367.
498. D. C. Olson, S. R. Bysouth, P. K. Dasgupta and V. Kuban, *Process Contr. Qual.* **5** (1994) 259.
499. J. S. Cosano, A. Izquierdo, M. D. Luque de Castro, M. Valcarcel, C. Aguilar and G. Penelas, *Anal. Chim. Acta* **308** (1995) 187.
500. J. F. van Staden, *Fresenius Z. Anal. Chem.* **328** (1987) 68.
501. N. W. Barnett, D. G. Rolfe, T. A. Bowser and T. W. Paton, *Anal. Chim. Acta* **282** (1993) 551.
502. T. J. Cardwell, R. W. Cattrall, G. J. Cross, G. R. O'Connell, J. D. Petty and G. D. Scollary, *Anal. Chim. Acta* **308** (1995) 197.
503. P. MacLaurin, K. S. Parker, A. Townshend, P. J. Worsfold, N. W. Barnett and M. Crane, *Anal. Chm. Acta* **238** (1990) 171.
504. J. A. Sweileh, P. K. Dasgupta and J. L. Lopez, *Mikrochim. Acta* **III** (1996) 175.
505. K. N. Andrew, N. J. Bundell, D. Price and P. J. Worsfold, *Anal. Chem.* **18** (1994) 917A.
506. K. M. Pedersen, M. Kümmel and H. Søeberg, *Anal. Chim. Acta* **238** (1990) 191.
507. J. Alonso-Chamarro, J. Bartroli and R. Barber, *Anal. Chim. Acta* **261** (1992) 219.
508. R. L. Benson, P. J. Worsfold and F. W. Sweeting, *Anal. Chim. Acta* **238** (1990) 177.
509. L. W. Forman, B. D. Thomas and F. S. Jacobson, *Anal. Chim. Acta* **249** (1991) 101.
510. K. Schügerl, L. Brandes, T. Dullan, K. Holzhauer-Rieger, S. Hotop, U. Hübner, X. Wu and W. Zhou, *Anal. Chim. Acta* **249** (1991) 87.
511. H. Jürgens, R. Kabuss, T. Plumbaum, B. Weigel, G, Kretzmer, K. Schügerl, K. Andres, E. Ignatzek and F. Giffhorn, *Anal. Chim. Acta* **298** (1994) 141.
512. H. Jürgens, L. Brandes, R. Joppien, M. Siebold, J. Schubert, X. Wu, G. Kretzmer and K. Schügerl, *Anal. Chim. Acta* **302** (1995) 289.
513. R. W. Min, V. Rajendran, N. Larsson, L. Gorton, J. Planas and B. Hahn-Hägerdal, *Anal. Chim. Acta* **366** (1998) 127.
514. R. L. C. Chen and K. Matsumoto, *Anal. Chim. Acta* **308** (1995) 145.
515. K. Nikolajsen, J. Nielsen and J. Villadsen, *Anal. Chim. Acta* **214** (1988) 137.
516. U. Brand, B. Reinhardt, F. Rüther, T. Scheper and K. Schügerl, *Anal. Chim. Acta* **238** (1990) 201.

517. N.-G. Hundeck. A. Sauerbrei, U. Hübner, T. Scheper, R. Koch and G. Antranikian, *Anal. Chim. Acta* **238** (1990) 211.
518. K. B. Male, P. O. Gartu, A. A. Kamen and J. H. T. Luong, *Anal. Chim. Acta* **351** (1997) 159.
519. B. A. A. Dremel, W. Yang and R. D. Schmid, *Anal. Chim. Acta* **234** (1990) 107.
520. M. Tservistas, B. Weigel and K. Schügerl, *Anal. Chim. Acta* **316** (1995) 117.
521. T. Scheper, W. Brandes, H. Maschke, F. Plötz and C. Müller, *J. Biotechnol.* **31** (1993) 345.
522. M. Carlsen, C. Johansen, R. W. Min, J. Nielsen, H. Meier and F. Lantreibecq, *Anal. Chim. Acta* **279** (1993) 51.
523. R. Freitag, C. Fenge, T. Scheper, K. Schügerl, A. Spreinat, G. Antranikian and E. Fraune, *Anal. Chim. Acta* **249** (1991) 113.
524. T. Ding and R. D. Schmid, *Anal. Chim. Acta* **234** (1990) 247.
525. K. Schügerl, *J. Biotechnol.* **31** (1993) 241.

Chapter 9

Sequential and Batch Injection Techniques

FIA methods, so widely accepted for many years in analytical laboratories by research people and analysts in routine laboratories, have an important functional advantage. At large effectiveness of determinations they usually require a small volume of the sample and they do not consume large volumes of reagents. This is an advantage compared to conventional wet analytical procedures or titration methods. This is, however, not sufficiently advantageous compared to numerous modern discrete analyzers used mainly in clinical analysis [1, 2].

In order to compete with these techniques the newest methodology of flow injection meaurements, so-called *sequential injection analysis* (SIA), has been developed. This term is already widely accepted in the literature, although perhaps more appropriate and informative would be *sequential flow injection analysis*, as the most important feature of this method is a flow of the sample solution through the detector during a detection. Continuous delivery of carrier solution and solutions of reagents in typical systems with segmented flow or FIA has been replaced by a technique employed in numerous discrete analysers. Using mostly syringe pumps and a multiport rotary selection valve, to the tubing of the measuring system in an appropriate sequence segments of the sample and reagents are introduced. After changing the direction of the flow and switching one or more selecting valves, solutions are delivered to the detector. On their way segemnts of solutions overlap and mix, and then a steady-state equilibrium signal is measured in the detector. This concept was utilised in several clinical analysers equipped with one pump and valve or several such elements, e.g. [3, 4]. A similar concept of solution handling with automatic burette and multiport valve has been employed in a photometric titrator with the use of a syringe pump also as a flow-through photometric detector, where a reaction takes place, which is involved in the detection principle [5].

1. Principle of Measurement and Basic Instrumentation of SIA

The principle of sequential injecion flow measurements invented by Ruzicka and Marshall [6] is that an analytical signal is measured in a flow-through detector in the course of the flow of series of liquid segments of sample and reagents through a detector. Therefore it is a typical flow measurement with essentially reduced consumption of reagents compared to other flow techniques, including FIA, even in a very economical version of merging zones. In the latter, there are continuously delivered at least two carrier solutions, whereas in SIA there is only one. Sequential injection analysis is the newest generation of flow measurement techniques in chemical analysis [7], and it quickly gains numerous applications in various fields of chemical analysis [8].

Sources of the concept of methodology of SIA can be found in flow injection works with the use of sinusoidal flow pumps [9, 10], or even earlier in works on flow systems with reversed or oscillating flow [11–15]. Twice or multiple pumping of the sample segment through the same detector at different times and with different dispersions of the sample allows information to be obtained about the kinetics of the chemical system involved as well as performing determinations based on kinetics. Oscillating flow has been found useful in improving the contact of the reactants in heterogeneous flow injection systems. The use of a cam-driven piston pump allows one to obtain a flow of the solution in the system, where the flow rate and its direction in the function of time follows a sinusoidal pattern. The integration of the sinusoidal flow pump with a computer-controlled multi-point injection valve allows simultaneous and synchronised sample zone injection and stream switching [9]. The use of piston pumps ensures the extraordinary smoothness of all flow injection profiles and the stability of the base-line. Such a system can be successfully used for performing stopped-flow or flow-reversed measurements.

In a typical SIA system both piston and peristaltic pumps can be successfully used (Fig. 1A). Sequential injection uses a selector valve (instead of an injection valve), through which given measured volumes of the carrier stream, sample solution and reagent solution are aspirated into a holding coil by means of a pump, which is capable of forward-reverse movement. As a result of this the sample and reagent zones are stacked in the holding coil adjacent to each other. When the valve is switched into the detector position and the flow is reversed, the stacked zones are pumped to the detector through the valve and reactor. During this flow segments of sample and reagents interdisperse and a detectable product P is formed in the overlapped zone (Fig. 1B). The

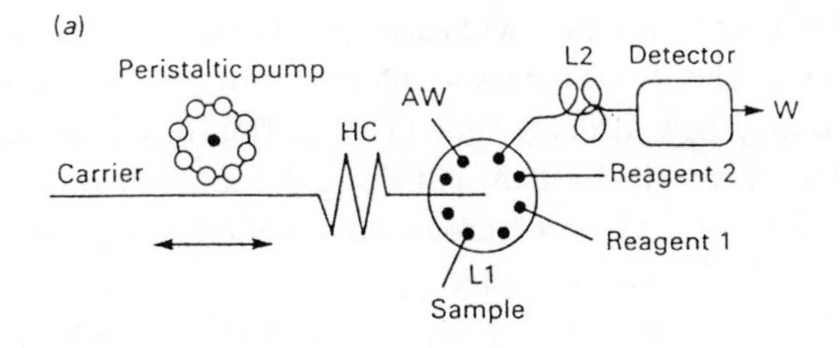

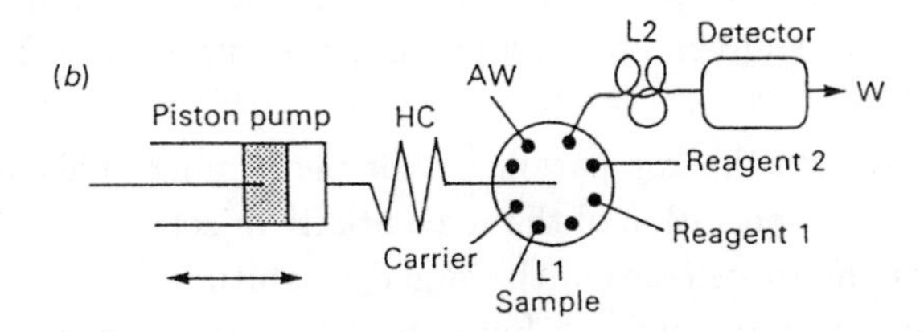

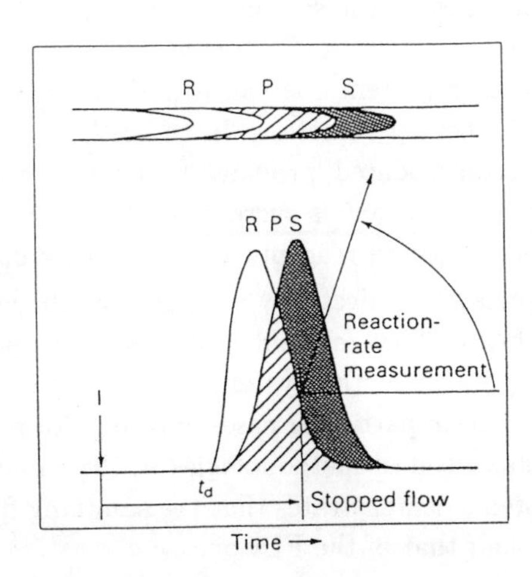

Fig. 1. Principle of sequential injection analysis concept [16]. (A) Flow schemes in the systems with (a) peristaltic pump or (b) piston pump; HC — holding coil; L1 — sampling line; L2 — flow-through reactor; W — waste; AW — auxiliary waste. (B) Structure of injected zones and concentration profiles as seen by the detector; R — reagent; S — sample; P — composite region where the analyte is transformed into a detectable product. (*Reprinted by permission of copyright owner.*)

peak observed at continuous flow will correspond to overlapped parts of sample and reagent peaks. Besides the measurement of transient signal by continuous flow also by stopped flow at delay time t_d a reaction-rate measurement can be carried out, as is schematically shown in Fig. 1B, too. Contrary to typical multichannel FIA systems a SIA measurements are carried out usually in a single channel setup, with the use of a single pump and single valve. As the precision and accuracy of determination depends on the accuracy of introducing a fixed volume of the sample and reagent solutions into a holding coil, a system has to be equipped in a complex and precise selector valve, in a pump with a precise function of sudden stop and go, and the system should be controlled by a computer.

In an SIA system the key parameter for performing analytical determination is a degree of penetration of the sequentially injected zones [10]. Similarly to the FIA system an increase of the signal magnitude can be achieved by an increase of the injected sample volume; however, an increase in sensitivity is possible only when a sufficient reagent excess is available in the overlapped zone. A simultaneous increase of the sample volume and reagent above a certain value in a given system is not an effective way to increase the sensitivity of a measurement. For each system one can experimentally adjust the optimum magnitude of both these parameters. In the SIA system a two-reagent chemistry can also be accommodated, provided that the sample volume is sufficiently small, that the sample zone is surrounded by the reagent zones, and that the concentrations of injected reagents are sufficiently large.

The zone penetration is more effectively promoted by the length of reversed flow than by the number of flow reversals [10]. As multiple flow reversals increase the overall dispersion in the system and the time to complete the measurement, they are used in particular cases only, e.g. for mixing of zones of different viscosity. Comparison of methodologies of the same determinations in the FIA and SIA systems demonstrates that the sampling frequencies of the SIA systems are about half that of the FIA ones.

In assembling of the SIA system a high precision piston pump can be replaced not only by a peristaltic pump [16] but also by a mechanised burette actuated by a stepper motor [17]. In both cases precision at level 1% RSD can be assumed as satisfactory.

A particular version of SIA instrumentation is the one where all conduits are capillaries and the fluid driving system is an electroosmotic pump [18]. In a system the high voltage source is used for capillary electrophoresis and a system contains a modified in-house electropneumatically operated selector valve. A schematic arrangement of this system for spectrophotometric determination of nitrite or ammonia is shown in Fig. 2. In a conventional multiport selector valve a rotor slot volume was changed from about 10 μl to about 50 nl. The pumping rate can be adjusted by the applied voltage and also by the number of pump capillaries in parallel. In determination of ammonia in the system with three reagents the time of determination was about 2.5 min, and precision at 0.2 mM level expressed by a peak height RSD of 0.8%. A similar system with a membrane sampling interface for gaseous analytes has been developed for determination of ammonia employing a miniature diffusion scrubber with a porous polypropylene tube [19].

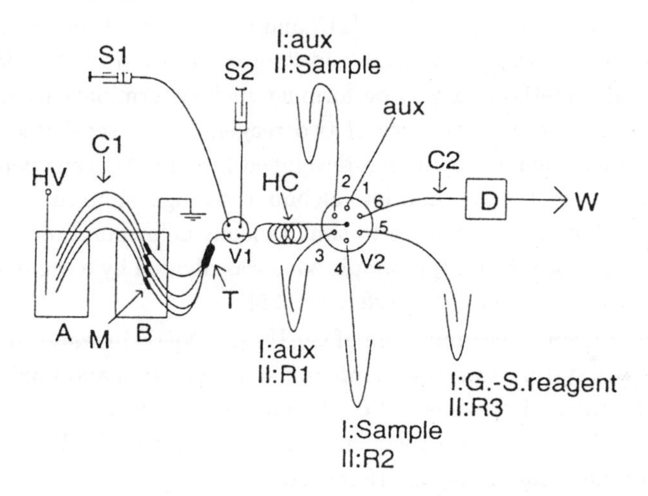

Fig. 2. Schematic diagram of the SIA system in capillary format with an electro-osmotic pump [18]. HV — high voltage power supply; A, B — pumping electrolyte solution containers; M — membrane joint; C1 — pumping capillary; T — connector; HC — holding coil; V1 — four-way valve; S1 and S2 — syringes; V2 — six-way selector valve; R1, R2, R3 — reagents; aux — unused auxiliary solution port. I — sample/reagent arrangements for the nitrite determination system. II — arrangements for ammonia determination. (*Reprinted by permission of copyright owner.*)

2. Measurement Techniques and On-Line Sample Pretreatment in SIA

Like in FIA systems, in sequential analysis systems the most often utilised technique is measurement at continuous flow in systems similar to that shown in Fig. 1 and the analytical signal for a single sample is recorded as a peak. The height of a signal is proportional to the analyte concentration in the sample. Calibration of the system is based on injection of several standard solutions.

As a way to simplify a conventional procedure the sample was proposed as a carrier in the SIA system [20]. This was employed for spectrophotometric determination of iron (II) and phosphate. Determinations were carried by the method of standard addition; however, the measuring system requires a thorough optimisation in order to eliminate the carry-over errors between samples.

In different ways in SIA systems a simultaneous determinations of e.g. two analytes can be carried out. In the simplest case, schematically illustrated by a manifold in Fig. 3A, two independent detectors can be incorporated into the system. Such setups with two spectrophotometric detectors were developed for waste water quality monitoring [21], and with a flow-through cell with two ISE's for determination of chloride and fluoride in waters [22]. More characteristic for SIA methodology is performing such determinations utilising the sandwich technique with the use of two reagent zones separated by a large sample volume. Such a system was developed for simultaneous determination of nitrite and iron (II) [23], and for calcium and magnesium in waters [24]. In the application of this technique to simultaneous determination of phosphate and silicate in waste water, the sample zone was divided by a segment of oxalic acid to eliminate the mutual interference [25].

For simultaneous determination of Co(II) and Ni(II) in water and soil samples in the SIA system with spectrophotometric detection and mixing chamber the different rates of dissociation of the citrate complexes were utilised [26]. The detection is based on a subsequent rapid reaction with PAR in the presence of EDTA to mask the major interferences.

This methodology with stopped flow can be successfully utilised for kinetic measurements. The principle of such a measurement is shown by the schematic diagram in Fig. 1B. On arrival at the detector of the product-containing zone, it is captured with the observation field of the detector and the changes of signal are observed in a given period of stopped flow. An example of this application can be determination of the activity of amyloglucosidase, which is widely used in the food industry [28]. The assay is based on three consecutive enzymatic reactions catalysed by analyte enzyme, mutarotase and glucose hydrogenase. In the developed SIA system a mixing chamber and detection of NADH at 340 nm were used (Fig. 4). This mode of SIA measurement was also employed in trace radiometric determination of ^{99}Tc in nuclear waste, where it improved the

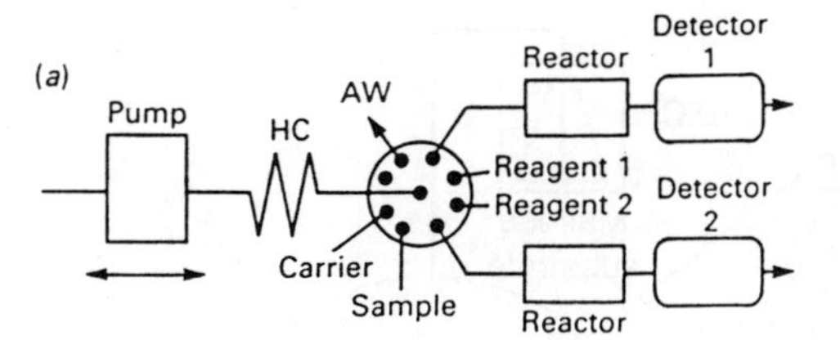

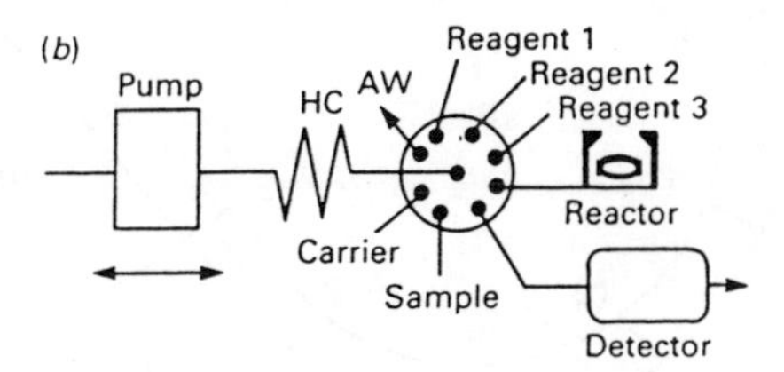

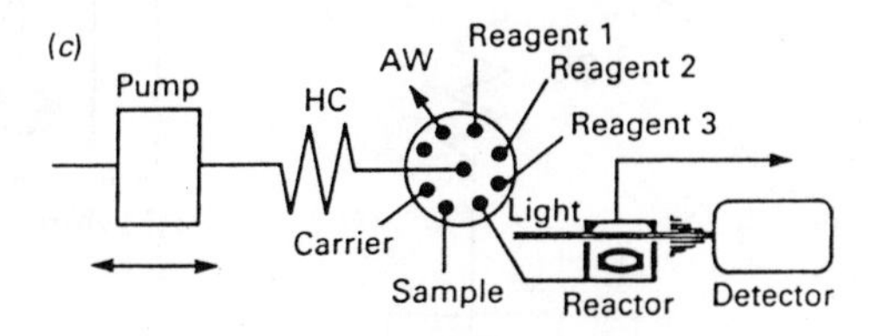

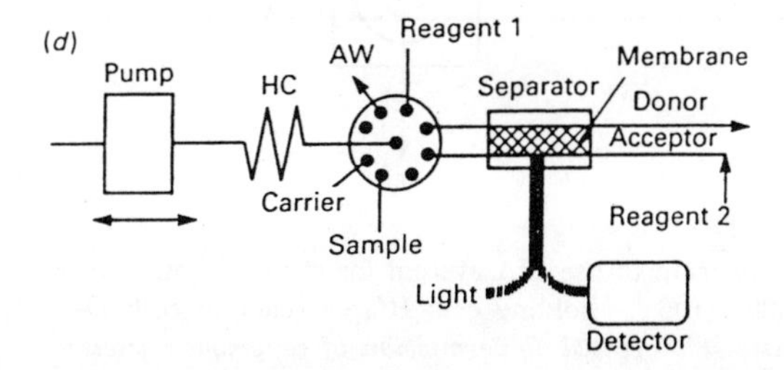

Fig. 3. Schematic diagrams of sequential injection systems developed for various applications [16]. (a) Two-detector systems for simultaneous determination of two analytes; (b) multi-reagent setup with several reagents to be mixed with the sample zone in a mixing chamber; (c) as for (b), where integrated reactor and detector allows reaction-rate measurements with no delay; (d) system with on-line gas diffusion or dialysis. (*Reprinted by permission of copyright owner.*)

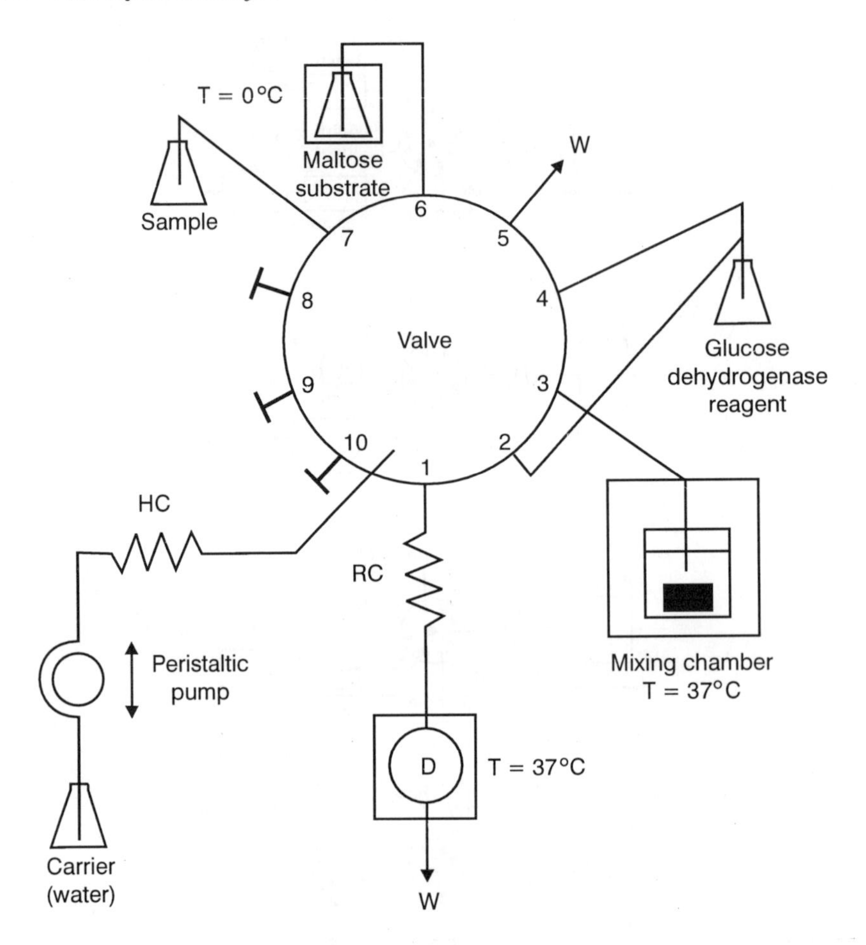

Fig. 4. Schematic diagram of the SIA system for stopped-flow determination of amyloglucosidase activity [28]. HC — holding coil; RC — reaction coil; D — spectrophotometric detector; W — waste. (*Reprinted by permission of copyright owner.*)

sensitivity of on-line radioactivity detection [9]. The stopped-flow method was also used for simultaneous determination of chloride and pH in waste waters with potentiometric detection [30]. The method uses a separate flow cell to measure pH by stopping the flow during the determination of chloride.

Besides experimental optimisation of measuring parameters reported in most papers on SIA determinations for the same purpose also advanced chemometric procedures can be applied. An expert system was developed which was able to make a self-configuration of the SIA system that was demonstrated on an example of Ca and Mg determination in natural waters with spectrophotometric diode array detection [31]. The SIA system for spectrophotometric determination of iron (III) in water samples was optimised by using jointly the neural network and genetic algorithms [32]. Artificial neural networks have also been applied to diagnose multivariate spectrophotometric responses in the SIA system for simultaneous determination of Ca and Mg with spectrophotometric detection [3].

Among other methodologies of flow measurements in SIA systems satisfactory results were reported for titrations. In the setup with a mixing chamber which contained a generating electrode and served as the spectrophotometric flow cell, a coulometric titration for bromine number determination was carried out [34]. Titrations of various olefins and samples of petroleum were performed with the mercuric chloride catalyst and an RSD of 2.1% was obtained for a set of ten determinations. A novel concept of titration was also developed with the use of a 35 μm diameter agarose bead loaded with a pH indicator [35]. They served in a non-aqueous sample stream as a pH-sensing material swollen with aqueous NaOH titrant. In a developed system the transmittance measurements were made by fibre optics through optical windows (Fig. 5). The indicator monitored the remaining titrant within the beads during perfusion with sulphuric acid in samples with titrated 1-butanol. The irreversible reaction can be applied due to automated packing and disposal of beads in a flow cell.

In numerous developed SIA systems also various on-line sample pretreatment methods were applied. The preconcentration of pertechnate on a resin column was reported in the above-mentioned radiometric determination of ^{99}Tc [29]. Preconcentration of iron (III) on a microcolumn packed with Chelex 100 was used in determinations of Fe (III) with spectrophotometric [36] and AAS [37] detections.

In sequential injection flow analysis systems also flow-through reactors with immobilised enzymes were already applied. In the determination of D-lactic acid with spectrophotometric detection a reactor with D-lactate dehydrogenase co-immobilised with L-alanine aminotransferase on porous glass was employed [38]. Then in the determination of D-glucose by chemiluminescence an open tubular enzyme reactor was employed with glucose oxidase immobilised on the inner wall of a nylon tube [39].

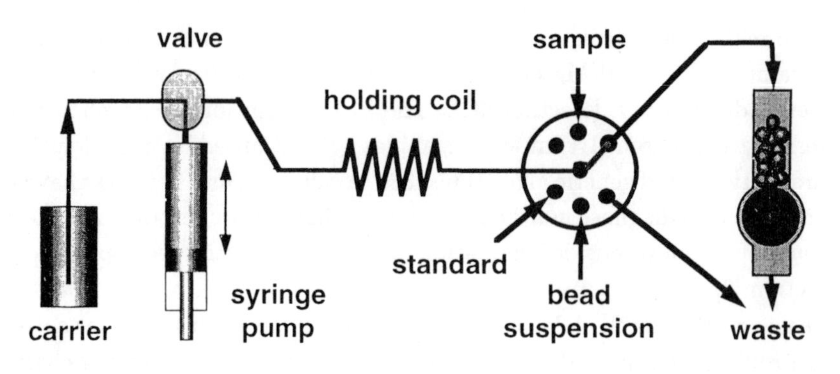

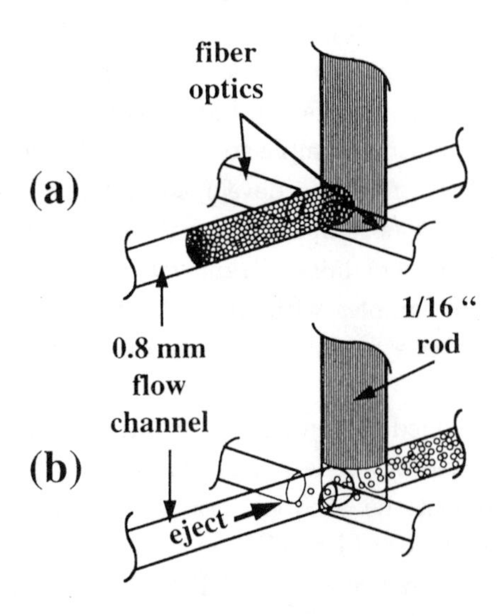

Fig. 5. Schematic diagram of the SIA system (A) and transmittance flow cell (B) used for titration of H_2SO_4 in 1-butanol using agarose beads with internally bound pH indicator and spectrophotometric detection [35]. (a) Perfusion with H_2SO_4 and monitoring of transmittance of entrapped beads; (b) ejection of spent beads. (*Reprinted by permission of copyright owner.*)

Preconcentration of trace analytes can be carried out not only on solid sorbents, but also on wetting films on the inner walls of an open tubular reactor. In the determination of Cr (III) and Cr (VI) the reaction product of

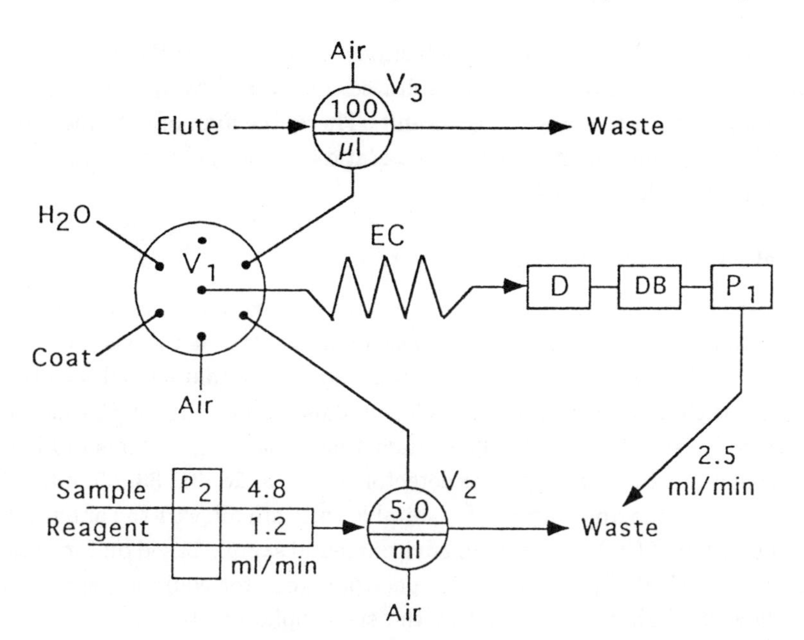

Fig. 6. Schematic diagram of the SIA system for spectrophotometric determination of Cr (VI) employing preconcentration by a wetting film extraction [40]. P1, P2 — peristaltic pumps; V1 — selection valve; V2, V3 — injection valves; EC — extraction coil; D — detector; DB — displacement bottle; Coat — coating solvent; Elute — eluting solvent. (*Reprinted by permission of copyright owner.*)

Cr (VI) with diphenylcarbazide was ion-paired with perchlorate and extracted into film consisting of octanol and 4-methyl-2-pentanone on the wall of a Teflon tube [40]. For spectrophotometric detection the wetting film was eluted with acetonitrile. In the system shown in Fig. 6 an enrichment factor of 25 was achieved at a sampling frequency of 17 h^{-1}. The solvent extraction in SIA systems can be effectively carried out also in a different way. An aqueous sample and organic solvent are injected sequentially into an extraction coil, then mixed and separated due to the different flow velocities of the aqueous and organic phases resulting from hydrophobic interaction of organic solvent with the walls of the extraction coil. Such a system was hyphenated with the HPLC setup for determination of barbiturtes in urine [41].

On-line gas diffusion, widely employed in flow analysis, was also successfully used in SIA measuring systems. Two examples of such application were developed for selective determination of ammonia in waters with conductivity

[42] and spectrophotometric [43] detections. In both cases the sample and an alkaline solution are sequentially aspirated and mixed by flow reversal while they are propelled to a gas-diffusion unit. There, the ammonia formed diffuses through a hydrophobic membrane to a static acceptor solution, which is then transported to the detector.

3. Detection Methods in SIA Systems

As shown in Table 1 listing the practical applications of SIA systems, the most frequently used detection is spectrophotometry in the visible range of radiation. This was also employed with the use of commercial spectrophotometers with flow-through cuvettes for fundamental studies of SIA methodology with dye solutions [6, 9, 10]. In numerous measuring systems diode array spectrophotometers were used as detector, e.g. [20, 28, 31, 33, 35, 43, 44]. In the latter case the application of a suitable software allowed the simultaneous determination of Ca and Mg in an environmental sample based on formation of complexes with PAR. A fibre-optic spectrophotometer equipped with a sandwich flow-through cell was used in the spectrophotometric determination of calcium with *o*-cresolphthaleine complexone [45]. In measurements with spectrophotometric detection the concept of a renewable optical sensor can also be utilised, which is based on the use of minute amounts of beads on which the reagent is adsorbed and the analyte becomes preconcentrated and monitored [46]. The principle of this measurement is illustrated in Fig. 7. In a special jet-ring cell a defined volume of suspension of beads with an immobilised appropriate reagent is retained on the surface of the detector window. The carrier solution flows over to waste through a circular gap between the sensor body and jet-ring circumference. When the sample is injected the analyte perfuses the retained bead and formation of the coloured product is carried out on particles. Finally the flow is reversed and beads are ejected from the cell. Using a fibre-optic sensor this concept was developed for spectrophotometric determination of chromium (VI) using C18 styrene-divinylbenzene-based beads loaded with diphenylcarbohydrazide [47].

For the first time this concept of measurement on a renewable solid support was demonstrated with fluorimetric detection where as detector was used a microscope with epifluorescence attachment and halogen source [46]. Model studies were carried out with sorption of substituted fluorescein and also utilising staining of BHK cells grown on Cytodex beads with propidium iodide. In immunochemical application of this measurement with fluorimetric detection

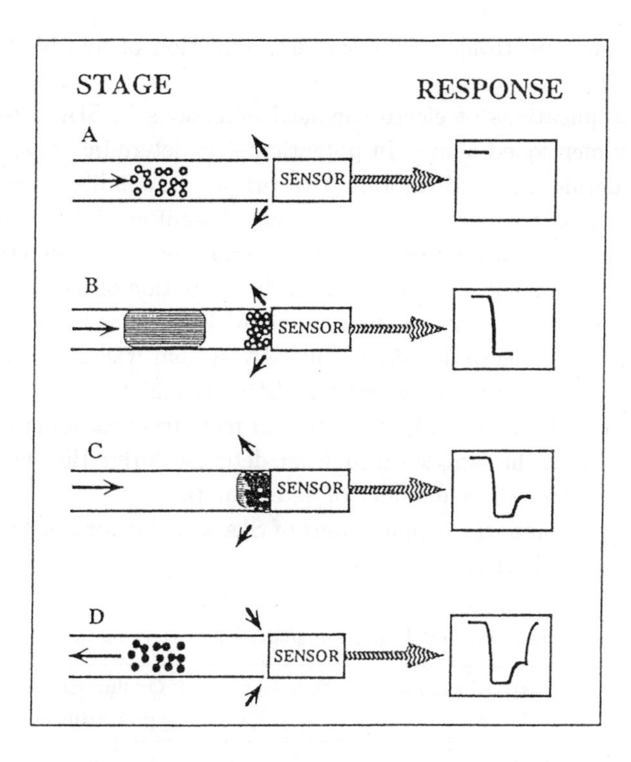

Fig. 7. Schematic diagram of four states of the measuring cycle and observed response for the renewable chemical sensor [47]. A — charging the system with the beads; B — injection of the analyte zone; C — perfusion; D — monitoring of the signal and discharge of the spent reactive layer. (*Reprinted by permission of copyright owner.*)

anti-mouse IgG1-coated agarose beads were used for IgG1 determination in both a competitive and a non-competitive format [48]. Second different configurations of fluid driving systems were applied for the use of this concept for perfusion studies of adherent cells [49].

A fountain cell where fluid entering via a central inlet was directed normal to a flat optical surface forming a fountain-like pattern was successfully applied for chemiluminescence detection in the SIA system of hydrogen peroxide and glucose using luminol [50]. In another SIA system with chemiluminescence detection a home-made photodiode detector was used [39].

Among other optical detectors in SIA systems also AAS detection with flame atomisation was used for determination of Fe (III) [37] and direct plasma atomic emission spectrometry for the determination of hydride-forming elements [51].

Several applications of electrochemical detections in SIA determinations were already mentioned above. In potentiometric determinations, besides fluoride and chloride [22], and chloride with pH [30], the indirect procedure was also reported for determination of glycerol and 2,3-butanediol in wine [52]. Amperometric detection in the SIA system was employed for the determination of phosphate in waters and wastes based on the formation of molybdophosphate [53], and glucose and lactate in the systems with enzymes immobilised at the surface of working electrodes [54]. The SIA system was also designed with anodic stripping voltammetry detection where plating of the mercury film was carried on-line, which reduced production of mercury-containing waste [5]. A repeated passing of the sample through the detector during the deposition step was used to enhance the sensitivity of determination.

In Table 1 the reported applications of SIA systems for determinations in real samples are collected.

Table 1. Applications of sequential injection analysis for determinations in real samples.

Analyte	Analyzed sample	Detection	On-line sample pretreatment	Reference
Ammonia	Wastes	Conductivity	Gas diffusion	42
	Wastes, aerosols	Photometry	Gas diffusion	43
Amyloglucosidase activity	Fermentation media	Photometry	–	28
As	Biological SRMs, sea water	DCP-AES	Hydride generation	51
Barbiturates	Urine	HPLC	Wetting film extraction	41
Bromine member	Olefins, petroleum distillates	Photometry	Coulometric titration	34
2,3-butanediol	Wine	Potentiometry	SPE of interferences	52
Ca	Water	Photometry	–	44
	White water	Photometry	–	45
Chemical oxygen demand	Wastewater	UV	–	21
Chloride	Water	Potentiometry	–	22
	Wastewater	Potentiometry	–	30

continued

Table 1 (*continued*)

Analyte	Analyzed sample	Detection	On-line sample pretreatment	Reference
Co (II)	Soil extracts, water	Photometry	–	26
Cr (III), Cr (VI)	Water	Photometry	Wetting film extraction	40
Fe	Water	Photometry	–	32
	Sea water, tap water	Photometry	SPE	36
	Tap water	AAS	SPE	37
Fluoride	Water	Potentiometry	–	22
Ge	Sea water	DCP-AES	Hydride generation	51
Glucose	Fermentation media	Chemiluminescence	–	39
		Amperometry	–	54
Glycerol	Wine	Potentiometry	SPE of interferences	52
Hg	Sediment, biological materials	Cold vapour AAS	Gas-liquid separation	57
D-lactic acid	Pork meat	Photometry	Solvent extraction	38
Mg	Water	Photometry	–	44
Ni(II)	Soil extracts, water	Photometry	–	26
Nitrate	Wastewater	UV	–	21
Nitrate, nitrite	Waste water, sea water, aerosols	Photometry	–	56, 59
pH	Wastewater	Potentiometry	–	30
Phosphate	Water, waste water	Amperometry	–	53
	Waste water	Photometry	–	25, 58
Silicate	Waste water	Photometry	–	25
^{99}Tc	Nuclear waste	Radiometry	SPE	29

*DCP-AES — direct current plasma atomic emission spectrometry; SPE — solid-phase extraction.

4. Batch Injection Analysis

Yet another sample handling has been invented in flow measurements where detection is carried out during the sample flow at the sensing surface of the detector, which is named "batch injection analysis" (BIA). Injection is performed in this case directly onto the surface of the detector and transient signal is recorded during the sample flow of the injected sample segment over the detector surface. The term "batch injection analysis" was introduced into the analytical literature by Wang and Taha [60], proposing for this purpose a simple amperometric large-volume wall-jet detector with upside located a flat surface of a working electrode (Fig. 8). Although the authors of this work have described the proposed measurements as a "non-flow injection-based technique", as recording of the analytical signal occurs during the flow of the sample segment over the detector surface it is without doubt flow measurement. Contrary to other methodologies of flow measurements, in the BIA system there is no transport of the sample or reagent solutions through a tubing. This simplifies a measuring setup; however, it makes impossible — at least in configurations reported so far in the literature — on-line steps of sample pretreatment, which is a very convenient feature for mechanisation and automation of analytical procedures commonly exploited in FIA and SIA systems. However, there are two exceptions. One of them is an amperometric measurement with enzymes immobilised at the surface of the working electrode, providing a conversion of the analyte into electroactive products detected by the sensor used [60, 61]. As another example of the BIA system with sample pretreatment, measurements with anodic stripping voltammetry detection [62–66] or adsorptive stripping voltammetry detection [67] can be considered, where one separate stage of the procedure is accumulation of the analyte at the electrode surface and another one anodic stripping with recording of the analytical signal. It seems then that one of the subjects for the future developments of this measuring technique might be development of unique methods of sample treatment in the BIA system prior to its delivery to the detector surface.

As was shown by Tur'yan in the review about micro-cells for voltammetry and stripping voltammetry [68], the concept of batch-injected analysis with sample injection by a micro-pipette into a Teflon capillary with the working carbon paste electrode (Fig. 8B) was proposed much earlier [69]. A capillary format of BIA with a micro-disk working electrode in a large-volume cell was reported much later [70]. Rotating disk working electrodes were used in a wall-jet arrangement [71], and also in thin-layer cell design shown in Fig. 8C [66]. The important advantage of the design with rotating working electrodes is the possibility of automatic removal of the sample.

In most cases the sample after contact with the surface of the working electrode is washed out by the large volume of the supporting electrolyte in the

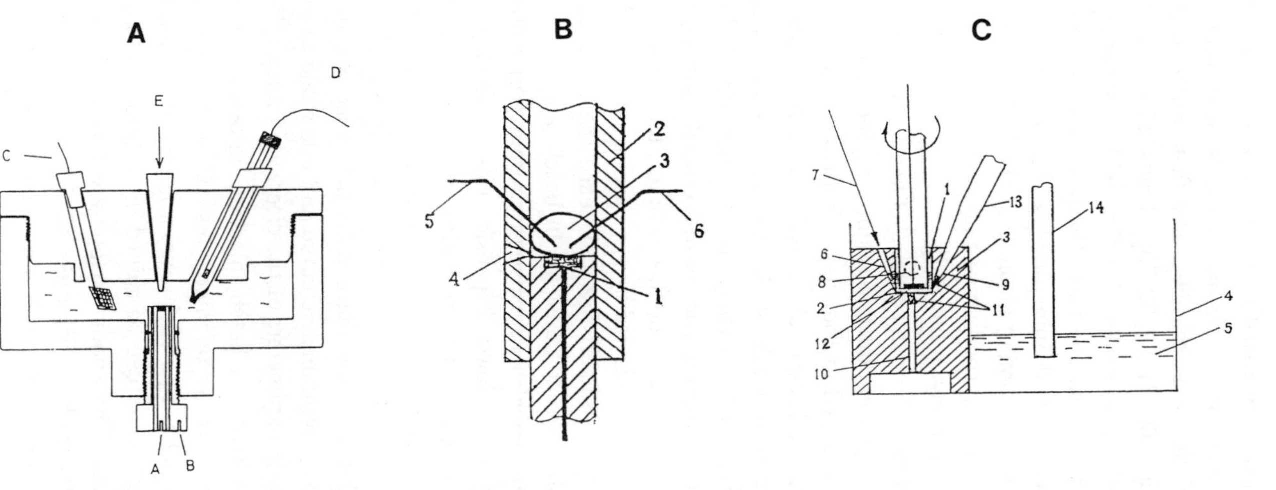

Fig. 8. Schematic diagrams of cells used in batch injection analysis with amperometric detection. (A) Large-volume wall-jet cell [74]; (B) a capillary cell with a paste electrode in a Teflon tubing [69], and a thin-layer cell with a working rotating disk electrode [66]. In (A): A — disk electrode; B — ring electrode; C — auxiliary electrode; D — SCE reference electrode; E — micropipette tip. In (B): 1 — working carbon paste electrode; 2 — Teflon capillary; 3 — sample solution; 4 — injection port for micropipette; 5 — a platinum wire auxiliary electrode; 6 — a silver wire reference electrode. In (C): 1 — inner microcompartment; 2 — sample solution; 3 — body of microcompartment; 4 — outer large-volume compartment; 5 — supporting electrolyte; 6 — channel for sample and washing solution; 9 — channel for auxiliary electrode; 10 — channel for contact with reference electrode; 11 — ceramic plugs; 12 — rotating disk working electrode; 13 — auxiliary electrode; 14 — reference electrode. (*Reprinted by permission of copyright owner.*)

detector compartment. As a considerable dilution of the sample takes place in this case, a large number of analyses can be carried out without replacing the supporting electrolyte in the detection cell [60]. The first works on BIA employed amperometric detection [60, 61, 70, 72–75]. Besides platinum or glassy carbon disk electrodes as working electrodes, thin-film gold electrodes prepared by cathodic sputtering using a plastic sheet as substrate were used for detection of hydrogen peroxide with pulsed amperometric detection [72]. For the electrocatalytic amperometric detection of alcohols in the BIA system, glassy carbon electrodes were modified by conductive polymeric nickel(II) tetrakis (3-methoxy-4-hydroxyphenyl) porphyrin film [73].

BIA is similar to conventional flow injection systems in that an injected sample zone is transported in a reproducible manner to the sensing surface of a detector. This results in signal in the shape of a sharp peak. A large-volume wall-jet cell configuration is especially suitable for this purpose where a micro-pipette tip functions as a nozzle of the wall jet. In such a case the incident sample zone is only active in creating the analytical signal. Both the sample volume and the electrode-tip distance have a crucial effect on the signal magnitude. Similarly to the effect of the flow rate in FIA measurements a peak width in BIA measurements is affected by the stirring rates in a detector which influences the wash-out characteristics and hence the peak sharpness (Fig. 9) [60]. The form of the current transients in batch injection measurements was analysed and compared with a steady-state response of wall-jet disk electrodes [74]. Differences found at lower dispension rates were attributed to radial diffusion effects. It was found that the optimum distance between the micro-syringe tip and the electrode surface is 3 mm and that maximum sensitivity requires a minimum injection volume of 14 μl. It was also shown that consecutive injections during a slow linear potential sweep permit one to record the pseudo-steady-state voltammetric curve and allow one to measure the thermodynamic and kinetic parameters of the electrode process [75].

The enhancement of sensitivity in batch injection amperometry can be achieved by the use of square-wave voltammetry instead of the measurement of maximum limiting current in fixed-potential amperometric detection [75]. A further improvement of detectability can be gained in stripping voltammetry batch injection measurements. The anodic stripping determinations can be carried out both with a static working electrode [62–65] and with a rotating disk electrode [66]. The measurement consists of three phases: injection of Hg (II) solution in order to form a mercury layer on a glassy carbon working

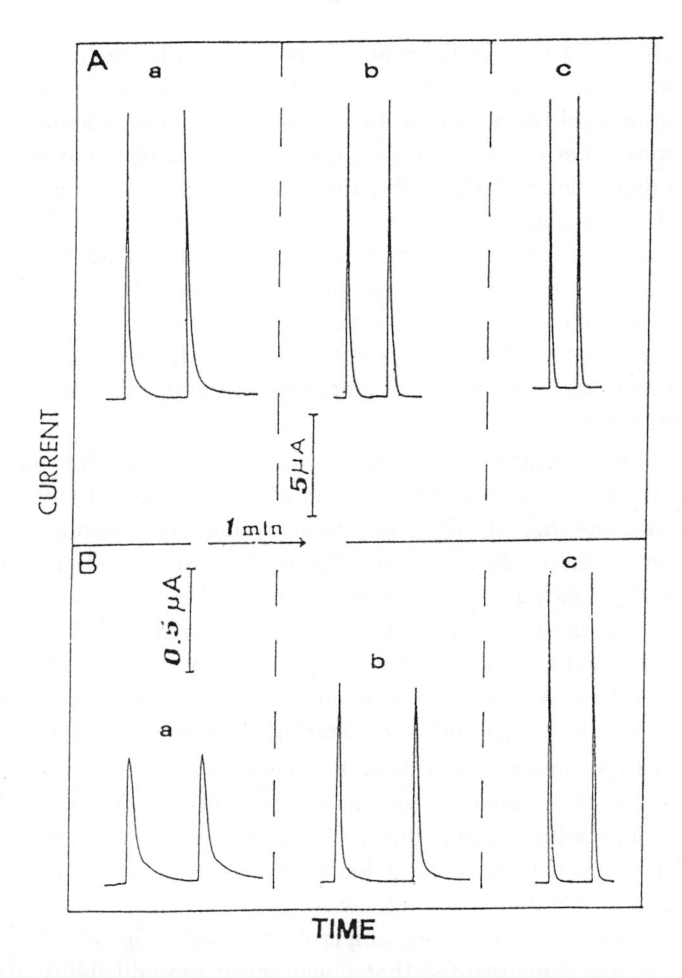

Fig. 9. Detection peaks recorded in BIA (A) and FIA (B) measurements for 0.1 mM ferricyanide, using different stirring rates (A: 0 (a), 250 (b) and 500 (c) rpm) and flow rates (B: 0.2 (a), 0.5 (b) and 1.0 (c) ml·min^{-1}) [60]. Conditions: operating potential, +0.9 V; sample volume, 20 μl; electrode tip distance, 2 mm; electrolyte and cell solutions, 0.1 M KCl. (*Reprinted by permission of copyright owner.*)

electrode, injection of the sample solution for preconcentration of analytes in the film and a stripping of determined analytes from the film. Covering of the electrode surface with a mercury film can be made outside of the BIA cell prior to a series of measurements [62, 66], or in the detection cell used for determinations [63–65]. In determinations with anodic square-wave voltammetric scan simultaneous measurements of several trace metals in the μg l^{-1} concentration level, at a rate of 48 h^{-1} with a relative standard deviation of 1.9%, were reported [62]. Increasing sensitivity arising from continuing to apply the deposition potential after the end of the injection period has been observed [63]. This allows one to obtain better detectability for BIA compared to conventional flow injection measurements. The use of Nafion-coated mercury thin film electrodes is advantageous in reducing electrode contamination by components of complex matrices [65]. A glassy carbon disk mini-electrode and cylindrical carbon fibre microelectrode substrates were used for the determination of zinc in zinc-insulin complexes by square-wave anodic stripping voltammetry [64].

The BIA with square-wave stripping voltammetric detection was also used in adsorptive mode for the determination of traces of nickel and cobalt [67]. In a large-volume wall-jet cell configuration a mercury thin-film electrode on a glassy carbon electrode was used. The method was based on the use of a nioxime (1,2-cyclohexanedione dioxime) ligand for adsorptive preconcentration of analytes. The practical detection limit was estimated as 5 nM for nickel and 2 nM for cobalt for injection volume 50 μl. Similarly to anodic stripping measurements in the BIA system it was found that the continuing adsorption after the end of the injection period is maximising the sensitivity of determination.

Besides amperometry, other detection methods are also developed for BIA measurements. Potentiometry with membrane ion-selective electrodes seems to be especially advantageous due to its ability in several cases to provide selective response without a need for sample pretreatment. Measurements are carried out also in a large-volume wall-jet cells with inverted electrodes similarly to most commonly used amperometric cells (Fig. 10). For pH measurements it was demonstrated that much better reproducibility of response in the BIA system is obtained for planar than for spherical electrodes [76]. In BIA pH measurements it was also found that the higher the buffer capacity of the background solution, the more effective the sample removal. pH response exhibited a super-Nernstian slope of 65 mV/decade in the pH range of 2–11 [76]. In several papers the use of chloride ISE in BIA systems was demonstrated [60, 76, 77]. No noticeable carry-over was observed between the 5 and 100 mM chloride solutions [60], and between 50 μM and 1 mM fluoride solutions [76]. In these two cases the sensitivity of response was slightly lower than Nernstian, namely 57.8 and 56.8 for chloride and fluoride, respectively [76]. The respective RSD values were 2.4 and 1.4% for these analytes.

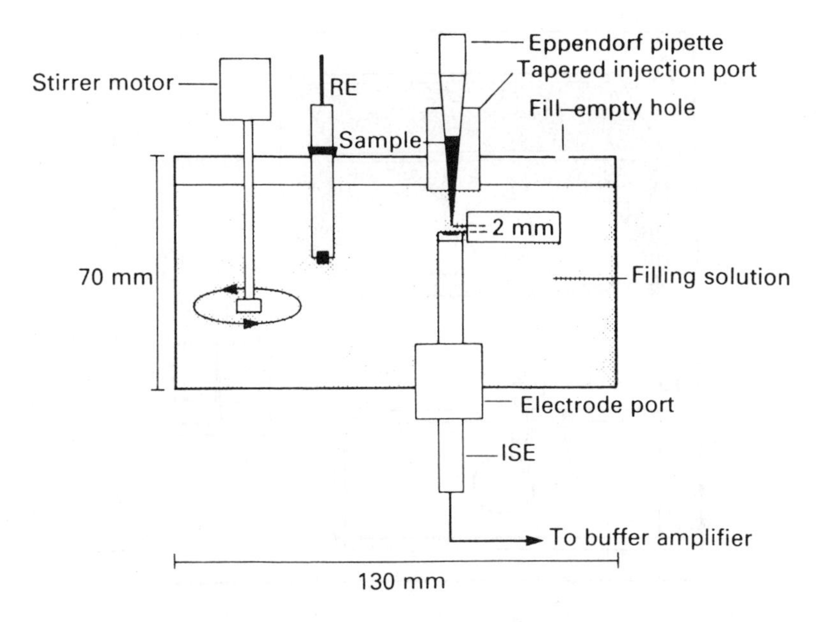

Fig. 10. Schematic diagram of a cylindrical batch injection cell for potentiometric detection with ion-selective electrode (ISE) and reference electrode (RE) [78]. (*Reprinted by permission of copyright owner.*)

In batch injection systems measurements with ion-selective electrodes with plasticised membranes were also carried out. In determination of sodium and potassium it was found that the peak response is primarily controlled by the active transport towards and away from the membrane provided by the wall-jet effect, whereas baseline recovery is controlled by the dynamics of diffusion into the bulk-filling solution [78]. The slope of response observed was considerably smaller than Nernstian, namely 49.3 mV/pM decade for potassium and 35.7 mV/pM decade for sodium. The same type of membrane electrodes with PVC membranes were used for multicomponent BIA with a multi-channel pipette and an array of three ion-selective electrodes (calcium, potassium and sodium) [79]. Measurements were carried out with background solution 0.1 M lithium chloride, with no cross-response being observed in any case. Again, the slope of response observed was sub-Nernstian in all cases.

An interesting practical application of BIA with potentiometric detection is determination of acetylsalicylic acid in tablets with a salicylate ion-selective electrode [80]. With sampling rate 90 determinations per hour there was no

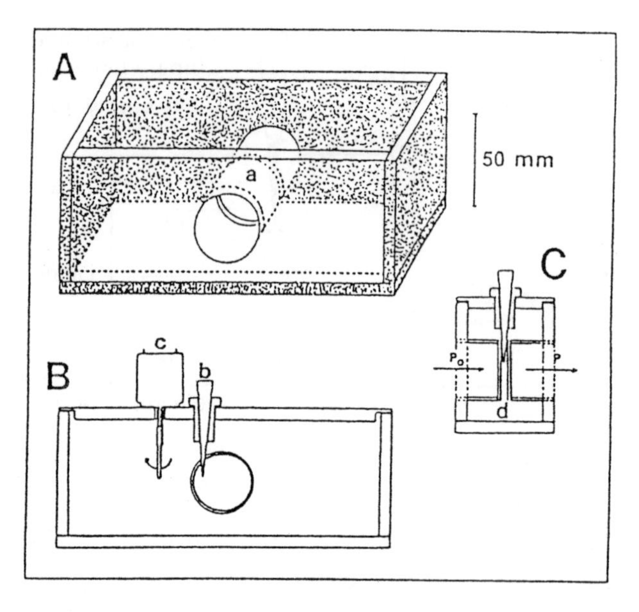

Fig. 11. Schematic diagram of a batch injection spectroscopic cell [81]. (A) View without cover; (B), (C) cross-section views (length and side, respectively). (*Reprinted by permission of copyright owner.*)

significant difference between BIA results and those obtained by the standard Pharmacopea method.

Spectrophotometric batch injection measurements can be carried out in a flow cell schematically shown in Fig. 11 [81]. It was based on the modification of the slope of the sample compartment of a commercial spectrophotometer in such a way that a sample solution is injected into an optical pathway between two fixed Perspex windows. The whole cell volume was 300 ml and the injected sample solutions 10-20μl. Reported preliminary results were obtained for the injection of coloured solutions of potassium dichromate, potassium permanganate and iron (II), which was detected as its 1,10-phenanthroline complex by using the preceding derivatisation reaction. In the study of the analyte build-up it was found that the presence of $KMnO_4$ in the 300 ml cell solution at a level corresponding to 250 and 500 injections resulted in 1.3 and 2.7% decreases of the 1 mM $KMnO_4$ response.

Thermal batch injection measurements employed for enzymatic, redox and acid-base reactions were performed with a differential measuring unit consisting of two thermistors, one sensing the heat, the other acting as a reference [82]. For enzymatic detection catalase and glucose oxidase were immobilised on the thermistor surface. A catalase-bound thermistor was used for the detection of hydrogen peroxide, and a glucose-oxidase-bound one for the detection of glucose. For determinations based on redox or acid-base reactions the reagent was present in the cell solution. Tin (II) was determined with potassium permanganate in the BIA cell, while nitric acid was determined with the cell solution containing NaOH. In all examined cases there was no observable carry-over and the precision was typically 2% (RSD).

The way these last two discussed detections were employed indicates the possibility of performing in BIA systems determinations based on the use of a suitable reaction with great effectiveness. Appropriate design of the measuring cell may allow the use in these measurements of the most common spectrophotometric detection in the visible range.

5. References

1. M. Valcarcel and M. D. Luque de Castro, *Automatic Methods of Analysis* (Elsevier, Amsterdam, 1988).
2. M. Trojanowicz, *Automatyzacja w analizie chemicznej* (Wydawnictwa Naukowo-Techniczne, Warszawa, 1992).
3. C. A. Burtis, W. F. Johnson and T. O. Tiffany, *Anal. Lett.* **7** (1974) 591.
4. T. O. Tiffany, J. M. Jansen, C. A. Burtis, J. B. Overton and C. D. Scott, *Clin. Chem.* **18** (1972) 829.
5. L. Andersson, A. Granel and M. Strandberg, *Anal. Chim. Acta* **103** (1978) 489.
6. J. Ruzicka and G. D. Marshall, *Anal. Chim. Acta* **237** (1990) 329.
7. J. Ruzicka, *Anal. Chim. Acta* **261** (1992) 3.
8. G. D. Christian, *Analyst* **119** (1994) 2309.
9. J. Ruzicka, G. D. Marshall and G. D. Christian, *Anal. Chem.* **62** (1990) 1861.
10. T. Gübeli, G. D. Christian and J. Ruzicka, *Anal. Chem.* **62** (1991) 2407.
11. J. Ruz, A. Rios, M. D. Luque de Catro and M. Valcarcel, *Talanta* **33** (1986) 199.
12. D. Betteridge, P. D. Oates and A. P. Wade, *Anal. Chem.* **59** (1987) 1236.
13. J. Ruzicka and E. H. Hansen, *Anal. Chim. Acta* **214** (1988) 1.
14. A. Rios, M. D. Luque de Castro and M. Valcarcel, *Anal. Chem.* **60** (1988) 1540.
15. J. Toei, *Analyst* **113** (1988) 475.
16. A. Ivaska and J. Ruzicka, *Analyst* **118** (1993) 885.
17. A. Cladera, C. Tomas, E. Gomez, J. M. Estela and V. Cerda, *Anal. Chim. Acta* **302** (1995) 297.
18. S. Liu and P. K. Dasgupta, *Talanta* **41** (1994) 1903.
19. S. Liu and P. K. Dasgupta, *Anal. Chim. Acta* **308** (1995) 281.
20. F. Mas, A. Cladera, J. M. Estela and V. Cerda, *Analyst* **123** (1998) 1541.

21. O. Thomas, F. Theravlaz, V. Cerda, D. Constant and P. Quevauviller, *Trends Anal. Chem.* **16** (1997) 419.
22. J. Alpizar, A. Crespi, A. Cladera, R. Forteza and V. Cerda, *Electroanalysis* **8** (1996) 1051.
23. J. M. Estela, A. Cladera, A. Muñoz and V. Cerda, *Int. J. Env. Anal. Chem.* **64** (1996) 205.
24. E. Gomez, C. Tomas, A. Cladera, J. M. Estela and V. Cerda, *Analyst* **120** (1995) 1181.
25. F. Mas-Torres, A. Muñoz, J. M. Estela and V. Cerda, *Analyst* **122** (1997) 1033.
26. R. E. Taljaard and J. F. Van Standen, *Anal. Chim. Acta* **366** (1998) 177.
27. G. D. Christian and J. Ruzicka, *Anal. Chim. Acta* **261** (1992) 11.
28. E. H. Hansen, B. Willamsen, S. K. Winther, H. Drabøl, *Talanta* **41** (1994) 1881.
29. O. Egorov, M. J. O'Hara, J. Ruzicka and J. W. Grate, *Anal. Chem.* **70** (1998) 977.
30. J. Alpizar, A. Crespi, A. Cladera, R. Forteza and V. Cerda, *Lab. Rob. Autom.* **8** (1996) 165.
31. A. Rius, M. P. Callao, R. X. Rius, *Anal. Chim. Acta.* **316** (1995) 27.
32. J. de Carcia, M. L. M. F. S. Sarasia, A. N. Araujo, J. L. F. C. Lima, M. del Valle and M. Poch, *Anal. Chim. Acta* **348** (1997) 143.
33. I. Ruisanchez, J. Lozano, M. S. Larrechi, F. X. Rius and J. Zupan, *Anal. Chim. Acta* **348** (1997) 113.
34. R. H. Taylor, C. Winbo, G. D. Christian and J. Ruzicka, *Talanta* **39** (1992) 789.
35. D. A. Holamn, G. D. Christian and J. Ruzicka, *Anal. Chem.* **69** (1997) 1763.
36. E. Rubi, R. Forteza and V. Cerda, *Lab. Rob. Autom.* **8** (1996) 149.
37. E. Rubi, M. S. Jimenez, F. B. de Mirabo, R. Forteza and V. Cerda, *Talanta* **44** (1997) 553.
38. H.-C. Shu, H. Håkanson and B. Mattiasson, *Anal. Chim. Acta* **283** (1993) 727.
39. X. Liu and E. H. Hansen, *Anal. Chim. Acta* **326** (1996) 1.
40. Y. Luo, S. Nakano, D. A. Holman, J. Ruzicka and G. D. Christian, *Talanta* **44** (1997) 1563.
41. K. L. Peterson, B. K. Logan, G. D. Christian and J. Ruzicka, *Anal. Chim. Acta* **337** (1997) 99.
42. M. T. Oms, A. Cerda and V. Cerda, *Electroanalysis* **8** (1996) 387.
43. M. T. Oms, A. Cerda, A. Cladera, V. Cerda and R. Forteza, *Anal. Chim. Acta.* **318** (1996) 251.
44. E. Gomez, C. Tomas, A. Cladera, J. M. Estela and V. Cerda, *Analyst* **120** (1995) 1181.
45. J. Nyman and A. Ivaska, *Anal. Chim. Acta.* **308** (1995) 286.
46. J. Ruzicka, Cy. H. Pollema and K. M. Scudder, *Anal. Chem.* **65** (1993) 3566.
47. O. Egorov and J. Ruzicka, *Analyst* **120** (1995) 1959.
48. Cy H. Pollema and J. Ruzicka, *Anal. Chem.* **66** (1994) 1825.
49. P. J. Baxter, L. Hallgren, Cy H. Pollema, M. Truka and J. Ruzicka, *Anal. Chem.* **67** (1995) 1486.

50. D. J. Tucker, B. Toivola, C. H. Pollema, J. Ruzicka and G. D. Christian, *Analyst* **119** (1994) 975.
51. P. Ek, S.-G. Hulden and A. Ivaska, *J. Anal. At. Spectr.* **10** (1995) 121.
52. G. C. Luca, B. F. Reis, E. A. G. Zagatto, M. C. B. S. M. Montenegro, A. N. Araujo and J. L. F. C. Lima, *Anal. Chim. Acta* **366** (1998) 193.
53. A. R. Crespi, R. Forteza and V. Cerda, *Lab. Robot. Autom.* **7** (1995) 245.
54. W. Schumann, H. Wohlschläger, J. Huber, H.-L. Schmidt and H. Stadler, *Anal. Chim. Acta* **315** (1995) 113.
55. A. Ivaska and W. W. Kubiak, *Talanta* **44** (1997) 713.
56. M. T. Oms, A. Cerda and W. Cerda, *Anal. Chim. Acta* **315** (1995) 321
57. F. M. Banza de Mirabo, A. Ch. Thomas, E. Rubi, R. Forteza, and V. Cerda, *Anal. Chim. Acta* **355** (1997) 203.
58. A. Muñoz, F. Mas Torres, J. M. Estela and V. Cerda, *Anal. Chim. Acta* **350** (1997) 21.
59. A. Cerda, M. T. Oms, R. Forteza and V. Cerda, *Anal. Chim. Acta* **371** (1998) 63.
60. J. Wang and Z. Taha, *Anal. Chem.* **63** (1991) 1053.
61. A. Amine, J.-M. Kauffmann and G. Palleschi, *Anal. Chim. Acta* **273** (1993) 213.
62. J. Wang, J. Lu and L. Chen, *Anal. Chim. Acta* **259** (1992) 123.
63. C. M. A. Brett, A. M. O. Brett and L. Tugulea, *Anal. Chim. Acta* **322** (1996) 151.
64. R. M. Barbosa, L. M. Rosario, C. M. A. Brett and A. M. O. Brett, *Analyst* **121** (1996) 1789.
65. C. M. A. Brett, A. M. O. Brett, F.-M. Matysik, S. Matysik and S. Kumbhat, *Talanta* **43** (1996) 2015.
66. Ya. I. Tur'yan, E. M. Strochkova, I. Kuselman and A. Shenhar, *Fresenius J. Anal. Chem.* **354** (1996) 410.
67. C. M. A. Brett, A. M. O. Brett and L. Tugulea, *Electroanalaysis* **8** (1996) 639.
68. Ya. I. Tur'yan, *Talanta* **44** (1997) 1.
69. M. Karolczak, R. Dreiling, R. N. Adams, L. J. Felice and P. T. Kissinger, *Anal. Lett.* **9** (1976) 783.
70. U. Backofen, F.-M. Matysiak, W. Hoffmann and H.-J. Ache, *Fresenius J. Anal. Chem.* **362** (1998) 189.
71. L. Chen, J. Wang and L. Agnes, *Electroanalysis* **3** (1991) 773.
72. A. M. O. Brett, F.-M. Matysiak and M. T. Vieira, *Electroanalysis* **9** (1997) 209.
73. A. Ciszewski and G. Milczarek, *J. Electroanal. Chem.* **413** (1996) 137.
74. C. M. A. Brett, A. M. O. Brett and L. C. Mitoseriu, *Electroanalysis* **7** (1995) 225.
75. C. M. A. Brett, A. M. O. Brett and L. C. Mitoseriu, *Anal. Chem.* **66** (1994) 3145.
76. J. Wang and Z. Taha, *Anal. Chim. Acta* **252** (1991) 215.
77. J. Wang, *Microchem. J.* **45** (1992) 219.
78. J. Liu, Q. Chen, D. Diamond and J. Wang, *Analyst* **118** (1993) 1131.
79. D. Diamont, J. Lu, Q. Chen and J. Wang, *Anal. Chim. Acta* **281** (1993) 629.

80. J. C. B. Fernandes, C. B. Garcia, L. A. Grandin, G. D. Neto and O. E. S. Godinho, *J. Braz. Chem. Soc.* **9** (1998) 249.
81. J. Wang and L. Agnes, *Anal. Lett.* **26** (1993) 2329.
82. J. Wang and Z. Taha, *Anal. Lett.* **24** (1991) 1389.

Chapter 10

Commercially Available Instrumentation for FIA

The evidence of advanced development of a given measuring methodology and its acceptance by routine analytical laboratories is an interest of manufacturers of measuring instruments. Practically only apparatus for which a wide range of accessories and spare parts is available together with the manufacturer's authorised service and consulting plus a possibly large list of applications may have a chance to compete for applications in routine analytical laboratories. In spite of the strong competition from various other instrumental techniques, flow injection methods have gained this level of advancement on the market of analytical instrumentation. Besides the broad literature available in thousands of original scientific papers, numerous monographs and computer databases, for many years several manufacturers from all over the world have offered continuously modernised instrumentation for FIA, although one has also to realise that most of the research work in this field is carried out with laboratory-built setups. Such prototypes then are very often utilised for manufacturing routine commercial instrumentation and its continuous improvement.

The market of chemical analytical instrumentation is very dynamic and competitive. It is connected with the necessity of permanent improvement in the performance of analytical instruments, their reliability and degree of mechanisation and automation. Various manufacturers have different ranges of distribution of their products, and some local producers are not widely known and hence they may not be included in this review of commercial instrumentation. Manufacturers with a given offer may appear on the market and then suddenly disappear when their offer has not been successful. The instrumentation presented below is only a partial illustration of the commercial offer of flow injection instruments available on the market at the end of the nineties.

Table 1. Manufacturers of flow injection instrumentation.

Company	Address	Phone No.	Fax No.
Alitea USA	Alitea Instuments USA LLC, Post Office Box 26, Medina, Washington 98039-0026, USA	(206) 453-4235	(206) 688-1565
Alpkem	OI Analytical Headquarters, 151 Graham Rd., PO Box 9010, College Station, Texas 77842-9010, USA	(800) 653-1711	(409) 690-0440
Burkard	Burkard Scientific Ltd., PO Box 55, Uxbridge, Middlesex UB8 2RT, UK	(01895) 230-056	(01895) 230-058
Eppendorf	Eppendorf-Netheler-Hinz Postfach 650670 2000 Hamburg 65, Germany	(040) 53801-0	(040) 538-01556
Hitachi	Hitachi Instruments Inc., 882 Ichige, Katsuta-shi, Ibaraki, 312 Japan		
Ismatec	Feldeggstrasse 6, PO Box CH-8152, Glattbrugg-Zürich, Switzerland	(01) 810-3040	(01) 810-5292
Lachat	Zellweger Analytics Inc., Lachat Instruments Division, 6645 West Mill Road, Milwaukee, WI 53218-1239, USA	(414) 358-4200	(414) 358-4206
Perkin-Elmer	761 Main Av., Norwalk, CT 06859-0012, USA	(800) 762-4000	(203) 762-4222
Sanuki	Sanuki Industry Co., 3-4-13 Midorigaoka, Hamura-shi, Tokyo, 190-11 Japan		
Tecator	Foss Tecator Box 70, S-263 21 Höganäs, Sweden	046(42) 361-500	046(42) 340-349
Zhaofa	Zhaofa Institute of Automatic Analysis, 72 San Hao Street, Shen Yang, China 110003	394-833	

Already in the first monograph on FIA a system for FIA from Bifok AB of Sweden has been mentioned [1]. In another monograph, published several years later by Valcarcel and Luque de Castro [2], one can find remarks on several companies producing FIA instruments (Tecator, Lachat, FIAtron, Hitachi). In the second edition of the monograph by pioneers of FIA Ruzicka and Hansen [3], one more detailed description of commercial instruments manufactured by the mentioned producers can be found.

A valuable review of flow injection instruments has been published in 1996, in the journal *Analytical Chemistry* [4]. It provides a comparison of functional parameters of eight instruments from seven companies. There were compared instruments of general laboratory application, as well as instruments dedicated e.g. for process analysis (from Eppendorf and Ionics), for environmental and agricultural applications (from Lachat and Perstorp), or for sample preparation, preconcentration and hydride generation in atomic absorption spectrometry (Perkin–Elmer).

For many years also instrumentation for sequential flow analysis (SIA) has been commercially available, which has been used for research applications in several laboratories, e.g. [5–8].

The aim of this chapter is to present the commercial products of several manufacturers supplying complete instrumentation for FIA and for SIA. The addresses of these manufacturers are listed in Table 1.

1. FIA

A laboratory system for performing flow injection determinations can be assembled in each laboratory from various elements available from numerous manufacturers (pumps, flow-through detectors, valves); however, routine applications are practically non-realistic without the existence of commercial instrumentation. Several such systems can be purchased in complete configuration and often with a large library of developed application notes. Several producers of this equipment are listed below in alphabetical order. Data cited about instruments were taken from technical information supplied by the manufacturers.

1.1. *Alitea USA*

For FIA, Alitea USA offers two complete solution handling systems not containing a detection system. **FIAlab®-2000** is a manually operated, portable, low cost system including a four-channel peristaltic pump, an injection valve, and a complete kit of connectors, pump tubings and tools.

FIAlab®-3500 is a compact modular solution handling system with the open architecture design of both hardware and software. It incorporates a microsyringe pump, a four-channel peristaltic pump and a 6–10-position selector valve, so it can be used for both FIA and SIA measurements. It also includes FIAlab® for Windows control software that makes FIAlab-3500 a command centre for custom control of samplers, detectors and other peripheral units. The software includes capabilities such as response peak height, peak locations and real-time plotting.

The list of available sensors and detectors from the same company obeys:

— a PC-1000 fibre-optic diode array spectrophotometer mounted directly on an A/D card of the computer;
— an optical detection system consisting of a modified Brinkman optic filter photometer and five interference filters, interfaced to an Alitea Universal Membrane Flow Cell. This system can also be used as a detector for chemiluminescence measurements;
— a micro-flow-cell designed for use with a wide range of electrodes, including pH, ion-selective and voltammetric.

It is also worth mentioning that through Alitea's web page one can have access to Hansen's Comprehensive Flow Injection Bibliography, which has been continuously updated since 1974. The bibliography consists of over 8000 titles and is available at: http://www.flowinjection.com/search.html.

1.2. *Alpkem*

Alpkem, since 1996 a division of OI Analytical, has for nearly 30 years manufactured automated ion analysers. Its **Flow Solution® 3000** was designed as a flow injection analyser, which also retains the ability to perform segmented flow analysis. A single channel unit includes a sampler, a cassette with plug-in modules containing a peristaltic pump, an injection valve, a chemistry manifold and an expanded range photometric detector. Electrochemical reactions can be monitored by plug-in amperometric or ion-selective detectors. Up to four channel systems can be configured. Advanced expanded range photometric detection allows single calibration ranges of 3–4 orders of magnitude. Analytical test cartridges are designed to conform to regulatory agencies such as the US EPA and DIN. A Windows-based software WinFLOW™ provides real-time peak editing, automatic absorbance scaling, automatic saving data and automatic quality control coupled with recalibration. It also corrects carry-over and base-line drift. The EnviroFlow software package allows random access sampling, automatic recalibration and real-time results. Applications developed for FlowSolution 3000 include about 60 compounds or parameters in such matrices as drinking water, waste water, soil extracts/digests, sea water, plant extracts and fertilisers.

1.3. *Burkard*

The **FIAflo** analyser in a basic offered configuration is composed of a flow injection module (with an automatic single or dual valve system, a peristaltic pump of up to 10 channels, a chemistry tray and a heating bath), a carousel or serpentine chain sample changer, a spectrophotometric detector, a chart recorder or a dedicated microprocessor data analyser monitor. Optional detectors include a flame photometer and an ion-selective detector with flow cell. The monitor 286 Processor can handle four independent channels, store data, analyse peaks, display all four channels simultaneously and print out results. In automatic operation mode throughputs up to 240 h^{-1} are possible. The present offer of application methods includes 19 procedures for water analysis, 5 for effluents, 10 industrial and 5 for agriculture analysis.

1.4. *Eppendorf*

Eppendorf Variable Analysers EVA^{TM} are flow injection systems of a modular design with a variety of standard interfaces, method-specific reaction manifolds and a wide range of interchangeable detector drawers. They incorporate several modules that can be programmed to act as a "Master" controlling all other EVA modules except the Manifold, as well as all peripheral devices. The EVA-Master module incorporates three functions: data processor, controller and detector. Changing the detection requires one to change the drawer in the Master module. Available detectors include: filter photometer, one- or two-electrode system, gas-sensing electrode detector, conductivity detector and adapter for other analysers. The EVA-Pump module is a peristaltic pump with up to six channels and with an integrated microprocessor, which can also operate as a controller for the other modules. The EVA-Injector module contains an eight-port rotary injection valve where the injection volume, load/inject intervals and dual injection of different volumes can be programmed. This module can synchronise the activities of other units or it can be controlled by other EVA modules or by the host computer. The EVA-Selector module is a user-programmable eight-port valve for multi-point selection for on-line analysis, switching from samples to standards and introduction of different reagents. The EVA-Manifold is the module for specific determination that provides mixing and reaction of samples and reagents, controls dispersion, and performs gas diffusion or extractions. Methods can be changed by changing from one application-specific drawer to another. The system is also equipped with the EVA-Sampler. Each module has software packages and connections to a host computer. The list of EVA application notes includes 4 procedures for the biotechnology and pharmaceutical industry, 5 for the chemical industry, 10 for environmental analysis and 11 for beverage and food analysis.

AMKO/Eppendorf model CS-2000 is an on-line process analyser that measures H_2S and CO_2 loadings in amine solutions. The analyser utilises an FIA method. The operating principle of the analyser is the detection and quantification of chemically bound H_2S and CO_2 in an ethanolamine matrix by acidifying the amine and driving the liberated gases to two detectors: ultraviolet for H_2S and infrared for CO_2. Possible wide applications of this analyser include natural gas treating, refinery systems, ammonia plants, hydrogen plants and cryogenic systems. An on-board microprocessor automates the entire system that can monitor up to four amine streams.

1.5. *Hitachi*

Flow Injection Analyser Model K-1000 is a single piece instrument containing a two-channel piston pump providing a high reproducibility of results, a four-channel peristaltic pump for delivery of reagents, an automatic 16-port valve, a reaction bath and a reaction coil. The analyser allows one to perform measurements manually for each sample, at a constant interval set by a timer or continuously in combination with a sampler. The analyser can be equipped in a solvent extraction unit and connected to several Hitachi spectrophotometers and fluorimeters. With the use of a multichannel pump and a 16-port valve a sample plug together with several zones of reagents can be introduced into the carrier stream (Fig. 1). This so-called "sandwich method" of FIA is a simplified version of the later-developed sequential injection analysis. Contrary to SIA, in this method there is no preliminary loading of the sequence of injected zones into a holding coil, but this sequence is formed by sequential injection by fast switching of the valve directly into a carrier stream without

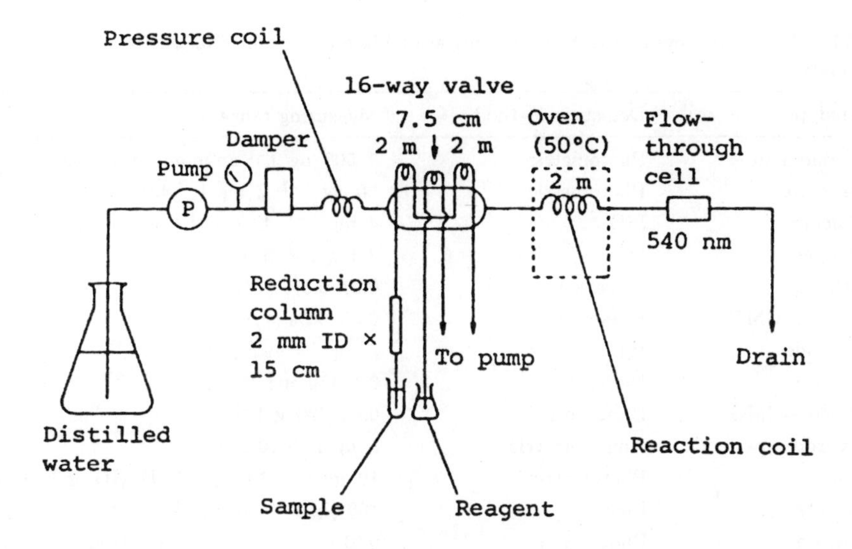

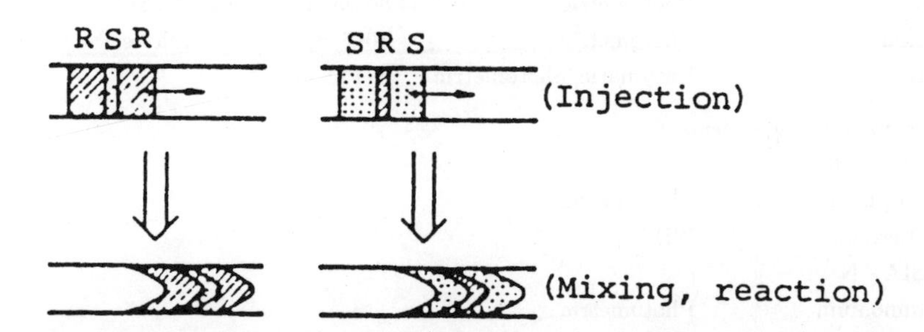

Fig. 1. Manifold for spectrophotometric determination of nitrate with a Hitachi model K-1000 flow injection analyser in a sandwich mode of operation and schematic illustration of sample sandwich and reagent sandwich mode of operation. Figures reproduced from manufacturer's technical brochure. (*Reprinted by permission of copyright owner.*)

flow reversal. The analyser allows also a simple performance of the merging zones injection technique. A multi-port valve is made of alumina ceramics which is resistible against chemicals and has a long service life.

Table 2. List of flow injection procedures available for a commercial ASIA analyser from Ismatec.

Analyte	Detection method	Measuring range
Ammonium	Photometric	5–500 $\mu g\ l^{-1}$ ammonium nitrogen
Bromate	Photometric	20 $\mu g\ l^{-1}$–2 $mg\ l^{-1}$ bromate
Calcium	ISE	4 $mg\ l^{-1}$–40 $g\ l^{-1}$ Ca
Copper	ISE	3 $mg\ l^{-1}$–6 $g\ l^{-1}$ Cu
Ethanol	Biosensor	0.01–0.5% (v/v)
Ethanol-GMC	Biosensors	0.05–50% (v/v)
Fluoride	ISE	1 $mg\ l^{-1}$–9.5 $g\ l^{-1}$ F
Glucose	Biosensor	0.1–150 $mg\ l^{-1}$
Glucose-GMC	Biosensors	0.03–300 $g\ l^{-1}$
Hydrogen peroxide	Amperometric	5 $\mu g\ l^{-1}$–10 $mg\ l^{-1}$
Iron	Photometric	10 $\mu g\ l^{-1}$–10 $mg\ l^{-1}$ Fe (II) or Fe (III)
Nitrate	ISE	160 $\mu g\ l^{-1}$–60 $g\ l^{-1}$ nitrate
Nitrate	Photometric	0.01–10 $mg\ l^{-1}$ nitrate nitrogen
Nitrite	Photometric	1–200 $\mu g\ l^{-1}$ nitrite nitrogen
pH	ISE	1–13
Phenol	Photometric	0.05–10 $mg\ l^{-1}$
Phosphate	Photometric	2 $\mu g\ l^{-1}$–10 $mg\ l^{-1}$ orthophosphate P
Potassium	ISE	20 $mg\ l^{-1}$–4 $g\ l^{-1}$ K
Silica	Photometric	50 $\mu g\ l^{-1}$–5 $mg\ l^{-1}$ Si
Sulphate	Photometric	30–500 $mg\ l^{-1}$ sulphate
Urea	Enzymatic/photometric	5–150 $mg\ l^{-1}$
Preprogrammed systems		
ASIA - PK		
Phosphate	Photometric	
Potassium	ISE	
ASIA - N		
Ammonium	Photometric	
Nitrate	Photometric	
Nitrite	Photometric	
ASIA Bio		
Glucosed	Biosensor	
C1-C4 Alcohols	Biosensor	

*ISE — ion-selective electrode; GMC — gradient mixing chamber.

1.6. *Ismatec*

ASIA (Automated Sample Injection Analyser) is a modular and compact system where the master rack has seven slots for plug-in modules. It includes a liquid crystal display, a keyboard and ASIA Windows PC Software, power supply, three A/D converters and a connection for external modules. Results can be printed out on a printer via a serial interface. As plug-in modules there are available a four-channel fixed or variable speed peristaltic pumps, different injection valves with or without gradient mixing chamber, a LED photometric detector, a potentiometric wall-jet detector for ion-selective electrodes, an amperometric biosensing detector with immobilised enzyme cartridges, a manifold carrier, a heater and a diffusion chamber. Pre-programmed and preplumber configurations are supplied with a sampler for 50 samples and with PC-Software (Windows 3.1). The list of available procedures with concentration ranges is shown in Table 2. The available throughput is from 30 to 120 samples h^{-1}, depending on the procedure. As additional detection systems any flow-through detector with analogue signal ± 2 V can be coupled to the ASIA system.

1.7. *Lachat*

Lachat Instruments shipped its first FIA instrument in 1980. Its QuikChem systems have since undergone a significant evolution and offer nowadays a list of application notes for various matrices that is the largest one available from manufacturers of flow injection instrumentation. Already in 1987 11 Lachat QuikChem methods have been approved by US EPA for National Pollutant Discharge Elimination System monitoring programs [9].

The most modern and advanced version of the instrument **QuikChem™ 8000** Automated Ion Analyser combines simultaneous operation of flow injection and chemically suppressed high-performance ion chromatography. The system is run by QuickChem 8000's Omnion software based on Windows 3.1. Omnion combined with the XYZ Sampler and Automated Dilution Station, automates preparation of working standards and detectors and dilutes off-scale samples into the working range of the method. The Shared Peripheral System™ incorporated into the Omnion enables one to run FIA and ion chromatography in any combination, simultaneously and independently. In FIA determinations, typically between 90 and 120 samples h^{-1} can be processed. Volatile analytes can be digested and distilled in-line. Individual modules for

FIA are mounted to perform parallel, up to seven analytical channels. FIA can be used as complementary sample preparation techniques for ion chromatography to pre-dilute samples and perform both pre- and post-column derivatisation. In addition, ion chromatography can be used as a matrix modifier prior to FIA injection. Ion chromatography is performed with conductivity detection, and the flow injection part of the system with a photometric detector. Model QuikChem equipped with an XYZ sampler has a full random access sampling and optional dilutor. Model 1000 has semiautomated sampling, whereas model 4000 samples with sequential sampling. Lachat offers over 400 analytical methods for a variety of matrices, from the very simple to the most complex — environmental waters, brackish water, brines, soils, plants, food, etc.

1.8. *Perkin-Elmer*

Perkin-Elmer was the first instrument maker to launch flow injection apparatus as accessories to earlier-produced atomic absorption spectrometers and UV/VIS spectrophotometers.

The **FIAS**™ series of flow injection systems are designed to cowork with flame atomic absorption spectrometers or spectrophotometers equipped with a flow cell. The latter is supplied as the Lambda FIA-System. FIAS systems include one or two independently controlled peristaltic pumps for carrier, reagent and sample solutions. The pumps use exchangeable tubing cassettes which are designed to secure pump tubing against the pump rollers. The modular design of the manifold section offers high flexibility for different applications such as a sample dilutor, adding reagents or modifiers, removing interfering matrices or concentrating trace elements. They provide a means to analyse microlitre sample volumes or sample solutions with a high level of dissolved solids without burner clogging. The literature offered by manufacturers for flow injection flame AAS includes 12 application notes for various elements, matrices and operations. For on-line sorbent extraction with preconcentration time 60 s for different volumes the developed procedures yield 14 to 174-fold sensitivity enhancement. FIAS systems can also allow the mechanisation of analyses requiring more complex sample treatment, such as cold vapour mercury determinations and hydride generation for determination of such elements as As, Se, Te, Bi, Sb and Sn. Each FIAS system can be connected to an autosampler.

For mercury trace determination below $\mu g\ l^{-1}$ concentrations Perkin-Elmer offers a compact **Flow Injection Mercury System FIMS**™, which is an atomic absorption dedicated spectrometer based on the flow injection technique. It uses a high-performance single-beam optical system with a low-pressure mercury lamp and solar-blind detector. FIMS-100 has one persitaltic pump, and FIMS-400 has two pumps. They can work together with samplers and other accessories, such as an amalgamation system for preconcentration or with microwave unit for on-line sample digestion. FIMS systems can be controlled from a personal computer using Windows-based AA WinLab™ software.

1.9. *Sanuki*

Sanuki produces a wide range of flow injection analysers in various configurations. A design of series **FI-3000** analysers is based on the use of a computer-controlled double-plunger pump suitable for precise microlitre delivery of solutions and complete in-line mixing [10]. Systems are equipped with a six-port valve where ceramic and fluorine resin being used for the valve's contacting surface to solvent makes possible its use for any kind of analysis. The system can be connected to spectrophotometric or fluorimetric detectors as well as to a sampler. It is also equipped in accessories to perform on-line solvent extraction. The list of supplied application notes includes numerous procedures for environmental analysis, biotechnology, industrial and food analysis.

1.10. *Tecator*

Foss Tecator is a manufacturer with the longest tradition in production of flow injection instruments for routine applications. At the end of the nineties the company offers two different instruments — one designed for the different applications **FIAstar® 5012** system, and another one the strictly dedicated **Aquatec®** system.

The FIAstar 5012 is a two-channel flow injection system for colorimetric analysis of water, environmental and agricultural samples. It is controlled by a SuperFlow Duo software package from an external computer. With the addition of a sampler, it is a fully mechanised system with unattended operation. Results for both channels are displayed on screen in the real-time regime, with automatic absorbance scaling, base-line correction and true random access sampling. The analyser is equipped with two peristaltic pumps with eight channels, with a dual injection valve, and can be operated manually or under PC control. It is supplied with a filter photometer but all types of flow-through detectors may be interfaced with the system. Sample size in recommended procedures is from 40 to 400 μl and sample throughput 60–180 sample h^{-1} per channel. Universal Chemifolds for mixing of reagents and samples in both standard and solvent resistant versions are available as well as dialysis and extraction modules for in-line sample preparation. The long list of application notes for the FIAstar system includes about 80 methods for water analysis for different ranges of concentrations, 15 methods for fertilisers, 14 for soil and plants and a similar number of methods for beverage, food and feed analysis.

Aquatec is a dedicated flow injection system for determination of ammonia, nitrate, nitrite, phosphate and chloride in water. The fully mechanised system includes an analyser with cassettes for given analytes, a sampler and laptop computer for control of the whole system, a display of recorded changes of signal and processing of results. Cassettes contain Chemifolds and containers solutions of reagents for one day of analyser operation. Determinations can be carried out with frequency 40–70 sample h^{-1}.

1.11. *Zhaofa*

The **LZ-2000 Flow Injection Processor** is a microprocessor-controlled compact system where parameters and status of operation are monitored with a 6-figure digital display. It is equipped with a 16-port multifunctional valve and dual low-pulsation variable speed, and 7-channel peristaltic pumps. It can be connected to different detection systems. For assembling of different manifolds the company supplies a wide range of accessories, including connectors, a gas-diffusion separator, ion-exchange columns and a complete hydride generator LZ-1200 as a separate independent instrument.

2. SIA

As was mentioned above, although the analyser **Hitachi K-1000** is described as a flow injection analyser with the use of a unique sandwich flow path, it is in fact a flow injection analyser with the sequential injection of sample and reagent segments directly into the carrier solution.

Two different SIA systems with typical loading into holding coil and flow reversal are offered by Alitea USA. **FIAlab®-3000** is a fully portable microsyringe system, designed primarily for SIA, but also capable of FIA. It features a six-port selector valve, a motor-driven piston pump and a two-channel persitaltic pump. An advanced model of this instrumentation was already mentioned above, **FIAlab®-3500**. It is equipped with a six to ten-position selector valve and a four-channel peristaltic pump. Both systems can be controlled with FIAlab® for Windows control software which can also control, external devices such as pumps, valves and samplers.

3. References

1. J. Ruzicka and E. H. Hansen, *Flow Injection Analysis* (Wiley, New York, 1981), p. 100.
2. M. Valcarcel and M. D. Luque de Castro, *Flow-Injection Analysis: Principles and Applications* (Ellis Horwood, New York, 1987), p. 388.
3. J. Ruzicka and E. H. Hansen, *Flow Injection Analysis*, 2nd Edition (Wiley, New York, 1988), pp. 292–295.
4. A. Newman, *Anal. Chem.* **58** (1996) 203A.
5. E. H. Hansen, B. Willumsen, S. K. Winther and H. Drabøl, *Talanta* **41** (1994) 1881.
6. J. Nyman and A. Ivaska, *Anal. Chim. Acta* **308** (1995) 286.
7. X. Liu and E. H. Hansen, *Anal. Chim. Acta* **326** (1996) 1.
8. O. Egorov, M. J. O'Hara, J. Ruzicka and J. W. Grate, *Anal. Chem.* **70** (1998) 977.
9. C. B. Ranger, *Am. Lab. September* (1988) 35.
10. T. Korenaga, X. Zhou, T. Moriwake, H. Muraki, T. Naito and S. Sanuki, *Anal. Chem.* **66** (1993) 73.

Chapter 11

Current Trends in Developments of Flow Analysis

All the above-presented developments and instrumentation convincingly demonstrate that at the end of the nineties the measurements in flow conditions play a significant role in chemical analysis. Although commonly it is not accepted to include chromatographic methods into flow analysis, all chromatographic detectors in liquid and gas chromatography operate in flow mode. The same is applied to electromigration capillary methods. A solid position in analytical instrumentation is held by methods where flow detection is not directly connected with chromatographic or electromigration separation of analytes. The latter methods are commonly considered as flow analysis.

Obviously, for various technical reasons, the availability of commercial routine instruments, economic or legislative factors, the role of various kinds of instrumentation in routine applications and research laboratories is different. Well-tested and verified concepts with advanced instrumentation are less the subject of scientific research, but are rather manufactured as commercial products and find a place in routine applications.

The instruments (analysers) for continuous monitoring of various components of gases and liquids are important and widely applied measuring devices in process and environmental analysis. They are manufactured by numerous specialised makers and they are basic instrumentation for chemical analysis in various branches of industry and environmental protection. They are produced with a broad spectrum of optical and electrochemical detections.

Flow determinations with segmented stream and the use of steady-state signal for many years have been practically eliminated from clinical laboratories by more efficient and economic discrete analysers. They are still, however, produced for routine environmental applications, for agricultural and industrial laboratories. Very many of these methods are recommended as standard ones, as they allow one to mechanise various accepted methods, mostly of wet analysis.

A large available and constantly expanding scientific literatue together with the increasing offer of commercial instrumentation allows one to predict the rising role of flow injection methods in the near future in various routine applications. The role of this methodology and trends in its development are discussed in the literature [1, 2].

1. Miniaturisation

In vividly developing areas of flow injection methodology several main streams of further progress can be distinguished. One of them is miniaturisation of both hydraulic systems and detectors employing micromechanics, electronics of large scale of integration and optoelectronics. The main purpose of this development activity is to construct portable devices which besides small size might utilise all advantages of measurement of transient signal. This combined with essential reduction of the required sample volume and consumption of reagents might allow competition with constructions of discrete analysers.

Several different concepts of miniaturisation of flow injection systems have already been employed and cited in earlier chapters starting from microconduit systems with etched channels in a micromachined substrate [3]. Two different approaches can be used for this purpose. One of them is scaling down some elements of instrumentation but employing the same devices as those found in conventional flow injection systems. As an example a capillary flow injection system with conventional peristaltic pumping [4] or using pressure-induced flow [5] can be given. In flow injection mode also one can operate a microbore flow system based on controlled pumping with the piston pumps in the range of 10–30 μl and the sample volume 3–4 μl [6]. There are also undertaken studies on the design of miniaturise systems with piezoelectrically driven micropumps [7, 8]. Satisfactory integration of miniature elements might lead in future to the development of a system capable of performing almost all steps of the real analytical procedure, which is discussed in the literature as a so-called *micro total analysis system.*

Another concept on miniaturisation of flow injection and sequential injection systems, at least in their hydrodynamic part, is the application of an electro-osmotic flow as the liquid propulsion mechanism with fused silica capillary tubing with an inner dimension of 75 μm as the manifold tubing [9–1]. In such a system might be used hydrodynamic injection, similar to that used in capillary electrophoresis, or an injection valve with fixed volume internal loops. In capillary systems a mixing of sample and reagent can be achieved utilising the difference in their electrophoretic mobility [12]. A suitably designed capillary electrophoretic system allows one to obtain a setup for so-called *electroinjection analysis*, where sample and reagent are injected from the opposite end of the capillary tube [13].

An additional element that might potentially increase the attractiveness of miniaturised flow injection analysers is the use of closed-loop circulation of reagents within the system. Only a few examples of this concept have been used so far, in conventional FIA systems. This seems to be a promising direction for further developments.

Portable flow injection analysers of different scales of miniaturisation and integration have already been reported for some applications [14–17], but this is open area for much more advancement.

2. Multicomponent Detection

The flow injection methods were developed as a response to the need for mechanisation of various operations with samples in different stages of the analytical procedure. They successfully simplify various operations of sample pretreatment and make them more precise. A specially valuable and attractive feature of routine analytical instruments is their ability to determine several components or to perform speciation analysis of a given element of the analysed sample in a single run.

The flow procedure itself does not provide a possibility of carrying on a multicomponent determination. This is, however, only an apparent impression. Numerous examples in the previous chapters show that this technique offers a possibility of very flexible design of multidetector systems, as well as a construction of flow manifolds with a single detector capable of multicomponent determinations or speciation. This is for sure an area for further exploration and creative inventions.

A separate aspect that is increasingly recognised by analysts is the use of flow injection sample processing for mechanisation of sample handling and pretreatment prior to measurements with typical multicomponent methods, such as voltammetry, spectrophotometry with diode array detectors or with scanning a wavelength, emission atomic spectrometry or mass spectrometry. The literature on this subject is already quite wide, e.g. [17]. In one of the recent, spectacular applications flow injection sample handling in an electrospray ionisation mass spectroscopy system was employed for a three-dimensional spectral mapping of 96-well combinatorial chemistry racks in search of new drugs [18]. Hyphenation of the FIA system to a large, specialised instrument often requires essential construction adjustment, but then as a profit it offers improvements in sample pretreatment, calibration, derivatisation or preliminary separation of sample components. It may be concluded that in the near future more and more instrument makers will include in their offer of accessories certain designs of flow injection devices.

3. Hyphenation to High-Performance Separation Systems

Increasing interest has also been recently observed in hyphenation of flow injection systems with column chromatography and capillary electrophoresis. In a similar manner much earlier flow systems with segmented stream were combined with the HPLC system. Some examples of FIA-HPLC combined setups were shown in previous chapters. Several of them have been reviewed [20]. This hyphenation allow one to achieve three goals: preconcentration of analytes, their derivatisation, or the clean-up sample. In the first two cases this allows one to enhance the sensitivity of determination in chromatographic determination; in the latter, to improve the selectivity and effectiveness. Practically for the same purposes FIA can be hyphenated with gas chromatography, which was demonstrated above for speciation of organometallic compounds. As examples of FIA-HPLC combination one can find determination of sugars as borate complexes [21], determination of bile acids in serum [22], or transition metal ions [23]. Due to the wide use of chromatography in various areas of analytical chemistry and the carrying-out of many pretreatment operations manually, further studies on the development of the hyphenated systems GC-FIA and HPLC-FIA are very desirable.

Much more difficult technically is the combination of flow injection systems with capillary electrophoresis. Because it provides similar advantages like in the case of column chromatography techniques, and as capillary electrophoresis gains increasing interest, several research groups are working in this areas. A satisfactory combination of these two techniques yields remarkable improvements. It was already demonstrated that in chiral separation of intermediate enantiomers in chloroamphenicol synthesis significant improvement of the precision and sample was achieved [23]. In determination of common ions the FIA-CE interface enables a high sampling throughput of up to 150 samples h^{-1} [28]. Efficient preconcentration was shown in determination of pseudoephedrine [26] and inorganic ions in waters [27]. Finally, an efficient sample clean-up was also shown in determination of polyphenols in wine [28]. It can be concluded that not only the use of electromigration techniques in the design of new generation flow injection systems but also their hyphenation with conventional FIA systems can be profitable and interesting for further development.

4. References

1. E. H. Hansen, *Anal. Chim. Acta* **308** (1995) 3.
2. J. Ruzicka and E. H. Hansen, *Trends Anal. Chem.* **17** (1998) 69.
3. J. Ruzicka and E. H. Hansen, *Anal. Chem. Acta* **161** (1984) 1.
4. D. M. Spence and S. R. Crouch, *Anal. Chem.* **69** (1997) 165.
5. D. M. Spence and S. R. Crouch, *Anal. Chim. Acta* **366** (1998) 305.
6. K. Carlsson, H. S. Jacobsen, A. L. Jednsen, T. Stenstrøm and B. Karlberg, *Anal. Chim. Acta* **354** (1997) 35.
7. B. H. van der Schoot, S. Jeanneret, A. van der Berg and N. F. de Rooij, *Anal. Methods Instrum.* **1** (1993) 35.
8. P. Graveston, J. Branebjerg, O. Søndergard Jensen, *J. Micromech. Microeng.* **3** (1993) 168.
9. S. Liu and P. K. Dasgupta, *Anal. Chim. Acta* **268** (1992) 1.
10. S. Liu and P. K. Dasgupta, *Anal. Chim. Acta* **283** (1993) 739.
11. S. Liu and P. K. Dasgupta, *Talanta* **41** (1994) 1903.
12. B. J. Harmon, D. H. Patterson and F. E. Regnier, *Anal. Chem.* **65** (1993) 2655.
13. V. P. Andreev, A. G. Kamenov and N. S. Popov, *Talanta* **43** (1996) 909.
14. K. Sonne and P. K. Dasgupta, *Anal. Chem.* **63** (1991) 427.
15. K. N. Andrew, N. J. Bundell, D. Price and P. J. Worsfold, *Anal. Chem.* **66** (1994) 917A.
16. T. Dimitrakopoulos, P. W. Alexander and D. B. Hibbert, *Electroanalysis* **8** (1996) 438.
17. L. T. Di Benedetto, P. W. Alexander and D. B. Hibbert, *Anal. Chim. Acta* **321** (1996) 61.
18. M. D. Luque de Castro and M. T. Tena, *Talanta* **42** (1995) 151.
19. E. Görlach, R. Richmond and I. Lewis, *Anal. Chem.* **70** (1998) 3227.
20. M. D. Luque de Castro and M. Valcarcel, *J. Chromatogr.* **600** (1992) 183.
21. H. Ohura, T. Imato, S. Yamasaki and N. Ishibashi, *Anal. Sci.* **6** (1990) 777.
22. A. Membiela, F. Lazaro, M. D. Luque de Castro and M. Valcarcel, *Anal. Chim. Acta* **249** (1991) 461.
23. P. Richetr, J. M. Fernandez-Romero, M. D. Luque de Castro and M. Valcarcel, *Chromatographia* **34** (1992) 445.
24. Z.-S. Liu and Z.-L. Fang, *Anal. Chim. Acta* **353** (1997) 199.
25. P. Kuban, A. Engström, J. C. Olsson, G. Thorsen, R. Tryzell and B. Karlberg, *Anal. Chim. Acta* **337** (1997) 117.

26. H.-W. Chen and Z.-L. Fang, *Anal. Chim. Acta* **355** (1997) 135.
27. L. Arce, A. Rios and M. Valcarcel, *J. Chromatogr. A* **791** (1997) 279.
28. L. Arce, M. T. Tena, A. Rios and M. Valcarcel, *Anal. Chim. Acta* **359** (1998) 27.